C0-AWZ-906

THE PRIMARY BATTERY

THE ELECTROCHEMICAL SOCIETY SERIES

THE CORROSION HANDBOOK
Edited by Herbert H. Uhlig

ELECTROCHEMISTRY IN BIOLOGY AND MEDICINE
Edited by Theodore Shedlovsky

ARCS IN INERT ATMOSPHERES AND VACUUM
Edited by W. E. Kuhn

MODERN ELECTROPLATING, SECOND EDITION
Edited by Frederick A. Lowenheim

FIRST INTERNATIONAL CONFERENCE ON ELECTRON AND ION BEAM SCIENCE AND TECHNOLOGY
Edited by Robert Bakish

THE ELECTRON MICROPROBE
Edited by T. D. McKinley, K. F. J. Heinrich and D. B. Wittry

CHEMICAL PHYSICS OF IONIC SOLUTIONS
Edited by B. E. Conway and R. G. Barradas

VAPOR DEPOSITION
Edited by C. F. Powell, J. H. Oxley and J. M. Blocher, Jr.

HIGH-TEMPERATURE MATERIALS AND TECHNOLOGY
Edited by Ivor E. Campbell and Edwin M. Sherwood

ALKALINE STORAGE BATTERIES
S. Uno Falk and Alvin J. Salkind

THE PRIMARY BATTERY (*In Two Volumes*)
Edited by George W. Heise and N. Corey Cahoon

THE CORROSION MONOGRAPH SERIES

R. T. Foley, N. Hackerman, C. V. King, F. L. LaQue, H. H. Uhlig, Editors

HIGH-TEMPERATURE OXIDATION OF METALS, *by Kofstad*
THE STRESS CORROSION OF METALS, *by Logan*
CORROSION OF LIGHT METALS
by Godard, Bothwell, Kane, and Jepson

THE PRIMARY BATTERY

VOLUME 1

Edited by

GEORGE W. HEISE

RESEARCH LABORATORY
CONSUMER PRODUCTS DIVISION
UNION CARBIDE CORPORATION (RET.)

and

N. COREY CAHOON

ELECTROCHEMICAL DEVELOPMENT LABORATORY
CONSUMER PRODUCTS DIVISION
UNION CARBIDE CORPORATION

Sponsored by

THE ELECTROCHEMICAL SOCIETY, INC.
New York, New York

JOHN WILEY & SONS, INC.
NEW YORK · LONDON · SYDNEY · TORONTO

Library of Congress Catalogue Card Number: 73–121906

ISBN 0 471 36899 7

Printed in the United States of America

10 9 8 7 6 5 4 3 2

Contributors

N. COREY CAHOON, *Senior Scientist, Electrochemical Development Laboratory, Consumer Products Division, Union Carbide Corporation, Cleveland, Ohio*

THEDFORD P. DIRKSE, *Professor of Chemistry, Calvin College, Grand Rapids, Michigan*

JOHN M. FREUND, *Kodak Apparatus Division, Eastman Kodak Company, Rochester, New York*

WALTER J. HAMER, *Chief, Electrochemistry Section, National Bureau of Standards, Washington, D.C.*

GEORGE W. HEISE, *Consultant; Associate Director of Research (ret.), Parma Research Center, Consumer Products Division, Union Carbide Corporation, Parma, Ohio*

HARRY W. HOLLAND, *Electronics Division, Union Carbide Corporation, Greenville, South Carolina*

DEMETRIOS V. LOUZOS, *Consumer Products Division, Research Laboratory, Union Carbide Corporation, Parma, Ohio*

JOHN N. MRGUDICH, *United States Army Electronics Command, Fort Monmouth, New Jersey (ret.)*

RALPH ROBERTS, *Director, Power Program, Office of Naval Research, Department of the Navy, Washington, D.C.*

SAMUEL RUBEN, *Director Ruben Laboratories, New Rochelle, New York*

ERWIN A. SCHUMACHER, *Consultant; Group Leader (ret.), Parma Research Center, Consumer Products Division, Union Carbide Corporation, Parma, Ohio*

EUGENE P. SCHWARTZ, *Associate Professor of Chemistry, DePauw University, Greencastle, Indiana*

WILLIAM C. SPINDLER, *Head, Electrochemistry Branch, Research Department, Naval Weapons Center, Department of the Navy, Corona, California*

ERNEST B. YEAGER, *Professor of Chemistry, Case-Western Reserve University, Cleveland, Ohio*

Advisory Board

To all technologists who have labored to convert the battery art to a science

Foreword

The Battery Division of The Electrochemical Society, Inc., was organized at the 1947 Fall Meeting of the Society in Boston. Its purpose included the stimulation of research, publication, and the exchange of information relating to primary or secondary cells for the production of electrical energy. What propitious timing! The 1967 Census of Manufactures indicates that the total annual product of the battery industry in the United States has steadily increased to the present annual value of about one billion dollars, about triple that reported for 1947.

The growth in volume of the battery industry has been accompanied by a demand in military and consumer markets to extend the utility of batteries. In the field of primary batteries, the one of particular interest here, the batteries must operate usefully over a wide range of ambient temperatures and environmental conditions, must have a wide range of energy and power densities, and often must be available in unusual packaging arrays. On one hand, development and application of transistorized circuits have led to miniaturization as in the case of hearing aids and their battery requirements. On the other hand, there is a trend toward larger sizes for stationary batteries used in railway signaling installations.

It should be no wonder then that all possible materials and their combinations are scrutinized and examined constantly in striving to create and invent improved cells. The technology of manufacturing at sustained high quality economically and the requisite knowledge become increasingly complex with time. There are no reasons at this time to justify any belief that the tasks in this field will become any easier.

This conclusion is indicated with certainty by the results of rather extensive research and development programs over the past decade. In particular, there may be cited the work in the fields of fused salt, solid-state electrolyte, organic electrolyte, and fuel-cell systems.

A partial answer to the problem of an explosively expanding literature is its periodic review and appraisal. Such compilations will be of great value to those concerned with all aspects of the industry, to the users and consumers who must understand the battery and its capabilities for its proper application and the avoidance of maintenance, and to those who have and seek a general background and interest in scientific and technologic achievements. The scope of the primary battery field was found to be broad enough to justify more than one published volume. This book is an authoritative review of primary battery technology.

ARTHUR FLEISCHER

Orange, New Jersey
January 1970

Preface

During this century, particularly since the introduction of commercial radio in the early 1920's, the growth of the primary battery has been spectacular. Its range of usefulness has increased continuously—from microelectronic applications to multiwatt outputs. The resulting improvement of old systems and the development of new cell types have been accompanied by rapid advances in theory as well as in technology. The work in hand was undertaken at the request of the Executive Committee of the Battery Division, The Electrochemical Society, to meet the need for a comprehensive and authoritative text in this important field.

In view of the continuing expansion of battery technology, it would have been exceedingly difficult for one author to do justice to the subject matter. Therefore experts in the field were asked to discuss individual cell systems. Though each author's contribution can stand by itself, every effort has been made to coordinate and integrate the separate chapters into a coherent whole and, through the format of the book, to minimize repetition and overlap.

Furthering this plan, the introductory chapter furnishes the historical background, not only of the cells in current use, but also of those that have influenced the course of battery development but are no longer of sufficient commercial importance to warrant separate chapters. This, together with a chapter on the basic electrochemistry involved in battery science and technology, is designed to furnish a basis for the subsequent, detailed discussions of individual cell systems. The latter, in the current volume, include the various alkaline batteries, continuous-feed and fuel cells, and standard cells, with separate chapters on low-temperature nonaqueous systems and on solid electrolyte batteries.

Because of the fluid state of the battery art, emphasis has been placed on general principles rather than on details of current construction and composition which may rapidly become obsolete. This is in keeping with our aim to make the current work a contribution to electrochemistry, of interest to the student and general reader as well as to the technologist.

We thank the many individuals and organizations that facilitated the compilation of this monograph: the authors, advisory board, and reviewers who gave of their time to the development and critical appraisal of individual chapters; the industrial and governmental organizations which released for publication the chapters written by staff members; the publishers, as acknowledged in the text, who permitted the use of copyrighted material.

GEORGE W. HEISE
N. COREY CAHOON

April 1970
Cleveland, Ohio

Contents

. . . un appareil . . . qui jouiroit, en un mot, d'une charge indéfectible, d'une action sur le fluide électrique, ou impulsion, perpétuelle . . .

Volta, 1800

THE PRIMARY BATTERY

1

Historical Introduction

George W. Heise

Few important discoveries and inventions of the last 200 years have surpassed the primary battery in impact on the progress of science and technology. It furnished the basis for the development of electrochemistry and, until the advent of the dynamo in the last half of the nineteenth century, constituted the only important source of continuous current. It is still the outstanding means for direct conversion of chemical to electrical energy and, commercially, is one of our most important electrochemical products.

THE PRE-VOLTA PERIOD (1678–1800)

By the end of the eighteenth century, the study of electricity, limited almost entirely to electrostatic phenomena (8), had reached something of an impasse. The only available sources of electrical power were of the static or frictional type, capable of producing momentary impulses of very high voltage but of almost negligible coulombic current.[1] Even with these, van Troostwijk and Deiman (221) accomplished the electrolysis of water in 1789. However, indicative of the handicaps suffered by early investigators, in 1797 Pearson (177) reported that 14,600 discharges of a "double plate electrical machine 24 inches in diameter" were required to "produce, at least, one-third of a cubical inch of gaz," i.e., of a mixture of hydrogen and oxygen.

The first source of continuous current, the primary battery, though

[1]A lightning flash, whose electrical nature was demonstrated in 1752 by Benjamin Franklin's famous kite experiment (80), may have a tension of a billion volts and a total energy content of several thousand kilowatt-hours. Yet it lasts perhaps a millisecond and involves current of the order of only 20–50 coulombs (246), an amount barely sufficient to operate a conventional flashlight for 2 minutes.

long foreshadowed,[2] was slow to develop. As early as 1678, the Dutch naturalist Jan Swammerdam made the following significant observation (216):

> Let there be a cylindrical glass tube in the interior of which is placed a muscle, whence proceeds a nerve that has been enveloped in its course with a small silver wire, so as to give us the power of raising it without pressing it too much or wounding it. This wire is made to pass through a ring bored in the extremity of a small copper support and soldered to a sort of piston or partition; but the little silver wire is so arranged that on passing between the glass and the piston the nerve may be drawn by the hand and so touch the copper. The muscle is immediately seen to contract.

The electrochemical significance of this phenomenon was not recognized, it was not studied further, and it appears to have been overlooked entirely in subsequent investigations.

In 1762, a Swiss, Johann Sulzer (215), noted that pieces of lead and silver, separately tasteless, developed a taste "similar to that of iron sulfate" when connected and applied to the tongue (see Fig. 1.1). This observation, too, remained unexplained and unexploited, and another opportunity to develop a battery was lost.

Apparently unaware of Swammerdam's earlier discovery, various investigators independently noted the contraction of frog legs under electrical stress. Among them were du Verney (228) in 1700, Stuart (212) in 1732, Caldani (29) in 1757, and Cotugno (46) in 1780. However, it remained for Galvani to correlate the various observations and, in experiments performed in 1786 which confirmed and enlarged upon the findings of Swammerdam, to show that the effect of dissimilar metals on the muscles or nerves of frogs was actually electrical in character (86). Galvani reached the conclusion that animal tissue was in some manner the source of electricity, a hypothesis which he held, despite all evidence to the contrary, to the time of his death in 1798.

[2]There are numerous references in recent literature [(87), (89), (98)] to a 2000-year-old "battery" discovered in the vicinity of Baghdad in 1932 by a German archaeologist, Wilhelm Koenig. The mere fact that the artifact was a small clay vase containing an iron rod inside a metal cup ("a thin copper cylinder about four inches long") is hardly conclusive evidence of its use as a galvanic cell. And the slight electrical capacity obtainable from a cell of this size and composition offers no support for the supposition that it "was most probably used by Arabian silversmiths . . . for electrogilding their wares."

The art of gilding nonmetallic as well as metallic surfaces, without the aid of current, is of ancient origin, known to Heredotus (c. 484–425 B.C.) and described by Pliny the Elder (A.D. 23–79) in his *Historiae Naturalis*. It was generally accomplished directly with gold foil or, alternatively, by the application of amalgams, or of mixtures of lead and gold, from which the base metal could be driven off by subsequent heating. Without more substantial evidence, the authenticity of the "ancient battery" cannot be accepted as proved.

NOUVELLE
THEORIE
DES
PLAISIRS,

Par Mr SULZER, *de l'Académie Royale des Sciences & Belles-Lettres de Berlin;*

AVEC
DES RÉFLEXIONS
Sur l'Origine du Plaiſir,

Par Mr. KAESTNER *de la même Académie.*

Vero fruere non ſuperbus gaudio.
MARTIAL.

M. DCC. LXVII.

DES PLAISIRS. 155

vement de vibration, qu'ils conſervent pendant un temps ſenſible; les nerfs ne ſont point des cordes tendues, ni des corps rigides. Car dans ce cas, une ſeule impreſſion momentanée feroit durer les ſenſations, ce qui répugne à l'expérience. En effet dès qu'on ferme l'œil, dès qu'on bouche l'oreille, les ſenſations ceſſent. Au lieu qu'elles continueroient, ſi les nerfs avoient un mouvement ſenſible de vibration (*)

(*) Cette ſuppoſition paroît confirmée par une expérience aſſez curieuſe. Si l'on joint deux pieces, l'une de plomb & l'autre d'argent, de ſorte que les deux bords faſſent un même plan, & qu'on les approche ſur la langue on en ſentira quelque goût, aſſez approchant au goût de vitriol de fer, au lieu que chaque

156 NOUVELLE THEORIE

Sur ces remarques ſe fonde mon ſecond principe général, ſavoir : *que toute ſenſation totale eſt compoſée d'un grand nombre de ſenſations momentanées qui ſe ſuccedent avec une rapidité à ne point laiſſer entrevoir les momens de temps qui s'écoulent d'un coup à l'autre.*

Partant maintenant de ces deux principes, il me ſemble qu'il n'eſt plus fort difficile d'ex-

piece à part ne donne aucune trace de ce goût. Il n'eſt pas probable que, par cette jonction des deux métaux, il arrive quelque ſolution de l'un ou de l'autre, & que les particules diſſoutes s'inſinuent dans la langue. Il faut donc conclure, que la jonction de ces métaux opere dans l'un ou l'autre, ou dans tous les deux, une vibration dans leurs particules, & que cette vibration, qui doit néceſſairement affecter les nerfs de la langue, y produit le goût mentionné.

THE FIRST SUGGESTION OF THE VOLTAIC CELL
[Fac-simile pages from Sulzer's book, 17[illegible]7. See foot note (*).]

Figure 1.1 Sulzer's "Theorie" (facsimile). From (15).

THE SIMPLE VOLTA BATTERY

Allessandro Volta, professor of natural philosophy at the University of Pavia, became interested in Galvani's theory but eventually discarded it (232), assuming instead that electricity was caused by the mere contact of dissimilar metals. Indeed, as early as 1792, he seems to have forecast the electromotive series of metals and begun work on cells or batteries capable of delivering continuous current (231). However, his work was not generally known until 1800 when, in a letter to Sir Joseph Banks, President of the Royal Society of London, Volta made the sensational announcement of the "construction of an apparatus . . . of unfailing charge, of perpetual power . . ." (233). The description of his "pile," freely translated from the original French, reads as follows (cf. Fig. 1.2):

> I provide myself with several dozen small round plates or disks of copper, brass or, better, of silver, an inch in diameter, more or less (e.g., coins) and with an equal number of plates of tin or, better, of zinc of the same or nearly the same shape and size . . . . Further, I prepare a sufficient number of (slightly smaller) disks of pasteboard ["carton"], of leather, or of some other spongy material, capable of absorbing and retaining as much as possible of the water or the liquid on which the success of the experiments depends . . . . On a table or base, I then lay one of the metallic plates, for example, one of silver, and on top of the first I fit a second, of zinc; on the second I place one of the moistened disks, then another plate of silver followed immediately by another of zinc,

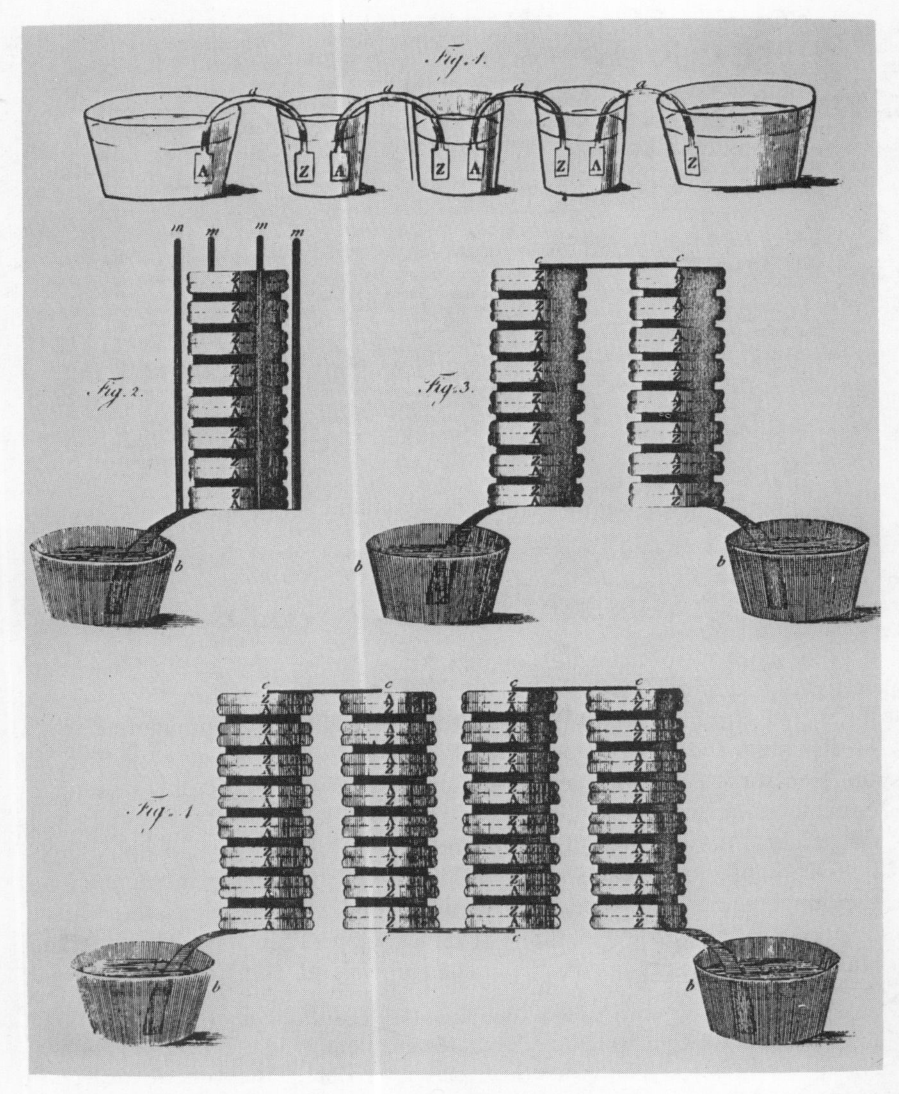

Figure 1.2 The Volta pile and "couronne de tasses." Copy of original plate (233).

succeeded again by a moistened disk . . . . I thus keep on, to form a column as high as can maintain itself without falling.

Alternatively, Volta described his "couronne de tasses," a battery of wet cells (Fig. 1.2) in which small vessels containing solutions of salt or lye were connected by means of "metallic arcs, one branch of which, *Aa*, or only the submerged extremity *A*, is of red or yellow copper or, better yet, silvered copper, and the other, *Z*, immersed in the adjacent goblet, is of tin or, better, of zinc."

EARLY APPLICATIONS

Volta's letter was dated March 20, 1800; there was a paragraph in the *Morning Chronicle* of May 30, 1800, giving an inaccurate account of the invention, and describing the general effects as exhibited by Dr. Garnet in the lecture at the Royal Institution (166); the letter itself was not formally read before the Royal Society until June 26. Meanwhile, it had been shown upon arrival to various British scientists who, with a source of continuous current finally available, immediately started the investigations that furnish much of the basis of modern electrochemistry. Thus it was that, as early as May 2, 1800, Nicholson and Carlisle accomplished the electrolysis of water with the aid of a Volta pile. Indeed, their report (167) in the July issue of *Nicholson's Journal of Natural Philosophy, Chemistry and the Arts* seems actually to have antedated the publication of Volta's letter.

Between 1800 and 1802, work on the electrodeposition of metals was carried on independently by Cruickshank and his associates (50), (168), by Ritter (192) in Bavaria, and by Brugnatelli (24), one of Volta's own colleagues, in Pavia, Italy. Sir Humphry Davy produced an electric arc as early as 1800, and two years later stated that "when, instead of metals, pieces of well-burned charcoal were employed, the spark was still larger and of a vivid whiteness" (61). Gautherot's observation (1801) of the difference in potential acquired by similar metal plates immersed in sulfuric acid when current was passed between them (93) formed the basis for Ritter's development about 1804, of a "charging or secondary pile" (193), which might be considered the forerunner of the modern storage battery. Davy accomplished the isolation of alkali metals by the decomposition of molten salts or oxides in 1807 (63). And it was with Volta cells that H. C. Oersted in 1819 made his momentous discovery of the magnetic effect of current (173), which laid the foundation for the great role of electromagnetic phenomena in science and technology.

Electrochemical telegraphs dependent on the electrolysis of water and capable of transmitting messages over distances of the order of two miles were proposed by Salva in 1804 (200) and by Soemmering in 1809 (210). Soemmering's telegraph originally had 35 separate circuits corresponding to the letters of the German alphabet and the requisite numbers, so that hydrogen evolution could be caused at any desired terminal at the receiving end (Fig. 1.3). Later a code was adopted and only

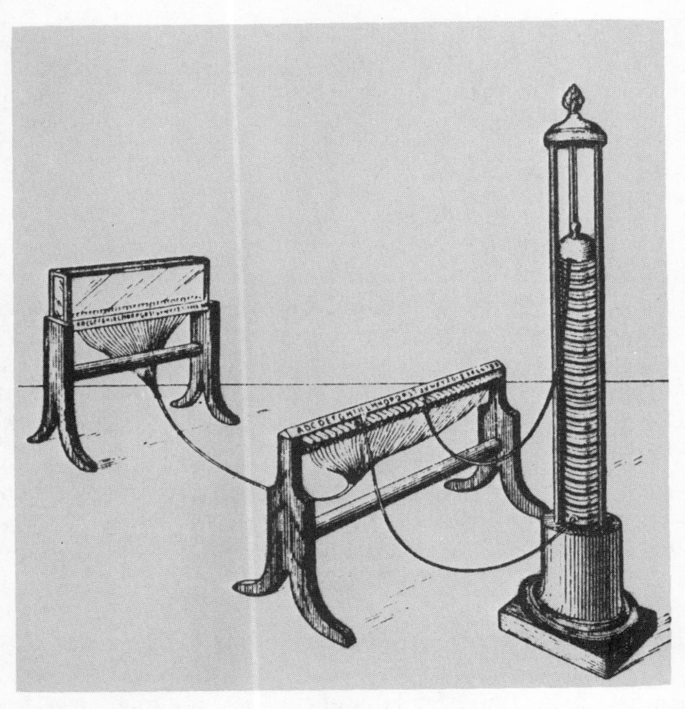

Figure 1.3 The Soemmering telegraph. From (217), courtesy of Houghton-Mifflin Co., Boston, Mass.

a single circuit was required. In England, the project was refused governmental support because "telegraphs of any kind are wholly unnecessary and no other than the kind now in use [i.e., the semaphore] will be adopted." [Cf. Mottelay (159), p. 439.]

Early theory

Volta's work demolished Galvani's theory of "animal electricity." His own widely accepted assumption that the current of a battery originated simply from the (dry) contact of dissimilar metallic conductors,[3] "Electricité excitée par le simple contact mutuel des métaux de différente espèce, et même par celui des autres conducteurs," (233) may actually

[3]In consequence, the early Volta batteries were made up of *pairs* of the electrode metals in dry contact, separated from the adjacent pairs by the electrolyte member (see Fig. 1.2), so that each battery had two superfluous metal pieces, zinc at the *cathode* and copper or equivalent at the *anode*. The fact that contact with the zinc *terminal* actually meant contact with the copper *electrode,* and vice versa, has caused some confusion in the interpretation of early electrochemical work (111).

have hindered rather than stimulated the rational development and improvement of cell systems. Indeed, in the absence of a more satisfactory explanation for the action of a cell, subsequent development of batteries was relatively slow and, before 1820, the enthusiasm for and interest in electrochemistry seems at least temporarily to have lost intensity. Thus a book by Bostock on galvanism published in 1818 (20) contains the following (p. 102):

> Although it may be somewhat hazardous to form predictions respecting the progress of science, I may remark that the impulse which was given in the first instance by Galvani's original experiments, was revived by Volta's discovery of the pile, and was carried to the highest pitch by Sir Humphry Davy's application of it to chemical decomposition, seems to have, in great measure, subsided. It may be conjectured that we have carried the power of the instrument to the utmost extent of which it admits; and it does not appear that we are at present in the way of making any important additions to our knowledge of its effects, or of obtaining any new light upon the theory of its action.

Fortunately, the foundations for a more rational interpretation of battery action were laid at an early date. Even before the publication of Volta's paper, Fabroni (73), in amplication of a 1792 memoir to the Florentine Academy, wrote:

> inquiring whether the phenomenon of galvanism is not solely due to chemical affinities of which electricity may be one of the concomitant effects and also ascribing the violent convulsions in a frog to a chemical change which is produced by the contact of one of the metals with some liquid matter on the animal's body, the latter decomposing and allowing its oxygen to combine with the metal.

Similarly, Haldane (106), in agreement with the views of Fabroni, said of the Volta pile:

> The whole operation can be received [sic] only on a combustion similar to that which arises from the combination of sulfur, and iron filings with water.

And in the same vein, Wollaston (248) in 1801 expressed the belief that the power of the battery was due to the oxidation of the anodic metal and "proportional to the disposition of one of the metals to be oxidated by the fluid interposed." Van Marum and Pfaff (152), Teed (218), and Parrot (176) are credited with similar observations.

Nevertheless, the contact theory was slow to be discarded. As late as 1840, Faraday (76), in his *Experimental Researches* (Ser. 16, No. 1797), includes Becquerel, Oersted, de la Rive, Ritchie, Pouillet, and Schoenbein among those responsible for the development of the theory of chemical action, but he notes that "a host of philosophers . . . all bright stars in the exalted regions of science," Pfaff, Zamboni, Matteucci, among others and, "as to the excitement of the power, even Davy had sustained Volta's theory of contact." The inclusion of Davy in this list may be open to question, in view of that investigator's statement made in 1800 (58):

> The oxidation of the zinc in the pile and the chemical changes connected with it are *somehow* the cause of the electrical effects it produces.

Faraday in 1840 (76) stated his own views as follows (Ser. 17, Nos. 2038 and 2071):

> Where no chemical action occurs no current is produced. The contact theory assumes, in fact, that a force which is able to overcome powerful resistance . . . can arise out of nothing; that without any change in the acting matter or the consumption of any generating force, a current can be produced which shall go on forever against a constant resistance, or only be stopped, as in the voltaic trough, by the ruins which its exertion has heaped up in its own course. This would indeed be *a creation of power*, and is like no other force in nature.

By no means the least of the contributions to battery science and technology were Faraday's studies of the mechanism of electrochemical processes, in the course of which he developed the nomenclature which is still in use. He rejected "the term *pole*, with its prefixes of positive and negative, and the attached ideas of attraction and repulsion," and stated this belief:

> The *poles*, as they are usually called, are only the doors or ways by which the electric current passes into and out of the decomposing body . . . .

On this basis, in a paper read before the Royal Society early in 1834, he proposed the terms *electrode* (from "odos," a path or way), *anode* (a way up), *cathode* (a way down), *electrolyte, electrolysis, ion, anion* (that which goes up), *cation* (that which goes down). He considered the anode "where oxygen, chlorine and acids are evolved" as negative, the cathode as positive. He did this, he says [(76), Ser. VII, No. 11, pp. 661 ff.; cf. also (75), Vol. II, 1758, pp. 272 ff.]:

> . . . merely in accordance with the conventional, though in some degree tacit, agreement entered into by scientific men, that they may have a constant, certain, and definite means of referring to the directions of the forces of that current.

Mechanical development and improvement

The original Volta cells, dependent entirely on the anode reaction as their source of power, were, of course, very feeble by modern standards, and by no means as "indéfectible" or "perpétuelle" as optimistically proclaimed by their inventor. A logical first step toward more powerful batteries, other than mechanical improvement on the pile itself,[4] was development of the wet cell, thus ensuring an adequate supply of electrolyte and correspondingly greater capacity. As early as 1800, Cruickshank (51) developed the "trough" battery (Fig. 1.4) in which pairs of zinc and copper electrodes fastened back to back were cemented into grooves in the walls of the container to form the partitions between cells. Wilkinson (245), using a compartmented trough, attached the metal electrodes to a rod so that the cell assemblies could be lifted from the electrolyte when not in service, an expedient widely used in subsequent battery technology to minimize shelf deterioration.

Another important step was taken by Wollaston (249), who in 1815 devised a cell assembly comprising a flat zinc anode with the copper cathode bent around it the form of a U (Fig. 1.5). The electrochemical utilization of both sides of the zinc doubled the active electrode surface without substantial increase in size of cell, a construction simplified by Oersted's use of copper boxes to serve not only as cathodes but also as cell containers (172).

Despite these and similar efforts at improvement, the early cells, deriving their power solely from the dissolution of the zinc or equivalent anode, and quickly polarizing because of the cathodic liberation of hydrogen, remained weak and incapable of sustained current output. To meet the growing demands for greater power and capacity, some fantastic increases in size were recorded in both electrode area and

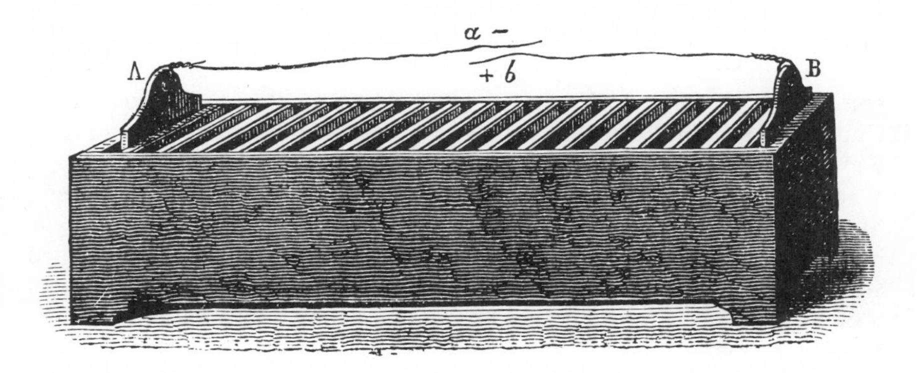

Figure 1.4 Cruickshank trough battery. From (165), Fig. 4, p. 12.

[4]By way of example, Volta, and subsequently others, noted that deterioration could be reduced by coating the pile with wax or equivalent to prevent loss of moisture (and wasteful corrosion), and Ritter turned up the edges of the metal electrodes "to prevent the liquid pressed out of the separators from flowing away" (193).

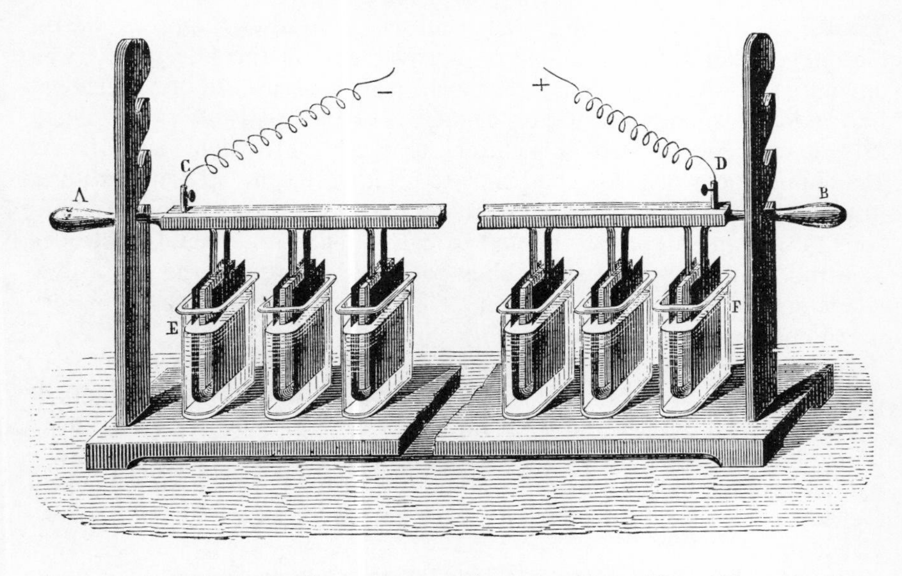

Figure 1.5 Wollaston battery – plunge type. From (34), p. 101.

volume of electrolyte. Thus Pepys (179) in 1802 is reported to have built a 60-cell battery of the Cruickshank type with electrodes 6 feet square. A plunge battery of the Wilkinson type built for Sir Humphry Davy's use had 2000 4-inch square "double plates" of zinc and copper. It was with this battery, paid for by the Royal Society of London through member subscriptions, that Davy in 1809 (64) gave a public demonstration of an electric arc "of nearly 3 inches in length and of dazzling splendor." A 20-cell Wollaston-type plunge battery constructed by Children (36) had electrodes 6 feet by 2 feet, 6 inches in size and took 945 gallons of electrolyte.

In 1819, Robert Hare, in the United States, described helical assemblies to obtain cells of large electrode area (108) (Fig. 1.6). A plunge-type battery constructed about 1822 on this principle by Pepys (180) contained strips of copper and zinc 50 to 60 feet long by 2 feet wide, spirally wound around a wooden cylinder and separated from one another by three horsehair ropes.

It was quickly recognized that polarization due to liberation of hydrogen at the cathode reduced the power output of the simple Volta cell, and efforts to overcome this difficulty were quick to follow. As early as 1800, Haldane (106) and others discovered that exposure to air or oxygen improved the performance of a Volta pile, and it was soon observed that, in a wet cell, polarization could be reduced or eliminated

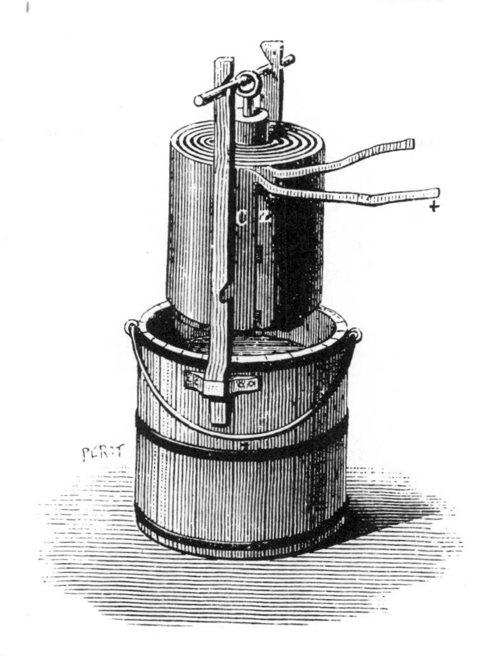

Figure 1.6 Helical battery assembly. From (165), p. 16.

by briefly withdrawing the cathode from the electrolyte. Even stirring of the electrolyte was found to have beneficial effect on the behavior of a wet cell, and Faraday (76) reported [Ser. 8, No. 1039 (1834)] that simply agitating the sulfuric acid electrolyte "by the introduction of a feather" between the zinc and copper electrodes "more than doubled the power of the battery."

Mechanical depolarization

Since even the proponents of the chemical theory considered only the anode reaction a source of power, observations such as the foregoing led to physical ways of reducing cathodic polarization, rather than to the search for active cathodic materials capable of contributing energy to the system. As a result, mechanical methods for overcoming hydrogen polarization [cf. Tommasi (219), pp. 96 ff.], in which cathodes were shaken, brushed, alternately raised and lowered, or mounted axially to permit rotation, partly in air, partly in electrolyte, persisted throughout the nineteenth century. Some were dependent on manual operation, some on clockwork, and some, according to the optimistic claims of their inventors, on part of the battery's own energy output (127) (Fig. 1.7).

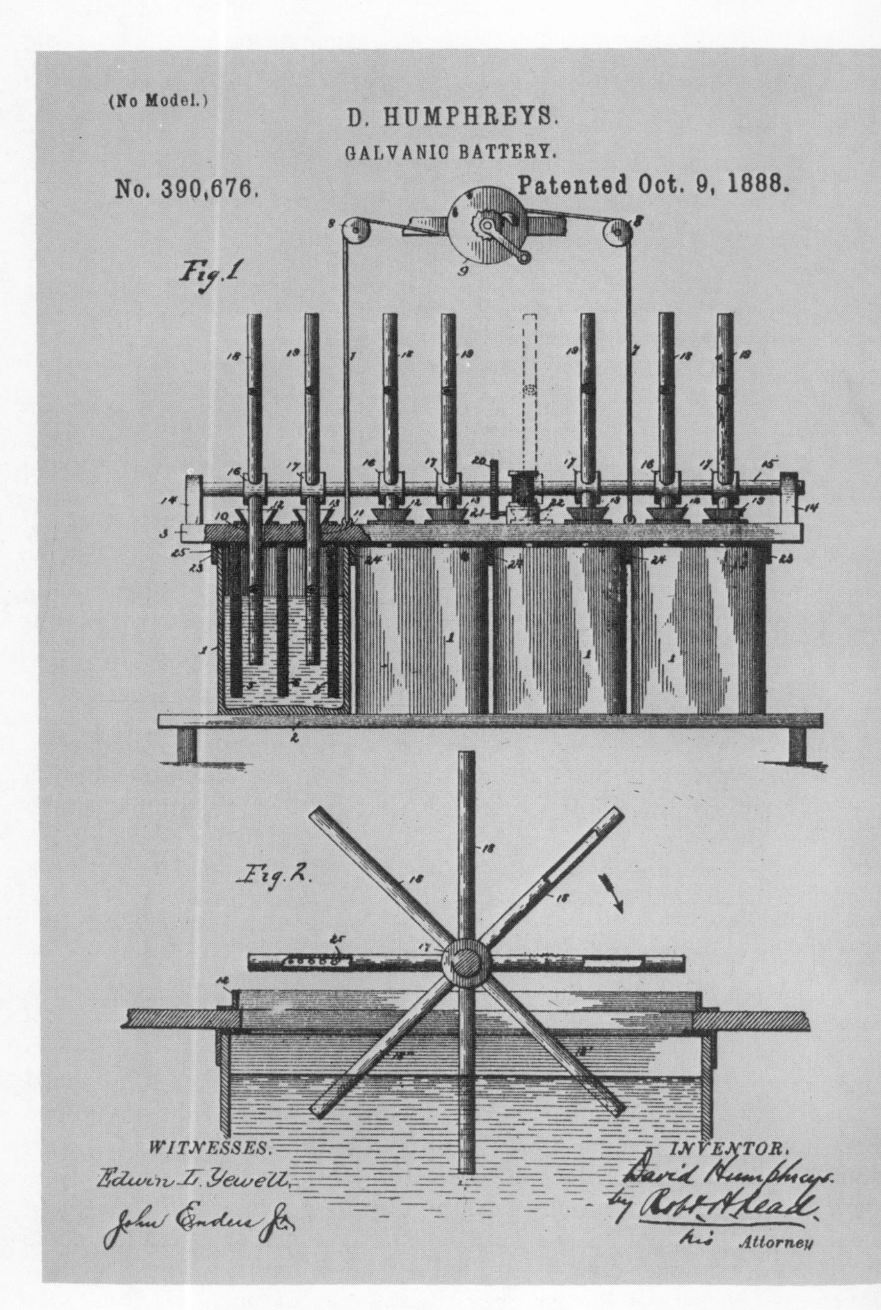

Figure 1.7 Mechanically depolarized battery. From (127).

With the introduction of efficient cathodic reactants, such methods, never of greater technical importance, have become matters of historic rather than of practical interest. Indeed, the unique advantage of the primary battery is its ability to function as a source of electrical energy without moving parts and, except for systems, e.g., those of the continuous-feed type (cf. Chap. 9), which demand forced movement of electrolyte or fluid reactants, there is little or no need for auxiliary equipment in primary battery technology.

In view of the state of electrochemical knowledge in the early decades of the nineteenth century, it is hardly surprising that a logical step in the improvement of the Volta cell, the reduction of polarization through the use of cathodes of highly developed surface and intrinsically low hydrogen overvoltage, should have been slow in coming. However, in 1840 Smee (209) proposed a cell capable of sustained operation, in which a vertical cathode of platinized platinum or platinized silver was suspended between two amalgamated zinc plates with sulfuric acid as electrolyte. Various modifications and improvements followed in due course, notably Walker's substitution in 1857 of platinized carbon for the platinum or silver previously used (237). Although such cells gave only about ½ volt on normal operating load, they were widely used at least as late as 1875 for plating, telegraph (224), and even, according to Cazin (34), for the operation of Transatlantic cables. Good results also were claimed for a later modification, the "Velvo-Carbone" cell described by Barnett (5). Cathodes for this unit were made by cementing velvet or fustian to carbon rods, and subsequently carbonizing the cloth by heat to retain the pile and high surface area of the fabric.

Despite its many defects, the simple voltaic wet cells, e.g., zinc: carbon elements with ammonium chloride electrolyte, proved useful for many years, particularly in applications such as doorbell and annunciator service, not requiring continuous or heavy-drain operation.

RISE OF THE COMMERCIAL BATTERY (1830–1880)

Although there are numerous references to the early use of oxidizing substances such as nitric acid [(58), (218), (108)] or ferric sulfate (64) in the catholyte, or manganese dioxide[5] at the cathode, the prevention of hydrogen polarization by chemical means[6] was slow to find acceptance

[5]Volta himself noted the high cathodic potential of manganese dioxide at least as early as 1802 (234), and Zamboni incorporated this material in his "dry pile" of 1812 (251).

[6]It is most unfortunate that oxidizing reactants should have been considered loosely, even in recent times, as "depolarizers" acting by the oxidation of *liberated* hydrogen. They are actually the reactants determining the electrode potential and operating at voltages

in battery practice. When it did emerge, in the 1830s, cells became available in which cathodic as well as anodic reactions contributed to the energy output of the system, thus establishing the basis of modern battery technology. Still the sole source of continuous current, the primary battery, rapidly became an important article of commerce. The new, reasonably efficient batteries were used in such diverse, rapidly expanding fields as electrometallurgy (electroplating, electrotypy), communication, the operation of motors, and even the charging of storage batteries.

By 1838, the Russian physicist M. H. Jacobi had developed and demonstrated the production of "copies of engraved copper plates by voltaic action" (131). The graphitizing of nonmetallic surfaces, developed by Murray in 1840, made possible the use of matrices of nonconducting materials such as wax or gutta percha [cf. Foerster (79), p. 483]. Samuel F. B. Morse began his work on the telegraph in 1832; a 40-mile experimental line between Washington, D.C., and Baltimore, Md., was completed in 1844; by 1851, the year of the laying of the first submarine cable between Dover and Calais, some 50 telegraph companies were in operation in the United States alone. The telephone was invented by Alexander Graham Bell in 1876 and came into public use a year later. The first electric motors of significant size were produced independently by Thomas Davenport (185) in the United States and by Jacobi (130) in Russia, in 1834. In 1838, Jacobi operated a battery-powered, motor-driven boat on the Neva river at St. Petersburg, and "in 1839 a (Davenport) motor weighing one hundred pounds, horsepower unknown, was driving a large printing press in New York City" [cf. Taylor (217), p. 700]. The lead storage battery (Pb: H_2SO_4: PbO_2) was developed—the Planté system with solid lead electrodes in 1859 (183), the Faure prototype of the modern "pasted grid" cell between 1870 and 1880 (78)—while the only convenient source of current for recharging purposes was the primary battery.[7]

Even the science fiction of the day gave full recognition to the importance of the primary battery. Jules Verne (227), for example, equipped

---

precluding the discharge of hydrogen (57). A more rational nomenclature would be "cathogen" or "cathodate" for the cathodic species, "anogen" or "anodate" for the anodic, but the term "depolarizer" seems too well entrenched to make substitution practicable at this time.

[7]An amusing sidelight on recharging is furnished by Faraday's speculation [(76), Ser. XV, 1792 ff.] that "restoring the powers" of a "Gymnotus or Torpedo" (electric eel) might be accomplished by the "sending of currents . . . in the same [sic] direction as those he sets forth"; conversely, he asked, "Would sending currents through in the contrary direction exhaust the animal rapidly?"

Prof. Hardwigg with a battery to light his way to the center of the earth. The power source of the submarine Nautilus, the creation of the same author, was also a primary battery; Captain Nemo expressed preference for the system of Bunsen over that of Ruhmkorff for his undersea travels.

Some bizarre applications have been suggested at various times, such as the "electrogenesis" of insects (49) and the production of "metallic mummies" by the electroplating of corpses (169). Various phases of electrotherapy were also formulated (164); the electric belt has been used extensively as a cure-all, even in recent times. In its original form, as described (218) by its inventor in 1801, this was a "galvanic belt or chain . . . consisting of fifteen small square plates of zinc and connected with two links of plates [sic] copper wire . . ." which afforded relief from "a constant pain in the small of my back and loins." In an improved modification, duly reported in 1857 to the French Academy of Sciences by Pulvermacher (188), two wires, one of zinc and one of copper, were spiral-wound, without touching one another, around small wooden cylinders to form the individual cell assemblies, which were connected in series and mounted on a flexible belt, which could be wrapped around the body (Fig. 1.8). This was activated by immersing the battery in dilute vinegar, the wetted wood cylinders "playing the role of the cloth sheets in the columnar piles" [(34), p. 108]. A means of extracting poisons from the human body (31) by means of a battery is shown in Fig. 1.9.

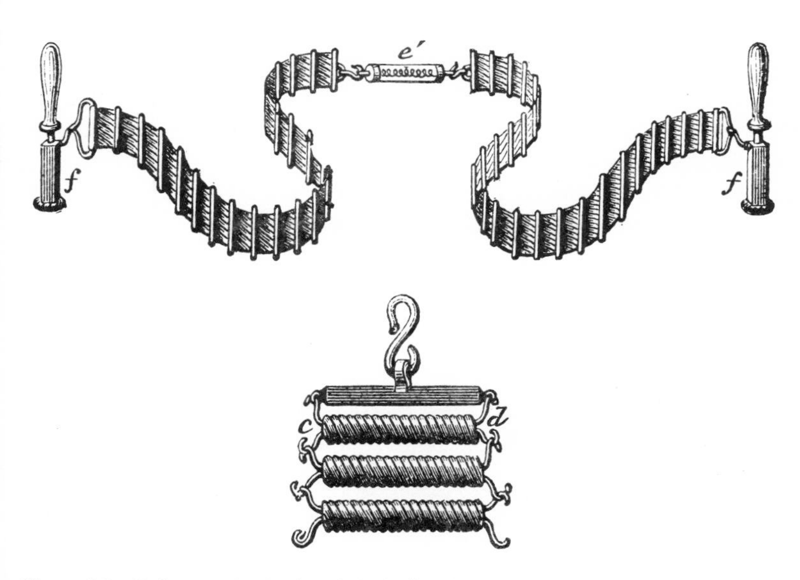

Figure 1.8 Pulvermacher's electric belt ("chaine galvanique"). From (34), p. 107.

606,887. ELECTRIC EXTRACTION OF POISONS. JOHN B. CAMPBELL, Cincinnati, Ohio. Filed Oct. 5, 1896. Serial No. 607,955. (No model.)

Claim. – The herein-described method of extracting poisons from the human body, consisting in providing the negative electrode of an electric battery with interchangeable receivers of animal, vegetable or mineral substance, to extract a poison from the body of a nature or kind corresponding with or to the negative electrode employed, as set forth.

Figure 1.9 Electrical extraction of poisons. From (31).

The primary battery was the only commercial source of continuous current until 1882, when the Edison system of central power production was installed. Few, if any possible combinations of available battery elements seem to have been overlooked during this period. Tommasi's treatise on batteries, for example, published in 1889 (219), has over 400 entries under the heading "piles diversés," many of them, to be sure, only minor variants of basic types. A comprehensive listing of these is beyond the scope of the present work, and only those which have had major influence on the course of battery development will be considered here.

Two-fluid cells

The first cathodic reactants to receive wide acceptance were water-soluble; used in single-fluid cells, they caused accelerated corrosion of the anode metal and self-discharge of the system. This defect was of minor importance for batteries that could be assembled just before use and subjected to immediate discharge, preferably on continuous load and in a comparatively brief period. In normal practice, however, some means of preventing contact between anode and cathodic reactant had to be provided to ensure adequate battery life and power output. Perhaps the simplest method, and one widely used (*vide supra*) was the removal of the anode metal from the electrolyte during periods of idleness, which obviously limited the usefulness of the battery. Better, and historically more important, was the development of two-fluid cell systems in which the catholyte, containing the cathodic species, was separated from the anolyte, perhaps by means of a diaphragm or porous pot.

Daniell Cells. Among the most important of the two-fluid cells is the system which, though definitely anticipated by Becquerel (1829) (9) and Wach (1830) (235), is generally credited to Daniell (1836) (53). An early modification (Fig. 1.10) comprised a copper cup (the cathode) (A) and a centrally located amalgamated zinc rod (2), with an animal membrane (gut) (B) separating anode and cathode compartments. The anolyte was dilute sulfuric acid, the catholyte a saturated solution of copper sulfate. During discharge, zinc dissolved anodically

$$Zn \rightarrow Zn^{++} + 2e$$

while copper in equivalent amount was deposited at the cathode

$$Cu^{++} + 2e \rightarrow Cu$$

the over-all reaction being expressed by the equation

$$Zn + CuSO_4 \rightarrow ZnSO_4 + Cu$$

To prevent premature depletion of copper sulfate, crystals of this salt were placed in an annular grid (G) at the top of the cathode compartment. As the anolyte became saturated with zinc sulfate, fresh acid could be added at the top (E) of the anode compartment, while the heavier, exhaused solution was removed through a siphon (T).

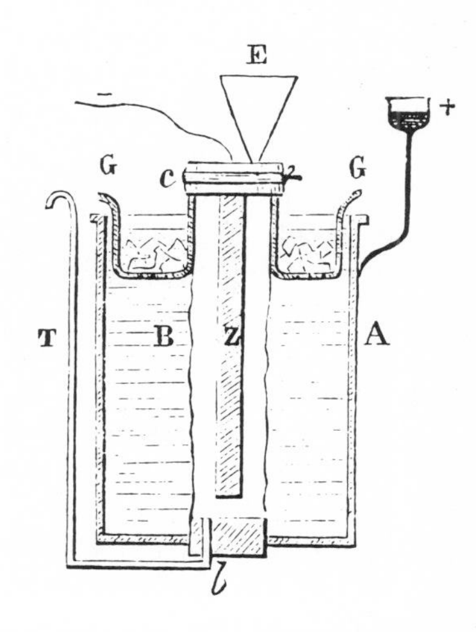

Figure 1.10 Early form of Daniell cell. From (34), p. 169.

Because of its simplicity and its good operating characteristics, the Daniell cell subsequently appeared in a number of forms, now of historic rather than technical importance. Porous ceramics soon replaced animal membranes (90), and anode and cathode compartments were interchanged, as shown in Fig. 1.11. Methods for increasing the amount of copper

Figure 1.11 Daniell cells with ceramic diaphragms. From (165), p. 78.

sulfate available for service were devised as in the "*pile à ballon*" of Fig. 1.12. Trough-type multicell batteries with porous plate separators were developed; the sulfates of zinc or magnesium were substituted for sulfuric acid; and various methods designed to reduce interdiffusion of anolyte and catholyte through the separators were introduced.

In the so-called gravity cell, introduced in the 1850s and variously credited to Varley (225), Callaud (30), and Meidinger (155), the diaphragm was entirely omitted. In this modification the electrolyte was in two layers, with the heavier, saturated copper sulfate solution and copper sulfate crystals at the bottom, the lighter anolyte, containing zinc sulfate or equivalent, at the top. This became the most popular form of the Daniell cell (Figs. 1.13 and 1.14).

Eventually there was even a dry type, credited to Trouvé (223), in which the space between horizontal zinc and copper electrode plates was occupied by disks of blotting paper, the uppermost wet with zinc sulfate, the lower with copper sulfate solution (Fig. 1.15). Commercial batteries containing as many as 500 such cells were marketed in the 1880s.

The Daniell cell had an open-circuit voltage of 1.08. Since there was no significant polarization on normal drain, and cell composition changed relatively little during service, operating voltage was well sustained throughout the battery's useful life. Copper ion migrated into the anolyte during periods of idleness, attacking the zinc electrode, a tendency which could be minimized by maintaining a low-drain short-circuit while the cell was not in use. Despite this handicap and an inherently high internal resistance, the gravity-type Daniell cell found wide application in

Figure 1.12 "Pile à ballon" (Breguet). From (165), p. 93.

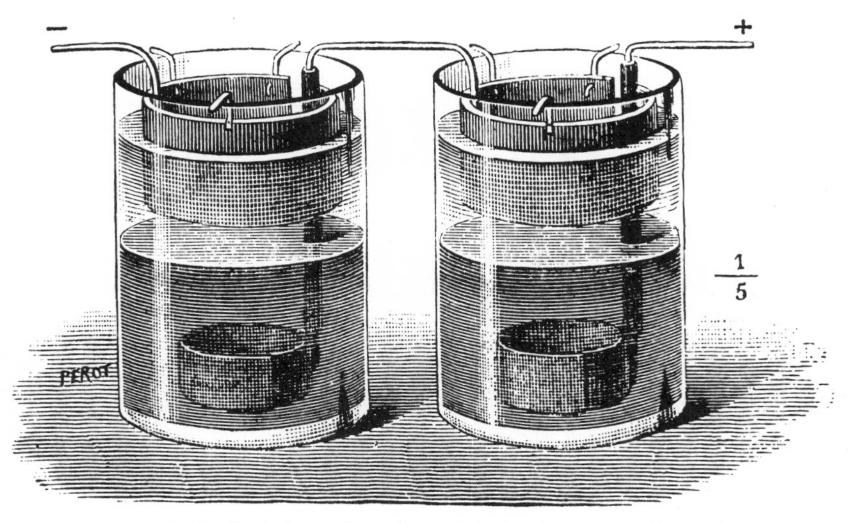

Figure 1.13 Early form of gravity cell (Callaud). From (165), p. 109.

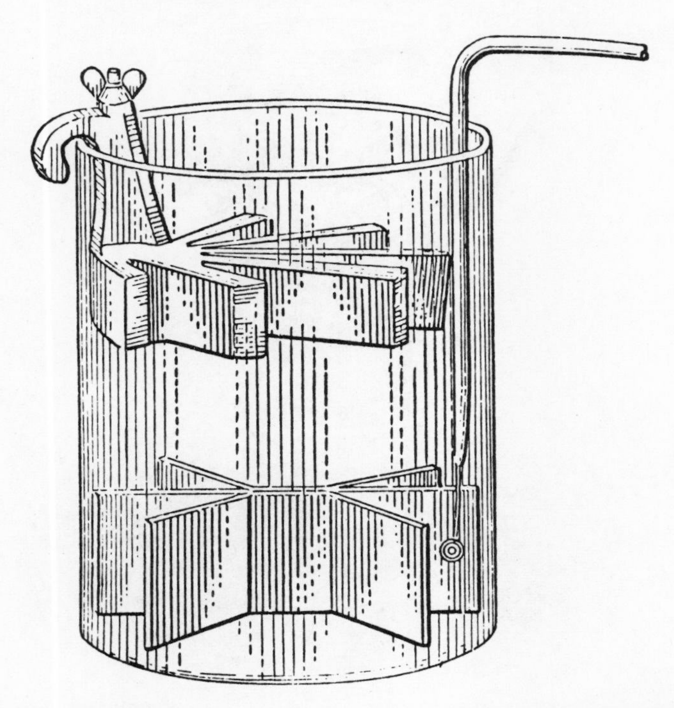

Figure 1.14 "Crowfoot"-type gravity cell. From W. A. Koehler Principles and Applications of Electrochemistry, Vol. II, Applications. p. 5.

telegraph and railway signal installations, until well into the twentieth century, when it was largely superseded by alkaline batteries of the Lalande-Chaperon type (Zn : NaOH : CuO) (140) (cf. Chap. 4).

Nitric Acid Cells. The Grove cell (101), which made its appearance in 1838, marked another important milestone in the development of the primary battery. As the element was originally constructed (Fig. 1.16), strong nitric acid inside the bowl of a clay pipe served as catholyte; the anolyte, ranged from acid to alkali but usually sulfuric acid, occupied the outer portion of the cell-containing vessel. The anode was amalgamated zinc, the cathode collector sheet platinum. Though nitric acid had previously been used in batteries, this seems to have been the first time an oxidizing agent in solution was recognized as a cathodic reactant rather than primarily as electrolyte. A commercial Grove cell is shown in Fig. 1.17.

Because of its high voltage (approximately 1.9) and its high power output, the Grove cell was soon widely used, and many modifications in

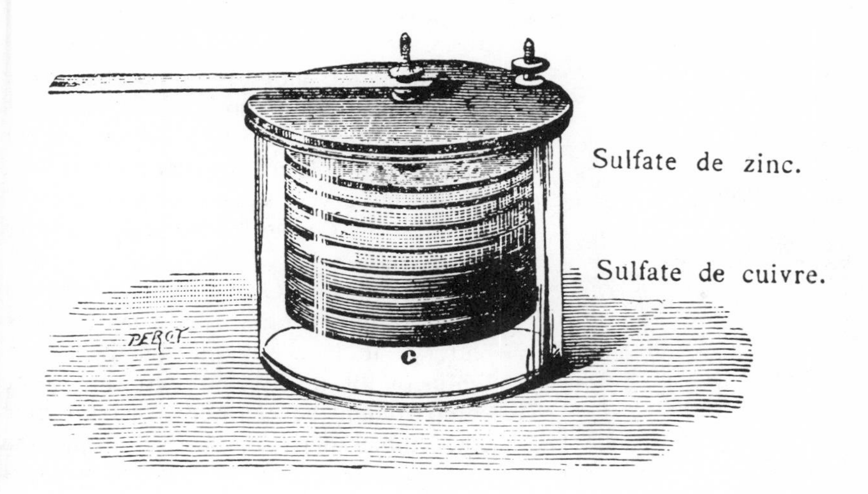

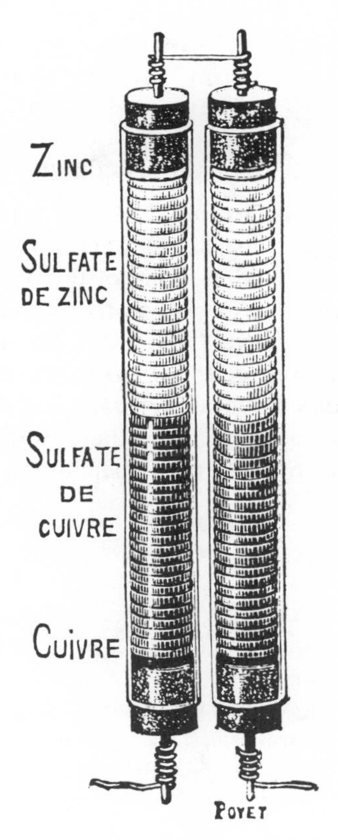

Figure 1.15 Trouvé (Daniell-type) dry batteries. From (219), p. 251.

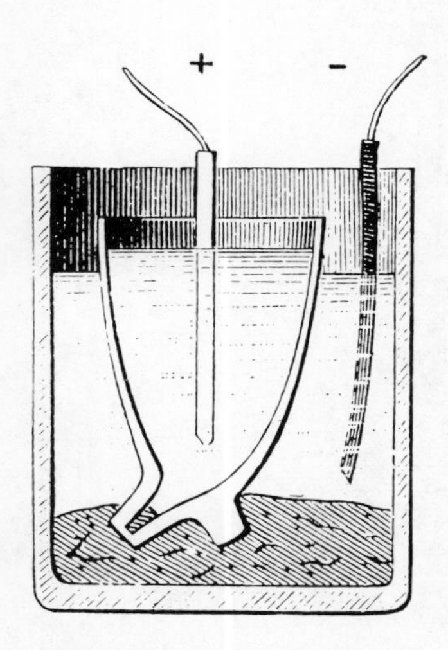

Figure 1.16 Original form of Grove's nitric-acid cell (1838). From (34), p. 209.

structure and anolyte composition were introduced. Most important was Bunsen's (27) substitution, in 1841, of carbon[8] for platinum, which greatly reduced the cost of the battery. One of the commercial forms of the Grove-Bunsen cell is shown in Fig. 1.18.

Although the Grove-Bunsen element found wide acceptance as a heavy-drain battery, it had several disadvantages. Diffusion of nitric acid into the anode compartment accelerated wasteful zinc corrosion. Indeed, the use of strong acid was itself objectionable due to noxious fumes and could even create a hazard in the absence of adequate ventilation during cell discharge, through reduction of nitric acid to nitric oxide gas (NO), and conversion of the latter, by atmospheric oxidation, to dinitrogentetroxide (N_2O_4).

Bichromate Cells. The substitution (1842) of chromic for nitric acid, generally credited to Poggendorff (184) but apparently considered by others [(27), (239)] as well, avoided many of the difficulties encountered in the Grove-Bunsen cell and gave rise, in its day, to the most important of the heavy-duty batteries. Like its Grove-Bunsen prototype, the original Poggendorff unit was two-fluid cell, with zinc anode and dilute sulfuric acid on one side of a porous diaphragm, a carbon electrode and a bichromate dissolved in sulfuric acid on the other. Sodium bichromate was generally preferred over the potassium salt because of its greater solubility.

The over-all cell reaction was

$$3Zn + Na_2Cr_2O_7 + 7H_2SO_4 = 3ZnSO_4 + Cr_2(SO_4) + Na_2SO_4 + 7H_2O$$

Cell voltage was 2.0–2.1, slightly above that of the Grove-Bunsen system.

[8]The use of various forms of carbon or graphite as cathode collector was by no means new, having been anticipated by Volta, Davy, and, according to Cazin (34), practiced independently by Chevreusse (35) in France in 1823, by Cooper (44) in England in 1840, by Schoenbein (204) in Germany in 1840, and by Silliman (206) in America in 1842. Apparently it remained for Bunsen to introduce a commercially practicable electrode in the form of shaped, cylindrical, element of molded and baked mixtures of carbons such as coke and charcoal, bonded with sugar, coal tar or the like.

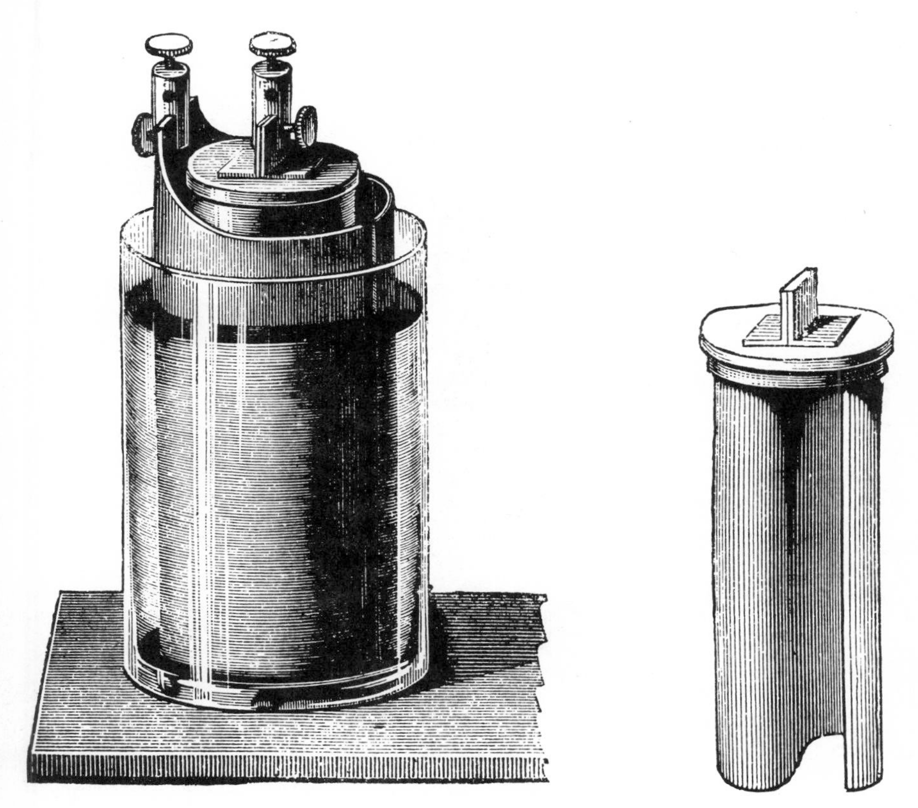

Figure 1.17 Grove cell with platinum cathode. From (165), p. 155.

Variations in construction were numerous. In the popular Fuller cell (84) (Fig. 1.19) anolyte, zinc anode and a small amount of liquid mercury were in the inner compartment, the cathode system in the outer. The Baudet "*pile siphoide*" system (6) (Fig. 1.20), which was developed to increase ampere-hour output during continuous discharge, provided means for introducing fresh and removing exhausted electrolyte.

In some modifications, the diaphragm was omitted (138). Perhaps the most popular of these was the type introduced in 1856 by Grenet (99), in which the zinc anode could be raised above the electrolyte to minimize wasteful corrosion during periods of idleness (Fig. 1.21). In the cylindrical Trouvé type cell (222), used also for other battery systems, the same effect could be achieved by turning the container upside-down (Fig. 1.22).

Because of its simplicity of construction, its relatively low cost of

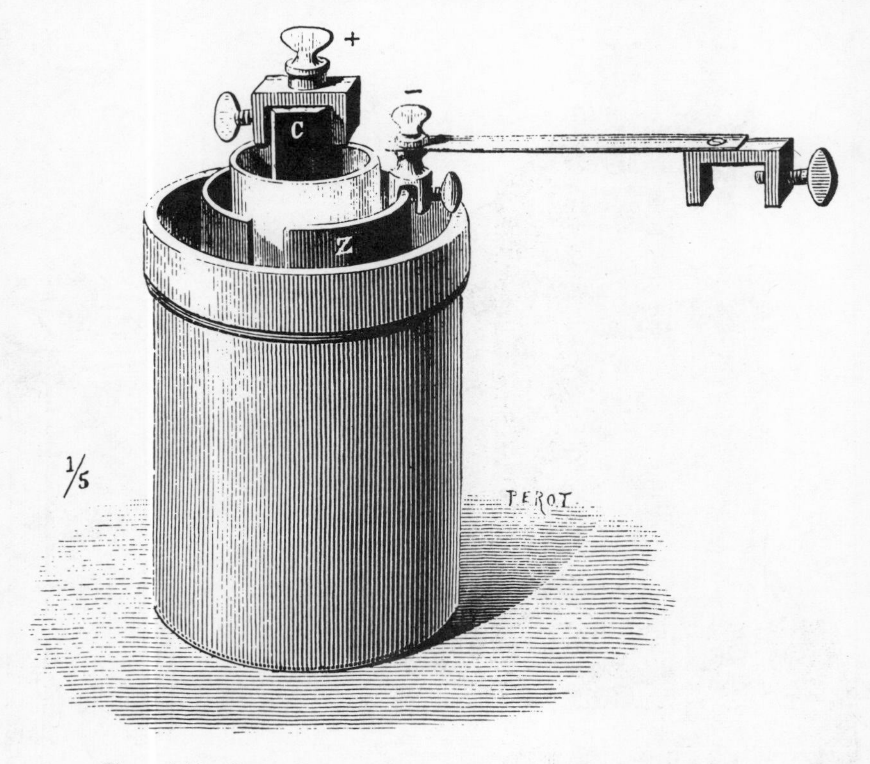

Figure 1.18 Grove-Bunsen cell (carbon cathode). From (165), p. 158.

operation, and its excellent service characteristics, the Poggendorff cell found many uses, in telegraphy (66), in electroplating, and in general applications for which high voltage and large ampere-hour capacity were important considerations. In 1906, Crocker (48) proposed a bichromate battery as a power source for automobiles, and as late as 1916, Cooper [(45), p. 423] concluded that £20 would cover the annual battery cost for lighting the principal rooms of a small country house, "which in many cases, would not be an exorbitant price to pay."

Bichromate cells have also served well throughout the years in military applications. The French developed portable forms (181) used in the Franco-Prussian War for the explosion of mines, in the field and during the siege of Paris. A reserve-type battery was used as recently as World War II as power source for the radio component of the proximity fuze (56).

The French were among the first to recognize the advantages of providing the electrolyte ingredients (mixtures of sodium sulfate, sulfuric acid, and bichromates) in dry form, easily shipped and handled, from

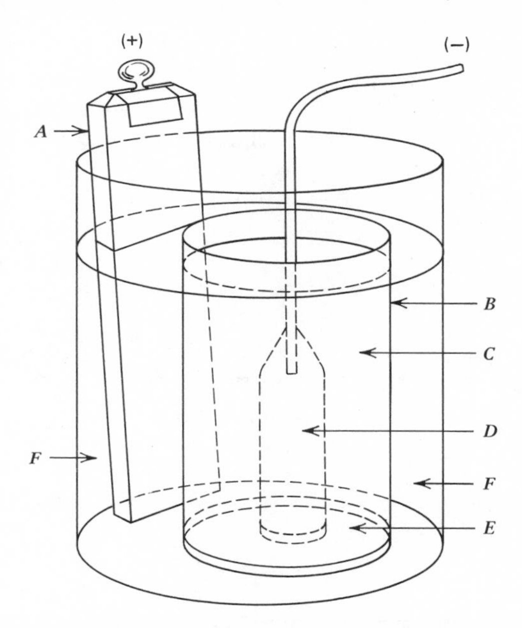

Figure 1.19 Fuller bichromate cell. (*A*) Carbon (cathode) collector plate; (*B*) porous pot; (*C*) anolyte: dilute H_2SO_4; (*D*) zinc anode; (*E*) mercury layer; (*F*) catholyte: alkali bichromate and dilute H_2SO_4.

which the electrolyte itself could be prepared by the simple addition of water. The sale of such "exciting salts" persisted well into the twentieth century under various trade names, e.g., "sel de Voisin et Drosnier" [(219), p. 139] in France, "Chromolyte" in Great Britain, and "Chromac" in the United States.

Miscellaneous Cells. As might be expected, many soluble oxidants other than nitric or chromic acids were used over the years as cathodic reactants,[9] the list including such diverse materials as permanganates, hypochlorites, chlorates, and even aqua regia (144). As late as the

[9]In the system

$$Pb : HClO_4\,(aq) : PbO_2$$

developed in the 1940s (205), perchloric acid served not as depolarizer but as electrolyte in which, as with fluoboric or fluosilicic acids, both anodic and cathodic species were soluble:

$$Pb + 4HClO_4 + PbO_2 \rightarrow 2Pb(ClO_4)_2 + 2H_2O$$

Owing to the instability of the system, cells were of the reserve type.

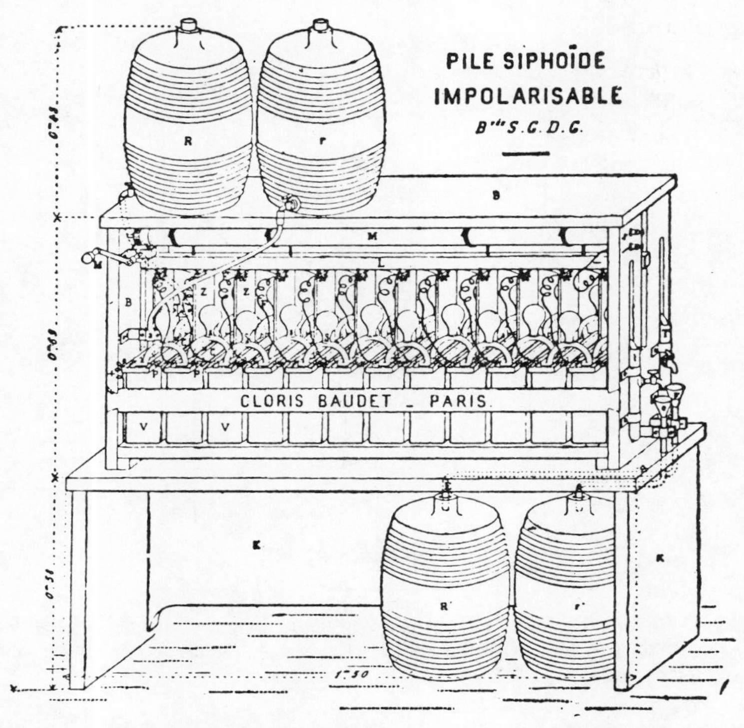

Figure 1.20 "Pile siphoide" (Baudet). From (219), p. 207.

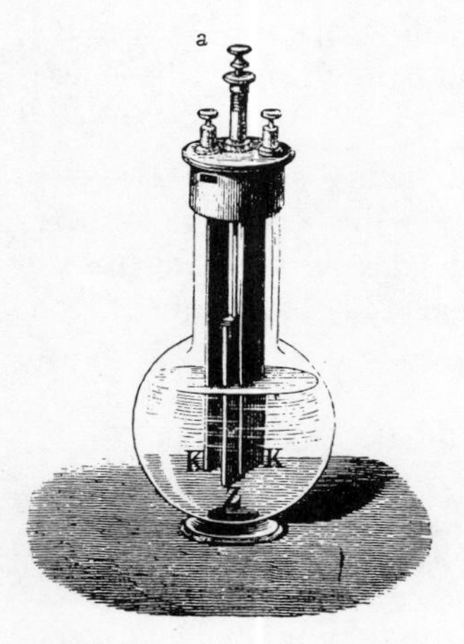

Figure 1.21 Single-fluid bichromate cell (Grenet). From (165), p. 213.

mid-1920s, a two-fluid diaphragm-type Darimont cell (55) using a ferric salt

$$Zn : NaCl\,(aq) :: FeCl_3\,(aq) : C$$

was in commercial production.

Because of the inherent difficulty of preventing excessive corrosive attack on the anode, cells of this type have found only limited acceptance. At the present time they may still have merit as "one-shot" reserve cells which can be activated just before being put into operation, and then used up in a single service discharge.

SINGLE FLUID CELLS

Cells with Solid "Depolarizers." The use of solid cathodic reactants, though long foreshadowed, was slow to be realized in battery technology. Volta (234) was aware of the high cathodic potential of manganese dioxide, and Zamboni(251) incorporated this material in his "dry pile" of 1812, a stack of dry paper disks, tinned on one side and coated with pulverized MnO_2 on the other. Faraday [(76), Ser. 17, No. 2041 ff, 1840] discussed the studies of A. C. Becquerel(10), de la Rive(196), and Muncke(160) on MnO_2 and PbO_2, as well as his own experiments with the oxides of iron and nickel.

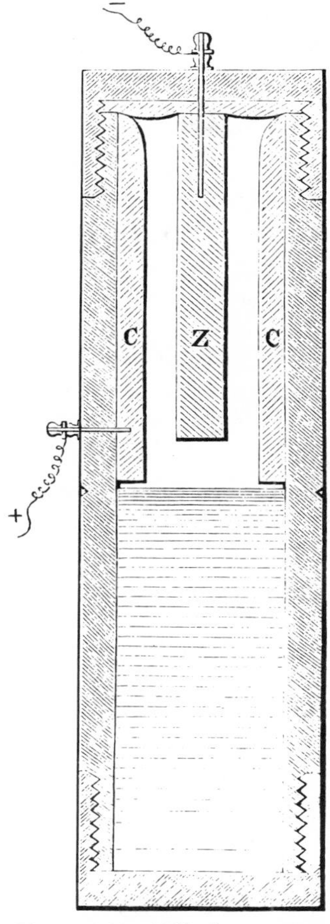

Figure 1.22 "Pile à renversement" (Trouvé). From (34), p. 157.

Of the many metallic oxides that might be considered cathodic reactants, a relatively small number found large-scale application in commercial batteries; notably the oxides of manganese, copper Chap. 4, mercury Chap. 5, silver Chap. 6, lead, and nickel, the last two for the most part in secondary cells.

The work of de la Rive, which probably was primarily responsible for the utilization of metallic oxides, culminated, in 1843, in his own development of a cell with zinc anode, sulfuric acid electrolyte, and a cathode assembly consisting of a porous cup containing granular lead peroxide packed around a platinum collector(197). The discharge reaction of this cell may be represented by the equation

$$Zn + PbO_2 + 2H_2SO_4 \rightarrow ZnSO_4 + PbSO_4 + 2H_2O$$

Because of its high voltage (2.5) and excellent heavy drain characteristics, this system attracted wide interest and appeared in various forms and under different names (Wheatstone, Reynier, Main, Harrison, etc.) over the years. A reserve cell of the added electrolyte type was developed for military purposes by Schlotter(202) as late as 1940, and a similar contemporary unit with cadmium anode was described by Pucher

(187). The cell reaction is reversible but efforts made to utilize the system in secondary batteries have met with indifferent success, owing to the difficulty under battery conditions of replating zinc uniformly and in massive, adherent form. Nevertheless, batteries of this type were used briefly, about 1887, to operate street cars in New York (236).

A cell in which the difficulty of replating zinc was avoided was in production as a radio "A" battery as late as the mid-1920s. In this instance, the cathode was a preformed PbO_2 plate as used in storage batteries, to be exchanged for a freshly recharged unit when exhausted. The amalgamated zinc anode was greatly overweight and was capable of operating for several discharge cycles.

De la Rive also made the attempt, duplicated in 1853 by Payerne [(219), p. 278] to substitute MnO_2 for PbO_2 with sulfuric acid electrolyte, but with little success.

Leclanché Wet Cell. One of the most important of the batteries with solid cathodic reactants, the prototype of the modern dry cell, was the system developed by Georges Leclanché (157), (146) in the 1860s.[10] This was a wet cell, with an amalgamated zinc rod as anode, a conducting mix of manganese dioxide and carbon in a porous pot as cathode, a central carbon rod or plate as current-collector, and a saturated solution of ammonium chloride as electrolyte (Fig. 1.23). Here, at last, was a battery that was rugged and cheap, which contained no highly corrosive acid or alkali, and which showed a minimum of shelf deterioration or self-discharge. Operating life on telegraph service of "from one to three years or even more" was expected, and the general excellence of the cell even at the time of Leclanché's first paper was attested by the "more than twenty thousand cells actually in service on different railroads, in France and abroad."

Indicative of the "state of the art," Leclanché recorded cell voltage as 1.382 times that of the Daniell cell, approximately 1.5 volts on the modern scale. For a "large model" cell he claimed an output of electrical work over a 1-year period as represented by the electrodeposition of 100–125 grams of copper (85–105 ampere hours). Internal resistance was given as about that of 450 meters of iron wire of 4 millimeter diameter. In some of the later models, cloth bags or wrappings were substituted with good results for the porous ceramic containers holding the cathode mix [(45), pp. 210 ff].

[10]The first reference to the Leclanché system seems to have been in a patent (145) issued in 1866, disclosing zinc anode, ammonium chloride electrolyte, and a cathode member of powdered manganese dioxide packed around a carbon collector plate and enclosed in a porous sac or equivalent "diaphragme" material. The crushed carbon or graphite of the commercial cell is not mentioned as component of the cathode mix.

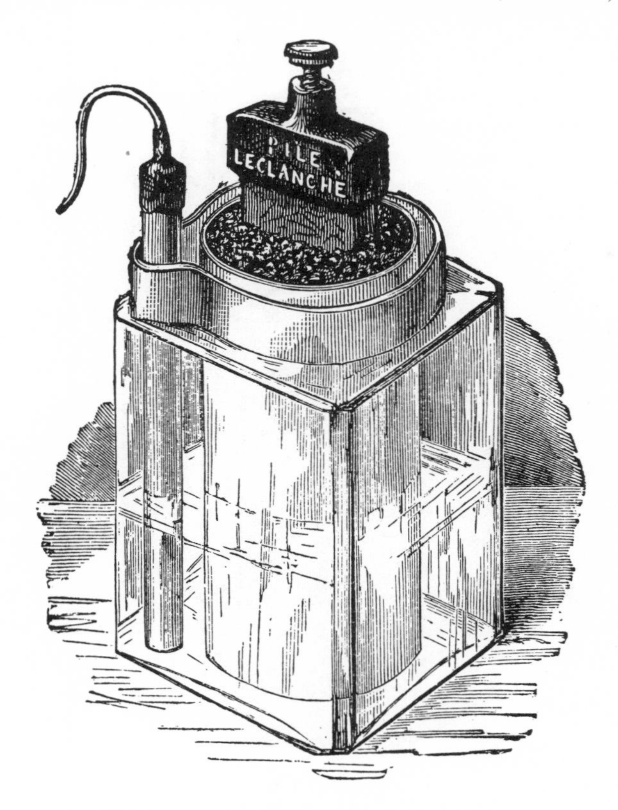

Figure 1.23 Leclanché wet cell. From (146).

Leclanché also developed agglomerated cathodes (146) resin-bonded blocks of manganese dioxide and carbon fitted against carbon collector plates (Fig. 1.24), thus eliminating the need for porous containers or envelopes. In the subsequent Leclanché–Barbier modification, probably the most important commercial type, there was no separate carbon plate; the agglomerate, in annular form, served as both cathode and current collector (Fig. 1.25). Leclanché's electrodes were described as compacted at temperatures of the order of 100°C, but commercial agglomerates, according to Drotschmann (68), were often baked at temperatures high enough to reduce MnO_2 and minimize its effectiveness.

Because of inefficient utilization of MnO_2, the Leclanché wet cell never became a satisfactory heavy-duty unit, and even on light and intermittent service, diffusion of electrodeposited hydrogen away from the cathode surface and the role of atmospheric oxygen as cathodic

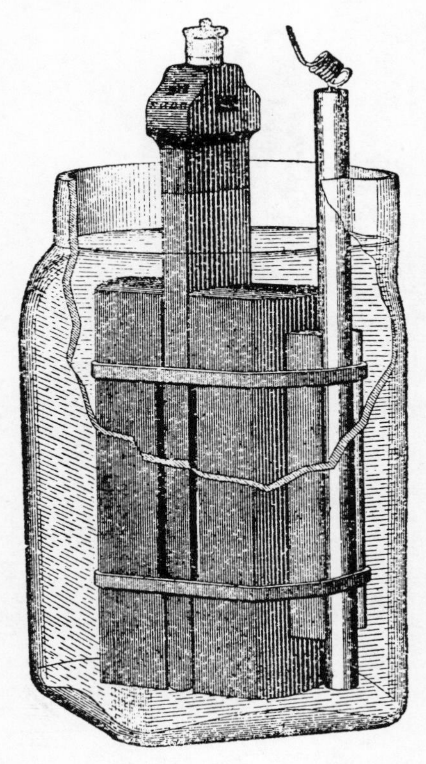

Figure 1.24 Leclanché cell with agglomerated cathode block. From (45), p. 202.

reactant played an important part in the operation of the battery. Even with the original construction, Leclanché noted that, for best results, the battery jar should be filled only partially with electrolyte: "Il faut avoir soin que cette dissolution ne baigne le vase poreux que jusqu'à la moitié de sa hauteur" (146). Nasarischwily (161), summarizing the work of previous investigators (170), (32), noted that it had been

> determined experimentally that in the Leclanché wet cell only perhaps half the [cathodic] oxidation is due to manganese dioxide.... With MnO_2:C electrodes of improper manufacture, carbon has been shown to be the only active electrode material.... For light service applications the MnO_2... [is unnecessary and]... can be omitted entirely.

All this helps to explain why it became common practice to use the cathode repeatedly, only zinc and electrolyte requiring replacement when a cell was exhausted, and why MnO_2 content was reduced or even entirely eliminated in late-model cells.

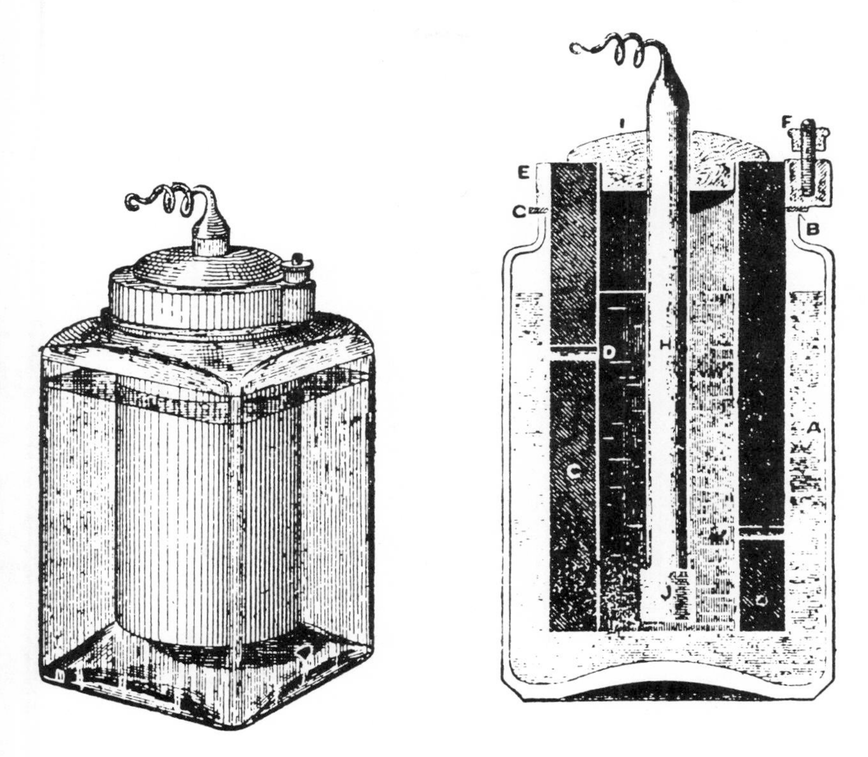

Figure 1.25 Leclanché-Barbier cell (146). From (45), 1916, p. 207.

In light, intermittent service such as telephone, doorbell, or annunciator operation, the Leclanché wet cell was satisfactory and was widely used until well into the twentieth century. Today, its chief importance lies in its impact on modern battery technology as prototype of the conventional dry battery.

Copper Oxide Cells. Copper oxide was used in batteries with neutral or acid electrolytes at least as early as 1870, e.g., in the Denys cell [Tomassi (219), p. 279], but did not become a commercially important cathodic reactant until about 1880, when Lalande and Chaperon (140) developed the system zinc:alkali hydroxide electrolyte:cupric oxide (Fig. 1.26).

In the early batteries, the copper oxide was usually in powdered form, carried at the bottom of the cell container or packed into perforated iron or copper baskets, but agglomerated plate-cathodes predominate in modern practice.

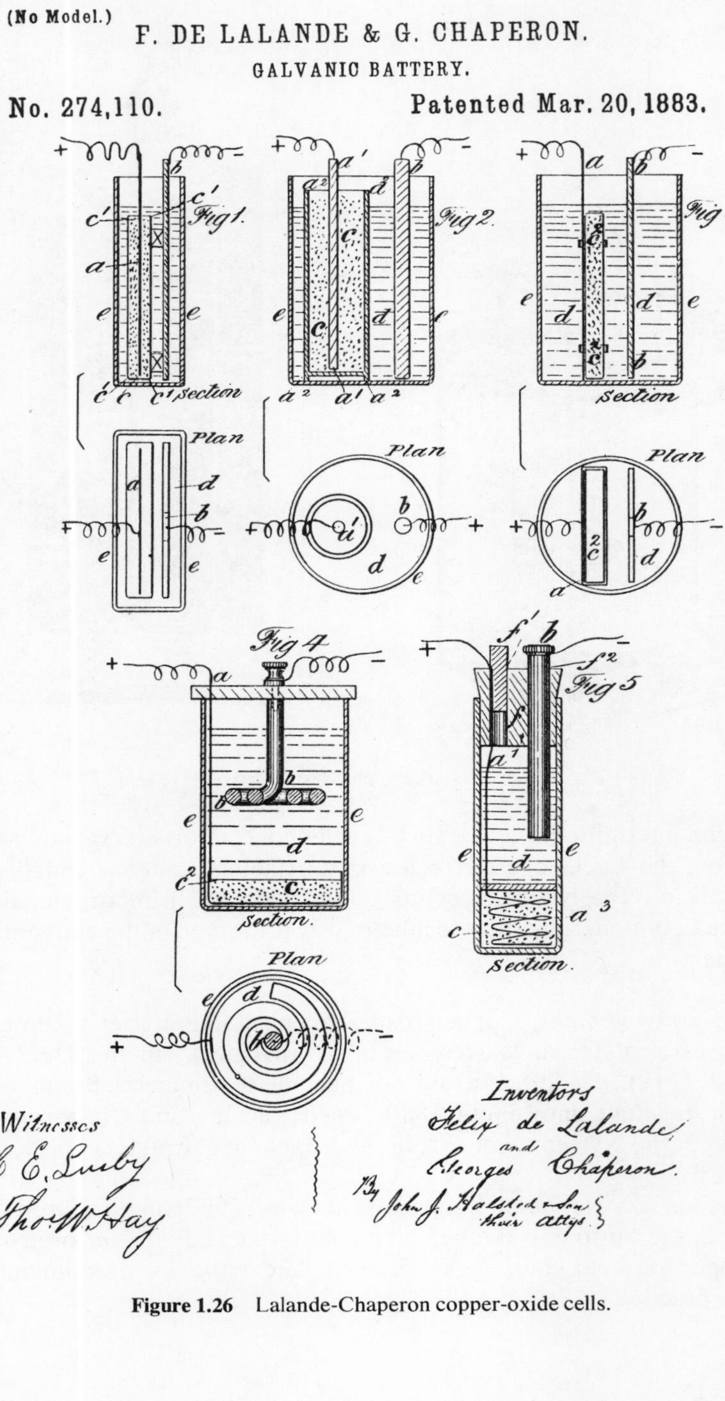
(No Model.)
F. DE LALANDE & G. CHAPERON.
GALVANIC BATTERY.
No. 274,110.
Patented Mar. 20, 1883.
Fig 1.
Section
Plan
Fig 2.
Plan
Fig. 3.
Section
Plan
Fig 4
Section.
Plan
Fig 5
Section.
Witnesses
C. E. Lusby
Tho. W. Hay
Inventors
Felix de Lalande,
and
Georges Châperon.
By John J. Halsted & Son
their attys

Figure 1.26 Lalande-Chaperon copper-oxide cells.

This battery, described in detail in Chapter 4, has an open circuit voltage of 1.06 and an operating voltage under normal service conditions of 0.5–0.7. It has almost completely displaced the Daniell cell and is still widely used where long service life and well-sustained voltage, even under heavy load, are important considerations, as in railway signal or track circuits.

Miscellaneous Cells. Other than oxides, one of the first of the solid cathodic reactants to be used appears to have been lead sulfate. As early as 1837, the elder Becquerel (11) projected a cell with a canvas-covered zinc anode and a cathode terminal or carbon of copper in a suspension of lead sulfate in saturated salt solution. According to Cazin [(34), p. 134], batteries of this type, with cell volumes up to 3000 liters, were used in the treatment of silver or lead ores and, in particular, for the "reduction of large quantities of lead sulfate originating in the lead chambers of sulfuric acid manufacture." Though cell voltage was low—only of the order of 0.5–0.6—Cazin noted that "six cells in series would give quite a strong spark."

In a modified version of the same system, introduced about 1860 by the younger Becquerel (14), a porous cylindrical cathode member comprising a paste of the lead sulfate and saturated salt solution was molded and allowed to harden around a lead rod, the whole coated with plaster to prevent disintegration. Salt solution or dilute sulfuric acid served as electrolyte.

Marié-Davy (151) is credited with dry-type flat cells—disks of tin plate and zinc separated by electrolyte members of blotting paper and lead sulfate paste—assembled in meter-high columns, 40 to a battery "pour les usages médicaux." Lead chloride was mentioned as a possible substitute for lead sulfate.

Sulfates of mercury (Hg_2SO_4, $HgSO_4$, or the basic turpeth $HgSO_4 \cdot 2HgO$) were first used by Marié-Davy in 1859 in cells of the type zinc:sulfuric acid:mercury sulfate paste (carbon). Grenet (100) and others developed commerical mercurous sulfate cells, once widely used, e.g., on the French telegraph lines, because of relatively high voltage (1.4–1.5) and capacity. The system is of interest as the prototype of the Clark standard cell (38) described in Chapter 12.

Marié-Davy (1860) also described the cell Zn : H_2O : AgCl (fused), with a silver crucible serving as cell container and cathode terminal.[11] To quote the inventor [Niaudet (165), p. 141]:

[11]This may well be considered the prototype of the water-activated reserve cells of the 1940s utilizing magnesium anodes, with silver chloride (143) or cuprous chloride (1) as cathodic reactants.

Its internal resistance, at first very large, diminishes gradually as the zinc chloride formed dissolves in the water. If this salt is dissolved beforehand, the cell immediately delivers a strong current.

Similar batteries of varying design, with common salt and ammonium chloride as well as with zinc chloride solutions as electrolyte, were subsequently developed by de la Rue (199), Pincus (182), and Gaiffe (85), and the system Zn:KOH:AgCl is ascribed to Skrivanow (208).

Other insoluble chlorides used in batteries at various times include lead chloride, proposed by Marie-Davy (151), mercurous chloride by Heraud (119) in 1879, and cuprous chloride by Wensky (241) in 1891.

Cells with gaseous reactants

Observations on the use of oxygen as cathodic reactant can be traced back as far as 1800. In June of that year, Haldane (106) reported that the operation of a Volta pile depended to a large extent on access of air, that pure oxygen intensified cell action, and that in both cases oxygen was absorbed during discharge. These findings were confirmed and amplified by Davy (58), Pepys (178), and Nauche, Graperon, and Baget (163).

It was not until 1839, however, that Grove (102) gave the first description of a cell with two gas-energized electrodes of plain platinum, so arranged that

> ... over each piece of platinum was inverted a tube of gas, four-tenths of an inch in diameter, one of oxygen, the other of hydrogen, acidulated water reaching a certain mark of the glass, so that about half of the platina was exposed to the gas, and half to the water. The instant the tubes were lowered so as to expose part of the surfaces of platinum to the gases, the galvanometer needle was deflected .... I hope, by repeating this experiment in series, to effect decomposition of water by means of its composition.

By 1842, Grove (102) had constructed a battery of 50 hydrogen-oxygen cells, this time with platinized platinum electrodes. He was then able to fulfill his prediction, "a beautiful instance of the correlation of natural forces," even though, with the cell construction used, "twenty-six pairs were the smallest number which would decompose water" (Fig. 1.27).

Grove's work is of great importance not only as the beginnings of the gas electrode and the fuel cell (Chap. 9) but also as representing an anticipatory step toward the doctrine of the conservation of energy.[12]

Grove himself was quick to discover (103) that gases other than oxygen,

[12]According to Helmholtz (118), "the first who saw truly the general law here referred to, and expressed it correctly, was a German physician, J. R. Mayer of Heilbronn, in the year 1842." Grove's lecture "On the Correlation of Physical Forces," printed in 1846 (104), anticipated by a year Helmholtz's own first publication on the same subject (117).

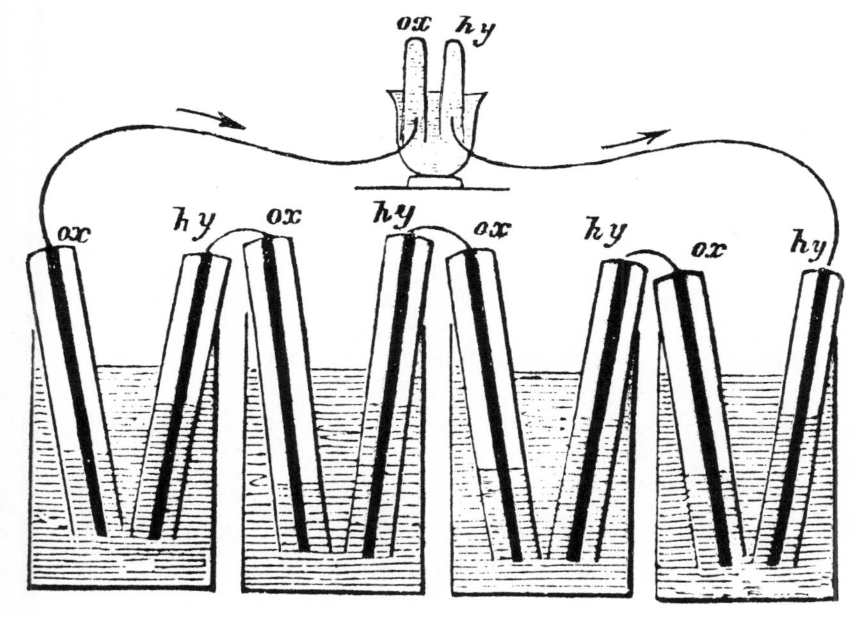

Figure 1.27 Grove gas battery (1839–1842). From (102).

e.g., chlorine or nitrogen peroxide, acted effectively as cathodic reactants, and that the hydrogen: chlorine as well as the hydrogen: oxygen system could be used reversibly. i.e., as accumulators. As early as 1853, a serious effort seems to have been made to exploit a chlorine-depolarized cathode in the "Pile de Le Roux" [Tommasi (219), p. 298]. Nevertheless, the Grove gas cells, and their analogues, despite their fundamental importance, had little immediate effect on battery technology. Both Cazin (34) and Niaudet (165), in the early 1880s, characterized Grove's work as of scientific interest but of no practical value.

Anodic reactants

From the very beginning of battery development, zinc was almost exclusively used as battery anode because of its availability and low cost, high anodic potential, usefulness over a wide pH range, and the solubility of its salts. Amalgamation, generally practiced to reduce wasteful corrosion of zinc in batteries, is usually credited to Kemp (135) (1828) and Sturgeon (213) (1830), but this expedient[13] seems to be about as old as the

[13]Sturgeon (214) acknowledged the priority of Kemp but indicated that the latter had worked with semi-liquid amalgams rather than with surface-amalgamated anodes. Sturgeon

battery itself, having been used as early as 1801 by Wollaston (248) (who believed amalgamated zinc or tin to be more readily "oxidated" than the unamalgamated metal), by Brugnatelli in 1802 (24), by Davy (65) in 1826, and no doubt by others. According to King (136), de la Rive (195) in 1830 "was the first to examine and explain the relation between amalgamation and local action."

The search for anodes other than zinc was actively pursued, in some instances even before the materials being investigated were available in commercial quantities or at moderate cost. Tin and iron, in addition to zinc, were studied from the very beginning of battery development. Sodium and potassium, in the form of amalgams, were used at least as early as 1847 (96), (190), and references to cadmium (142) and indium (72) appeared in the 1880s.

Aluminum as battery electrode found mention in 1855 (126), though as cathode, after passivation in nitric acid, rather than as anode. Due to its surface oxide film, this metal is reasonably stable, generally speaking, only in neutral electrolyte, and has no voltage advantage over zinc in such media. When the oxidic film is dissolved as in acidic or basic solutions, or removed by surface amalgamation (175), it may show higher anodic potential but corrodes too rapidly to be applicable to conventional cells. Such factors, plus scarcity and high cost, may have limited the efforts of investigators to exploit aluminum, and it was not until well into the twentieth century that substantial progress toward the utilization of this metal as anode was made (116), (211), (95), (28).

The first utilization of magnesium as anode seems to have been made about 1865, when Bultinck (26) substituted filaments of that metal for the zinc wires in a Pulvermacher battery (*vide supra*), noting that the resultant cells, moistened only with water, "without addition of salt or acid," gave the same effects as the chain with zinc anodes, for which acidulated water is required [(34), p. 132]. However, it was not until the 1880s, possibly because of considerations of cost and availability, that serious efforts were made to use magnesium anodes in conventional cells (109).

The investigation of anodic reactants was by no means restricted to metals, and much of it foreshadowed the drive toward the fuel cell (Chap. 9). As early as 1802, Davy (60) had used carbon as anode, with nitric acid as catholyte. Grove's work on the hydrogen:oxygen system has already been mentioned, and this experimenter himself was quick to

---

himself seems to have had some doubts regarding the practicability of amalgamation: "Were it not on account of the brittleness and other inconveniences occasioned by the incorporation of the mercury with the zinc, amalgamation of the zinc surfaces would become an important improvement." [Quoted by Faraday (76), Ser. VII No. 999, 1837, fn.]

discover that ethylene and carbon monoxide could be substituted for hydrogen (103). Gaugain (92), a contemporary of de la Rive, studied the behavior of the vapors of alcohol and ether as well as hydrogen at carbon electrodes, against air or oxygen cathodes, with molten glass as electrolyte. Efforts to utilize illuminating, producer, or water-gas were made by Maiche and several others [(219), pp. 410 ff.; (22), pp. 299 ff.].

Considerable work was also done, at comparatively early dates, with liquid reactants.[14] Thus in 1852, Matteucci (153) described cells with solutions of ferrous sulfate, sodium thiosulfate, or the sulfides of potassium as anolytes, separated by diaphragms from nitric or chromic acid catholytes, the same metal, platinum, serving as terminal electrodes. Though not very stable, "such cells have high electromotive force . . . a single couple of nitric acid and potassium pentasulfide will decompose water" [(34), p. 261].

Uncommon systems

In the normal course of battery development, few available avenues of exploration were overlooked. The acid:base system, Pt:HNO_3:KOH:Pt was suggested by Becquerel *père* (12) in 1837. Fused salt electrolytes were employed at least as early as 1855 (13) and in subsequent attempts to make carbon-consuming fuel cells (129).

There are numerous references to photogalvanic cells [(219), pp. 578 ff.; (22), pp. 295 ff.) beginning with observations by A. E. Becquerel (1840) and Grove (1846) (104). A cell with tin electrodes (one irradiated, the other dark) was credited to Minchin (1880) and to Fleming (1882). Kitching (1882) worked with electrodes coated with light-sensitive silver chloride, Saur (1882) with silver sulfide. An early attempt to produce a regenerative system is represented by the Case cell (33) of 1886, shown in Fig. 1.28. In this cell, sealed to prevent access of air, horizontal carbon plates served as electrodes, the upper provided with a porous ceramic coating, the lower covered with a layer of granulated tin; the electrolyte was a solution of chromic chloride. Little current was obtainable at room temperature, but output rose as the cell was heated in a water bath, reaching a maximum at about 95°C. The cell reaction

$$Sn + 2CrCl_3 \rightleftarrows SnCl_2 + 2CrCl_2$$

was reported to be entirely reversible, proceeding, on discharge, with the anodic solution of tin, whereas, with lowering of temperature to about

[14] As early as 1793, Robison (198) noted that porter drunk from a pewter pot produced "an agreeable relish" and a "more brisk sensation" than that obtained from a glass or ceramic container, this effect being attributed to the contact of the metal with two dissimilar fluids (saliva and porter).

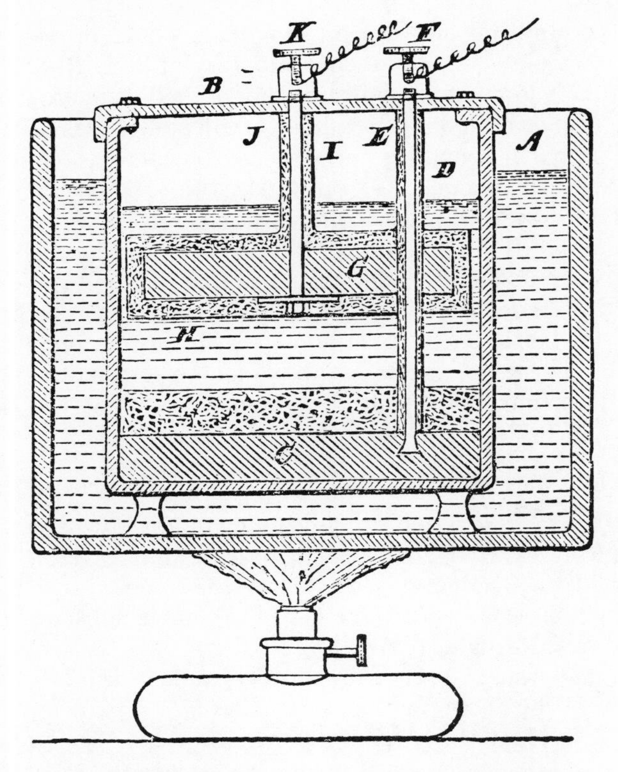

Figure 1.28 Case regenerative cell. From (219), p. 260.

15°C, tin was precipitated on the lower carbon plate and the original cell composition was restored.

A survey would hardly be complete without some mention of the bizarre systems reported in reputable technical journals and perpetuated in textbooks of the period. Baconio (3) is credited with a Volta-type pile made entirely of vegetable substances—disks of beetroot and walnut wood separated by leaves of scurvy grass, the latter serving as electrolyte members. In 1804, Dyckhoff (69) described a Volta pile with an air-space instead of an electrolyte between anode and cathode and, with the same technique, Watkins (240), not to be outdone, used only one metal—in the form of plates of zinc, scraped and polished on one side, left untreated on the other—as electrode material. Tommasi immortalized the zinc: carbon *pile au urine* ($E = 0.76$ volt) of Coiffier (43), in which the biological product served merely as electrolyte, not, in anticipation of recent developments (Chap. 9), as anodic reactant in biological fuel cells.

Competing energy-conversion systems

Thermopiles. In bringing the discussion of this period to a close, it should be noted that the supremacy of the primary battery was not maintained without challenge from other potential converter systems. Following the work of Seebeck, about 1821, on the thermoelectric effect, and thus long before the dynamo imposed a threat, serious efforts were made throughout the years to develop thermocouples as power sources, some of the later models of quite imposing size.

A cylindrical unit credited to Clamond (37) (Fig. 1.29) is described as 2.5 meters tall and 1 meter in diameter. It contained 2 thermopiles each of 3000 iron:zinc-antimony alloy couples arranged in 30 "chains" of 100 elements. The voltage of a single pile was given as 109, its resistance 15.5 ohms, and its current output was "capable of producing a source of electric light of *40 becs Carcel*" (380 candle power). Heating was done with coke, at a rate of expenditure of 10 kilograms per hour. "These results" says Cazin [(34), p. 293)] "are quite remarkable, and show clearly that the thermoelectric piles can already compete with liquid cells." Evidently, low thermal efficiency—probably not exceeding one per cent—and the advent of the dynamo postponed the acceptance of the thermopile as a commercial power unit.

Rise of the storage battery

As long as secondary batteries or accumulators could be recharged only with current from primary cells, there was little incentive for their intensive development. However, the primary battery was neither a cheap nor convenient source of sustained charge, and efforts were not lacking to find a better means of large-scale current generation.

Both motor and generator had their origins in observations made by Faraday in 1821 [(75), Vol. 1, pp. 499 ff.] and his subsequent studies of "the evolution of electricity from magnetism" and "electricity from revolving copper plates . . . and . . . moving conductors" [(76), Ser. 2, No. 27 ff., 1831]. Unlike the battery powered motor which appeared in the 1830s (185), (130), the generator was slow to develop. The first models were capable of producing only alternating current, long considered to be of no practical value. The first converter or commutator is credited to Sturgeon (1834) (213), but despite this and subsequent improvements, the direct current generator long remained inefficient and unreliable, and it was not until 1882 that the introduction of the Edison system of power production from central stations ushered in the era of cheap electricity and made possible the large-scale exploitation of the storage battery.

Figure 1.29 Clamond thermopile. From (34), p. 292.

Though it continued to gain in usefulness and importance, the primary battery had to yield to the accumulator some service applications, particularly those of sustained high current demand.

PRIMARY BATTERIES IN THE ELECTRICAL AGE (1880–1920)

By the time dynamo and storage battery had become firmly established, an expanding field of usefulness for small power units had developed which ensured the continuance of the primary battery as an important article of commerce. Electrical means of communication powered by batteries were undergoing rapid growth. The Morse system of telegraphy, first introduced on a commercial scale in the United States in 1844, had spread to Europe in the 1860s and was still expanding. In the United States alone, the number of telephones, which had risen from 26,000 in 1877 to 227,000 in 1890, was over 1,350,000 at the turn of the century and was growing at a rate of over 500,000 installations per year. Other uses—bell ringing annunciator, alarm, railway signal and track circuit, energizing of "medical coils," and a wide variety of miscellaneous applications requiring small direct currents—helped to swell the demand for primary batteries.[15]

Dry cells of the Leclanché type

The Leclanché-type dry cell, which made its appearance in the 1880s, is generally credited to Gassner (91). However, the concept of a non-spillable, portable battery, in which the electrolyte was immobilized, was of long standing, and most of the factors that contributed to the success of the conventional dry cell appeared at an early date. The original Volta pile, in which the electrolyte was held in a bibulous disk or membrane, was of this type. Hachette and Desormes (105) described the use of starch paste in 1809. Sand (or earth) as electrolyte absorbent was employed in the 1840s by Bagration (4) (and by Jacobi) in the Zn: NH_4Cl: Cu system and by Minotto in his modification of the Daniell cell (1863) [(34), p. 186]. It was also used by Leclanché himself, who described this expedient in his 1866 patent (145) covering the use of manganese dioxide and sal ammoniac:

> I then take fine sand or sawdust, in short, any substance capable of producing a paste with the liquid electrolyte, and with it fill the jar .... Finally I moisten the entire mass, sand and manganese dioxide, with concentrated ammonium chloride solution.

Other methods employed over the years [(2), (47), (219), (22)], involved such diverse immobilizing agents as kieselguhr, asbestos, glass

[15]It was with a home-made primary battery, constructed "out of all sorts of cups, tumblers and so on, with pieces of carbon in them" (122) that young Charles M. Hall, in 1886, obtained the current for his invention of the electrolytic process of making aluminum (107).

wool, paper, coconut fibre, cellulose, plaster of Paris, gelatin, and agar. According to Bottone (22), the Becquerel and Onimus cell of 1884 consisted

> . . .of zinc and carbon standing in a made paste with plaster of Paris, a solution of zinc chloride and sal ammoniac in water mixed with sesquioxide of iron and peroxide of manganese.

On the basis of the foregoing, Gassner probably should be considered a pioneer in the production of a commercially successful unit in its conventional form, i.e., with the zinc anode serving as cell container, rather than as inventor of the Leclanché-type dry cell.[16] Indeed, some of his early disclosures deal with ferric hydroxide rather than manganese dioxide as cathodic reactant, and his choice of plaster of Paris to immobilize the electrolyte was hardly a happy one. However, once manganese dioxide:carbon mixes were used, as originally suggested by Leclanché, and cell liners of bibulous pulpboard or paper, cereal pastes, and combinations of these (133) were substituted for plaster of Paris, the foundations of modern dry battery technology were laid. Unfortunately, the value of paste-coated paper as a means of improving zinc:electrolyte contact and reducing wasteful corrosion was not generally recognized, and it was not until about 1915 that this expedient was generally adopted in the United States for the conventional 6-inch dry cell.

The dry cell, from the very first similar in shape and size to its modern 6-inch cylindrical counterpart, found rapid acceptance, supplanting the conventional wet cells in many applications and paving the way to new areas of usefulness. Manufacture of miniature cells was stimulated by the introduction of the flashlight, an American invention (42), (156), introduced by Conrad Hubert (125) some 20 years after the development of Edison's incandescent lamp. Though capable at first of only a limited number of momentary flashes—hence its name—the flashlight enjoyed a rapidly increasing popularity which spurred battery manufacture and furnished the incentive for improvement in quality. Meanwhile, the larger "six-inch" dry battery was still finding new and important uses such as ignition service for internal combustion engines in the rapidly expanding automobile industry. Primarily because of the dry battery, the Leclanché system, with the proud record of 20,000 units in service in 1868 (146), could boast an annual production rate, in the United States alone, of two million unit cells in 1900, twenty times that number only ten years later, and an approach to two billion in the mid-1960s.

[16]Bottone attributes to Skrivanow a zinc:carbon cell of 1882 with immobilized electrolyte, with the zinc anode in the "form of a cylindrical pot, a plate, or a square box, at will. . . . This comes very close to the dry cells of today" [(22), p. 31].

Gas electrodes and fuel cells

This was the era of much study of gas electrodes (203) and serious efforts were made, though with limited success, to develop fuel cells (7), particularly the high-temperature types with molten electrolytes. The halogens, too, received some attention, serving as cathodic reactants [(219), pp. 298 ff.]. Chlorine appeared in Upward's (1876) battery:

$$Zn : ZnCl_2 \text{ solution} :: \text{diaphragm} :: Cl_2(C)$$

bromine (1884) in Koosen's cell:

$$Zn : \text{acid} :: \text{diaphragm} : Br_2 \text{ (Pt or C)}$$

iodine in the cells of Doat (67) or Laurie (1881):

$$Zn : KI \quad \text{or} \quad ZnI_2 : I_2 [C]$$

None of these seems to have been of any commercial significance.

Of far greater impact on battery technology were the comparatively early attempts to use atmospheric oxygen as cathodic reactant (226), (121), (201), (161). In the Maiche cell of 1878 (150) (Fig. 1.30), once extensively used in telegraph service, a glass container was topped by an ebonite lid to which a perforated, porous vessel was attached. This vessel contained pulverized, platinized retort carbon. An ebonite tube through the center of the lid supported a small cup or basin containing slugs of zinc and a small amount of mercury. The electrolyte was a solution of common salt or, preferably, of sal ammoniac, into which the porous vessel containing the cathodic carbon was immersed to a depth of about 2 centimeters.

For the Walker-Wilkins cell (238) of the 1890s (Fig. 1.31), at one time seriously considered for lighting purposes, the outer container was a perforated cylinder mounted on an earthenware basin. Inside, a porous jar of somewhat smaller diameter served as container or zinc electrode and caustic potash electrolyte. The annular space between jar and perforated cylinder contained an outer layer of coarse carbon, an inner layer of fine carbon or graphite, and, serving as cathode collector, a perforated nickel cylinder. Electrolyte slowly seeping through the porous jar kept the carbon moist, and any excess flow was caught in the supporting basin.

These cells and their analogues are now of only historic interest, as forerunners of today's commercial batteries utilizing atmospheric oxygen

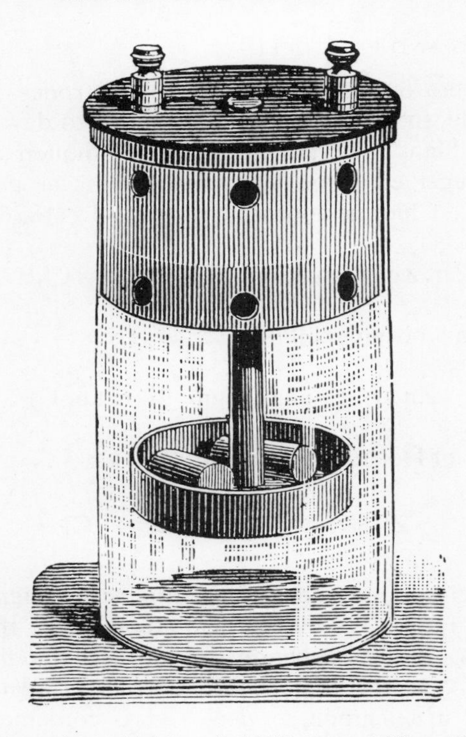

Figure 1.30 Maiche cell (1879). From (219), p. 102.

(Chap. 8). In their day, however, they represented a well-conceived drive for low-cost battery operation. (With zinc at 10 cents per pound, and neglecting battery construction and maintenance, the cost of power on the basis of expendable materials only—e.g., zinc and caustic—would be of the order of 20 cents per kilowatt-hour.)

Stimulated in large measure by interruption of supplies of manganese dioxide during World War I, much work was also done, particularly in France, on "air-depolarized" dry cells. Results were not impressive. The access of air that could be provided in batteries of conventional design, though limiting cell operation, caused desiccation and accelerated zinc corrosion, resulting in poor service maintenance and a generally inferior product.

Appraisal of the period

Despite the growing commercial importance of the primary battery during the period under discussion, technological progress, on the whole, failed to keep pace with industrial expansion. This was the era of Van't

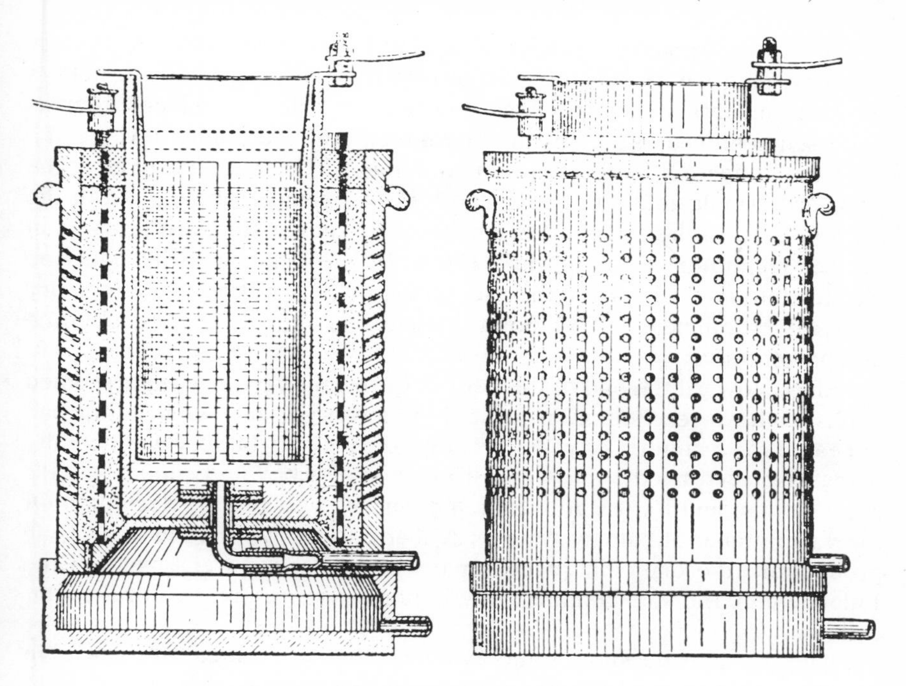

Figure 1.31 Walker-Wilkins cell. From (45), p. 183.

Hoff, Oswald, Nernst, Arrhenius, and the many others who laid the foundations of modern electrochemical theory, yet the primary battery derived relatively little immediate benefit from their efforts. In 1902, Bottone (22), though admitting the capabilities of dry cells for light, intermittent service (telephone, bell-ringing, signal operation and the like), noted:

> Neither Leclanché's nor dry cells are of any practical use, except only for very short intermittent lighting of low candle power . . . for a second or two at a time, to see the hour by a watch; or in a dark cupboard, to light up momentarily, to enable one to get at any article contained therein. For any lighting extending over more than a few seconds at a time they are absolutely useless.

As late as 1906 Crocker (48) emphatically, if not entirely accurately, voiced his opinion of primary batteries in general with the statement:

> One is at a loss to point out any radical or even important advance in primary batteries . . .; the types now on the market . . . are not substantially better than what might have been obtained 60 years ago.

Such views, though unduly pessimistic, reflect the fact that the battery industry, as a whole, was slow to underwrite the research on which its ultimate success was dependent. Further, in a field in which progress had been achieved largely by empirical methods, there may have been a tendency to ignore the teachings of contemporary science and even of the prior art. Amalgamation of zinc anodes, for example, though old in the art and used in the original Leclanché cell, was not standard practice in dry cell technology until the early 1900s. In one instance, its introduction was fortuitous, arising from the use, on the pulpboard liners of 6-inch dry cells, of a commercial "bill-board" paste containing mercuric chloride as a preservative.

Indeed, toward the end of the period, the primary battery again seemed almost to have reached an impasse. No new large-scale applications had appeared for a number of years, whereas the primary battery was actually being displaced by the small transformer in doorbell and similar service applications, and its use for ignition purposes was rapidly declining. Of the vast number of battery systems available, only two were in large-scale production: the dry cell, successor to the Leclanché wet type, and the Lalande cell, the sole commercially important representative of the wet-cell family.

There was, to be sure, some evidence that a brighter future was in prospect. Commercial production was still increasing, thanks largely to the growing utilization of dry cells, particularly in the smaller, flashlight sizes. Greater emphasis on quality—the continuous service of a flashlight battery had become a matter of hours rather than the "few seconds" noted by Bottone at the turn of the century—led to the development of standard specifications and tests, initiated in the United States by the Electrochemical Society (71) and subsequently taken over and kept up to date by the National Bureau of Standards (162). Laboratories organized within the industry were expanding their research activities, and their number was increasing.[17] But in a field that had developed as an art rather than as a science, much research effort was of necessity directed toward the solution of practical problems associated with manufacture, rather than to basic studies. Even when fundamental investigations were undertaken, their results all too frequently could not be correlated with actual cell performance. Electrochemistry continued to be a neglected area in the academic world, and contributions from that source were meager.

[17]C. F. Burgess, one of the most prominent electrochemists of his day, left the academic field (1913) to organize the industrial laboratory and battery manufacturing enterprise (1917) that bore his name, and devoted his talents in large measure to the advancement of battery science and technology (154).

BATTERIES IN THE ELECTRONIC AGE

The current period may be considered as beginning with the introduction of radio broadcasting in 1920, the year the pioneer station, KDKA, was established in Pittsburgh, Pa. by the Westinghouse Electric Co. Immediate public acceptance set off a chain reaction: accelerated growth of the battery industry; greater competition among manufacturers; fundamental research, which resulted in a better understanding of cell processes and led to improvement of conventional batteries and the introduction of new systems. Major developments continued to come from the industry itself but, as batteries grew in importance and aroused scientific interest, other agencies, academic and governmental, began to make significant contributions (54). The technical literature on batteries has shown a vast increase—a survey made in 1948 (18) yielded some 3600 references to Leclanché dry cells alone—and is still growing at an accelerated rate.

Since the history of this period is, in effect, the theme of the succeeding chapters, in which current concepts of battery science and technology are discussed, only some of the major trends influencing the course of primary battery development need be considered here.

Impact of radio

The first radio receivers were, of course, battery operated, with the "A" or filament current supplied by storage elements, the "B" or plate circuit almost invariably by batteries of conventional Leclanché-type flashlight cells. Whereas flashlight service, with its emphasis on heavy drain (0.2–0.5 ampere) and high operating voltage to ensure maximum light, had accounted for a relatively small portion of the built-in capacity of the cells, the low drains of plate circuits (usually from 0.004 to perhaps 0.020 ampere) permitted much greater utilization and correspondingly better correlation of cell composition and electrical output. Increase in capacity through development of improved mix formulas followed almost as a matter of course. Concurrently, better space utilization through the substitution of flat units for the conventional cylindrical cells, essentially a reversion to the type of construction used in the original Volta pile, helped to increase capacity per weight and volume of battery (189), (128), (82), (81).

Battery production for radio purposes received a severe setback in the late 1920s, when a-c sets, operating directly from the power lines, were introduced, seriously curtailing the production of battery-powered equipment. The counter to that threat, suggested by H. W. Kadell (134), who did much of the pioneer work, was the development of low-drain receivers designed especially for unwired homes, e.g., farms, and dependent on long-life primary batteries (23). An immediate consequence was stimulation of work on the utilization of atmospheric oxygen as cathodic reactant, long considered as replacement for the copper oxide of the Lalande cell. The development of a fully assembled, "Air Cell" battery, capable of 1000 hours, nominally a year's service, on filament circuits, followed in due course (115). In addition to its expansion of the area of usefulness of the primary battery, this, in turn, led to the production of the high-current-density oxygen electrodes (110), (132), (113), which furnished the basis for much of the subsequent work on continuous-feed and fuel cells (Chap. 9).

Other factors

The use of small cells in flashlight and analogous service has continued its steady growth over the years, accompanied by marked improvement in quality. In the Leclanché dry cell, for example, service capacity has more than tripled since 1920 (114). The alkaline dry-type [Zn : KOH (H_2O) : MnO_2(C)] system introduced in 1949 (120), and greatly improved since, has better heavy drain and secondary cell characteristics than the conventional dry battery (Chap. 7).

Modern military requirements gave great impetus to the progress of the primary battery, not only in the modification of and refinement of conventional cells, but also in the exploitation of old systems not previously developed because they lacked commercial application, and in the introduction of new systems. World War II boosted battery production in the United States to an annual rate of three billion units, even though civilian usage of conventional cell was seriously curtailed. Leclanché dry cells were used in enormous quantities – for flashlight, telephone, radio including the two way "walkie-talkie," the firing of bazookas and other guns, mine detector work, and the like. Air-Cell batteries of 600–800 ampere-hour capacity were used in large numbers for the lighting of troop kitchen, hospital and sleeping cars, and for telephone installations in the war zone. Zamboni piles (124) were used for certain types of night-viewing equipment. An A-B-C battery of bichromate cells was designed as power supply for the radio component of the proximity fuze, rugged enough, when shot from anti-aircraft guns, to withstand forces 20,000 times that of gravity, and spin rates of 26,000 r.p.m. This battery (Fig. 1.32), though small enough to fit into a $2\frac{1}{2}$-inch cylindrical space of $1\frac{1}{2}$-inch diameter, included a $1\frac{1}{2}$-volt "A", a 100-volt "B", and a 6-volt "C" section. It was of flat-cell, reserve-type construction, electrolyte being housed in a separate ampule which was broken on firing. Production had reached almost 2,000,000 per month at War's end.

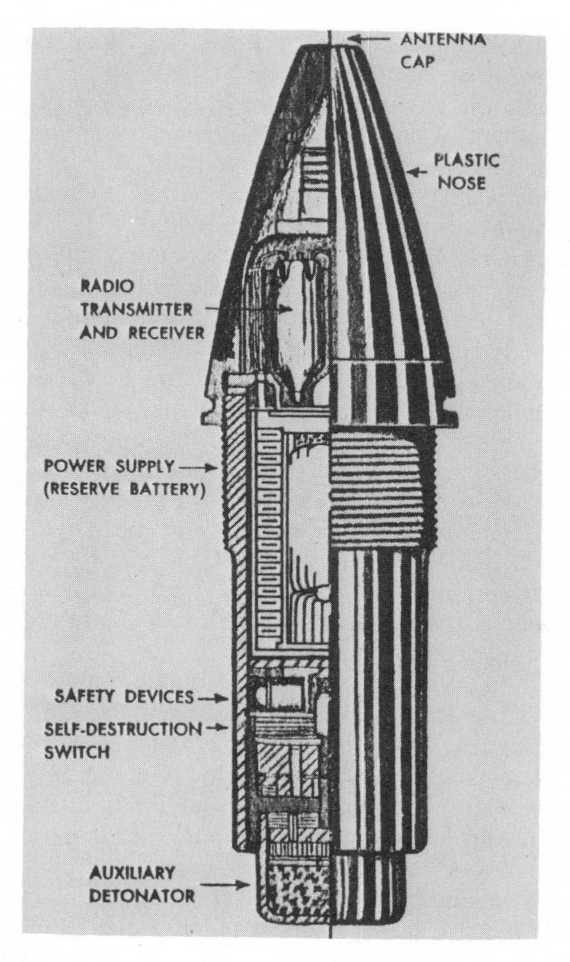

Figure 1.32 Proximity fuze with deferred-action battery. From (137) (courtesy of *Modern Plastic*, McGraw-Hill Publishing Co., Inc., New York).

Military needs, in peace as well as in war, have also been responsible for, or at least have greatly accelerated, the adoption of systems previously considered too esoteric or too expensive to warrant commercial exploitation e.g., those employing the oxides of silver (39) (Chap. 6) or mercury (40) (Chap. 5). Reserve or "deferred-action" cells have made possible the utilization of systems ordinarily too unstable for conventional batteries.

Another factor, perhaps even more responsible for the introduction and large-scale utilization of new systems, has been the reduction in power requirement of battery-operated equipment, brought about by general refinement of components, the introduction of the transistor, and the development of new fields of usefulness. For cells hardly larger than an aspirin tablet[18] such small quantities of active materials are required that few if any reactants need be barred because of cost.

Last but by no means least in its impact on battery technology has been the upsurge during the 1950s of interest in continuous-feed and fuel cells, stimulated in large measure by military needs and the requirements of space exploration. As a result (discussed in detail in Chap. 9), multiple-cell units have been made available (250), capable of uninterrupted operation for thousands of hours at high current densities, e.g., 9–10 amperes per square decimeter (100 amperes per square foot). The more advanced systems have been limited largely to hydrogen and oxygen as electrode reactants with caustic alkali solutions as electrolyte, and costs have been high. However, with simple and efficient reforming methods available for the procurement of hydrogen from cheap hydrocarbons, with the possibility of using air instead of oxygen, and with good prospects for the direct oxidation, with acid electrolyte, of hydrocarbons to carbon dioxide and water, it should still be possible to find large-scale peace-time applications for fuel cells.

The primary battery has come a long way since its invention at the end of the eighteenth century. From its beginnings as the feeble Volta pile it has become an efficient converter of chemical to electrical energy, in sizes ranging from miniature portable units to large stationary power plants. Its capability has progressed from light, intermittent service and limited capacity to continuous heavy-drain operation. It can be made to operate at temperatures ranging from well below freezing to those, e.g., of 1000°C or more, with molten salt electrolytes. With commercial production, currently valued, in the United States alone, at $300,000,000 per year and still increasing, and new areas of usefulness emerging, the immediate future of the primary battery seems assured.

Meanwhile, constantly expanding support in theoretical as well as applied research effort continues to accelerate technological progress, not necessarily confined to the exploitation of the battery as a power source. By way of example, E. Yeager (250) has pointed out that in the

[18]A mercury cell used in hearing aids is currently manufactured as a disk 0.755 centimeter (0.305 inch) in diameter and 0.34 centimeter (0.135 inch) thick, weighing 0.57 gram (0.02 ounce); a Leclanché dry cell which will keep a watch running for a year is 1.12 centimeter (0.442 inch) in diameter, 0.32 centimeter (0.125 inch) thick, and weighs about 1.4 grams (0.049 ounce).

production of chlorine and caustic soda by means of the mercury cell, the sodium amalgam formed might be used directly as anode in a continuous-feed battery with air or oxygen as cathodic reactant to recover part of the power wasted during brine electrolysis. Simpler, if practicable, would be the substitution of an oxygen electrode for the inert graphite electrode of the conventional brine cell, eliminating hydrogen overvoltage and actually contributing energy to the system, with a net reduction in over-all power requirement of the order of 20 per cent. With these and other applications of active electrodes (110), (132), (247) liquid or gaseous, anodic as well as cathodic – to electro-oxidations and reductions, both organic and inorganic electrosynthesis, and even to the electrowinning of metals – the primary battery should continue to have important influence on the future progress of the still "relatively neglected" (242) science and technology of electrochemistry.

REFERENCES

1. Adams, B. N., U.S. Pat. 2,322,210 (1943); cf. Pucher (186).
2. Anon., *The Electrician*, **42**, 185 (1889); cf. Tommasi (219), pp. 91, 101, 240, 253, 254, 275.
3. Baconio, J., *Phil. Mag.*, **23**, 283 (1805); *Nicholson's J.* **18**, 159 (1807).
4. Bagration, P. R., *Arch. de l'Electr.*, **4** (1844); cf. Cazin (34), p. 112; Tommasi (219), p. 107.
5. Barnett, H. T., *The Electrician*, **33**, 564 (1894); cf. Bottone (22), p. 105; Cooper (45), p. 150.
6. Baudet, C.; cf. Tommasi (219), p. 207.
7. Baur, E. and Tobler, J., *Z. Elektrochem.*, **39**, 169–180 (1933).
8. Beccaria, G. B., *Dell' elletricismo e naturale*, Turin, 1753; cf. Whittaker (243).
9. Becquerel, A. C., *Ann. Chim. Phys.*, **41**, 5–45 (1829).
10. *Id.*, *Ann. Chim.*, **60**, 164 171 (1835); cf. Faraday (76), Ser. XVII, 2041 (1840).
11. *Id.*, *Compt. rend.*, **4**, 35 (1837); cf. Cazin (34), p. 167.
12. *Id.*, *Compt. rend.*, **4**, 882 (1837); cf. Cazin (34), p. 133.
13. *Id.*, *Traité d'électricité et de magnétisme* (1855); cf. Cazin (34), p. 258.
14. Becquerel, A. E., *Compt. rend.*, **50**, 660–2 (1860); cf. Cazin (34), p. 134.
15. Benjamin, Park, *The Voltaic Cell*. John Wiley & Sons, New York (1893).
16. Biot, J. B., *Ann. Chim. Phys.*, **5** 191 (1803); *Ann. Chim.*, **47**, 15 (1803).
17. Blake, I. C., *J. Electrochem. Soc.*, **99**, 202 C (1952).
18. Bolen, M. and Weil, B. H., *Literature Search on Dry Cell Technology*, Georgia Inst. Tech., Atlanta, Ga. (1948).
19. Booe, J. E., *J. Electrochem. Soc.* **99**, 197–200 C (1952).
20. Bostock, J., *Account of the History and Present State of Galvanism*. London, 1818; cf. Mottelay (159), p. 443.
21. Boswell, T. L., U.S. Pat. 2,683,184 (1954); *J. Electrochem. Soc.*, **105**, 239–41 (1958).
22. Bottone, S. R., *Galvanic Batteries*. Whittaker & Co., London & New York, 1902.
23. Bowditch, F. T., "Battery Design Problems of the Air Cell Receiver." *Proc. Inst. Radio Eng.*, **20**, 215–27 (1932).

24. Brugnatelli, L. V., (1802); cf. Tommasi (219), p. 90.
25. *Id.*, cf. *Phil. Mag.*, **23**, (1805); cf. Mottelay (159), p. 361.
26. Bultinck, *Compt. rend.*, **61**, 585–6 (1865); cf. Cazin (34), p. 131.
27. Bunsen, R., *Pogg. Ann.*, **54**, 417 (1841); **55**, 265 (1842); **57**, 110 (1843) *Compt. rend.* **16** (1843).
28. Cahoon, N. C. and Korver, M. P., *J. Electrochem. Soc.*, **106**, 469 (1959).
29. Caldani, L. M. A.; cf. Mottelay (159), pp. 148–303; Fahie (74), p. 1344.
30. Callaud, A., French Pat. 36,643 (1858); cf. King (136); *Ann. Teleg.*, **1** 46 (1858), cf. Vinal (230); *Cosmos*, **19** (1861), **20** (1862), cf. Cazin (34), p. 193.
31. Campbell, J. B., U.S. Pat. 606,887 (1898).
32. Carhart, H. S., *Primary Batteries*, Allyn & Bacon, Boston, 1891; Carhart and Schoop, *Die Primärelemente*, Halle, a. S., 1895.
33. Case, W. E., U.S. Pats. 344,345-6-7 (1886); cf. Tommasi (219), p. 260; Bottone (22), p. 291.
34. Cazin, A., *Piles Électriques*, Gauthier-Villars, Paris, 1881.
35. Chevreusse, *Ann. Chim. Phys. II*, **29** (1825); cf. Cazin (34), p. 214.
36. Children, J. G., *Phil. Trans.*, **99**, 32–8 (1809); **105**, 363–74 (1815); *Phil. Mag.*, **42**, 144 (1813).
37. Clamond, *Compt. rend.*, **88**, 925 (1879).
38. Clark, J., Latimer, *Proc. Roy. Soc.* (London), **20**, 444 (1872); *Phil. Trans.*, **164**, 1–14 (1874).
39. Clarke, C. L., Brit. Pat. 1932 (1883); cf. Howard (123).
40. *Id.*, U.S. Pat. 298,175, (1884); cf. Booe (19).
41. Codd, M. A., *Practical Primary Cells*, Pitman & Sons, London (1929).
42. Cohen, J. L., U.S. Pat. 636,492 (1899).
43. Coiffier, cf. Tommasi (219), p. 113.
44. Cooper, J. T., *Phil. Mag. III*, **16**, 35 (1840).
45. Cooper, W. R., *Primary Batteries* (New Ed.), D. Van Nostrand Co., New York and London, 1916.
46. Cotugno, D., Letters to G. Vivenzio (10-2-1784); cf. Mottelay (159), p. 274; Fahie (74), p. 1344.
47. Cox, H. B., "Dry Battery," U.S. Pat. 350,294 (1886); cf. Ger. Pat. 35398 (1885).
48. Crocker, F. C., *Trans. Electrochem. Soc.*, **10**, 107 (1906).
49. Crosse, A., "On the Production of Insects by Voltaic Electricity," *Ann. Elect.* **1**, 242 (1837); cf. "G.W.H." (112); Gardner (88).
50. Cruickshank, W., *Nicholson's J.*, **4**, 187 (1800); cf. Wilkinson (244), Vol. II, 52–63; 96–99; Nicholson, Carlisle, and Cruickshank (168).
51. *Id.*, *Nicholson's J.* **4**, 254–64 (1800); *Gilbert Ann.*, **7** (1801).
52. *Id.*, "Description oı Mr. Pepys' Large Galvanic Apparatus." *Phil. Mag.*, **15**, 94–6 (1803).
53. Daniell, J. F., *Phil. Mag. III*, **8**, 421 (1836); *Phil. Trans.*, **126**, 106, 125–129 (1836); **127**, 141–150 (1837).
54. Daniels, F., *Trans. Electrochem. Soc.*, **53**, 47 (1928).
55. Darimont, *Engineer*, **137**, 636 (1924).
56. Darland, W. G., *et al.*, U.S. Pat. 2,985,702 (1961); cf. Freund (83); Kleiderer (137).
57. Davidson, A. W., *J. Chem. Educ.*, **25**, 536 (1948).
58. Davy, H., "On the Causes of the Galvanic Phenomena . . .," Nicholson's J., **4**, 338–41; 394–402 (1800); *Phil. Trans.*, **91**, 347 (1801).
59. *Id.*, *Phil. Mag.*, **11**, 326, 340 (1801).
60. *Id.*, *Nicholson's J.*, **5**, 2 (1802).
61. *Id.*, (1802) cf. Fahie (74), p. 1348.

62. *Id.*, (Second Bakerian Lecture), *Phil. Trans.*, **97**, 1–49 (1807).
63. *Id.*, *Phil. Trans.*, **98**, 1–44, 333–70 (1808).
64. *Id.*, *Phil. Trans.*, **100**, 16–74 (1810); cf. Mottelay (159), p. 338.
65. *Id.*, *Phil. Trans.*, **116**, 383–422 (1826).
66. Dehms, J., *Télég. internat.*, **1** (1869–71); cf. Niaudet (165) p. 204.
67. Doat, V., *Compt. rend.*, **42**, 769, 855 (1856).
68. Drotschmann, C., *Trockenbatterien* (3rd ed.), Becker & Erler, Leipzig (1945).
69. Dyckhoff, M., *Nicholson's J.*, 7, 303 (1804); cf. Mottelay (159), p. 387.
70. Edison, T. A., U.S. Pats. 704, 303–4 (1902).
71. The Electrochemical Society, *Trans. Electrochem. Soc.*, **24**, 275 (1912).
72. Erhard, Th. (1881); cf. Tommasi (219), p. 224. For later work on indium cf. Boswell (21).
73. Fabroni, V. G. M., *Nicholson's J.*, **3**, 308 (1799); **4**, 120 (1800); *J. Phys.*, **6**, 348 (1799); *Becquerel's Traité d'Électricité*, **1**, 81–91; cf. Haldane (106); Faraday (76), Ser. 16, Nos. 1796 ff. (1840).
74. Fahie, J. J., "Magnetism, Electricity and Electromagnetism up to the Time of the Crowning Work of Michael Faraday in 1831," *J. Inst. Elect. Eng.*, **69**, 1331–57 (1931).
75. Faraday, M., *Faraday's Diary, 1820–1862* (Thos. Martin, Editor), 8 Vols., G. Bell & Sons, Ltd., London (1932).
76. *Id.*, *Experimental Researches in Electricity*, 1831–55, Vol. I, 2nd ed., Richard and John Edward Taylor, London (1849); Vol. II, Bernard Quaritch, London (1844); Vol. III, Taylor and Francis, London (1855).
77. Faure, C. A., *Compt. rend.*, **73**, 890 (1871).
78. *Id.*, *Compt. rend.*, **92**, 951–3 (1881); cf. *Electrician*, **6**, 323; **7**, 122, 249 (1881).
79. Foerster, F., *Electrochemie wæsseriger Loesungen*, (3rd ed.), Barth, Leipzig (1922).
80. Franklin, Benj., Letter XII "Experiments and Observations on Electricity . . . ," Philadelphia (1754); cf. Mottelay (159), pp. 193 ff.
81. Franz, A. O., Martinez, J. M., and Koppelman, U.S. Pat. 2,416,576 (1947).
82. French, H. F., *Proc. Inst. Rad. Eng.*, **29**, 299–303 (1941); U.S. Pat. 2,272,969 (1942).
83. Freund, J. M., U.S. Pat. 2,981,779 (1961).
84. Fuller, (1871); cf. Cazin (34), pp. 251 ff; Niaudet (165), pp. 206 ff.; Cooper (45), pp. 161 ff.
85. Gaiffe; cf. Niaudet (165), p. 146; Tommasi (219), p. 293.
86. Galvani, A., "Aloysii Galvani de viribus Electricitatis in Motu Muscularis Commentarius," *De Bononiensi Scientarum et Artum Instituto atque Academia Commentarii* **7** (1791); cf. Magie (149). (English translation: M. Foley, Burndy Library Publ. 10, Norwalk, Conn., 1953.
87. Gamow G., *The Birth and Death of the Sun*, Viking Press, New York (1940).
88. Gardner, M., *Fads and Fallacies* . . . . (2nd ed.), Dover Publications, New York (1957) (p. 117).
89. Garrett, A. B., *Batteries of Today*, Research Press, Dayton, Ohio (1957).
90. Gassiot, J. P., *Phil. Mag. III*, **13**, 436–9 (1839).
91. Gassner, C., Ger. Pats. 37,748 (1886); 45,251 (1887); U.S. Pat. 373,064 (1887); cf. Krehbiel (139).
92. Gaugain; cf. Tommasi (219), p. 410; [English translation in Benjamin (15), p. 323].
93. Gautherot, N., *Phil. Mag.*, **24**, 183 (1806) (Proc.).
94. Gay-Lussac, J. L., *Ann. Chim. Phys.* **11**, 76–87 (1816).
95. Glicksman, R., *J. Electrochem. Soc.*, **106**, 457–64 (1959).
96. Goodman, J., "On a New and Practical Voltaic Battery of the Highest Power, in which Potassium Forms the Positive Element," Manchester (1847); *Phil. Mag.*, [3] **30**, 127 (1847).

97. Gore, G., *The Art of Metallurgy*, Longmans, Green & Co., London, 1877 (8th printing, 1906).
98. Gray, W. F., *J. Electrochem. Soc.*, **110**, 211 C (1963).
99. Grenet (1856); cf. Tommasi (219), p. 124.
100. *Id.*, *Cosmos*, **23**, (1863); cf. Cazin (34), p. 243.
101. Grove, W. R., *Phil. Mag.*, *III*, **13**, 430 (1838); **14**, 388 (1839); **15**, 287 (1839).
102. *Id.*, *Phil. Mag.*, [3] **14**, 127 (1839); **21**, 417 (1842); cf. Grove (104).
103. *Id.*, *Ann. Chim. Phys.*, **58**, 202 (1843).
104. *Id.*, On the correlation of physical forces (lectures in the London Institution), London (1846).
105. Hachette and Desormes, C. B., *Ann. Chim. Phys.*, **5**, 191 (1803); cf. Biot (16); Gay-Lussac (94).
106. Haldane, H., "Experiments made with the Metallic Pile of Signor Volta....," *Nicholson's J.*, **4**, 313–19 (1800).
107. Hall, C. M., U.S. Pat. 400,076 (1889).
108. Hare, Robert, *Am. J. Sci. (Silliman's J.)*, **1**, 413–23 (1819); **3**, 105 (1821); **4**, 201 (1822); *et seq.*
109. Heim, G., *Elektrotechn. Z.*, **8**, 472, 517 (1887); cf. Laurie (142); Morehouse (158).
110. Heise, G. W., *Trans. Electrochem. Soc.*, **75**, 147 (1939).
111. *Id.*, *J. Electrochem. Soc.*, **97**, 65 C (1950).
112. *Id.*, *J. Electrochem. Soc.*, **100**, 264 C (1953).
113. *Id.*, *J. Electrochem. Soc.*, **101**, 291 C (1954).
114. Heise, G. W., and Cahoon, N. C., *J. Electrochem. Soc.*, **99**, 179–187 C (1952).
115. Heise, G. W., and Schumacher, E. A., *Trans. Electrochem. Soc.*, **62**, 337–344 (1932); Heise, Schumacher, and Fisher, C. R., *Trans. Electrochem. Soc.*, **92**, 173 (1947).
116. Heise, G. W., Schumacher, E. A., and Cahoon, N. C., *J. Electrochem. Soc.*, **94**, 99 (1948).
117. Helmholtz, H., *On the Conservation of Force*. (Pamphlet, 1847); cf. Helmholtz (118), p. 149.
118. *Id.*, "On the interaction of natural forces" (1854), In *Popular Lectures on Scientific Subjects*" (Transl. E. Atkinson), pp. 137–171. Longmans, Green & Co., London, 1889.
119. Heraud, A., *Compt. rend.*, **88**, 124–6 (1879); cf. Cazin (34), p. 165.
120. Herbert, W. S., *J. Electrochem. Soc.*, **99**, 109 C (1952).
121. Highton, H., *Brit. Pat.* 1638 (1872); cf. Bottone (22).
122. Holmes, H. N., *Oberlin College Bull.*, **346** (1937) (includes quotation from a speech by Prof. F. F. Jewett, 1920).
123. Howard, P. L., *J. Electrochem. Soc.*, **99**, 200–1 C (1952).
124. Howard, P. L., *J. Electrochem. Soc.*, **99**, 333–7 (1952).
125. Hubert, C.; cf. McQueen (154), p. 48.
126. Hulot, *Compt. rend.*, **40**, 1148 (1855); cf. Cazin (34), p. 131; Tommasi (219), p. 93.
127. Humphreys, D., U.S. Pat. 390,676 (1888).
128. Huntley, A. K., *Trans. Electrochem. Soc.*, **68**, 219 (1935).
129. Jablochkoff, P., *Compt. rend.*, **85**, 1052 (1877); cf. Niaudet (165), p. 228.
130. Jacobi, M. H., *L'Institut*, **2**, 394 (1834), cf. King (136).
131. *Id.*, *Phil. Mag.*, **15**, 161–5 (1839).
132. Janes, M., *Trans. Electrochem. Soc.*, **77**, 411 (1940).
133. Johnson, H. T., U.S. Pat. 487,839 (1892).
134. KaDell, H. W. (Private communication, 1927).

135. Kemp, K. T., *Edinburgh New Phil. J.*, **6**, 70–7 (1828); *Ann Elec., Mag., Chem. (Sturgeon)*, **1**, 81–8 (1837); cf. Benjamin (15); King (136).
136. King, W. J., The Development of Electrical Technology in the 19th Century: 1. The Electrochemical Cell and the Electromagnet. Paper 28. *U.S. National Museum Bull. 228*, Smithsonian Inst., Washington, D.C. (1962).
137. Kleiderer, C., *Modern Plastics* (Nov. 1945), p. 133.
138. Kousmine (1890); cf. Bottone (22), p. 249.
139. Krehbiel, H., *Elektrotech. Z.*, **11**, 422 (1890).
140. de Lalande F., and Chaperon, G., French Pat. 143,644, (1881); U.S. Pat. 274,110, 1883, *Compt. rend.*, **97**, 164 (1883).
141. Latimer, W. M., *Oxidation Potentials*, (2nd ed.) Prentice-Hall, Englewood Cliffs, N.J. (1952), p. 236.
142. Laurie (1886); cf. Tommasi (219), p. 121.
143. Lawson, H. E., U.S. Pat. 2,428,850, (1947); cf. Blake (17).
144. LeBlanc, F., *Compt. rend.*, **83**, 904 (1871).
145. Leclanché, G., French Pat. 71,865 (1866).
146. *Id.*, *Les Mondes*, **16**, 532–5 (1868); *Compt. rend.*, **83**, 54–6, 1236–8 (1876); Leclanché and Barbier, French Pat. 124,108 (1878); cf. du Moncel (157).
147. Leeson, *Phil. Mag. III*, **20**, 262 (1842).
148. Ley, W.; cf. Garrett (89) (1957).
149. Magie, W. F., *A Source Book of Physics*, McGraw-Hill, New York (1935).
150. Maiche, L., French Pat. 127,069 (1878); cf. Tommasi (219), pp. 102 ff., 411; Benjamin (15), p. 323.
151. Marié-Davy E. H., *Compt. rend.*, **49**, 1004-6 (1859); *Ann. Telegr.*, **3**, (Jan.-Feb. 1860); cf. Niaudet (165), pp. 137, 141; Cazin (34), pp. 135 ff.
152. van Marum, M., and Pfaff, *Phil. Mag.*, **12**, 161–4 (1801).
153. Matteucci C., *Ann. Chim. Phys.*, III, **34** (1852); cf. Cazin (34), p. 260.
154. McQueen, A., *A Romance in Research. The Life of Charles F. Burgess*, Instruments Publishing Co., Pittsburgh, (1951).
155. Meidinger, H., *Ann. Phys.*, **108**, 602 (1859).
156. Misell, D., U.S. Pat. 603, 112, (1898); 617, 592 (1899).
157. du Moncel, T., *Dingl. Polytech. J.*, **186**, 270 (1867); **188**, 96 (1868).
158. Morehouse, C. K., *J. Electrochem. Soc.*, **99**, 187 C (1952).
159. Mottelay, P. F., *Bibliographical History of Electricity and Magnetism*, Chas. Griffin & Co., London (1922).
160. Muncke, *Bibliothéque Universelle*, **1**, 160 (1836), cf. Faraday (76), Ser. XVII, 2043, (Jan. 1840).
161. Nasarischwily, A., "Galvanische Elemente mit Luftsauerstoffdepolarisation", *Z. Elektrochem.*, **29**, 320 (1923).
162. National Bureau of Standards, *Handbook 71. Specifications for Dry Cells and Batteries* (Dec. 1959).
163. Nauche, Graperon and Baget; cf. *Phil. Mag.*, **24**, 184 (1806).
164. Neuffen, P., *Anleitung zum Aufbau der Galvanischen Saeule u. zur Anwendung derselben auf verschiedenen Krankheiten*, Ulm (1802).
165. Niaudet, A., *Pile Électrique* (2nd ed.), J. Baudry, Paris (1880). (Translated by L. M. Fishback): *Elementary Treatise on Electric Batteries*, John Wiley & Sons, New York, 1880.)
166. Nicholson, W., *Nicholson's J.*, **4**, 241 (1800); cf. Haldane (106).
167. *Id.*, "Account of the New Electrical or Galvanic Apparatus of Sig. Alex. Volta and Experiments performed with the same," *Nicholson's J.*, **4**, 179–87 (1800).

168. *Id.*, Carlisle, A., Cruickshank, W., et al., "Experiments in Galvanic Electricity", *Phil. Mag.*, **7**, 337-347 (1800).
169. Novalhier, E. T., and Prevost, J. B., Brit. Pat. (1857); cf. Gore, (97), p. 222.
170. Obach, E., *London Elect. Rev.* (May 15, 1891); cf. Carhart (32), pp. 155–6.
171. *Id., Elektrochem. Z.*, 220 (1909); cf. Nasarischwily (161).
172. Oersted, H. C., *Schweiggers J.*, **20**, 205 (1817).
173. *Id., Ann. chim. phys.*, **14**, 417–25 (1820); *Schweiggers J.*, **29**, 364 (1820).
174. Ostwald, W., *Klassiker der exakten Wissenschaffen*, Leipzig (1888), No. 114, p. 128; pp. 4–5.
175. Paine, A. J., *Electrician,* **82**, 258 (1913); cf. Cooper (45), p. 234.
176. Parrot, G. F., *Ann. Phys.*, **12**, 49 (1802).
177. Pearson, G., "Experiments and Observations made with the view of Ascertaining the Nature of the Gaz Produced by passing Electric Discharges through Water," *Phil. Trans. Roy. Soc.*, **87**, (I), 140 (1797).
178. Pepys, W., (1801); cf. Viard (229); Mottelay (159), p. 378.
179. *Id.*, (1802); cf. Cruickshank (52); Mottelay (159), p. 371.
180. *Id., Phil. Trans.*, **113**, 187–8 (1823).
181. Picardat, *Les mines dans la guerre de campagne,* Gauthier-Villars, Paris (1874); cf. Niaudet (165), p. 208.
182. Pincus, *Pogg. Ann.*, **135**, (1868); *Compt. rend.*, **67**, 1076 (1868); cf. Cazin (34), p. 142.
183. Planté, G., *Compt. rend.*, **49**, 402 (1859); **50**, 640 (1860).
184. Poggendorff, J., *Pogg. Ann. Phys.*, **57**, 101 (1842).
185. Pope, F. L., *Elect. Eng.*, **11**, 1 (1891), cf. Taylor (217), p. 699.
186. Pucher, L. E., *J. Electrochem. Soc.*, **99**, 203 C (1952).
187. *Id., J. Electrochem. Soc.*, 204 C (1952).
188. Pulvermacher, *Compt. rend.*, **45**, 1047 (1857); cf. Cazin (34), p. 107.
189. Reed, C. J., and Morrill, M. T., U.S. Pat. 690,772 (1902).
190. Regnauld, J., *Compt. rend.*, **43**, 917–22 (1857).
191. Ritter, J. W., *Beitraege zur naeheren Kenntnis des Galvanismus . . .* (2 vols.), Jena (1800-1805).
192. *Id., Ann. Phys.*, **6**, 468 (1800).
193. *Id., Phil. Mag.*, **23**, 51 (1805); cf. Mottelay (159), pp. 381–2.
194. de la Rive, A. A., *Arch. de l'Électr.*, **3**, 525 (1823).
195. *Id., Bibliothèque Universelle, Sciences et Arts*, **43**, 391–411 (1830); cf. Faraday (76), Ser VII, 863 (1834); King (136), p. 241.
196. *Id., Ann. Chim.*, **61**, 40 (1836); *Bibliothéque Universelle,* **1**, 152–62 (1836); Faraday (76), Ser. XVII, 2041–43 (1840).
197. *Id.*, "Note sur une nouvelle combinaison voltaique." *Arch. Elect.*, **3**, 112–15 (1843); also *Traité d'Elect.*, **2**, 60 (1856); cf. Niaudet (165), p. 178.
198. Robison, J., Letter in Fowler, R.: *Experiments and Observations relative to ..... Animal Electricity*, Edinburgh, 1793. cf. Mottelay (159), p. 308.
199. de la Rue, W., and Mueller, H., *Compt. rend.*, **67**, 794–8 (1868); *Les Mondes*, **16** (1868); **18** (1870) cf. Cazin (34), p. 137.
200. Salva, F., "Application du galvanisme à la telegraphie" (1804); cf. Mottelay (159), p. 318.
201. Sauvage, *Ann telegr.*, (July-August 1875); cf. Nasarischwily (161).
202. Schlotter, W. J., *J. Electrochem. Soc.*, **99**, 205 C (1952).
203. Schmid, A., "Die Diffusions Gaselektrode," *Ferd. Enke*, Stuttgart (1923); *Helv. Chim. Acta*, **7**, 370 (1924).
204. Schoenbein, C. F., *Pogg. Ann.*, **49** (1840), cf. Cazin (34), p. 214.

205. Schrodt, J. P., Otting, W. J., Schoegler, J. O., and Craig, D. N., *Trans. Electrochem. Soc.*, **90**, 405 (1946).
206. Silliman, B., *Am. J. Sci. Arts*, **44** (1842); *Arch. Electr.*, **3**, (1842); cf. Cazin (34), p. 214.
207. Simpson, G. C., *J. Inst. Elect. Eng.* (London), **67**, 1275–6 (1920).
208. Skrivanow; cf. Tomassi (219), p. 294; Bottone (22), pp. 175 ff.
209. Smee, A., *Phil. Mag. III*, **16**, 315 (1840); cf. Niaudet (165), p. 48; Cooper (45), p. 150.
210. Soemmering, S. T., "Ueber einen elektrischen Telegraphen," *Denkschrift Muencher Akad.*, Munich (1811); cf. Taylor (217); Mottelay (159) pp. 406–7.
211. Stokes, J. J., Paper read before Electrochem. Soc. (Oct. 1955); cf. Glicksman (95) Cahoon and Korver (28).
212. Stuart, A., "Experiments to Prove the Existence of a Fluid in the Nerves," *Phil. Trans. Roy. Soc.* 327, 585 (1732); cf. Fahie (74), p. 1344.
213. Sturgeon, W., "Experimental Researches on Electromagnetism" (1830); *Phil. Mag.*, **11**, 194 (1832); III **7**, 230 (1835); *Ann. Élect.*, **7** (1841); cf. Cazin (34), p. 109.
214. *Id.*, (5th memoir) *Ann. Elect.*, **5**, 120–35 (1840).
215. Sulzer, Johann, Georg. *Theorie der angenehmen und unangenehmen Empfindungen*, Berlin (1762); *Nouvelle Théorie des Plaisirs* (1767); cf. Benjamin (15).
216. Swammerdam, Jan, in Vol. II, *Biblia Naturae* (1737) (Boerhave), Leipzig, (English translation, 1752); cf. Mottelay (159), p. 202.
217. Taylor, L. W., *Physics*, Houghton Mifflin Co., Boston (1941).
218. Teed, R., *Phil. Mag.*, **12**, 105 (1801).
219. Tommasi, D., *Traité des Piles Électriques*, Paris (1889).
220. Trommsdorff, B., *Geschichte des Galvanismus*, Erfurt (1808).
221. van Troostwijk, A. P., and Deiman, J. R., *J. Phys.*, (November 1789) Pearson (177).
222. Trouvé; cf. Cazin (34), pp. 156–7; Tommasi (219), p. 286.
223. *Id.*, cf. Tomassi (219), p. 250; Niaudet (165), p. 103.
224. Tyer (modified Smee cell); cf. Niaudet (165), p. 50.
225. Varley C., Brit. Pat. (1855), cf. Niaudet (165), p. 108; *Am. Ry. Signaling*, chap. V, Signal Sec., Assoc. Am. Railroads p. 6 (1948); Vinal (230).
226. Vergne, M., U.S. Pat. 28,317 (1867).
227. Verne, Jules, *Voyage au centre de la terre* (1864); *Vingt mille lieues sous les mers* (1869).
228. duVerney, J. G., *Histoire de l'Academie des Sciences* (1700), p. 40; (1742), p. 187; cf. Mottelay (159), p. 148; Fahie (79), p. 1344.
229. Viard, *Phil. Mag.*, [4] **6**, 241–58 (1843); *Ann. Chim. Phys.*, **36**, 129 (1843).
230. Vinal, G. W., *Primary Batteries*, John Wiley & Sons, New York, 1950.
231. Volta, A., Letters XIII and XIV to van Marum (Aug. 30 and Oct. 11, 1792); Mottelay (159), p. 247; Ostwald (174), No. 114, p. 128, No. 118, p. 4.
232. *Id.*, *Phil. Trans. Roy. Soc.*, **83**, 10–44 (1793); *Phil. Mag.*, **4**, 59, 163, 306 (1799).
233. *Id.*, "On the Electricity Excited by the mere contact of Conducting Substances of different kinds," *Phil. Trans. Roy. Soc. London*, **90**, 403–31 (1800).
234. *Id.*, *Ann. Chim.*, **40**, 224 (1802); cf. Faraday (76), Ser. XVII, 2041 (1840).
235. Wach, *Schweigger J.*, **58** (1830); cf. Cazin (34) p. 168; Tomassi (219), p. 227.
236. Wade, E. J., *Secondary Batteries*, London (1902).
237. Walker, C. V., *Phil. Mag.*, [4] **18**, 73–7 (1859).
238. Walker, W., Wilkins, F. H., and Lones, J., U.S. Pats. 524,229 and 524,291 (1894).
239. Warrington, R., *Phil. Mag. III*, **20**, 393 (1842).
240. Watkins, *Ann. Chim. Phys.*, **38** (1828); cf. Cazin (34), p. 128; Tomassi p. 530; Benjamin (15) p. 310.

241. Wensky, W., Brit. Pat. 49 (1891); cf. Pucher (186).
242. Westheimer, F. H., et al., *Chem. Eng. News*, **43** (No. 48) 72 (1965).
243. Whittaker, Sir Edmond, *A History of the Theories of Aether and Electricity*, London (1931), pp. 227 ff.
244. Wilkinson, C. H., *Elements of Galvanism in Theory and Practice*, John Murray, London (1804).
245. *Id.*, *Nicholson's J.*, **8**, 1–5 (1804); *Phil. Mag.*, **29**, 243–4 (1807).
246. Wilson, C. T. R., *Phil. Trans. Roy. Soc., A*, **221**, 73 (1920); cf. Simpson (207).
247. Winslow, N. M., *Trans. Electrochem. Soc.*, **80**, 120 (1941); **88**, 81 (1946).
248. Wollaston, W. H., "Experiments on the Chemical Production and Agency of Electricity," *Phil. Mag.*, **11**, 206 (1801).
249. *Id.*, *Gilbert Ann. Phys.* **64** (1816).
250. Yeager, E., *Science*, **134**, 1178–1186 (1961).
251. Zamboni, G., *Pila Elèttrica a Secco*, Verona (1812); *Ann. chim. phys.*, **11**, 190 (1812); cf. Mottelay (159), pp. 420 *et seq.*

2

Fundamental Aspects

ERNEST B. YEAGER AND EUGENE P. SCHWARTZ

INTRODUCTION

Many reactions involving the gain or loss of electrons initially appear attractive for use in power-producing electrochemical cells. In practice, relatively few have proved capable of supporting appreciable currents at practical voltages in batteries. In fact, even the open-circuit voltages (or the electromotive forces) often do not correspond to the thermodynamically predicted values because of substantial irreversibility and competing reactions.

How have the chemical systems of presently important batteries been chosen? What factors have guided their development? Much to the dismay of many battery electrochemists, almost all batteries of importance today are the result principally of an Edisonian approach and not fundamental research. Even so, the era when the art has predominated over the science in battery development is rapidly approaching an end and perhaps has already ended. Increasing reliance is placed on a fundamental understanding of the thermodynamics and various dynamic characteristics of the electrochemical processes in the exploration of new battery systems and the optimization of existing systems. It is in this light that the present chapter attempts to summarize some of the more relevant aspects of the thermodynamics of electrochemical cells, the physical and chemical properties of electrode surfaces, and various factors contributing to losses within batteries.

This chapter reviews the background necessary for understanding the fundamental aspects of the specific batteries to be described in later chapters. It is not systematic or comprehensive in its approach nor a substitute for the more extensive treatments of various aspects of fundamental electrochemistry which are more or less required reading for a scientist entering into battery research. Frequent reference will be made to these more comprehensive treatments.

The identification of the component reactions occurring in various parts of the cell and the over-all cell reactions at various stages of discharge is a necessary first step in developing an insight into the electrochemistry of a battery. Such identification often presents a very difficult problem because of the complex oxides and other solid compounds—usually nonstoichiometric—that form. Modern instrumental methods such as x-ray and electron diffraction, absorption spectroscopy, and radiochemical methods are powerful tools for the identification and characterization of cell reactants and products, but even so, information concerning the specific reactions occurring within many of the more important batteries is amazingly incomplete. Perhaps the single most important fundamental advance that can be made in battery electrochemistry would be the development of more specific analytical and structural tools for *in situ* characterization of battery components during discharge.

THERMODYNAMICS

If thermodynamic properties of the reactants and products of the over-all cell reaction are known, thermodynamics permits the calculation of the open-circuit cell voltage or *electromotive force* (emf) provided competing reactions are not involved. This thermodynamic calculation is not dependent on a knowledge of the mechanism or steps involved in the over-all cell reaction. In practice, considerable difficulty is often encountered because of uncertainty of the nature and thermodynamic properties of the species involved in the over-all cell reaction at various stages of discharge (or charge in the instance of storage batteries). Nevertheless, even when the data for the thermodynamic properties (particularly activities) of the chemical components within the cell are insufficient for precise calculations, thermodynamics may still permit predictions of cell voltage to within 0.1 V. When extraneous competing reactions are known or suspected to cause difficulty under open-circuit conditions, the thermodynamically calculated electromotive force represents an upper limit.

A detailed development of the thermodynamics of electrochemical cells is beyond the scope of this chapter. The reader is referred to standard texts for such treatments [e.g., Lewis, Randall, Pitzer, and Brewer (84), Glasstone (52), Kirkwood and Oppenheim (71). Only aspects of thermodynamics of direct importance in batteries will be considered here.

THE NERNST EQUATION

The thermodynamic potential for an electrochemical cell may be calculated from the standard oxidation or reduction potentials for the anodic and cathodic electrode reactions and the activities of the reactants and products of the overall cell reactions. If the cell reaction is

$$aA + bB \rightarrow cC + dD \tag{a}$$

the thermodynamically predicted cell voltage is given by the Nernst equation

$$E = E^\circ - \frac{RT}{nF} \ln W \tag{1}$$

where n is the number of equivalents of charge transferred from the anode to the cathode per mole of reaction (i.e., when the reaction proceeds once in the indicated manner), F is the Faraday constant ($9.65 \cdot 10^4$ coul/equivalent), E° is the standard cell-reaction potential, and

$$W = \frac{(a_C)^c (a_D)^d}{(a_A)^a (a_B)^b} \tag{1a}$$

The quantities a_i are the activities of the entities indicated by the subscripts and the other symbols have their usual meaning. Equation 1 does not take into account liquid-junction potentials, which are relatively uncommon in battery systems. The thermodynamic cell reaction potentials E and E° in Eq. (1) are related to the Gibbs free energy change (ΔG) and the standard free energy change (ΔG°) for the reaction by the equations

$$\Delta G = -nFE \tag{1b}$$

and

$$\Delta G^\circ = -nFE^\circ \tag{1c}$$

The sign conventions used for electrode potentials in this book are the same as those adopted by The Electrochemical Society. The *electrode potential* indicates the sign and numerical value for the voltage of the electronic conductor of an electrode system (usually a metal) relative to the standard hydrogen electrode (SHE) in contact with the same electrolyte with any possible liquid-junction potential excluded. Thus the electrode potentials for metals more active than hydrogen gas are negative in sign. The term *oxidation potential* refers to the thermodynamically predicted potential for a half-cell reaction when the reaction is written as an oxidation process. A positive value for the potential of an oxidation reaction indicates that this half-cell reaction occurs more readily than the corresponding oxidation reaction for the standard hydrogen electrode, whereas a negative value indicates just the reverse.

In an analogous manner the *reduction potential* refers to the thermodynamic potential for a half-cell reaction when the reaction is written as a reduction. A positive value for a reduction potential indicates that this reaction proceeds more readily than the reduction process at the standard hydrogen electrode, and a negative value indicates the reverse. In comparing experimentally measured potentials with the thermodynamic oxidation or reduction potentials, it should be noted that the electrode potentials and the reduction potentials have the same sign. The standard cell-reaction potential E° is the sum of the standard oxidation potential for the anodic reaction and the standard reduction potential for the cathodic reaction at the temperature T.

The Nernst equation is sometimes written with a positive sign instead of the negative sign shown in Eq. (1). The negative sign, however, should be used exclusively whenever the quantity W is expressed as the product of the activities of the products over the corresponding product for the reactants. The potential E in Eq. (1) is then positive if the reaction proceeds spontaneously as written and negative if the reaction does not proceed spontaneously but must be forced to proceed in the indicated direction by the application of an external potential of sufficient magnitude. This is in accord with the fact that the Gibbs free energy change is negative for a spontaneous process such as in a power producing cell.

The Nernst equation can be applied to the calculation of the potentials of the individual half cells. For a reduction reaction of the type

$$aA + ne^- \rightarrow cC + uU^{-r} \tag{b}$$

the corresponding reduction potential is

$$E_{\text{red}} = E^\circ_{\text{red}} - \frac{RT}{nF} \ln W_{\text{red}} \tag{2}$$

where

$$W_{\text{red}} = \frac{(a_C)^c (a_{U^{-r}})^u}{(a_A)^a} \tag{2a}$$

Similarly, for an oxidation reaction of the form

$$bB \rightarrow vV^{+s} + ne^- \tag{c}$$

the corresponding equation for the oxidation potential is

$$E_{ox} = E^{\circ}_{ox} - \frac{RT}{nF} \ln W_{ox} \tag{3}$$

where

$$W_{ox} = \frac{(a_{V^{+s}})^{v}}{(a_B)^{b}} \tag{3a}$$

If the product D in reaction (a) corresponds to $[(u/d)U^{-r} + (v/d)V^{+s}]$, then Eq. (1) is equal to the sum of Eq. (2) and (3) provided there is no liquid-junction potential. A cell which corresponds to these reactions is

$$A \mid C, U^{-r} \mid \mid V^{+s} \mid B$$

The electrode on the left involving reactant A will be positive if the cell reaction proceeds in the direction indicated in the reaction (a)—(c) and the thermodynamic cell reaction potential will be positive.

The values for the standard oxidation and reduction potentials may be obtained in many instances from the literature [e.g., Latimer (82), deBethune et al. (15), deBethune and Loud (16), Pourbaix (100)]. A table summarizing some of the values of particular interest to the battery electrochemists is given in Appendix B. When the standard oxidation or reduction potentials are not available, the battery electrochemists may be forced to calculate or to determine experimentally his own values. The procedures involved in determining accurate values for standard oxidation and reduction potentials may prove quite complicated and the reader is referred to such texts as Lewis, Randall, Pitzer, and Brewer (84) and Glasstone (52) for information relating to such procedures.

In some instances the thermodynamic potential of the cell E may be reasonably well approximated by the value of the standard cell potential E°, and hence by the sum of the standard oxidation and reduction potentials for the appropriate individual electrodes. Such is the case in systems where both the reactants and products for the over-all cell reactions are in separate pure phases. Under these conditions the activities of the reactants and products are unity and then $E = E^{\circ}$.

Activities

The concept of thermodynamic activity and the definitions of the standard states must be thoroughly understood if the Nernst equation is to be used reliably for the relatively complicated systems often encountered in battery electrochemistry. When expressed in terms of actual concentrations, many thermodynamic equations such as the Nernst equation are strictly applicable only to perfect systems. In situations where activity data are lacking there is a temptation to assume that activities may be approximated by concentrations in the use of the Nernst equation. Most batteries involve very concentrated electrolytes as well as solid phases of nonstoichiometric composition, however, and such approximations are usually relatively poor. Errors of the order of 0.1 V in the calculated potentials are easily possible.

The definition of activity is facilitated by the concept of *fugacity*. For a gaseous component i the fugacity f_i is defined by the equation

$$\mu_i = \mu_i{}^0 + RT \ln f_i \tag{4}$$

where

$$\frac{f_i}{P_i} \rightarrow 1 \qquad \text{as} \qquad P_T \rightarrow 0 \tag{4a}$$

In these equations, P_i is the partial pressure of component i and P_T is the total pressure on the gaseous mixture. The quantity μ_i is the *chemical potential* of component i and is defined as the partial derivative of the Gibbs free energy G with respect to the number of moles n_i of the ith component with all other independent variables held constant:

$$\mu_i = \left(\frac{\partial G}{\partial n_i}\right)_{T,P_T,n_{j(j\neq i)}} \tag{5}$$

The quantity $\mu_i{}^0$ is the *standard chemical potential* of the ith component; It is a constant, independent of P_T n_i, and f_i, but depending on temperature. Equation (4a) incorporates into the definition of fugacity the condition that f_i should approach the partial pressure of the ith component as the gaseous mixture approaches ideal behavior, which is the case when the total pressure approaches a very low value.

The *fugacity* of the ith component in the solid or liquid state is defined as equal to the fugacity of this component in the vapor in equilibrium with the liquid or solid. This definition of fugacity is used even when the vapor pressure of the ith component in the liquid or solid state is far too low to be measured or to have any practical significance.

The *activity* a_i of the ith component in a given phase is defined as the ratio of the fugacity f_i of this component to the fugacity $f_i{}^0$ in the standard state:

$$a_i = \frac{f_i}{f_i^0} \tag{6}$$

For a gaseous mixture, the ratio of the fugacities in Eq. (6) can often be replaced by the ratio of partial pressures of the ith component to a good approximation. Similarly, for a solid or liquid phase, the ratio of the fugacities may be approximated by the vapor pressure of the ith component. Thus

$$a_i \cong \frac{P_i}{P_i^0} \tag{6a}$$

The error resulting from the use of the approximate Eq. (6a) in subsequent calculations of potential with the Nernst equation is usually less than a millivolt for cells operating at a total pressure of a few atmospheres or less.

Standard States for Gaseous Components. The standard state for a component in a gaseous mixture is taken as $f_i^0 = 1$ atm for whatever temperature the activity is desired. Thus the activity of the ith component in a gaseous mixture can be set equal numerically to the partial fugacity of this component or approximately to the partial pressure of this component. The partial fugacity and partial pressure are expressed in atmospheres but it is evident from Eq. (6) and (6a) that the activity is dimensionless.

The calculation of open-circuit cell voltages for an over-all cell reaction involving gaseous components may be illustrated by considering the following high-temperature cell:

$$O_2(Pt)|(ZrO)_{0.85}(CaO)_{0.15}|H_2, H_2O(Pt)$$

in which the electrodes are platinum and the electrolyte is an ionic conducting solid oxide. The over-all cell reaction is

$$2H_2 + O_2 \rightarrow 2H_2O \tag{d}$$

where all components are gases at the operating conditions of the cell (1015°C or 1288°K) and four electrons are involved for every two molecules of water formed. For this reaction at pressures of a few atmospheres or less, the activities may be closely appropriated by the partial pressures as indicated in Eq. (6a) with the values for P_i^0 for each component set equal to 1 atm. Equations (1) and (1a) then become

$$E = E^\circ - \frac{RT}{4F} \ln \frac{(P_{H_2O})^2}{(P_{H_2})^2(P_{O_2})} \tag{7}$$

Weissbart and Ruka (126) have compared the voltages calculated with this equation with experimentally measured values at 1015°C. The solid line in Fig. 2.1 represents the calculated cell voltages while the points correspond to measured volumes with the oxygen partial pressure (P_{O_2}) held constant at 0.0621 atm. Agreement between calculated and experimental values for the cell voltage are within the accuracy with which the values for P_{H_2} and P_{H_2O} are known.

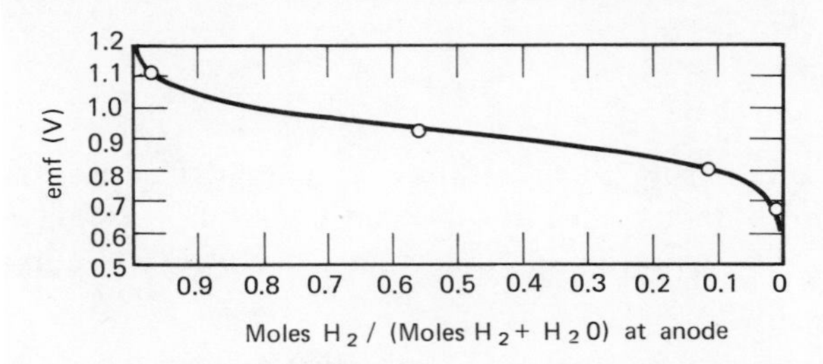

Figure 2.1 Comparison of measured voltage with theoretical emf of cell, $O_2(Pt)|(ZrO)_{0.85} \cdot (CaO)_{0.15}|H_2,H_2O(Pt)$ according to (126) (by permission of The Electrochemical Society). Solid line: theoretical; ○ experimental point; temperature 1015°C, O_2 pressure at cathode 731.2 mm.

Standard States for Pure Liquids and Pure Solids. The *standard state* for a pure liquid or pure solid at any particular temperature is the pure material at a total pressure of 1 atm and at that temperature. As a result, the standard state fugacity f_i^0 corresponds to the vapor pressure of the pure liquid or pure solid with the total pressure on the system equal to 1 atm. The fugacity f_i, however, also corresponds to that of the pure material although not necessarily at a total pressure 1 atm. If the total pressure on the system is 1 atm, then for a pure material a_i is equal exactly to 1. The vapor pressure and hence the fugacity of a pure material change very slowly with total pressure. Therefore, unless the total pressure is high (e.g., greater than 10^2 atm), the activity of either a pure liquid or pure solid phase is equal to unity to a very close approximation.

Standard States for Solid and Liquid Solutions. For the solvent in a solution, the *standard state* at a particular temperature is usually defined as the pure solvent at a total pressure of 1 atm and at that temperature. Therefore the standard state fugacity f_1^0 of the solvent[1] equals the fugacity of the pure solvent at a total pressure of 1 atm and the activity of the solvent a_1 is then equal to the ratio f_1/f_1^0, or to a good approximation P_1/P_1^0. If the system is ideal, Raoult's law can be applied and the ratio of the partial pressures is equal to the molefraction X_1, or $a_1 = X_1$. At the concentrations involved in most batteries, Raoult's law is not a good approximation and the activity of the solvent a_1 cannot be set equal to X_1 without incorporating a significant error in the final calculated potentials.

For solutes in solution the *standard state* is usually hypothetical, not an actual state that can be achieved. This hypothetical state is obtained by extrapolating the properties of an infinitely dilute solution to unit

[1]The subscript 1 will be reserved for the solvent unless otherwise indicated. The subscripts 2, 3, . . . , will be used for various solutes.

concentration as shown in Fig. 2.2. In this figure the fugacity of the ith component (or approximately the partial pressure) is plotted versus some measure of concentration such as molefraction, molality, or molarity at a total pressure of 1 atm. The standard state fugacity f_i^0 of the ith component is then obtained through a linear extrapolation of the infinitely dilute properties to unit concentration. The particular standard state and the value for f_i^0 are dependent on whether the extrapolation is to unit molefraction, unit molality, or unit molarity. In using the Nernst equation and activity data it is very important to know which particular unit concentration is involved in the definition of the standard state. For solid solutions and liquid metal solutions the molefraction is usually preferred. For solutes in liquid solutions other than liquid metals, molality and molarity are favored. *In this chapter activity data, standard electrode potentials, and other thermodynamic data involving standard states will be given for solutes in liquid solutions in terms of a standard state based on unity molality unless otherwise noted.*

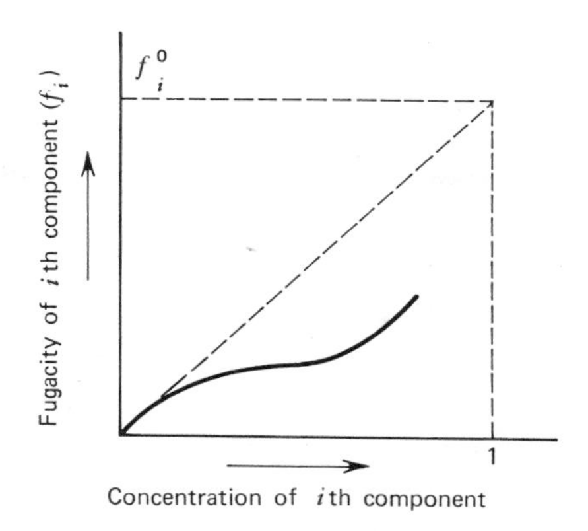

Figure 2.2 Procedure for defining the standard state fugacity f_i° of a solute in a solution.

In *sufficiently dilute* solutions, a_i may be approximated numerically by the molefraction, molality, or molarity, depending on the definition of the standard state. With the systems and the concentrations used in virtually all practical batteries, however, the deviation from ideality for the solution phases is far too extreme to permit the activities to be approximated by actual concentrations. The activities of solutes often deviate from the numerical values for the concentration by factors of 10^{+1} to 10^{-3} for concentrated electrolytic solutions and by even greater factors for alloys. Thus the battery chemist should not succumb to the temptation of replacing the activities for solutes in Eq. (1) to (3) with concentrations.

The errors which may occur upon substituting concentration for activity for an alloy can be illustrated by considering the cell

$$Na_XSn|glass|Na$$

which has been studied by Agruss (1) in conjunction with research on the use of electrochemical cells in closed cycles for the generation of electrical energy from thermal energy. This cell is an example of a concentration cell of the general form

$$A_{a_{II}}, B|A^+|A_{a_I}, B$$

where a_I and a_{II} are the activities of the active metal A in the two alloy electrodes and B corresponds to the nonactive component of the alloy. The cell reaction is

$$A_{a_I} \rightarrow A_{a_{II}} \tag{e}$$

and the thermodynamic cell reaction potential is given by the equation

$$E = -\frac{RT}{F} \ln \frac{a_{II}}{a_I} \tag{8}$$

with $E^\circ = 0$ and $n = 1$. If the standard state for component A corresponds to pure A, then $a_I = 1$. In Fig. 2.3 are plotted the cell reaction potentials calculated by means of Eq. (8) with activity data derived from literature data [Selected Values (110)]. Also shown are the cell reaction potentials calculated with the same equation but using the molefraction of Na rather than the activity. The experimental values for the electromotive force are close to the curve based on activities.

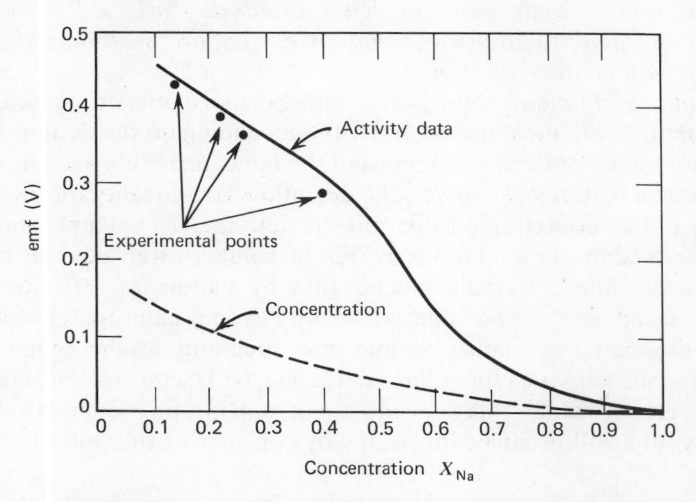

Figure 2.3 EMF versus concentration at 500°C for the cell Na_xSn|glass|Na. Solid line: calculated, taking activities into consideration with Eq. 8 and using literature data [Selected Values (110)]; dashed line: calculated with activity of Na replaced by molefraction of Na; points: experimental values [Agruss (1)] (by permission of The Electrochemical Society).

Activity Coefficients. The *activity coefficient* γ_i for either solute or solvent is defined by the relationship

$$\gamma_i = \frac{a_i}{c_i} \tag{9}$$

where c_i is the molefraction, molality, or molarity of the ith component. At infinite dilution γ_i approaches unity as the solution approaches ideal behavior.

Activities of Ionic Components. The definition of the *activity of individual ions* is the same as for ordinary solutes in solution:

$$a_+ = \frac{f_+}{f_+{}^0} \cong \frac{P_+}{P_+{}^0} \tag{10a}$$

and

$$a_- = \frac{f_-}{f_-{}^0} \cong \frac{P_-}{P_-{}^0} \tag{10b}$$

even though the fugacity and vapor pressures of the ionic constituents are far too small to measure. The values for $f_+{}^0$ and $f_-{}^0$ are obtained in principle by the same procedure as indicated in Fig. 2.2, extrapolation to unit molality or molarity. The *mean activity* for an electrolyte of the type V_vU_u with V^{+s} as the cation and U^{-r} as the anion is defined as

$$a_\pm = [(a_+)^v(a_-)^u]^{1/(v+u)} \tag{11}$$

Likewise, the *mean activity coefficient* $\gamma_\pm$ is

$$\gamma_\pm = [(\gamma_+)^v(\gamma_-)^u]^{1/(v+u)} \tag{12}$$

where $\gamma_+ = (a_+/c_+)$, $\gamma_- = (a_-/c_-)$, and $\gamma_\pm = (a_\pm/c_\pm)$, where the mean concentration $c_\pm$ is given by

$$c_\pm = [(c_+)^v(c_-)^u]^{1/(v+u)} \tag{13}$$

The mean concentration is usually expressed in terms of molality and is then represented by the symbol $m_\pm$.

For a strong electrolyte which is completely dissociated, the activity a_2 of the electrolyte as a whole can be shown to be

$$a_2 = (a_+)^v(a_-)^u = (a_\pm)^{v+u} \tag{14}$$

Thus in using the Nernst equation, it does not matter whether the strong electrolytes in the cell reactions are expressed in the form $(vV^{+s}+uU^{-r})$ or V_vU_u since the final results obtained using the individual ionic activities, the mean activities or the activity a_2 of the electrolyte as a whole all are the same. Most activity data presented in the literature are in terms of mean activities or mean activity coefficients.

Thermodynamics does not provide a means for the determination of individual ionic activities (a_+ or a_-) from experimental data. A knowledge of the individual ionic activities, however, is usually not necessary for the calculation of quantities of practical or experimental significance. The Nernst equation when written for individual half-cell reactions may include the activities of individual ions. When the Nernst equation for the half-cell reactions are combined to calculate the cell potential, however, the final result is found to be dependent on a_2 or $a_\pm$ and not just a_+ or a_- individually.

To illustrate this point, consider the cell

$$Br_2|Br^-, Zn^{+2}|Zn$$

for which the over-all cell reaction is

$$Zn + Br_2 \rightarrow ZnBr_2(aq) \tag{f}$$

where the Br_2 is in the form of pure liquid. The Nernst equations for the two half-cell reactions involve the single ion activities:

$$(E_{ox})_{Zn|Zn^{+2}} = (E^\circ_{ox})_{Zn|Zn^{+2}} - \frac{RT}{2F}\ln\frac{a_{Zn^{+2}}}{a_{Zn}} \tag{15}$$

and

$$(E_{red})_{Br_2|Br^-} = (E^\circ_{red})_{Br_2|Br^-} - \frac{RT}{2F}\ln\frac{(a_{Br^-})^2}{a_{Br_2}} \tag{16}$$

For a solution of $ZnBr_2$,

$$a_{ZnBr_2(aq)} = (a_{Zn^{+2}})(a_{Br^-})^2 = (a_\pm)^3 = (\gamma_\pm c_\pm)^3 \tag{17}$$

where $c_\pm$ is related to the concentration of the $ZnBr_2$ salt (c) by the expression $c_\pm = (4c^3)^{1/3}$ in accord with Eq. (13). The cell reaction potential is then

$$E = (E_{ox})_{Zn|Zn^{+2}} + (E_{red})_{Br_2|Br^-} = E^\circ - \frac{RT}{2F}\ln\frac{(a_\pm)^3}{(a_{Zn})(a_{Br_2})} \tag{18}$$

or at 25°C with $a_{Zn} = 1$ and $a_{Br_2} = 1$,

$$E = 1.828 - 0.0296 \log 4c^3\gamma_\pm{}^3 \qquad (18a)$$

where

$$E^\circ = (E^\circ_{ox})_{Zn|Zn^{+2}} + (E^\circ_{red})_{Br_2|Br^-} = 0.763 + 1.0652\text{ V}$$

according to Latimer (82).

The zinc-bromine cell has been considered by Barnartt and Forejt (5) as a secondary battery. After charging, these workers found an open-circuit cell voltage of approximately 1.83 V or close to E°, which means that the activity term in Eq. (18) was rather close to unity—a somewhat fortuitous situation.

Theoretical attempts have been made at the calculation of individual ionic activities as well as mean activities with the Debye-Hückel theory and modifications of this theory [see e.g., Lewis et al. (84), Glasstone (52), Harned and Owen (64), Robinson and Stokes (104)]. The results of such treatments are valid only for dilute solutions (concentrations less than 10^{-1} m) and are not applicable even as coarse approximations at the concentrations involved in virtually all battery electrochemistry.

In polyvalent electrolytes, very substantial ionic association is involved at the concentrations used in batteries. For example, most of the zinc ions in the zinc-bromine cell just considered are in the form of various bromo-complexes. For such electrolytes most of the literature data for activity coefficients are in the form of *stoichiometric mean activity coefficients*. The stoichiometric activity coefficients and the stoichiometric mean activity coefficients are defined by the equations

$$(\gamma_i)_s = \frac{a_i}{(m_i)_s} \qquad (19a)$$

and

$$(\gamma_\pm)_s = \frac{a_\pm}{(m_\pm)_s} \qquad (19b)$$

where $(m_i)_s$ is the molal concentration of the ith ion of the electrolyte assuming complete dissociation of the electrolyte into simple ions. The stoichiometric mean molality $(m_\pm)_s$ is defined by an equation similar to Eq. (13) with the ionic concentrations corresponding to the molalities $(m_i)_s$ calculated on the basis of complete dissociation. For example, for a solution of $ZnBr_2$ of molality m, the values are

$$(m_{Zn^{+2}})_s = m \qquad (19c)$$

$$(m_{\mathrm{Br}^-})_s = 2m \tag{19d}$$

and

$$(m_{\pm})_s = (4m^3)^{1/3} \tag{19e}$$

For calculations involving the Nernst equation, stoichiometric mean activity coefficients are quite convenient since the mean activities may be calculated directly from the stoichiometric mean activity coefficient and the over-all concentrations of the electrolyte without a knowledge of the extent of ion association.

Activity data for some electrolytes of special significance to the battery electrochemist are given in Appendix D. For a further discussion of activities and particularly their evaluation, the treatise on electrolytes by Harned and Owen (64) and Robinson and Stokes (104) are useful.

Thermodynamic characteristics of batteries at various stages of discharge

With some batteries the open-circuit voltage decreases only slightly during the discharge of the cells until the exhaustion of one of the electrochemical components, while with other batteries rather pronounced changes in the open-circuit voltages may occur. In batteries whose open-circuit voltages decrease only slightly during discharge, the activities of the reactants and products remain relatively unchanged during the useful life of the cell. This occurs when the reactants and products are pure components or the major components of some phase of relatively constant structure (e.g., water or a major ionic component in the electrolyte). Thus it is not surprising that the open-circuit voltage of the mercury cell,

$$\mathrm{Hg, HgO|KOH, ZnO(sat'd)|Zn}$$

is very constant during discharge. The over-all cell reaction is

$$\mathrm{Zn + HgO} + x\mathrm{H_2O} \rightarrow \mathrm{ZnO} \cdot x\mathrm{H_2O}(s) + \mathrm{Hg} \tag{g}$$

with each component of the over-all cell reaction in a separate phase; hence the activity of each component remains constant during discharge. The number of moles of water involved in the reaction is open to some question and therefore has been designated x. The cell voltage of this battery remains very constant during almost all of its useful life as is evident from Fig. 2.4, which shows the time dependence of the cell voltage with a very light load (large resistor across cell).

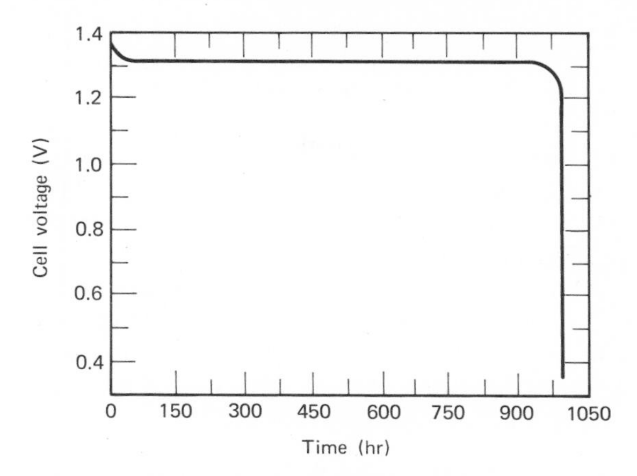

Figure 2.4 Cell voltage of the mercury battery [Hg,HgO|KOH, ZnO(sat'd)|Zn] as a function of time with a constant load of 1350 ohms at 70°F. Nominal capacity: 1000 ma-hr. From (33) (by permission of Union Carbide Consumers Products Co.).

The constancy of the activities of the components represented in the *over-all* cell reaction, however, is not an adequate criterion for constancy of open-circuit voltage during the useful life of a battery. For example, with the alkaline zinc-mercury cell, the half-cell reactions may be represented as

$$Zn + (x-1)H_2O + 2OH^- = ZnO \cdot xH_2O + 2e^- \qquad \text{(j)}$$

$$HgO + H_2O + 2e^- = Hg + 2OH^- \qquad \text{(k)}$$

During discharge, the OH^- concentration decreases in the solution adjacent to the zinc anode and increases in the solution adjacent to the HgO cathode. Convective transport is virtually nonexistent in such batteries because of their physical construction. Consequently, if the OH^- concentration in the battery was initially relatively low, marked changes in the activity of the OH^- at the anode and cathode would be expected and these might persist for very long times even on open-circuit after partial discharge. In the alkaline zinc-mercury battery, however, the KOH concentration is very high (35 to 40 weight percent) and the local changes in the activity of the OH^- are not sufficient to produce a substantial change in the cell voltage. It is to be noted that a tenfold change in the hydrogen ion concentration at either electrode is needed to produce a 0.06 V change in the thermodynamic half-cell potential of either electrode. With batteries such as the Leclanché cell [Zn|$ZnCl_2$, NH_4Cl|MnO_2] which are not as well buffered against *p*H changes, the activities of the hydronium and hydroxyl ions in the vicinities of the anode and cathode may shift by several powers of ten after partial discharge. These changes can produce very significant changes in the open-circuit voltage.

In cells involving the production or consumption of ionic species, the open circuit voltage will remain relatively constant during discharge if the activities of the ionic constituents do not change more than several fold. For example, in the zinc-bromine cell cited earlier, a tenfold charge in $a_{Zn^{+2}}$ would produce only a change of 0.04 V. By using relatively concentrated solutions (e.g., several molar or higher) or saturated solutions, the decrease of cell voltage associated with changes in ionic activities can be kept quite small (less than a few hundredths of a volt).

Metal oxide cathodes often present a relatively complicated situation from a thermodynamic viewpoint if the oxide can exist in both higher and lower valency states. If the higher and lower valency oxides are exactly *stoichiometric*[2] with no mutual solubility of one oxide in the other, the thermodynamic reduction potential should follow the dashed line shown in Fig. 2.5, providing the activities of other components undergo no major changes. Most oxides which exist in both higher and lower valency states usually exhibit mutual solubility in each other, which is the equivalent of nonstoichiometric behavior. Thus the activity of the high valency oxide will deviate considerably from unity as the concentration of the lower valency oxide builds up, and, likewise, the activity of the lower valency oxide will also undergo continuous change. Under

[2]The term *stoichiometric* is used to mean that the metal and oxygen are in the exact proportions represented by the usual formula for the oxide.

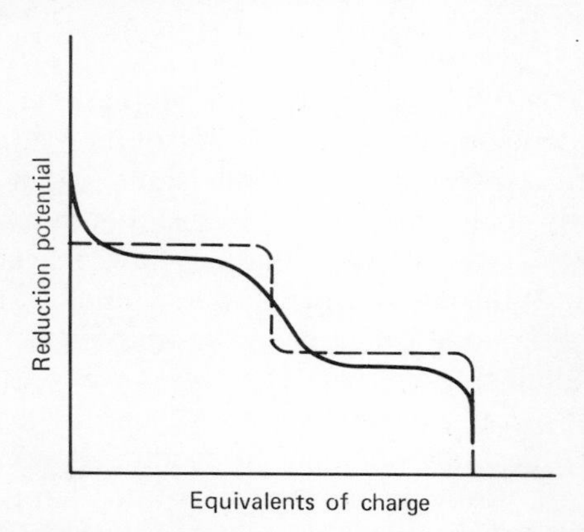

Figure 2.5 Thermodynamic reduction potential for an oxide capable of existing in two valency states. Solid line: behavior when the higher and lower oxides have considerable mutual solubility; dashed line: behavior when the higher and lower oxides are not soluble in each other.

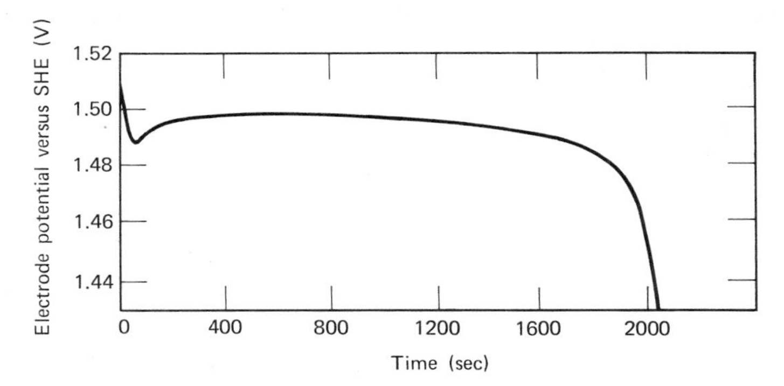

Figure 2.6 Discharge of electroplated β-PbO_2 electrode at constant current in H_2SO_4 (1 mole/liter) at 1 ma/cm^2 at 30°C. From (2) (by permission of The Electrochemical Society).

such circumstances the thermodynamic reduction potential will usually depend on the state of discharge in a manner similar to that represented by the solid curve in Fig. 2.5. Vetter (118, 121, 122) has presented a thermodynamic treatment of the electrochemistry of nonstoichiometric oxides and hydroxides.

Examples of oxide electrodes which discharge at relatively constant potential are the mercuric oxide electrode of the alkaline zinc-mercury cell discussed earlier and the PbO_2 electrode of the acid lead storage cell. The constancy of the potential of the PbO_2 electrode in sulfuric acid even on load is evident in Fig. 2.6, which shows the dependence of potential on time for electrodeposited β-PbO_2 at a discharge current density of 1 ma/cm^2 in 1 M H_2SO_4 of constant concentration. [Note the greatly expanded voltage scale.]

Behavior approximating that of the dashed line in Fig. 2.5 is exhibited by the silver oxide electrode for which the charging as well as discharging curves are shown in Fig. 2.7. The potential plateaus correspond approximately to the potentials found for the Ag_2O, $AgO|OH^-$ and Ag, $Ag_2O|OH^-$ electrodes, e.g., those by Dirkse (25).

Behavior somewhat similar to that shown by the solid curve in Fig. 2.5 has been found for the MnO_2 electrode in alkaline solutions at low discharge rates. The potential versus time curves, however, are quite contingent on the nature of the original MnO_2. The reduction of electrodeposited MnO_2 (curve B in Fig. 2.8) appears to proceed through two stages [Kozawa and Yeager (77)], the first involving the reduction of Mn^{IV} to Mn^{III} and the second from Mn^{III} to Mn^{II}. The decrease of potential during discharge during the first stage may be interpreted as indicative of a continuous increase in the ratio of the activity of Mn^{III} to that of Mn^{IV}, probably because of a range of solid solutions involving a

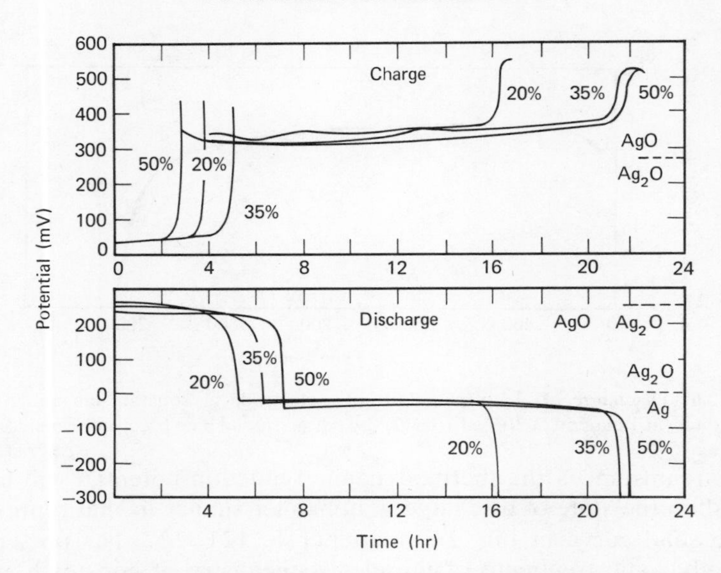

Figure 2.7 Typical 100-ma constant current charges and discharges of silver electrodes in 20, 35, and 50 per cent KOH at 25°C. Potentials are given with respect to a Ag, $Ag_2O|OH^-$ reference electrode in the same solution. Reversible potentials of the Ag_2O, $AgO|OH^-$ and Ag, $Ag_2O|OH^-$ electrodes are shown on the right-hand ordinates. From (125) (by permission of The Electrochemical Society).

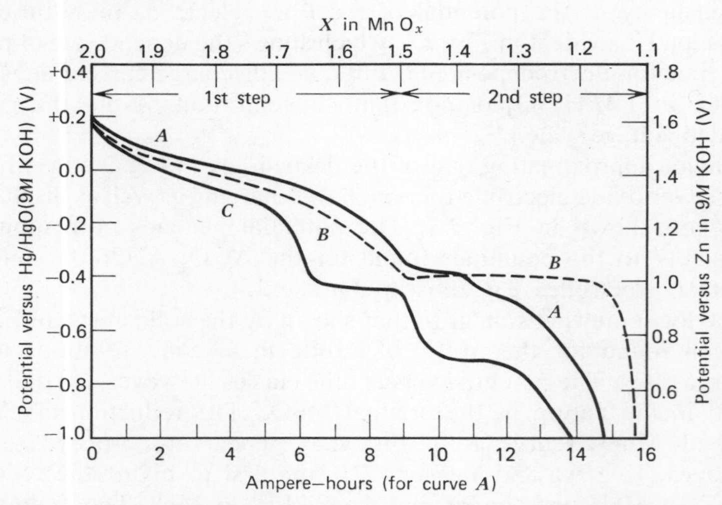

Figure 2.8 Discharge curves for manganese dioxide electrodes as a function of x in MnO_x. Curve A. At 0.33 ma/g of MnO_2 [from (76)]; B. At 11 ma/g of MnO_2 electrodeposited on a graphite rod [from (77)]; C. At 0.5 ma/g of MnO_2 [from (6)] (by permission of The Electrochemical Society).

higher (MnO_2) and a lower (likely MnOOH) valent oxide. At the end of the first stage, all of the Mn^{IV} appears to have been reduced to Mn^{III}. As the Mn^{III} is then further reduced to Mn^{II}, the potential remains relatively unchanged until most of the Mn^{III} is reduced. This behavior indicates that the activities of the Mn^{III} and Mn^{II} remain relatively constant and hence that two separate solid phases are involved, one controlling the Mn^{III} activity and the second the Mn^{II} activity in the solid phase and also in the electrolyte phase if the reduction process involves liquid-phase manganese species as has been suggested by some workers [e.g.,Kozawa and Yeager (77)]. In Fig. 2.8 also are shown the curves for MnO_2 electrodes of the type involved in alkaline zinc-manganese dioxide batteries. Substantial differences are evident. The behavior of the MnO_2 electrode is quite complex. As of the preparation of this chapter, a comprehensive quantitative explanation remains to be developed for the potential-current-time behavior of the MnO_2 electrode in either alkaline zinc-manganese dioxide cells or the Leclanché cell ($MnO_2|NH_4Cl, ZnCl_2|Zn$).

Temperature dependence of thermodynamic cell potentials

The temperature coefficient of the thermodynamic cell potential and of the standard value may be calculated by means of the Gibbs-Helmholtz equation:

$$\left(\frac{\partial E}{\partial T}\right)_p = \frac{\Delta S}{nF} = \frac{[(\Delta H/nF)+E]}{T} \tag{20}$$

$$\left(\frac{\partial E^\circ}{\partial T}\right)_p = \frac{\Delta S^\circ}{nF} = \frac{[(\Delta H^\circ/nF)+E^\circ]}{T} \tag{21}$$

where ΔH and ΔS are the molar enthalpy and entropy changes associated with the cell reaction for the particular concentrations which exist in the cell and ΔH° and ΔS° are the corresponding standard state values. The standard heats of formation are available for many of the reactants and products encountered in batteries [see, e.g., Latimer (82) and, Bichowsky and Rossini (8)], and therefore ΔH° for the cell reaction can be easily calculated from the difference in the sums of the standard heats of formation for the products and the reactants.

The values for ΔH can be calculated from ΔH° but for this calculation the reader is referred to one of the more complete thermodynamic texts [Lewis, Randall, Pitzer and Brewer (84); Kirkwood and Oppenheim (71)]. Often the actual heats of formation are available in the original literature from which the compilations by Latimer (82) and Bichowsky and Rossini (8) were obtained. While ΔH and ΔH° are often approximately equal, care must be exercised in substituting ΔH° for ΔH in Eq. (20) since the difference between two relatively large terms may be involved in this equation. In some instances standard entropies of formation may also be available in the literature [Latimer (82), Bichowsky and Rossini (8)] for both products and reactants and these data may be used to calculate ΔS° for the cell reaction.

Heat generation in electrochemical cells

The heat generated per Faraday of charge passed through the cell can be calculated by means of the equation

$$Q = \frac{T\Delta S}{n} - F(E_r - E) = \frac{\Delta H}{n} + FE \tag{22}$$

where E_r is the thermodynamically reversible potential and E is the actual terminal cell potential. A negative value for Q corresponds to heat liberated. The values for ΔS and ΔH may be obtained from either the literature or calculated by means of Eq. (20) from the observed temperature dependence of the open-circuit potential of the cell provided the latter corresponds to the thermodynamically reversible value. The conditions for which the open-circuit values do not correspond to thermodynamic values will be discussed later in this chapter.

The value for ΔS for an electrochemical cell can be either positive or negative but for almost all useful battery systems ΔS during discharge is positive. Even when the operating terminal voltage is substantially the same as the thermodynamic value, heat corresponding to $T(\Delta S)$ will be evolved within the cell. For some cells, particularly those involving gaseous reactants, the entropy increase may be quite substantial and the heat liberated appreciable even when there are no significant losses of any type within the cell. For example, for a cell of the type

$$H_2(P = 1\text{ atm})|H^+, Cl^-(a_\pm = 1)|Cl_2(P = 1\text{ atm})$$

for which ΔS is -29 cal/°K–Faraday, the heat $(-Q)$ liberated at 25°C under reversible conditions would be 8.7 kcal/Faraday or per mole of HCl formed.

Equation (22) takes into account all heat losses within the cell arising from *i-r* drops and various types of electrode losses. When the electrodes are shorted, $E = 0$ and the heat liberated then corresponds to $Q = \Delta H$. Under these circumstances the heat liberated is the same as would be generated in a direct, nonelectrochemical reaction at constant pressure and constant temperature.

EFFICIENCIES

The *overall efficiency* ϵ of a cell used as a source of energy is defined as

$$\epsilon = -\frac{\int_{\sigma=0}^{\sigma=q} E\,d\sigma}{\Delta G} \tag{23}$$

where ΔG is the molar Gibbs free energy change and σ is the charge (coulombs) transferred through the external circuit. The symbol q is the total charge transferred through the external circuit per mole of cell reaction. The upper limit for q is nF. Departures of q from this value may occur because of internal chemical reactions within the cell which yield no charge transfer through the external circuit. Typical of such reactions are (*a*) chemical shorting associated with the diffusion of cathodic reactants to the anode or vice versa, (*b*) anodic corrosion with the liberation of hydrogen gas, and (*c*) reaction of an oxide cathode with water or a presumably "inert" cell member such as a separator or immobilizing agent added to the electrolytes. In many cells, q approaches very closely to nF.

The deviation of the cell voltage E from the limiting thermodynamic value occurs because of *i-r* drops within the cell in the electrolyte and in the bulk of the electrodes as well as various types of electrode losses, which will be discussed shortly.

The *coulombic efficiency* Φ of an electrode or half-cell process is defined as

$$\Phi = \frac{q}{nF} \tag{24}$$

where q now represents the number of coulombs obtained with a given anode or cathode per mole of electrode reaction and n is the number of electrons in the reaction. The term "coulombic efficiency" is usually somewhat ambiguous when applied to the cell as a whole and is best reserved for describing the individual electrodes of the cell. The number of equivalents of the various reactants at the anodes and cathodes of a battery are usually unequal even initially for various practical reasons—to prevent the liberation of undesirable gases within a cell, to prevent puncture of the metal can when the can is a consumable electrode, and so on.

Some nonelectrochemists have attempted to compare the efficiencies of electrochemical cells with the efficiencies of energy converters such as thermoelectric devices that convert heat to electrical energy. Such comparisons are artificial since electrochemical devices utilize directly chemical energy and are not thermal converters. The only satisfactory basis for comparison is the electrical output per unit amount of a given fuel. For a given fuel the energy per unit weight will usually be far higher for a battery than a heat engine because the Carnot cycle limitations[3] imposed on any heat engine are not applicable to a battery.

A *thermal efficiency* (Θ) will be defined for an energy producing electrochemical cell in order that a number may be available for comparison with the thermal efficiency of heat converters:

$$\Theta = -\frac{\int_{\sigma=0}^{\sigma=q} E\,d\sigma}{\Delta H} \tag{25}$$

where ΔH is the molar enthalpy change at the operating temperature of the cell. The thermal efficiency has an upper limit which corresponds to the ratio $\Delta G/\Delta H$. Theoretical ΔG may be greater or less than ΔH, depending on the sign of ΔS and hence may have an upper limiting value greater or less than unity. Since ΔS is almost always positive for batteries, the upper limit for Θ is almost always less than unity. Even under open-circuit conditions the value for Θ may fall short of the theoretical upper limit by a significant amount.

Undue weight should not be given to the values for Θ in comparing the relative merits of various batteries since other factors usually are more important. For example, two hydrogen-oxygen fuel cells operating at a given temperature could, by chance, have identical operating voltages and hence identical values for the numerator in Eq. (25). If one of the cells (e.g., an ionic membrane cell) produced pure liquid water and the other water vapor at the cell temperature, the value for ΔH for the former would be approximately 10 kcal/mole larger numerically than that for the latter. The value of Θ for the cell producing liquid water would be 17 percent lower than that of the cell producing water vapor despite the fact that both cells consume the same fuel and have the same terminal voltage. The difference in ΔG in Eq. (23) for the two cells would only be a few percent and hence the difference in the values for the overall efficiency ϵ would only be a few percent. Thus it is evident that Θ may prove quite misleading as a basis for comparing cells.

[3]For a thermal converter operating between the two temperatures T_2 and T_1, thermodynamics imposes an upper limit on the thermal efficiency Θ:

$$\Theta = \frac{w}{Q_2} = \frac{T_2 - T_1}{T_2} \tag{25a}$$

where w is work done, Q_2 is heat *in* at the higher temperature T_2, and T_1 is the lower temperature at which heat is rejected from the device or cycle.

Standard thermodynamic data ($\Delta G°$, $\Delta H°$, and $\Delta S°$) for various batteries are given in Appendix C.

Battery chemists rate batteries in terms of *charge capacity* (e.g., ampere-hour capacity) and also *energy capacity* (e.g., watt-hour capacity). The end of the useful life of a battery is set by the decrease of the cell terminal voltage below an acceptable limit, a cut-off voltage. Consequently, the cut-off voltage must be specified for each type of battery for the charge or energy capacity to be fully meaningful. The generally accepted cut-off voltages as well as types of loads to be used in establishing ampere-hour capacities are stated in the National Bureau of Standards Handbook 71 (1959) as approved by the American Standards Association.

THE ELECTRICAL DOUBLE LAYER[4]

The chemical reactions that produce electrochemical power within a battery for the most part are heterogeneous processes occurring at the electrode surfaces rather than at a distance from the electrode surface. To understand the factors controlling these heterogeneous processes, it is necessary to consider the structure of the interface. Both the electrolyte and the electrode sides of the interface differ substantially from the bulk phases in their electrical properties. The present understanding of the electrical properties of the electrode side of the interface is limited and reflects a general lack of knowledge concerning the surface states of electronic conductors. The electrical properties of the solution side of the interface are better, although still far from adequately understood.

The potential difference across a metal-solution interface disturbs the ionic distribution within the solution in the vicinity of the interface and an ionic double-layer results (see Fig. 2.9). Immediately adjacent to the electrode surface there exists a compact layer of ions of charge predominantly opposite to that of the electrode. This layer is called the Helmholtz layer and the plane through the electrical centers of the ions is referred to as the Helmholtz plane. This plane corresponds to the distance of closest approach of the ions to the electrode surface for the particular type of ions which predominate in the Helmholtz layer.

The distance of closest approach, however, is contingent on the type of ion, the potential of the electrode relative to the ionic solution, and strength of the interactions of the electrode with the solvent dipoles. Most cations have sufficiently strong interaction with the water molecules in the inner coordination sphere that this layer remains essentially intact as the ion approaches the electrode surface, pending an electro-

[4]The reader will find more extensive discussions of the double layer in Parsons (93), (94); Delahay (19); and Conway (12).

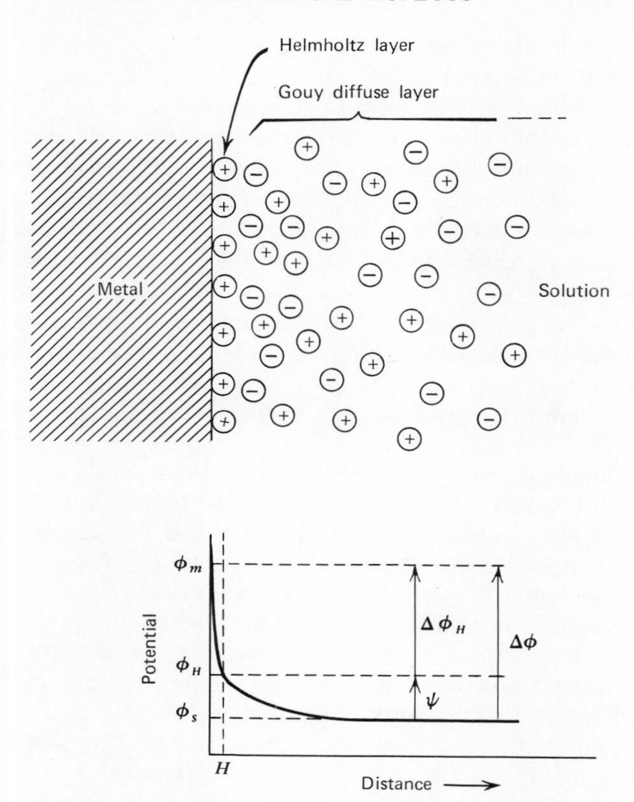

Figure 2.9 The electrical double layer at a metal-solution interface in the absence of specific adsorption.

chemical interaction of the ion with the electrode. Cations of relatively large size and hence low charge density [e.g., cesium, $(CH_3)_4N^+$] have weak interactions with the solvent molecules even immediately surrounding them. Consequently, the water molecules adjacent to that part of these ions facing the electrode surface can be relatively easily displaced and the ions can be pulled into positions where they make direct contact with the electrode surface when the potential of the electrode is negative to that of the solution. This is also possible with most anions when the electrode is positive relative to the solution since the interaction of anions with solvent molecules is usually relatively weak. The fluoride ion is a notable exception. When solvent molecules are not interposed between the electrode and the ion, relatively strong interactions of a chemical nature, specific to the ion and the electrode, may occur between the atomic orbitals of the ions and the surface orbitals of the electrode.

The ions may be described as specifically adsorbed when no solvent is interposed between the electrode and ions.

In some instances it is possible to have two Helmholtz planes, an inner and an outer Helmholtz plane. The inner plane corresponds to the electrical centers of ions which are in direct contact with the electrode surface while the outer plane may consist of ions that have water of hydration interposed between them and the electrode surface. Such a situation is encountered with most electrolytes over some range of potentials.

Extending from the solution side of the Helmholtz layer is the Gouy diffuse layer. The potential in this layer decreases nonlinearly and asymptotically approaches that of the bulk of the solution. Mathematical techniques similar to those used in the Debye-Hückel theory have been applied to the calculation of the ionic distribution in the diffuse layer. These treatments of the diffuse layer introduce the concept of an effective thickness which corresponds to the distance from the Helmholtz plane (outer Helmholtz plane if both inner and outer are involved) to a plane in the diffuse layer representing the electrical center of the excess charge in the diffuse layer. For the concentrated electrolytes involved in most battery systems, the effective thickness of the diffuse layer is of the order of 10^{-7} cm or, at most, only a few ionic (hydrated) diameters. Thus the diffuse layer for the solutions used in conventional batteries is quite thin.

The potential difference between the metal and the bulk of the electrolyte is denoted in Fig. 2.9 by

$$\Delta\phi = \phi_m - \phi_s = \Delta\phi_H + \psi \tag{26}$$

where the subscripts H, m, and s refer to the Helmholtz plane, the metal, and the bulk of the solution, respectively. Only relative values for $\Delta\phi$ can be measured directly since the potential of the electrode under study must be measured relative to some reference electrode whose potential does not correspond to the potential of the bulk of the solution.

Of particular interest is the *point of zero charge* (pzc) at which the excess charge on the metal electrode relative to the solution is zero. At the pzc in the absence of specific ionic adsorption the ionic double layer essentially disappears. Even so, the potential difference between the metal electrode and the bulk of the solution ($\Delta\phi$ in Fig. 2.9) does not necessarily become zero at the pzc because of surface potential effects within the surface layer of the metal and solvent dipole effects at the interface [see, e.g., Ershler (30); Parsons (93)]. Under such conditions, the potential drop across the diffuse layer (ψ) is probably within a few millivolts of zero and $\Delta\phi_H$ within a tenth of a volt of zero if the penetration of the electric field into the metal is not considered.

At the point of zero charge the interfacial tension of the electrode-solution interface passes through a maximum. The interfacial tension of liquid metal electrodes such as mercury can be readily measured as a function of potential relative to a constant reference electrode. Similar information can be obtained for solid electrodes although with more difficulty and less accuracy [see, e.g., Smolders and Duyvis (113); Smolders (112)]. Consequently, the point of zero charge can be determined from interfacial tension measurements. Several other methods, including capacity measurements, hardness studies, and adsorption studies, also have been used to determine the pzc of various solid and liquid metal electrodes [see, e.g., Conway (12); Delahay (19)].

In Fig. 2.10 values for ψ are shown as a function of the rational potential as calculated by Parsons (93) from the data of Grahame (58, 59) for mercury in various NaF solutions. Neither Na^+ or F^- ions are specifically adsorbed to any significant extent over the range of potentials represented in this figure. The potentials on the abscissa correspond to the *rational potentials* (E_R) or the potentials of the mercury electrode

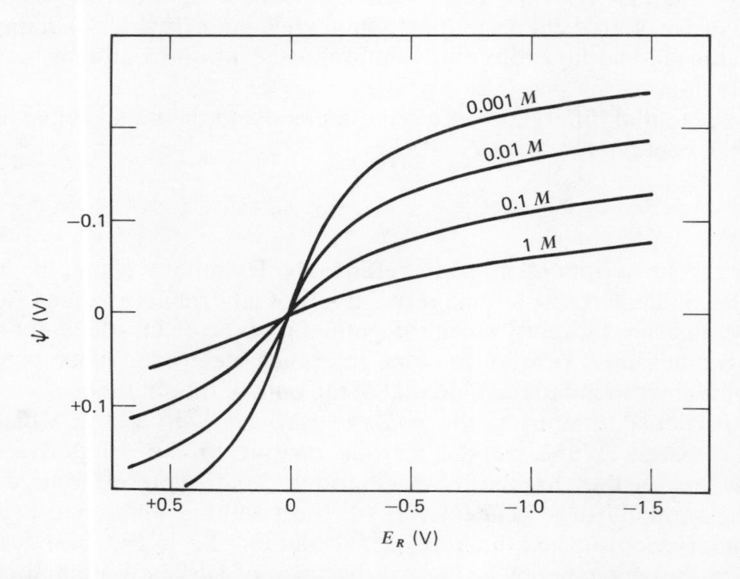

Figure 2.10 Potential of the Helmholtz plane (outer) relative to the bulk of the solution as a function of the rational potential in the absence of specific adsorption. Values calculated by Parsons (94) from the data of Grahame (58), (59) for mercury in NaF solutions at 25°C (by permission of Interscience Publishers).

relative to that for the pzc. Other uni-univalent electrolytes yield similar results if the ions are not specifically adsorbed.

For electrolytes with higher valent ions, the dependence of ψ on the rational potential should be similar if calculated at the same *ionic strengths*,[5] although the calculation would be considerably less reliable because of the difficulty of accounting for ion association in the double layer in such electrolytes. While the values for ψ are substantial in dilute electrolytes at potentials remote to the pzc, the values are quite small (numerically usually less than 50 mV) in one molar solutions. Thus in the concentrated electrolytes used in most batteries, the potential drops across the diffuse layers at the electrodes are only a small fraction of the over-all cell voltage. Nevertheless, the ψ potentials are still quite significant in kinetic studies of battery electrodes because the relatively small ψ potentials are still sufficient to produce large differences between the concentrations of the reacting ions in the plane of closest approach and the bulk of the solution.

The concentration $(c_i)_H$ of an ion of charge z_i in the outer Helmholtz plane can be calculated by means of the approximate equation

$$(c_i)_H = (c_i)_s \exp - \frac{z_i F \psi}{RT} \tag{28}$$

where $(c_i)_s$ is the bulk concentration of the ith type of ion with valence z_i. From this equation, it is evident that the concentration of a particular type of ion will be higher in the outer Helmholtz plane than in the bulk of the solution when z_i and ψ have opposite signs—a conclusion easily reached on the basis of simple electrostatic considerations. Although Eq. (28) is only an approximation, it has proved of great importance in interpreting the role of the double layer in electrode kinetics. Much of the pioneering work in this area was carried out by A. N. Frumkin in the Soviet Union beginning in the 1930s (36).

Some appreciation of the importance of the double layer properties in understanding electrode processes can be gained from Table 2.1, which lists the concentrations $(c_i)_H$ for the Na^+ and F^- ions in the outer Helmholtz plane or plane of closest approach for these ions for the same concentrations of NaF as involved in Fig. 2.10. The potentials of the mercury electrode are listed relative to the normal calomel reference electrode

[5] *Ionic strength* is defined by the expression

$$\chi = \tfrac{1}{2} \sum c_i z_i^2 \tag{27}$$

where c_i and z_i are the concentration and valence of the ith ion and the summation includes all of the ions in the solution.

(NCE). Equation (28) has been used to calculate values for $(c_{Na^+})_H$ and $(c_{F^-})_H$ from the values for ψ in Fig. 2.10. The potential corresponding to the pzc is ≈ -0.47 V versus normal calomel electrode. Table 2.1 shows that at potentials remote to the pzc the concentrations of the cations and anions in the outer Helmholtz plane may differ by many orders of magnitude in dilute solutions and by one order of magnitude even in solutions as concentrated as 1 *M*. In the highly concentrated electrolytes used in most batteries, the concentrations in the Helmholtz plane may still deviate by several fold from the bulk values particularly if the potential of the electrode is remote to the pzc as is often the case.

TABLE 2.1. CONCENTRATIONS OF Na^+ AND F^- IN THE OUTER HELMHOLTZ PLANE FOR MERCURY IN NaF SOLUTIONS AT 25°C IN THE ABSENCE OF SPECIFIC ADSORPTION[a]

Potential (V) re. NCE	Bulk Concentration (M)	ψ (V)	$(c_{Na^+})_H$ (M)	$(c_{F^-})_H$ (M)
−1.75	0.001	−0.241	11.8	8.46×10^{-8}
	0.01	−0.184	13.1	7.62×10^{-6}
	0.1	−0.128	14.8	6.76×10^{-4}
	1	−0.074	18.1	5.52×10^{-2}
−1.35	0.001	−0.22	5.27	1.90×10^{-7}
	0.01	−0.164	6.04	1.63×10^{-5}
	0.1	−0.109	7.03	1.42×10^{-3}
	1	−0.058	9.62	1.04×10^{-1}
−0.95	0.001	−0.187	1.45	6.89×10^{-7}
	0.01	−1.135	1.90	5.26×10^{-5}
	0.1	−0.0829	2.53	3.96×10^{-3}
	1	−0.039	4.56	2.19×10^{-1}
−0.55	0.001	−0.066	1.31×10^{-2}	7.64×10^{-5}
	0.01	−0.041	5.00×10^{-2}	2.00×10^{-3}
	0.1	−0.021	2.21×10^{-1}	4.42×10^{-2}
	1	−0.0085	1.41	7.18×10^{-1}
−0.45	0.001	+0.141	5.78×10^{-4}	1.73×10^{-3}
	0.01	+0.134	5.93×10^{-3}	1.69×10^{-2}
	0.1	+0.067	7.70×10^{-2}	1.30×10^{-1}
	1	+0.0024	9.11×10^{-1}	1.10

[a]From (19).

The ionic double layer at an electrode-solution interface gives rise to a considerable capacitance component. The *differential electrode capacity* associated with the double layer (the so called non-Faradaic capacity) is defined as

$$C_{DL} = \frac{d\sigma}{dE} \tag{29}$$

where E is the electrode potential relative to an invariant reference and σ is the excess charge on the metal electrode per unit area. This charge is equal numerically but opposite in sign to the total excess charge in the Helmholtz layer and the Gouy diffuse layers. The differential capacity so defined is the quantity normally determined in a–c impedance measurements and potential decay studies with individual electrodes. The over-all differential capacity is related to the contributions from the Helmholtz and diffuse layers, C_H and C_G, respectively, by the approximate equation

$$\frac{1}{C} = \frac{1}{C_H} + \frac{1}{C_G} \tag{30}$$

This equation is equivalent to assuming that the capacitive components of the Helmholtz and Gouy layers are connected in series. Except at the pzc, the component C_G is much larger than C_H and therefore the observed differential capacity is approximately equal to C_H.

In Fig. 2.11 the potential dependence of the observed differential capacity associated with mercury in various electrolytes is shown. In dilute solutions a minimum is usually discernable in the differential capacity at the pzc when the value of C_G decreases to a low value.

The differential capacity is quite sensitive to organic impurities in the electrolyte at potentials where these impurities are adsorbed on the

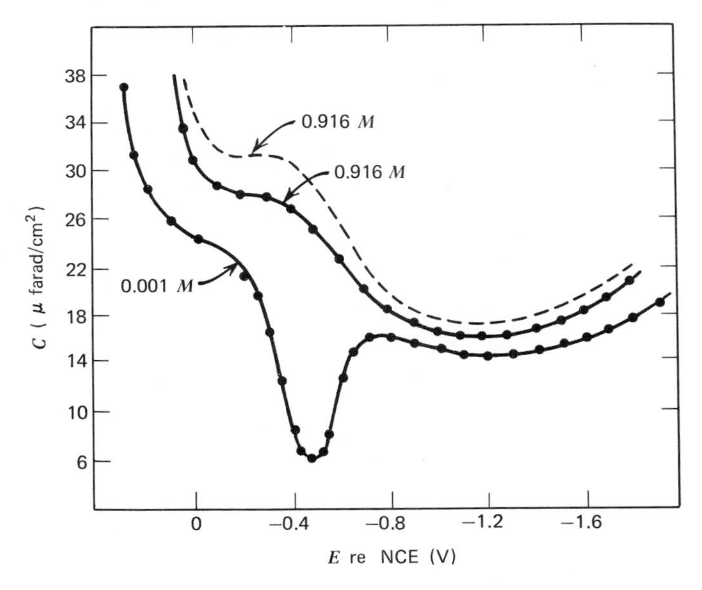

Figure 2.11 Differential capacity C of mercury in 0.001 and 0.916 M NaF at 25°C. Dashed line corresponds to C_H. From (58) (by permission of American Chemical Society).

electrode surface. The concentration of an impurity in the solution need be sufficient only to support the formation of a reasonable fraction of a monolayer on the electrode surface for the differential capacity to be depressed by a significant amount [see, e.g., Grahame (54); Delahay (19)].

Although the literature contains many double-layer data for mercury in various electrolytes, very few data are on hand for the solid electrodes and the very concentrated electrolytes used in most batteries. Furthermore, even if the experimental electrochemists undertake such measurements as double-layer capacities, interfacial tensions, and pzc potentials, the theoretical effort required to process these data is substantial and then only likely to yield semiquantitative information as to the activities of the reacting species at the electrode surface. Nevertheless, such semiquantitative information would make it possible to anticipate how changes in the nature and composition of the electrolyte will influence the kinetics of the various processes occurring at the electrodes, including such undesirable processes as hydrogen and oxygen discharge and anode passivation.

With a semiconductor as the electrode, the equivalent of the Gouy diffuse layer exists in the semiconductor phase in addition to the Helmholtz and Gouy layers in the solution phase. The diffuse layer within the semiconductor contributes to the overall capacity of the semiconductor-solution interface in the form of a series component. Measurements of the differential capacity associated with such interfaces as a function of the electrode potential (at least in principle) can be used to determine the carrier concentration in the semiconductor electrode. Trace elements in the semiconductor phase can produce major changes in the carrier concentration and can have a large effect on the differential capacity and also the reversibility of various electrochemical processes occurring at the semiconductor-solution interface. For a discussion of semiconductor electrodes, the reader is referred to the reviews by Green (60), Gerischer (45), and Holmes (68). The semiconductor properties of oxides are of major importance in interpreting the properties of oxide cathodes in various batteries, the oxygen cathodes of a number of fuel cells, and the passivation phenomena encountered with various metal anodes.

LOSSES IN BATTERIES

The flow of current through any electrochemical cell results in a departure of the cell terminal voltage from the value predicted thermodynamically. This deviation can be resolved into the following:

1. *i-r* losses in the bulk of the electrode and electrolyte phases as well as the leads

2. Losses associated with interfaces and in particular with the electrode-electrolyte interfaces.

The second represent deviations of the electrode potentials from thermodynamic expectations and are referred to as *overpotential* or *overvoltage*.

The resolution of the losses within a power producing cell into *i-r* drop and electrode losses is illustrated in Fig. 2.12, which shows the terminal voltage and various component voltages for a sodium amalgam-hydrogen cell with NaOH as the electrolyte [Dietrick, Yeager, and Hovorka (24)]. Sodium in a liquid amalgam is oxidized to sodium ions at the anode while water is reduced to H_2 gas at a platinized-platinum cathode. Other versions of this cell, particularly with an oxygen cathode, have been considered, although not yet proved practical, as a means for recovering power during the denuding of sodium amalgam in the caustic industry [Yeager (128)]. The interrupter method, to be described later, was used to establish the *i-r* drop free potentials of the anode and cathode under load and the *i-r*

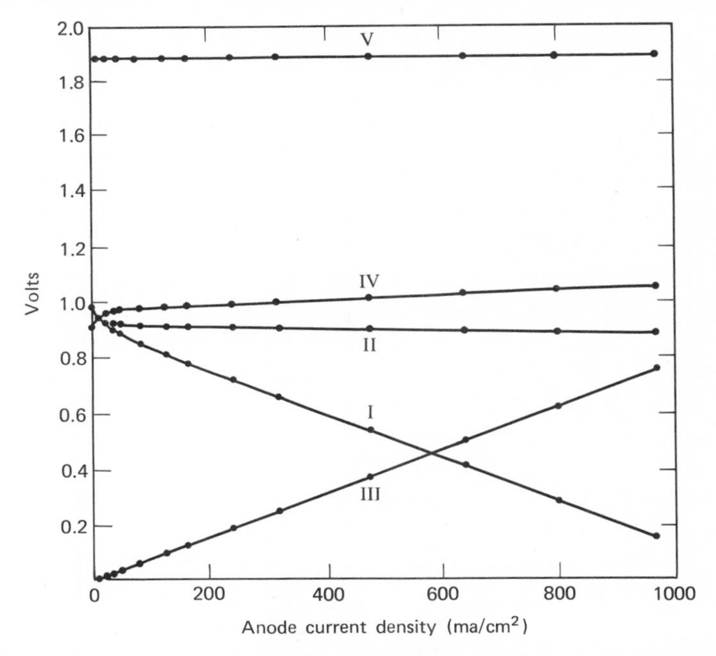

Figure 2.12 Cell terminal voltage, electrode potentials (*i-r* drop free), and *i-r* drop for the sodium amalgam–hydrogen cell Na-Hg|NaOH|H_2(Pt-Pt). Electrolyte: 5 *M* NaOH; amalgam concentration: 0.4 percent by weight Na; hydrogen pressure: 1 atm; temperature: 25°C. Curve I: cell terminal voltage; curve II: cell terminal voltage less *i-r* drop; curve III: *i-r* drop within electrolyte; curve IV: cathode potential re standard hydrogen electrode (SHE); curve V: anode potential re SHE. From (24).

drop within the electrolyte. Other *i-r* drops were negligible. From Fig. 2.12 it is apparent that most of the loss within the cell at high current densities was *i-r* drop within the cell, thus indicating the direction for further effort to improve the load characteristics. The anode potential was remarkably constant even at high current densities.

I-r LOSSES

The calculation of the *i-r* drop within various phases is relatively simple if the geometry of the cell is such that the current distribution is uniform throughout the phase. For a cell consisting of parallel electrode members (Fig. 2.13) the voltage drop in the electrolyte phase is given by the equation

$$E_{ir} = i \cdot r = i\left(\frac{x}{sa}\right) \tag{31}$$

where x is the thickness of the electrolyte phase, a is the cross-sectional area, s is the specific conductance of the electrolytic solution (in ohm^{-1} cm^{-1}), r is the resistance of this phase, and i is the current.

When the current distribution is not uniform, the calculation of the effective *i-r* losses within a given phase is considerably more complicated. For cells with substantial electrode polarization the current distribution is a function of the electrode polarization, the conductivity of the electrolyte, and the geometric characteristics of the cell. When the electrode polarization and *i-r* losses in the electrode phases are negligible, Eq.(31) may still be applicable with the term (x/A) replaced by an average or effective value. This effective value corresponds to the cell constant used in conductivity measurements and can be calculated for most cell geometries when porous electrodes are not involved.

In most batteries the electrode losses are not negligible compared to

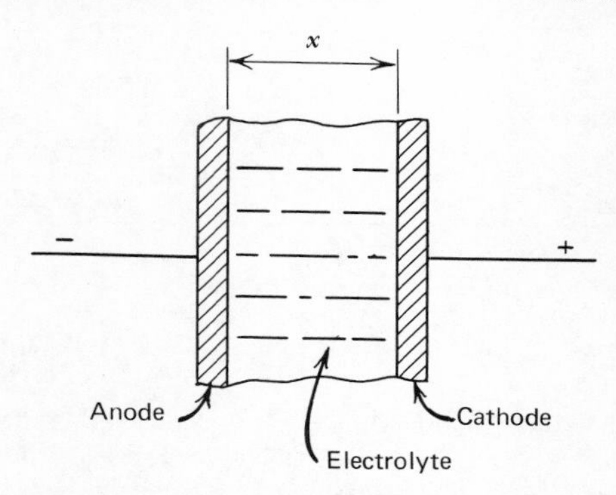

Figure 2.13 Cell with parallel electrodes.

i-r losses within the various phases. The electrode potential usually varies with location on the electrode surface because of the nonuniformity of the current distribution and mass transport problems, Under these conditions the potential in the electrolyte immediately adjacent to the electrode surface varies as the electrode surface is transversed. Thus a single value cannot be used to represent the *i-r* drop in the electrolyte or electrode phases for a given total current. Techniques for measuring the potential distribution in the solution immediately adjacent to the electrode surfaces will be discussed later in this chapter.

Cells with porous electrodes constitute a special problem. In such electrodes current flows within the solution-filled pores as well as the electrode phase. The calculation of the current distribution within such electrodes is difficult for batteries for which mass transport within the porous electrodes and other types of electrode losses must be considered simultaneously. Furthermore, experimental methods for characterizing the various types of losses associated with porous electrodes are not readily available. Consequently it may not be possible to resolve the losses in a cell with such porous electrodes clearly into *i-r* drops within the bath phases and losses associated with the electrode interfaces. The problem is particularly acute with the Leclanché cell in which the cathode consists of MnO_2 mixed with carbon with the void space filled with the NH_4Cl electrolyte. In this system at high currents most of the *i-r* drop is within the MnO_2-carbon bobbin and not a well defined value. Even so, many battery chemists choose to assign a numerical value to the internal resistance of this cell; the value then becomes quite dependent on the method of definition and measurement.

Losses associated with the electrode interface[6]

The *overpotential*[7] η is related to the actual electrode potential E and the reversible thermodynamic value E_r by the equation

$$\eta = E - E_r \tag{32}$$

where E and E_r may be measured relative to any convenient reference

[6]For more extensive discussions, see, e.g., Vetter (120), Conway (12), Delahay (19), Bockris (9), Kortum and Bockris (74).

[7]Some electrochemists use the term *overvoltage* or *electrode polarization* to represent the deviation of the electrode potential from the reversible thermodynamic value. The rigorous electrical definition of the term polarization, however, carries the concept of producing polarity and hence the "unpolarized" or reversible electrode would be expected to have no polarity according to this definition. Such is not the case and hence the authors prefer not to use the term polarization for η in (32).

electrode. The overpotential can be obtained directly by measuring the potential of the irreversible electrode against a reversible electrode of the same type. The values for η will be positive for an electrode at which oxidation is occurring and negative for an electrode at which reduction is occurring.

The factors contributing to the overpotential may be classified as follows:

1. *Concentration overpotential.* This type of overpotential is associated with the decrease in the concentration of reactants and increase in the concentration of products at the electrode interface because of the limited rate of transport of these components to and from the interface.
2. *Activation overpotential.* The limited rate at which some step or steps in the electrode reactions can proceed kinetically results in activation overpotential. An energy barrier and hence an activation energy are involved.
3. *Ohmic overpotential.* After the *i-r* drops within the bulk of the electrode and solution phases have been taken into account, some residual ohmic drop associated with the electrode-electrolyte interface may persist and is classified as a form of electrode overpotential. Ohmic overpotential is most commonly associated with some type of poor conducting film at the interface.

Each of these three types of losses at electrode interfaces will be discussed.

Concentration Overpotential. For a reaction of the type

$$bB \rightarrow vV^{+s} + ne^-$$

the concentration overpotential η_c can be calculated by means of the equation

$$\eta_c = E - E_r = \frac{RT}{nF} \ln\left[\frac{(a_{V^{+s}})^v}{(a_{V^{+s}})_e{}^v} \cdot \frac{(a_B)_e{}^b}{(a_B)^b}\right] \tag{33}$$

where E is the electrode potential, E_r is the reversible value of the electrode potential (i.e., the thermodynamic value), $(a_{V^{+s}})$ and (a_B) are the activities of the indicated components in the bulk of the electrode or electrolyte phases, and $(a_{V^{+s}})_e$ and $(a_B)_e$ are the corresponding quantities at the electrode-electrolyte interface. Equation (33) is only valid in the absence of appreciable activation overpotential. It may be derived from the Nernst equation on the assumption that the potential of the polarized electrode depends on the activities of the various components at the interface according to thermodynamic predictions.

Often in batteries the concentration of the principal ionic components remain relatively unchanged at the electrode interface when current is passing through the battery. Under such circumstances the activity coefficient of any reactants or products of the electrode reaction which may be minor components of the solution can be assumed to have constant activity coefficients and therefore the ratio of activities in the bulk to those at the interface can be replaced by the corresponding concentration ratio in Eq. (33). When activities of components in the solid state are involved, as in the case of oxide electrodes, the activity coefficients usually vary considerably with concentration. Therefore, for these components, the concentration ratio is usually not a good approximation for the activity ratio. Unfortunately, activity coefficients for solid state components are often unavailable and there may be no alternative to these poor approximations.

It should be noted that a relatively large change in the concentration of a component at the electrode interface is required to produce any significant amount of concentration overpotential. For example, a tenfold change in the concentration of a component at the interface will produce only ± 60 mV of concentration overpotential for a one-electron process at room temperature.

The transfer of components to and from an electrode surface may occur through the electrolytic solution, the electrode phase, and gas-filled pores within porous electrodes. In most batteries diffusion is the principal transport process.

In a fluid electrolyte, natural convection will usually arise as changes in the concentrations of reactants and products occur at the electrode surfaces. Under such circumstances the concentration profiles for reactants and products may be represented by curves of the type shown in Fig. 2.14. With increasing distance from the electrode surface, the concentration rapidly approaches that of the bulk of the solution. The steady-state conditions represented in Fig. 2.14 are usually achieved within a short time (fraction of a minute) at a particular current density in batteries with natural convection. For purposes of simplification the nonlinear concentration gradient can be considered in terms of the equivalent linear concentration gradient AB. The effective thickness of the concentration gradient may then be designated as δ.

According to Fick's first law, the diffusion flux J_i (moles/cm^2-sec) of a particular component i is related to the concentration gradient $(\partial c_i/\partial x)$ perpendicular to the electrode surface (one-dimensional gradient) by the equation

$$J_i = -D_i\left(\frac{\partial c_i}{\partial x}\right) \tag{34}$$

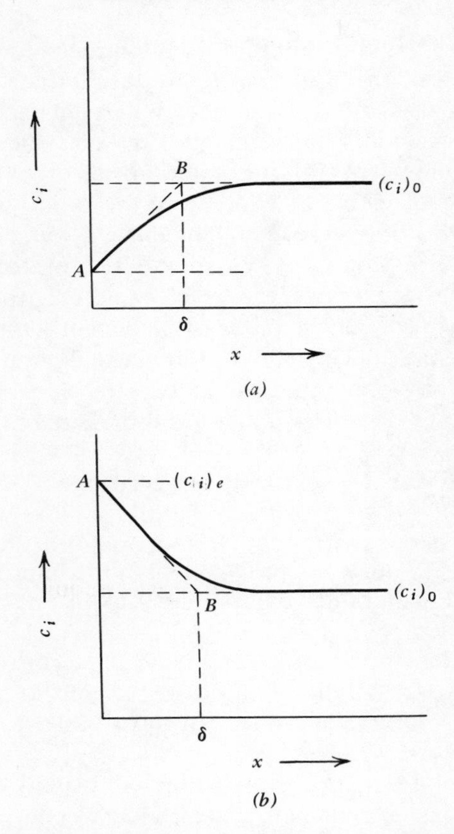

Figure 2.14 Concentration profiles at electrodes in systems with convection. Plane of electrode corresponds to $x = 0$. (*a*) For a reactant; (*b*) for a product.

where D_i is the diffusion coefficient in cm²/sec and c_i is the concentration in moles/cm³ of the diffusing species reaching or leaving the electrode surface. For the system indicated in Fig. 2.14 this equation may be used to calculate the relation between the current density and the equivalent linear gradient under steady-state conditions. Thus

$$I = n_i F J_i = -n_i F D_i \left(\frac{\partial c_i}{\partial x}\right) = \frac{n_i F D_i}{1000} \cdot \frac{(c_i)_0 - (c_i)_e}{\delta} \tag{35}$$

where $(c_i)_0$ and $(c_i)_e$ are the concentrations of the ith species in the bulk of the electrolytic solution and at the electrode surface, respectively, now expressed in moles/liter, and n_i is the number of equivalents of electrons involved in the electrode reaction per mole of diffusing species i. The algebraic sign of the concentration gradient and J should be discarded in using Eq. (35) since no special significance is attached to a plus or minus value for I in the present discussion.

Equation (35) is only applicable when transport by diffusion is very large compared to any transport which may occur because of electrolytic transference of ions. When transport of ionic constituents must be considered by both electrolytic migration as well as diffusion, Eq. (35) takes the form

$$I = \frac{n_i F D_i}{[1 \pm (n_i/z_i)t_i]} \frac{\partial c_i}{\partial x} = \frac{n_i F D_i}{1000} \cdot \frac{1}{1 \pm (n_i/z_i)t_i} \cdot \frac{(c_i)_0 - (c_i)_e}{\delta} \tag{36}$$

where t_i is the transference number of the migrating ion in the solution at the interface and z_i is the valence of the ion. The over-all sign of the term $[(n_i/z_i)t_i]$ should be minus if the electrolytic transference of the ion is in the same direction as the diffusion of the ion and plus if the electrolytic transference is in a direction opposite to the diffusion. The transference number t_i for an ion will approach ~ 0.5 only if its concentration is appreciable compared to that of the principal ionic components of the solution.

When the concentration gradient involves a reactant at the electrode surface (Fig. 2.14*a*), an upper limit is imposed on the current density I when $(c_i)_e$ approaches 0. This limiting current density I_L is given from Eq. (36):

$$I_L = \frac{n_i F D_i}{1000} \cdot \frac{1}{1 \pm (n_i/z_i)t_i} \cdot \frac{(c_i)_0}{\delta} \tag{37}$$

The value of t_i will approach zero if the solution contains another type of ion which has the same polarity and which is not discharged. Otherwise t_i will remain finite even though the concentration of the ith ion approaches zero at the electrode surface.

The value for the effective thickness of the concentration boundary layer δ depends on the type of convection and the configuration of the electrodes. For natural convection on vertical flat electrodes, δ is often of the order of 10^{-2} cm. For many simple solute molecules and ions in aqueous solutions, the diffusion coefficients fall in the range 10^{-6} to 10^{-5} cm²/sec. If the transference number of the diffusing species is small compared to unity, then I_L is usually in the range $10^{-2} n_i (c_i)_0$ to $10^{-1} n_i (c_i)_0$ where $(c_i)_0$ is expressed in moles/liter. With forced convection, the effective thickness of the diffusion boundary layer δ may be reduced to 10^{-3} cm and I_L will become approximately tenfold greater than with natural convection on vertical electrodes.

The limited rate of diffusion of products away from the electrode surface may also give rise to the equivalent of a limiting current density. The concentration of any cell product at the electrode interface cannot ordinarily exceed the solubility limit. If this limit is reached, the electrode may become covered with a nonconducting adherent solid film which will interfere with the passage of current through the cell. The current density at which such an effect may occur can be calculated from Eq. (36) by replacing $(c_i)_e$ with the concentration corresponding to the solubility of the particular component. However, many cases exist where the solubility of a reaction product does not impose an upper limit on the current density because the solid reaction product does not form as an adherent layer covering the electrode surface.

In the discussion of the diffusion boundary layer just presented, the electrode surface has been assumed to be very smooth compared to the dimensions of the boundary layer, i.e., surface irregularities are small compared to δ. Under such circumstances, regardless of the micro roughness of the surface, the area used in calculating the current density in (34 to 37) is the apparent area of the electrode. When the surface irregularities of the electrodes are comparable in dimensions to the thickness of the diffusion boundary layer, the calculation becomes complicated – probably prohibitively so for most applications.

The steady-state concentration overpotential η_c in a system with forced or natural convention can be calculated by means of the following equation obtained from Eq. (33) and (36):

$$\eta_c = \pm \frac{RT}{n_i F} \ln\left(1 + \frac{I}{Y_i}\right) \tag{38}$$

where

$$Y_i = \pm\left[\frac{n_i F D_i (c_i)_0}{1000\,\delta} \bigg/ \left(1 \pm \frac{n_i}{z_i} t_i\right)\right] \tag{38a}$$

Equation (38) is applicable when the concentration overpotential involves for practical purposes only one reactant or product and activation overpotential is negligible. The activity coefficient for the diffusing species has been assumed to be constant. In Eq. (38a) the plus sign is used outside the bracket when the concentration overpotential is caused by a product and the minus sign is used when a reactant is involved. The choice of signs in Eq. (38) should be made such that the concentration overpotential at a cathode (reduction) is negative and at an anode (oxidation) is positive. When the diffusing species i is a reactant, $Y_i = I_L$ according to Eq. (37).

In Fig. 2.15*a* is plotted the cathodic concentration overpotential accord-

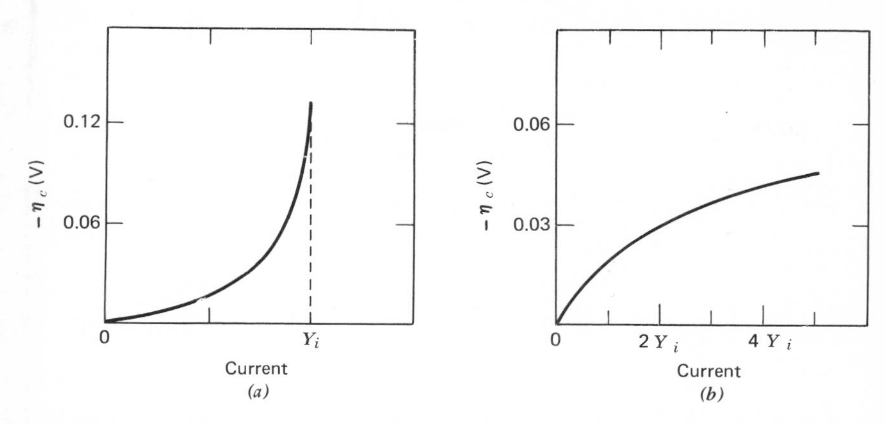

Figure 2.15 Cathodic concentration overpotential for a reduction process with natural or forced convection ($n = 1$ and $T = 298°\text{K}$). (*a*) Concentration overpotential caused by a diffusing reactant; (*b*) concentration overpotential caused by a diffusing product.

ing to Eq. (38) when only the depletion of a reactant is involved. A similar plot is given in Fig. 2.15*b* when the build-up of the concentration of a product is the cause of the concentration overpotential. The overpotential in both instances is linearly dependent on the current density only when I is small numerically compared to Y_i. The curve in Fig. 2.15*b* is based on the assumption that the solubility limit is not reached at the current densities shown in the figure.

In many batteries convection cannot occur within the electrolyte because the ionic conducting phase is in a form such as a gel, a paste, an ionic exchange membrane, or even a true solid. Natural convection is also missing in battery systems operating in the essentially gravitationless fields associated with orbital and space flight. Under such circumstances rather severe limitations are imposed on the maximum current densities which may be supported by the electrodes within the cells.

Three situations involving diffusion in the absence of any type of convection are represented in Fig. 2.16 for electrodes operating at a constant current density. The curves in Fig. 2.16*a* and 2.16*b* are for the concentrations of a soluble reactant and product, respectively, which are not involved at the other electrodes of the cell. In Fig. 2.16*c* the component is a reactant at one electrode and a product at the other electrode with no net change in the total amount of the component in the cell with time. In all three instances, transport is assumed to occur only through diffusion with electrolytic transference negligible. A steady state is never reached in either of the systems in Fig 2.16*a* and 2.16*b*. In the former, the concentration of the reactant at the electrode surface progressively

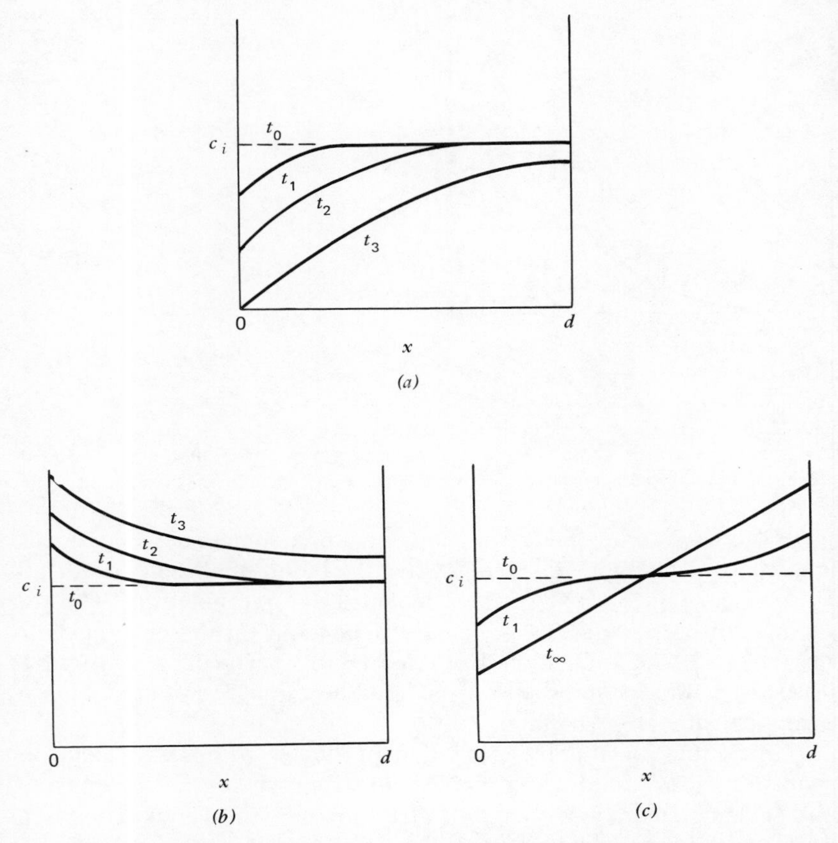

Figure 2.16 Concentration profiles for solution components without convection. The values $x = 0$ and $x = d$ correspond to the positions of the parallel electrodes. Time t_0 corresponds to zero time; t_1, t_2, t_3 to successively longer times. (a) Concentration of a component which is a reactant at only one electrode; (b) concentration of a component which is a product at only one electrode; (c) concentration of a component which is a reactant at one electrode and product at the other electrode.

approaches zero. When this state is reached, the current density through the electrode must decrease because of the limitations imposed by diffusion. In Fig. 2.16b the concentration at the electrode surface increases with time until the solubility limit (if any) is reached. Precipitation of the solid at the electrode surface may interfere with the passage of current. If this occurs, it may no longer be possible to maintain the same current density through the electrode and the current density may need to be decreased progressively to prevent the formation of a solid film over the electrode.

In Fig. 2.16*c* a steady-state condition represented by the curve designated t_∞ is approached for practical purposes after a sufficient length of time, which ranges from minutes to days depending on the electrode spacing and the characteristics of the solution. The gradient then becomes uniform between the electrodes. The upper limit for the current density may be set either by the depression of the concentration at one of the electrodes to zero or the reaching of a solubility limit at the other. Under *steady-state* conditions Eq. (36) may be applied but δ must be replaced by $\frac{1}{2}d$ where d is the distance between the two parallel electrodes. The quantity $(c_i)_e$ in Eq. (36) now becomes the concentration at the surface of either of the electrodes. The value for the limiting current densities imposed either by depletion of a reactant or the precipitation of a product, however, will be smaller in most batteries by a factor of 10^{-1} to 10^{-2} than the limiting current densities observed under comparable conditions with natural convection because of the much larger value for δ.

Examples of the various systems shown in Fig. 2.16 are as follows:

1. Hydrogen ions consumed at an oxide cathode in an immobilized electrolyte.
2. Metal cations (e.g., zinc) produced at a metal anode in an immobilized acid electrolyte.
3. Hydrogen ions in a hydrogen-oxygen cell with an immobilized electrolyte in which the transport number for the hydrogen ion is relatively small compared to unity.

Systems of the type shown in Figs. 2.16*a* and 2.16*b* are encountered relatively often in battery electrochemistry.

Concentration gradients involving reactants or products within the electrode phase often are responsible for substantial concentration overpotential in batteries. The most commonly encountered instances are oxide cathodes. Such cathodes are usually porous with much, if not all, of the pore structure filled with electrolytic solution. Other materials such as carbon may be added to the "electrode mix" at some stage in its formulation for such purposes as increasing conductivity and increasing electrolyte retention in the electrode mix. For the metal oxide to undergo complete reduction coulombically to a lower oxidation stage, either of the following must occur: (*a*) such species as hydrogen ions, oxide ions, and water molecules must diffuse through the solid phase; or (*b*) the oxide must slowly dissolve and recrystallize out on nearby sites in a lower valency state. In some oxide cathodes both may be involved. A reasonably complete mathematical description of the concentration overpotential for either of these mechanisms has not been accomplished. The problem is formidable since various concentration gradients within both the solid and liquid phases must be considered simultaneously in a system with nonuniform current distribution and usually appreciable *i-r* drops. For some important oxide electrodes such as the manganese dioxide cathode of the Leclanché cell, it is not even known with certainty which of these mechanisms and hence which concentration gradients in the solid or liquid phases are primarily responsible for the observed concentration overpotential.

Transport of gaseous reactants through gas-filled pores is encountered in many continuous-feed electrodes used in fuel cells; see, e.g., Justi and Winsel(70), Eisenberg(27), Yeager(128). The gaseous reactant enters the porous electrode from the rear and is transported to an electrochemical reaction zone within the electrode near the electrolyte side where the pores are partially or completely filled with electrolyte. Transport may occur in the gas filled process by:

1. Viscous flow under a pressure gradient.
2. Counterdiffusion of active and inert gaseous components in opposite directions.
3. Knudsen diffusion.
4. Surface migration of species adsorbed on the pore walls.

Within a porous electrode operating on a pure gaseous reactant such as H_2 or O_2, viscous flow under a pressure gradient should be the principal transport mechanism in gas-filled pores with radii large compared to the mean free path for the gas; i.e., pores with radii of 10^{-4} cm or greater for pressures of 1 atm or greater. As the gaseous reactant is consumed in the electrochemical reaction zone, the pressure in adjacent gas-filled pores will drop and a pressure gradient will develop within the electrode.

In most porous fuel cell electrodes, viscous flow is usually not controlling since such flow can support current densities orders of magnitude greater than those normally realized in such cells. Any nonreactive gas present in the oxygen supply, even in relatively small concentration ($> 10^{-3}$ molefraction), however, will accumulate within the pores and impede the transport of the reacting gaseous species. Under these conditions transport will become controlled by counterdiffusion of the gaseous reactant through a stagnant inert gas. From the theory of counterdiffusion through a stagnant layer [see, e.g., Treybal(117)], the relation between the current density and the partial pressure of oxygen at the rear of the electrode (P_R) and in the pores adjacent to the electrochemical reaction zone (P'_R) is

$$I = \frac{K}{L} \log \frac{P - P'_R}{P - P_R} \tag{39}$$

and the limiting current density I_L is given by

$$I_L = -\frac{K}{L}\log(1-X_R) = -\frac{K}{L}\log X_I \quad (39a)$$

where X_R is the molefraction of the reacting gas (R), X_I is the total molefraction of the nonreacting gases, L is the distance from the electrochemical reaction zone to the rear of the electrode, and K is a constant which depends on the pore structure of the electrode, temperature, and the molecular cross sections of the interdiffusing gases.

The limiting current density associated with counterdiffusion through a stagnant inert gas is quite evident in operation of fuel cell porous cathodes on air. In Fig. 2.17, curve A indicates the performance of a porous carbon cathode on air and curve B on pure oxygen. No limiting current was evident in operation on pure oxygen even at current densities tenfold greater than the highest represented in the figure. Equations (39) and (39a) indicate that fuel cell cathodes intended for operation on air should be made thin in order to minimize L and thus maximize the limiting current I_L.

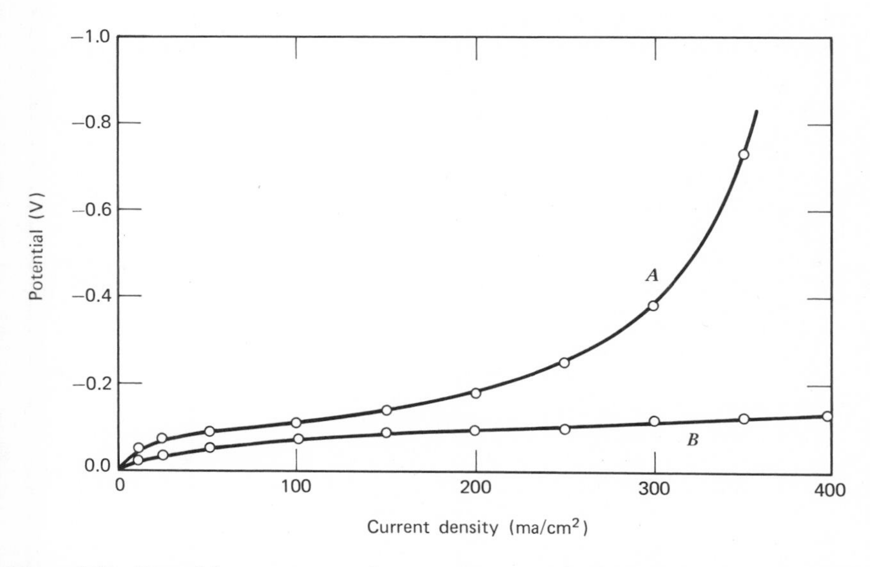

Figure 2.17 Potential-current curves for a porous carbon cathode on air (curve A) and pure oxygen (curve B). Electrode: active carbon layer (Nuchar C) on a graphite base; electrolyte: 5 M KOH + 0.01 M H_2O_2; temperature: 25°C; total pressure: 1 atm. Potentials on ordinate are relative to reversible $O_2|H_2O_2$, OH^- electrode in the same electrolyte and at the same partial pressure of oxygen. From (128) (by permission of Academic Press).

Knudsen diffusion predominates when the pore radii become smaller than the mean free path for the gas (10^{-5} cm for most gases at 1 atm). Collisions of the gas molecules with capillary or pore walls then predominate over collisions between gas molecules. Most porous electrodes operating on gases have a substantial fraction of the pore volume associated with pores with radii considerably greater than the mean free path and therefore Knudsen diffusion usually is not the principal transport mechanism.

No conclusive evidence exists presently to show that surface diffusion of adsorbed molecules contributes significantly to the transport of reactants or products within the pores of fuel cell electrodes. Surface migration of adsorbed species would not be an important transport mechanism at room temperatures because the surface concentration of physically adsorbed species such as O_2 or H_2 should be very small at room temperatures and the mobility of most chemiadsorbed species is low.

Activation Overpotential. Activation overpotential results because of the limited rate of some step in the over-all electrode reaction—a step which is characterized by an activation energy.[8] Typical of such steps are the transfer of charged species (ions or electrons) across the electrode interface, surface chemical reactions preceding or following the charge-transfer step, surface diffusion of various adsorbed species between different reactive sites and crystal nucleation. The irreversibility of such processes is the source of activation overpotential in many batteries. Activation overpotential in most batteries, however, usually is not very large and not the major factor contributing to the departure of the cell terminal voltage from the thermodynamic value.

Overpotential arising from the irreversibility of a charge-transfer step is one of the most common types of activation overpotential. The potential energy barrier associated with such a step may be represented by the curves shown in Fig. 2.18. An example of a charge-transfer step that may be described by such curves is the ionization of adsorbed hydrogen (H_{ad}) on an electrode surface:

$$H_{\mathrm{ad}} \rightleftharpoons H^+ + e_M^- \tag{1}$$

where e_M^- represents the electron in the valence or conduction bands of the metallic phase and H^+ is a proton in the hydronium ion residing in the Helmholtz layer. For this example, state I in Fig. 2.18 corresponds to H_{ad} and state II to the proton in a hydronium ion in the Helmholtz layer with

[8]Transport and conduction processes in the electrode and electrolyte phases also have activation energies but are not included with the factors contributing directly to activation overpotential.

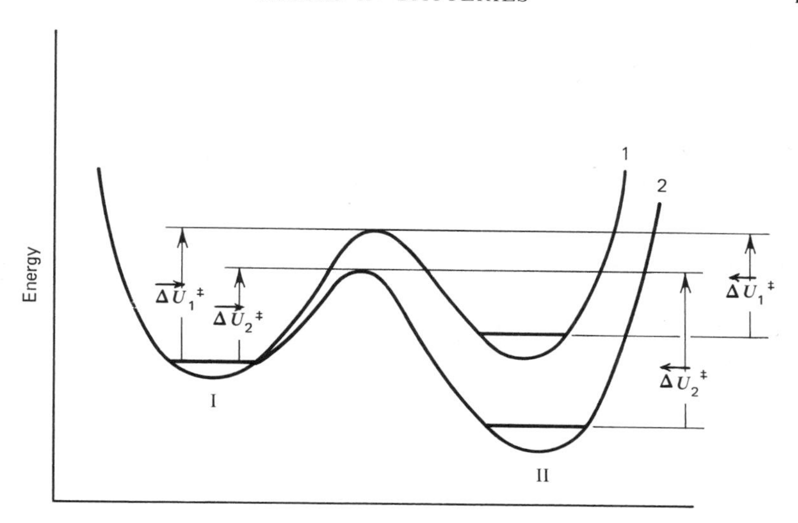

Figure 2.18 Energy-reaction coordinate diagram for charge-transfer step. For a process proceeding to the right, I corresponds to the prereaction state and II to the postreaction state; for a reaction proceeding to the left, vice versa.

the electron in metal. The ordinate in Fig. 2.18 represents the relative potential energy while the abscissa represents the reaction coordinate. The latter is a position coordinate which serves to specify the state of the chemical process. For (l) this coordinate can be the distance of the proton from the electrode surface. The system cannot reside at the minima in the potential energy curves in Fig. 2.18 but rather must have the zero-point energies indicated by the horizontal lines in states I and II for quantum mechanical reasons.

Let curve 1 in Fig. 2.18 represent the energy of the system at the reversible electrode potential for the reaction with the current and hence the net rate of reaction both equal to zero. The rate of transfer of species across the energy barrier from state I to state II is then equal to the rate for the reverse direction II to I. The rate in each direction per unit true area can be expressed in terms of an internal current density. Thus at the reversible potential $i_a = i_c = i_0$, where i_a and i_c are the internal current densities in the anodic and cathodic directions, respectively, and i_0 is the *exchange current density* which is equal to either of these internal current densities under reversible conditions.

When the electrode potential is shifted from the reversible value, the potential energies of states I and II also will shift relative to each other. Let curve 2 in Fig. 2.18 represent the potential curve when the electrode potential has been shifted from the reversible value in such a direction as to favor the forward process I to II and to retard the reverse process II to I. The rate of the process is determined by the height of the energy barrier in each direction. In polarizing the electrode so as to shift from curve 1 to curve 2, the height of the energy barrier for the forward process has been decreased ($\overrightarrow{\Delta U}_2^\ddagger < \overrightarrow{\Delta U}_1^\ddagger$) while the height of the energy barrier has been increased for the reverse process ($\overleftarrow{\Delta U}_2^\ddagger > \overleftarrow{\Delta U}_1^\ddagger$). For (l) with states I and II as defined earlier, the shift from curve 1 to curve 2 corresponds to polarizing the electrode to a more positive potential relative to the reversible electrode potential. If the electrode had been polarized to a more negative potential relative to the reversible value, the back reaction would have been favored.

When the electrode potential is shifted from the reversible value, the net rate of the reaction is no longer zero and a net current flows:

$$I = i_a - i_c \tag{40}$$

where I is the net current density with I positive for a net anodic current (oxidations) and negative for a net cathodic current (reduction). The electrode potential controls the relative height of the energy barriers in each direction and hence the net current I.

Absolute rate theory provides equations for the forward and reverse rates and hence the internal current densities in Eq. (40) [see, e.g., Glasstone, Laidler, and Eyring(53), Bockris(9)]. Thus

$$i_a = \kappa \exp \frac{\beta n F \eta_A}{RT} \tag{41}$$

$$i_c = \kappa \exp - \frac{\alpha n F \eta_A}{RT} \tag{42}$$

and

$$I = \kappa \left[\exp \frac{\beta n F \eta_A}{RT} - \exp - \frac{\alpha n F \eta_A}{RT} \right] \tag{43}$$

where

$$\alpha + \beta = 1 \tag{43a}$$

and n is the number of electrons involved in going between states I and II in either direction. The term η_A is the activation overpotential and represents the deviation of the actual electrode potential from the thermodynamic value when only this type of overpotential is involved. The activation overpotential is positive when it favors the anodic process (oxidation process) and negative when it favors the cathodic process (reduction). Only a fraction of the change in electrode potential is effective in modifying the height of the energy barrier for the process in either direction. The terms β and α are the *transfer coefficients* for the anodic and cathodic directions, respectively, and represent the fraction of the potential difference between the metal and the Helmholtz plane $(\phi_m - \phi_H)$ effective in changing the height of the energy barrier for the anodic and cathodic directions, respectively. The transfer coefficients are relative constant over surprisingly wide ranges of potentials for many electrochemical processes provided the Helmholtz layer does not undergo any major changes.

The quantity κ depends on the potential drop ψ across the diffuse portion of the ionic double layer and the nature and activities of the reacting species. For an electrochemical reaction of the form

$$A^{z_A} \longrightarrow B^{z_B} + ne^- \tag{m}$$

κ is given by the equation

$$\kappa = nFk_s(a_A)^{\alpha}(a_B)^{\beta} \exp - \frac{F(\alpha n + z_A)\psi}{RT} \tag{44}$$

where a_A and a_B are the activities of A and B in the bulk of the solution and k_s is a constant referred to by some authors as the *standard rate constant* [see, e.g., Delahay(19)]. The term ψ is the potential of the plane of closest approach of the reacting species to the electrode surface relative to the bulk of the solution. Equation (44) is fully valid only if the species A and B are not specifically adsorbed and the plane of closest approach can be assumed to correspond to the equivalent of the ordinary outer Helmholtz plane. From Eq. (44) it is evident that κ will be constant only if ψ is relatively constant for the range of potentials over which Eqs. (41) to (43) are being applied.

The exchange current density i_0 can be shown to be given by the following equation:

$$i_0 = nFk_s(a_A)^{\alpha}(a_B)^{\beta} \exp - \frac{F(\alpha n + z_A)\psi_r}{RT} = \kappa \exp - \frac{F(\alpha n + z_A)(\psi - \psi_r)}{RT} \tag{45}$$

where ψ_r is the reversible value for the potential of the outer Helmholtz plane relative to the bulk of the solution at the reversible electrode potential. From Eq. (45), it is evident that κ corresponds to i_0 when $\psi = \psi_r$. This is true to a reasonably good approximation in electrolytes of moderately high concentration ($> 0.1\ M$) if the reversible potential as well as the range of potentials over which Eq. (43) is being applied are on the same side and more than 0.1 V removed from the pzc. Otherwise κ may be considered as an *apparent exchange current density*.

For many processes α and β have been found to be approximately equal and therefore each equal to approximately 0.5. This should be expected if the potential energy curve is rather symmetrical in the vicinity of the barrier [see, e.g., Conway(12)].

Equations (41) to (43) are illustrated graphically in Fig. 2.19. For purposes of graphing, the internal anodic current density i_a has been assigned a positive value and the internal cathodic current density i_c a negative value.

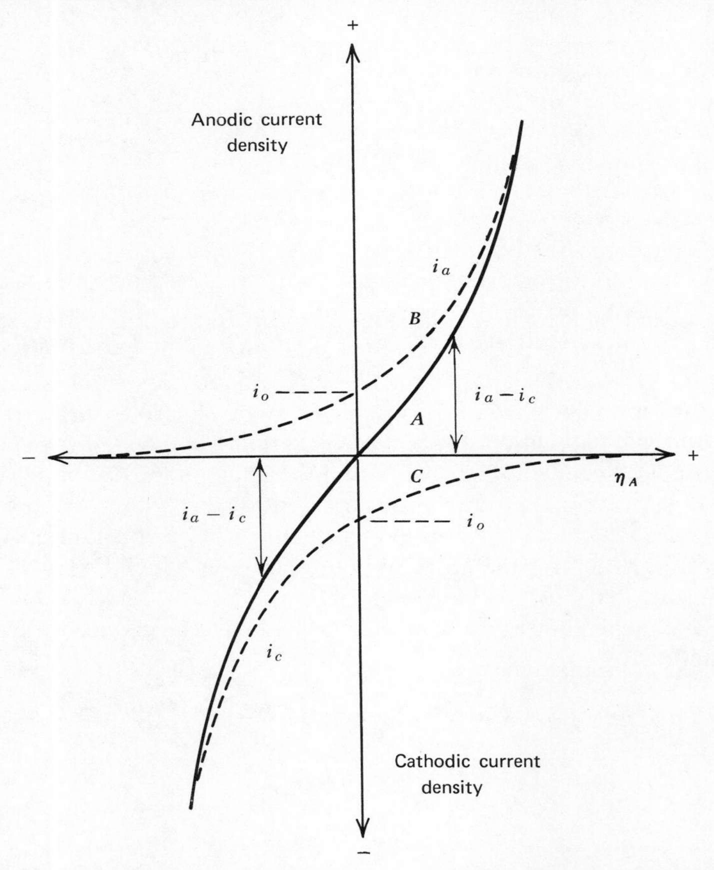

Figure 2.19 Current density versus activation overpotential. Curve A: net current density (Eq. 43); Curve B: internal anodic component (Eq. 41); Curve C: internal cathodic component (Eq. 42).

Curve A represents the net current while curves B and C represent the internal anodic and cathodic components, respectively. When the activation overpotential numerically[9] is less than $(30/n)$ mV at room temperatures, the overpotential curve (curve A in Fig. 2.19) is well approximated by a straight line which is described by the equation

$$\eta_A = \frac{RT}{nF}\frac{I}{i_0} \tag{46}$$

Equation (46) is obtained[9] by expanding the exponential terms in Eq. (43) with Maclaurin's expansion and dropping all terms past the first power. The quantity κ has been replaced by i_0 in Eq. (46).

As the overpotential η_A increases, the internal current in one direction becomes larger and that in the reverse direction becomes smaller until a potential is reached where the net current density is effectively equal to either the internal anodic or cathodic current density. This condition is reached for practical purposes when the activation overpotential numerically[10] is greater than $(120/n)$ mV at room temperatures. The relationship between the net current density and the activation overpotential is then adequately represented by either Eq. (41) or (42) with the net current density set equal to the internal current density in the appropriate direction (anodic or cathodic). Under these circumstances the net current density is related to the activation overpotential by the familiar *Tafel* equation:

$$\eta_A = A + B \log |I| \tag{47}$$

where for the anodic overpotential

$$A = -\frac{2.303\,RT}{\beta nF}\log i_0 \tag{47a}$$

and

$$B = \frac{2.303\,RT}{\beta nF} \tag{47b}$$

and for the cathodic overpotential

$$A = \frac{2.303\,RT}{(1-\beta)nF}\log i_0 \tag{47c}$$

and

$$B = -\frac{2.303\,RT}{(1-\beta)nF} \tag{47d}$$

[9]See, e.g., Potter(99), p. 128.

[10]Assuming that α and β are both approximately equal to 0.5.

Figure 2.20 presents an alternate graphical representation of Eqs. (41) to (43) in a form which more clearly illustrates the logrithmic aspects of the overpotential curves at higher overpotentials. The solid curve A corresponds to anodic net currents and the solid curve B to cathodic net currents. The dashed lines represent the internal components of the current densities with the assumption that κ in Eqs. (41) to (43) is constant.

Most electrochemical processes involving more than one electron appear to proceed by consecutive one-electron steps rather than in a single step involving two or more electrons. The preference for consecutive one-electron steps over a single multi-electron step is in accord with the idea that, other things equal, the potential energy barrier for steps

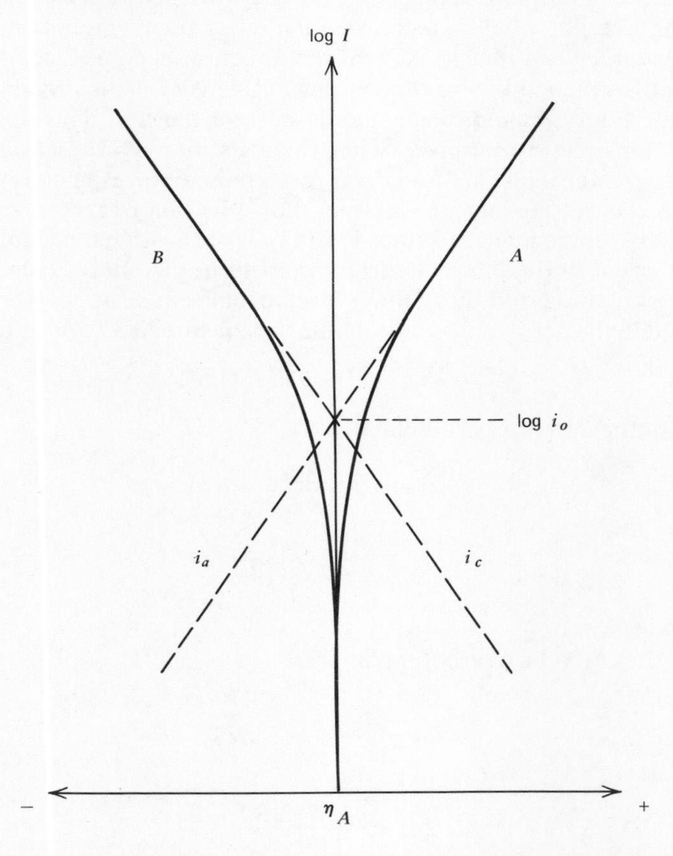

Figure 2.20 Logarithmic plot of current density versus activation overpotential. Curve A: anodic overpotential versus anodic net current density; B: cathodic overpotential versus cathodic net current density. NOTE: No algebraic signs are associated with i_a, i_c, and I.

involving two or more electrons simultaneously will be far higher than for one-electron steps.

As indicated earlier, the values for α and β often are both 0.5. Thus αn and βn in Eqs. (41) to (43) are often 0.5. In this case, the slope constant B in the Tafel equation (Eq. (47)) is ± 120 mV/decade at 25°C. In an experimental study, the finding of such a Tafel slope is considered by many workers to be evidence favoring a one-electron charge transfer process as the rate-determining step but by no means does it constitute unique evidence.

It is relatively uncommon to find a system that can be described exactly by the curves given in Fig. 2.20. Usually the pzc will fall somewhere in the anodic or cathodic range under investigation and will result in very substantial deviations from predictions based on constant κ using Eqs. (41) to (43). In Fig. 2.21 are shown the theoretical anodic curves for a process of the type

$$A^{+1} \rightarrow A^{+2} + e^- \tag{n}$$

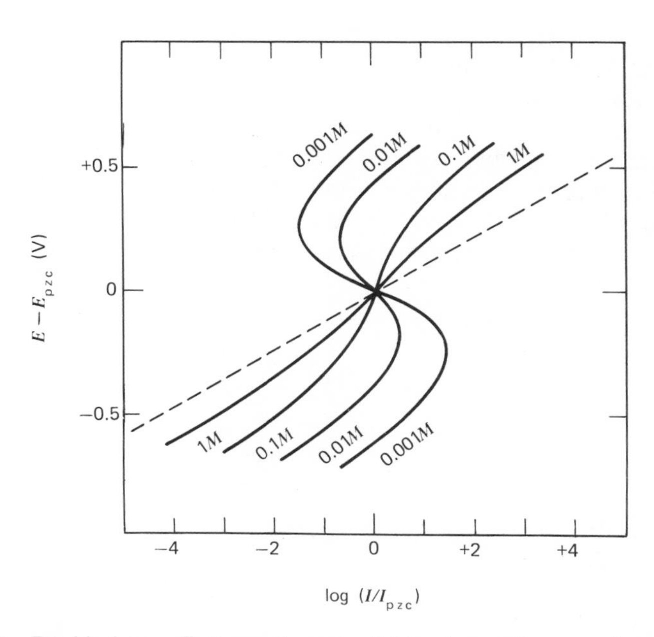

Figure 2.21 Double layer effects on the potential-log (current) curves for the process $A^{+1} \rightarrow A^{+2} + e^-$ using data for ψ on mercury at 25°C given in Fig. 2.10. Concentrations on solid curves refer to the supporting electrolyte NaF. The dashed line indicates the predicted behavior without a double layer correction ($\psi = 0$). The anodic transfer coefficient β has been taken to be $\frac{1}{2}$ and the back reaction not considered. From (94) (by permission of Interscience Publishers).

using Eqs. (41) and (44) with $\alpha = \frac{1}{2}$ and the data for ψ shown in Fig. 2.10. The back reaction has been assumed negligible. The ordinate corresponds to the rational potentials (electrode potentials relative to the pzc) while the abscissa corresponds to $\log(I/I_{\text{pzc}})$ where I_{pzc} is the current at the pzc. The numbers indicated on the curves correspond to the concentrations of the uni-univalent supporting electrolyte (NaF). The concentration of the reacting species A^{+1} has no effect on the graph because of the choice of $\log(I/I_{\text{pzc}})$ for the abscissa. Even in solutions as concentrated as 1 *M* NaF, the deviations from the simple Tafel dependence (dashed line) expected if κ is constant are very substantial. Thus electrochemical kineticists have insisted strongly that double layer effects must be considered even in the absence of specific adsorption in interpreting overpotential-current data. For a more complete treatment of the effects of the electrical double layer on electrode processes, see, e.g., Gierst(51); and Delahay(19)].

Even in concentrated solutions, specific adsorption can cause very serious deviations from the predictions of Eqs. (41) to (44) since α and β as well as κ may undergo substantial changes. An example is the discharge of hydrogen on lead in sulfuric acid, a system of interest in conjunction with the lead acid storage battery. In Fig. 2.22 the deviations from the linear Tafel behavior are associated with the specific adsorption of the SO_4^{-2} ions on the lead surface. A one-electron charge transfer step is believed to be the rate controlling step. On the lower curve for the overpotential in this figure, SO_4^{-2} adsorption is favored, whereas on the higher curve for the overpotential, this anion is desorbed. The fact that the adsorption

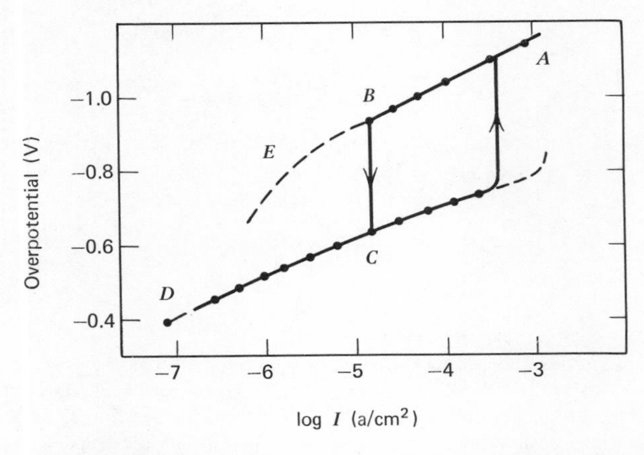

Figure 2.22 Hydrogen overpotential on lead in 0.5 *M* H_2SO_4 at 25°C. From (72) (by permission of the Faraday Society).

and desorption of the SO_4^{-2} anions is very slow leads to the hysteresis effects seen in the figure.

Equations (41) to (44) are applicable only when a charge-transfer step is rate determining and this step occurs only once when the over-all electrode process occurs once. Chemical reactions preceding or following the charge-transfer step, consecutive charge-transfer steps, and high surface concentrations of adsorbed intermediates can introduce substantial complications. An illustration of an electrode process with such complications is the oxidation of hydrogen gas on a platinum surface—a system of some importance because of its application in fuel cells. The rate-determining step responsible for the activation overpotential in this electrode system under some circumstances is believed to be the dissociative adsorption step:

$$H_2 \rightarrow 2H_{ad} \tag{o}$$

which precedes reaction (l). The hydrogen electrode reactions on platinum are far from fully understood, however [see, e.g., Frumkin and Aikasjan(38); Frumkin, Sobel, and Dmitrieva(40); Frumkin(37); Makowski, Heitz, and Yeager(85)].

Even when a step other than a simple charge-transfer step is rate determining, a logarithmic dependence of the activation overpotential on current density is often observed over an appreciable range of potentials (> 0.2 V). Although the dependence of η_A on log I may follow the general form of (47), the intercept A and the slope B will deviate from the values indicated by Eqs. (47a) to (47d). Corresponding equations for A and B have been worked out for various electrode mechanisms and rate-determining steps on a general basis as well as for specific systems. General mathematical discussions appear in Parsons(92), Bockris(9), and Vetter (119, 120). These general treatments indicate that a linear dependence of the activation overpotential on the logarithm of the current is to be expected for many systems and that when such Tafel type linearity is observed, the anodic slope B_a and cathodic slope B_c will be given by equations of the form

$$B_a = \frac{2.303\,RT}{\beta' F} \tag{48a}$$

and

$$B_c = \frac{2.303\,RT}{\alpha' F} \tag{48b}$$

with

$$\alpha' + \beta' = \frac{n}{\nu} \tag{48c}$$

provided the same mechanism and rate-determining step are involved for the regions where the anodic and cathodic slopes are evaluated and double layer effects are small. The terms α' and β' are referred to as the apparent transfer coefficients. The quantity ν in Eq. (48c) is the *stoichiometric number*, defined as the number of times the rate-determining step must occur for the over-all reaction to occur once. Table 2.2 indicates the

TABLE 2.2. STOICHIOMETRIC NUMBERS (ν) FOR VARIOUS HYDROGEN ELECTRODE REACTIONS

Mechanism	*a.* $H_2 \rightarrow 2H_{ad}$ *b.* $H_{ad} \rightarrow H^+ + e^-$		*c.* $H_2 \rightarrow H_{ad} + H^+ + e^-$ *b.* $H_{ad} \rightarrow H^+ + e^-$	
Rate-determining step	*a*	*b*	*c*	*b*
Stoichiometric number	1	2	1	1

stoichiometric numbers for the hydrogen electrode reactions with various steps rate determining. When available, the stoichiometric number can prove helpful in establishing or confirming the electrode reaction mechanism.

When the activation overpotential is small numerically [$< 30\,(\nu/n)$mV], general treatment indicates that η_A will be given by the equation

$$\eta_A = \frac{RT}{F}\frac{I}{i_0}\frac{\nu}{n} \tag{49}$$

which differs from Eq. (46) by the factor ν. Equation (49) can also be used to evaluate the stoichiometric number from overpotential measurements near the reversible electrode potential provided a reliable value for i_0 is available from studies in a range of potentials where linear Tafel dependence is observed.

Despite the extensive efforts of electrochemists over the past 125 years, many electrode systems exist for which there is no consensus of opinion among qualified workers as to the step or steps responsible for the activation overvoltage. This situation does not reflect a lack of effort but rather the difficulties associated with studying heterogeneous reaction kinetics at a solid-solution interface.

Many of the experimental data are not suitable for fundamental interpretation because of uncertainties concerning impurity effects. The heterogeneous character of most electrode surfaces represents an additional major complication in interpreting experimental data. Even if these two difficulties are circumvented or minimized, the role of the electrical double layer in electrode kinetics is still far from completely described mathematically. This problem is particularly acute in measurements in polyvalent electrolytes in the vicinity of the point of zero charge. Thus the battery electrochemist may not find it possible to establish electrode mechanism uniquely from the activation polarization data. He may be forced to content himself with just a knowledge of the activation polarization and its dependence on various experimental parameters.

Simultaneous Electrode Reactions. Consider a system in which two different electrochemical reactions can occur simultaneously on a given electrode surface, for example,

$$M \rightleftharpoons M^{+z} + ze^- \tag{p}$$
$$2H^+ + 2e^- \rightleftharpoons H_2 \tag{q}$$

where (p) corresponds to the anodic dissolution (or in the reverse direction the cathodic electrodeposition) of a metal while (q) corresponds to the discharge of hydrogen gas. The potential-current curves for an electrode at which both of these reactions are occurring simultaneously are shown in Fig. 2.23. The two dashed curves B and C represent the individual reactions (p) and (q), respectively, while the solid curve A represents the relation of the electrode potential to the over-all external current density through the electrode. The concentration of H_2 in the solution has been assumed to be small and the anodic currents corresponding to (q) in Fig. 2.23 are drawn as small.

The two processes can occur simultaneously in a uniform manner over the entire electrode surface or each process may occur on separate zones or patches on the electrode surface. In either instance, only in exceptional systems [see, e.g., Yeager et al. (130)] are the two curves independent of each other over any substantial range of conditions. Thus the rate of one process influences to some extent the current-voltage curve of the other process and vice versa. These interactions may occur because of changes in the concentration gradients, changes in the effective electrode area available for each reaction, direct competition for reaction sites, or changes in the surface crystal structure of the electrode.

When the total current through the electrode system is zero, the anodic current for one process must equal the cathodic current for the other. Thus in Fig. 2.23 the *rest* electrode potential corresponding to a net current of zero represents the potential at which the rate of dissolution of metal is exactly compensated by the discharge of hydrogen gas. This potential may differ significantly from the reversible potential for either process. If the system represented in Fig. 2.23 is the metal anode of a battery, then the anodic coulombic efficiency defined by Eq. (24) is the total anodic

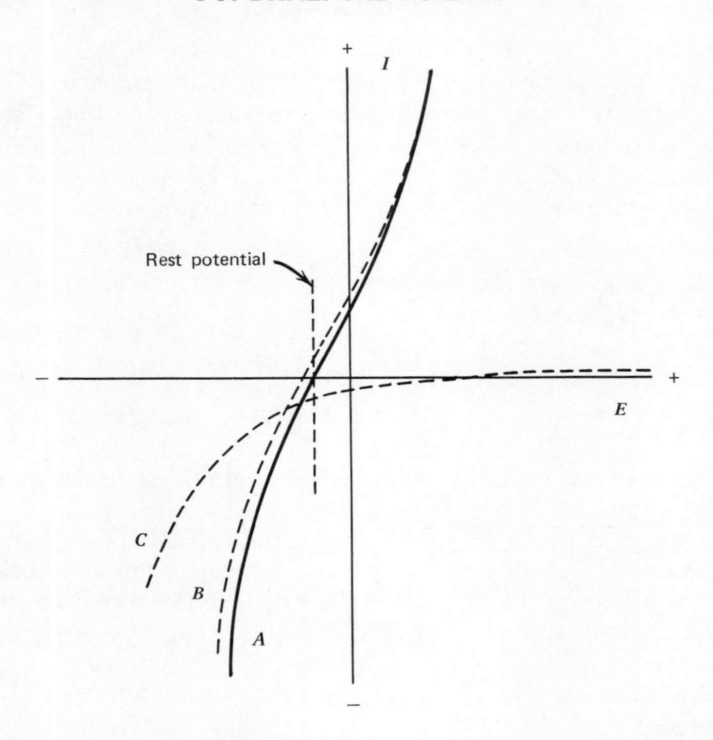

Figure 2.23 Electrode potential-current curves for an electrode with two simultaneous reactions. Curve A: potential versus external current density; B: potential versus current density for reaction p; C: potential versus current density for reaction q.

current (obtained from curve A) divided by the current density for (p) (curve B).

Another method of representing the current-potential behavior for a system of the type illustrated in Fig. 2.23 is shown in Fig. 2.24. Here again the two-simultaneous reactions are the cathodic evolution of hydrogen gas on a metal and the anodic dissolution of the metal. The reversible potential of the metal is considerably more negative than that of the reversible hydrogen electrode in the same solution. Note that each of the reactions has anodic [$(i_p)_a$, $(i_q)_a$] and cathodic [$(i_p)_c$, $(i_q)_c$] components like those shown for the single reaction in Fig. 2.20. When no net current flows, the anodic current for the metal dissolution (I_p) and the cathodic current for hydrogen gas formation (I_q) must be equal to each other and each corresponds to the *corrosion current* (I_C). For the system shown in Fig. 2.24, the open-circuit potential or rest potential (E_C) is remote to the reversible potentials of the metal and the hydrogen electrode. The curves

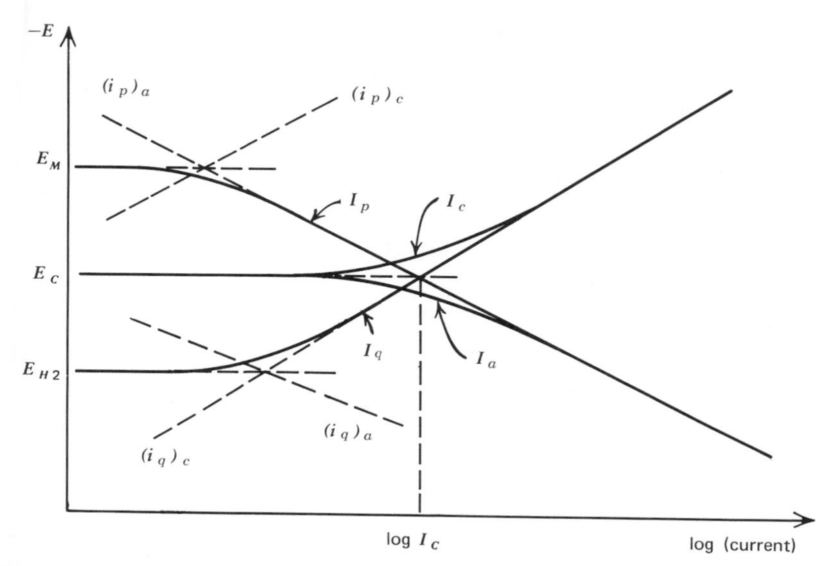

Figure 2.24 Potential-current curves for an active metal in an air-free acid solution. I_p: metal dissolution curve; I_q: hydrogen evolution reaction; I_c: potential versus external cathodic current; I_a: potential versus external anodic current; E_C: open-current potential; I_C: corrosion current; E_M: reversible potential of the metal-metal ion couple; E_{H_2}: reversible hydrogen electrode potential.

designed I_c and I_a represent the dependence of the electrode potential on the external cathodic and anodic current, respectively.

Aluminum and magnesium in acid solutions are illustrations of metals whose open-circuit potentials are remote to both their thermodynamic potentials and the reversible hydrogen electrode potential. When these metals are used as anodes in batteries with electrolytes yielding tolerable corrosion rates, however, it is not likely that the metal dissolution reaction can be even approximated by I_p in Fig. 2.24 because of the partial or complete coverage of the surface by oxides. In fact, with magnesium anodes in various batteries, the rate of formation of hydrogen gas actually increases when current is drawn from the battery in contradiction to the behavior indicated in Figs. 2.23 and 2.24. This situation arises because the passage of an appreciable current through the magnesium anode modifies the structure of the oxide on the surface.

With the zinc and lead anodes used in many batteries, the electrode potential is usually close to the reversible thermodynamic value on open circuit as well as reasonable loads. Under these circumstances the rate of corrosion of the anode on open circuit is controlled primarily by the hydrogen overpotential. Current-potential curves for such a system are similar to those represented schematically in Fig. 2.25.

In some batteries the anode is amalgamated in order to reduce the rate of anode corrosion on open-circuit. The hydrogen evolution reaction is less reversible on the amalgam than on the pure active metal. With some active metals such as zinc, the potential-current curve for the metal-metal ion couple is changed very little over the range of anode potentials and currents involved in the use of the battery. The behavior of the amalgamated electrode is represented by curves I'_c and I'_a in Fig. 2.25 with curves I_a and I'_a almost identical. The potential-current curve for the hydrogen evolution has been shifted from I_q to I'_q and the corrosion current reduced from I_C to I'_C. Various inorganic and organic species can also be added to the solution to shift the potential-current curve for hydrogen discharge to the left in Fig. 2.25 and hence to reduce the self-discharge, but these same agents may affect the potential-current curve for the metal dissolution in an adverse manner at high battery drain rates.

In Figs. 2.23 to 2.25, the discharge of hydrogen has been considered to be the only cathode process available to support the anodic dissolution of

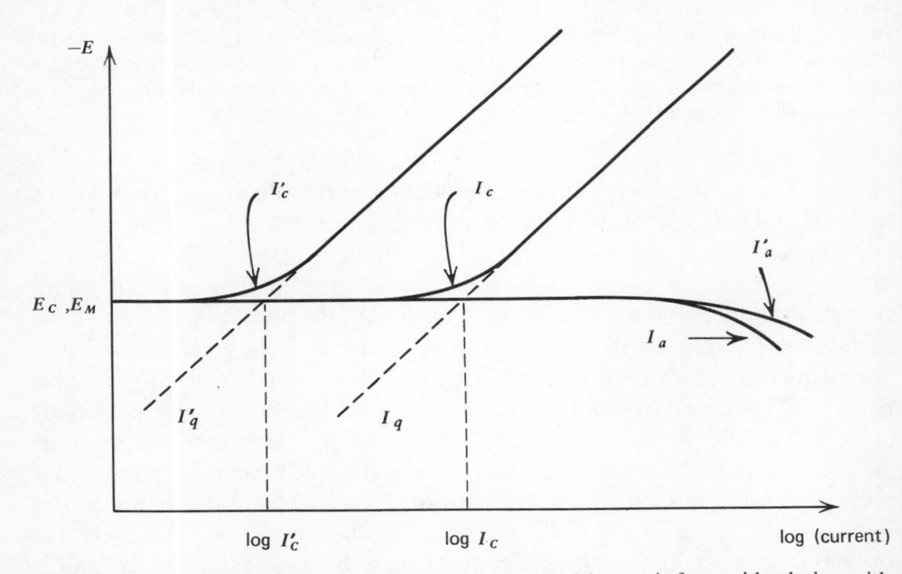

Figure 2.25 Potential-current curves for an active metal in an air-free acid solution with the open-circuit potential essentially equal to the reversible potential of the metal-metal ion couple. I_q, I'_q: potential versus cathodic current curves for hydrogen evolution reaction; I_a, I'_a: potential versus external cathodic current curve; I_c, I'_c: potential versus external cathodic current curve; E_c: open-circuit potential; I_C, I'_C: corrosion current. Curves with primes indicate effect of amalgamation (see text).

the metal. If the electrolyte contains a species that can be reduced at potentials more positive than those required for hydrogen evolution, then the dissolution of the metal may be increased substantially. An often encountered reducible species that can accelerate self-dissolution of an active metal is molecular oxygen. With anodes such as zinc, the open-circuit potential is sufficiently negative so that the oxygen reduction is transport-controlled and hence the rate of dissolution of the metal is also controlled by rate of transport of dissolved O_2 to the electrode surface.

Ohmic Overpotential. By definition the *ohmic overpotential* represents the residual ohmic losses after the *i-r* drops in the electrode and electrolyte phases have been taken into account. The usual procedure is to extrapolate the bulk properties of these phases to the interface in establishing the amount of *i-r* drop in each phase. Consequently, any abnormalities affecting the conductivity in either phase close to the interface can contribute to the ohmic overpotential. For example, if the electrolyte is depressed to a very low concentration in the Nernst diffusion layer at high current densities, the *i-r* drop in this layer may become appreciable and may be classified as one form of ohmic overpotential. Under most circumstances, however, this type of ohmic overpotential is quite negligible compared to the over-all electrode overpotential arising from other loss mechanisms.

Ohmic overpotential usually arises because of *i-r* drop associated with a layer or coating such as an oxide film on the electrode surface [see, e.g., Hoar(67)]. Such characteristics of the film as thickness and resistivity usually depend on current or potential. Thus the ohmic loss within such a film is often not a linear function of the current density since the resistance of the film need not be constant.

The treatment of charge transfer through such a film in terms of *i-r* drop concepts is valid provided the film is many monolayers thick. As the film approaches monomolecular dimensions, charge transfer involves the passage of the charged species over just a few specific potential energy barriers. In the limiting case of a monolayer, the charge transfer involves only one potential energy barrier and is usually considered in terms of activation overpotential. With films approaching monomolecular dimensions, the distinction between activation overpotential and ohmic overpotential becomes an artificial one. Even with films many layers thick, it is often difficult to establish any adequate experimental criteria for distinguishing between activation and ohmic overpotentials.

Passivation. Under some circumstances, as the potential of a dissolving metal electrode is shifted to more anodic values, the current passes through a maximum and then decreases to low values rather than increasing as would be ordinarily expected. In the range of potentials where the dissolution current is depressed, the electrode is described as passivated. Passivation, however, is not limited to corroding electrodes but can occur whenever some layer—monomolecular or multimolecular in thickness—forms on the electrode and inhibits electrochemical processes. The most common type of layer is an oxide which may range from the equivalent of a monolayer of chemiadsorbed oxygen to a layer of metal oxide of 1000 Å or thicker. With layers of oxide thick compared to molecular dimensions, the *i-r* drop within the film may be substantial. If the oxide layer is continuous over the surface, further dissolution of a metal electrode will require migration of the cations or oxide ions through the lattice and for most oxides such processes are very slow even with large voltage gradients at room temperatures.

The passivation of zinc in 6 *N* KOH is illustrated in the potential-current curves in Fig. 2.26. The passage of an anodic current through the

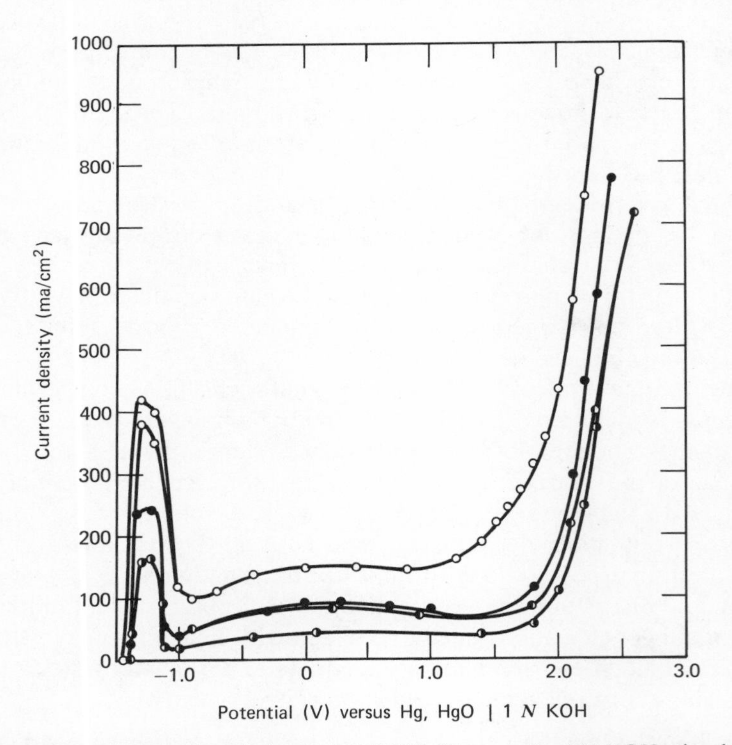

Figure 2.26 Anodic behavior of zinc in 6 *N* KOH. Clear circle: 6 *N* KOH stirred; black circle: 6 *N* KOH unstirred; half-black circle: 6 *N* KOH stirred saturated ZnO; cross: 6 *N* KOH unstirred saturated ZnO. From (105) (by permission of CITCE).

zinc electrode results in the formation of a hydrated zinc oxide on the electrode surface. Since the oxide has some appreciable solubility in the concentrated KOH, the thickness and permeability of the oxide layer depend on stirring and whether the electrolyte has been presaturated with ZnO. As the zinc electrode is polarized in the anodic direction, the current first increases, then passes through a maximum, and rapidly decreases to a relatively low value corresponding to a passive state in which the oxide layer is quite stable, adherent, and nonporous. Eventually the current again increases at anodic potentials when oxygen is discharged in the oxide.

The stability, adhesion, and resistivity of passivating oxide films usually depend on minor components and impurities present in the metal as well as the solutions. Passivation has been frequently encountered in alkaline batteries using zinc anodes and sets an upper limit to the currents that can be drawn from such cells in some instances.

Passivation is not always an undesirable phenomenon in batteries. In dry cells using magnesium anodes, the alloying agents in the anode as well as the electrolyte composition are chosen such that the anodes are passive under open-circuit conditions but rapidly become active when current is drawn from the battery. In this way self discharge through anode corrosion is minimized.

Maximum power point

For a battery in which i-r drops are the only significant losses, the *maximum power* is obtained from the battery when the load is sufficient to decrease the terminal voltage to one-half of the open-circuit value. For such batteries, the maximum power W_m is given by the equation

$$W_m = \frac{1}{2} E I_m = \frac{E^2}{4r} \tag{50}$$

where E is the open-circuit voltage, I_m is the current at the maximum power point, and r is the internal resistance of the battery. This equation is not applicable when appreciable electrode overpotential is involved. Losses at the electrode surfaces usually are not linear in current over any significant range of current densities and hence cannot be considered the equivalent of a simple resistive loss. Depending on the manner in which the overpotential varies with current, the maximum power point may occur at a terminal voltage either greater or less than that corresponding to one-half the open-circuit value. In such instances the maximum power point can be obtained readily by determining experimentally the voltage at which the product of the terminal voltage and the current is a maximum.

THE MEASUREMENT OF OVERPOTENTIAL

Objectives

In battery research, establishing the type of overpotential that is predominant at a given electrode can be a worthwhile objective of overpotential measurements. Such knowledge serves as a guide in optimizing the performance of a battery system. For example, if activation overpotential is predominant, efforts to improve cell performance in terms of lower electrode losses should be directed toward improving catalytic properties of the electrode surfaces, eliminating certain impurities that might be partially poisoning the electrode surfaces, and increasing the ratio of true to apparent surface area (e.g., by increasing the microporosity of the electrode). If mass transport in the solution or electrode phases is limiting, the research should take a quite different direction. Concentration overpotential may be decreased by such techniques as increasing natural and forced convection, increasing the concentration of reactants, and decreasing the viscosity of the solutions.

Classification of methods

The more important techniques for measuring overpotential are classified in Table 2.3 as steady-state and nonsteady-state methods. With the first class, measurements are made when the potential at a fixed current (galvanostatic method) or the current at a fixed electrode potential (potentiostatic method) has approached asymptotically a steady-state value which shows no further short time-duration variations. The steady-state galvanostatic techniques can be further subdivided into direct and indirect methods. With direct methods the potential is measured while a given current is flowing through the cell. With the indirect method the polarizing current is interrupted after it has flowed a sufficient time for steady-state conditions to be realized. The electrode potential is then measured before it has changed appreciably from the steady-state value. The polarizing current often is periodically interrupted. The indirect method ordinarily avoids the inclusion of any *i-r* drop within the potential measurements since the electrode potential is measured relative to a reference electrode with no current flowing.

The nonsteady-state methods provide for the study of the time dependence of the electrode potential or current during or following a disturbance of the electrode system. The step-function techniques involve the time dependence of the current (chronogalvanometry) or the potential (chronopotentiometry) following the application of a step function in the electrode potential or current, respectively. An interesting variation of the current-step function technique is the coulostatic or charge-step method

TABLE 2.3. ELECTRICAL METHODS FOR THE STUDY OF ELECTRODE PROCESSES

General Class	Specific Method	Maximum Values[a] Standard Rate Constant (k_s) (cm/sec)	References
Steady state[b]	direct potentiostatic	10^{-2}	(9), (97), (96), (28), (29), (4)
	direct galvanostatic	10^{-2}	(9), (97), (96), (28), (29), (4)
	indirect galvanostatic	10^{-2}	(109), (66), (107), (114), (26), (11), (102)
Nonsteady state	voltage-step function	1	(44), (50), (124), (48), (34), (7), (123)
	single current-step function	1	(20), (44)
	double current-step function	10	(46), (47), (49), (86)
	charge step function	1	(18), (12)
	linear sweep voltammetry and polarographic	10^{-1}	(17), (73), (10), (14), (65)
	a-c impedance measurements	1	(93), (17), (56), (101), (41), (42), (36), (55)
	Faradaic rectification	$10-100$	(17), (91), (3), (21), (87)

[a]Estimates indicate highest values of the standard rate constant k_s which can be determined for a reaction of the form $A \rightarrow B + e^-$ with approximately equal concentrations of reactant (A) and product (B). For relationship between k_s and i_0, see Eq. (45).

[b]Estimates of k_s for steady-state methods are applicable for forced convection of a conventional type.

in which a fixed charge is passed through the electrode. This technique permits the determination of the double layer capacity even in the presence of fast Faradaic processes.

In addition to rectangular step functions, the electrode can be perturbed from its steady-state open-circuit condition by means of a saw tooth or repetitive triangular potential function and the current measured as a function of potential and sweep rate with an oscilloscope. In both single and cyclic linear sweep voltammetry, the sweep rates may range from millivolts per second to kilovolts per second. Peaks are usually evident in the current-potential curves and are indicative of adsorption-desorption processes involving charged species at the electrode interface. Information concerning the kinetics of these adsorption-desorption processes as well as the kinetics of other Faradaic processes may be obtained from linear sweep voltammetry, although the quantitative interpretation of the voltammetry curves on a kinetic basis is more complex than for the rectangular step-function techniques. It should be noted that the polarographic technique of analysis is one form of linear sweep voltammetry in which the current is predominantly controlled by diffusion.

The a-c techniques involve the measurement of the complex impedance of the electrode system as a function of frequency, usually with an a-c bridge. From an equivalent circuit analysis the corresponding kinetic and mass transport parameters for the system are calculated. The Faradaic rectification technique involves the measurement of the d-c changes in the potential of an electrode system when an alternating current is passing through the electrode. These d-c potential changes result from the nonlinearity of the current-potential relationships for the electrode system and can be used to study the kinetics of the electrode reactions.

The various nonsteady-state measurements represented in Table 2.3 are more complicated experimentally, more difficult to interpret theoretically, and lack the direct significance of the steady-state overpotential measurements. The nonsteady-state methods, however, permit the study of activation overpotential for very reversible electrode systems for which the concentration overpotential otherwise would be predominant under steady-state conditions. This is evident in Table 2.3 from the values listed for the approximate upper limits for the standard rate constant[11] (k_s) at which the designated methods are ordinarily still adequate for the study of electrode kinetics. The steady-state techniques are of importance for the measurement of the over-all overpotential even with systems with much higher rate constants than listed in Table 2.3, but the measurements do not yield specific information con-

[11]See Eq. (45).

cerning activation overpotential and the kinetics of the electrode reactions. The steady-state techniques can be extended to the study of activation overpotential at somewhat higher values of k_s than listed in Table 2.3 by reducing the effective thickness of the boundary layer at the electrode interface to values lower than attainable with conventional agitation; e.g., through the use of high speed rotating electrodes [Koutecky and Levich (75), Levich (83), Riddiford (103), Gregory and Riddiford (62)] or ultrasonic cavitation [Penn, Yeager, and Hovorka (95), Yeager (127)].

Both the direct and indirect methods for measuring overpotentials usually can be applied to the actual battery system under investigation with only slight modification of the battery necessary; e.g., the introduction of a capillary tube (Luggin capillary) or a reference electrode. Thus it is possible to locate and to measure sources of losses within the actual battery with these methods. This is usually not true of the nonsteady-state and a-c methods. These methods ordinarily require that the measurements be made in specially designed cells which have been optimized for the study of one electrode (anode or cathode) at a time.

Only the steady-state techniques and the a-c technique will be discussed in detail in this chapter. The references indicated in Table 2.3 describe other nonsteady-state methods.

Direct methods

With the direct methods, the potential of a polarized electrode is measured relative to a reference electrode with current flowing through the polarized electrode. An attempt often is made to exclude *i-r* drop from the measurements through the use of one of the Luggin capillary arrangements shown in Fig. 2.27. The reference electrode against which the potentials are measured is placed in a separate compartment connected to the working cell by a solution bridge. The tip of the solution bridge is placed in close proximity to the particular electrode under study. The most commonly used arrangements are those shown in Fig. 2.27*a* for flat electrodes and Fig. 2.27*c* for spherical electrodes. The latter type electrode is preferred for fundamental laboratory investigations because it provides a more uniform current distribution. Many workers using the arrangements shown in Figs. 2.27*a* and 2.27*b* attempt to reduce the distance between the tip of the Luggin capillary and the electrode surface to a value such that any further decrease in distance produces no further decrease in potential, and hence *i-r* drop in the solution is presumed eliminated from the measurements.

An estimate of the *i-r* drop which may be tapped off by a Luggin capillary of the type shown in Fig. 2.27*a* can be obtained from the

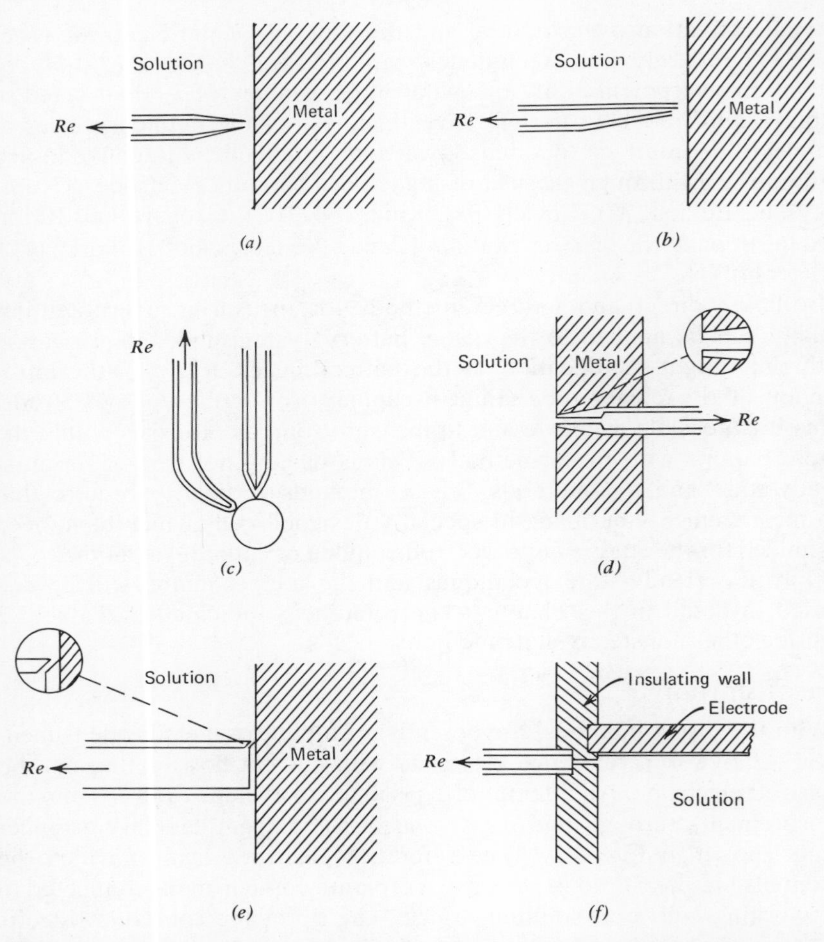

Figure 2.27 Luggin capillary arrangements for use in overpotential measurements by the direct method. Re = reference electrode.

equation $E = Ix/s$ where I is the current density in A/cm², x is the distance from the electrode surface to the Luggin tip, and s is the specific conductance of the electrolytic solution. For a Luggin capillary located at a distance of 0.2 cm and an electrolytic solution with a specific conductance of 0.2 mho/cm, this equation predicts $E \cong I$. Thus measurements at even the relatively moderate current density of 10^{-2} A/cm² under these conditions yield an error of 10^{-2} V. Any attempt to reduce the value of x to less than 10^{-1} cm usually will result in placing the capillary tip at a distance which is smaller than the outside diameter of the capillary and the current distribution on the electrode surfaces will be seriously disturbed. In addition, natural and forced convection may be adversely affected by the close proximity of the tip to the electrode surface.

The arrangement shown in Fig. 2.27*b* involves a beveled edge on the tip and is preferred to that shown in Fig. 2.27*a* or 2.27*c* for the case where the tip must be placed in close proximity to the electrode surface. It suffers from the same general defects, however, although perhaps to a slightly lesser extent. A preferable procedure [Bockris (9)] with Luggin capillaries of the type shown in Fig. 2.27*a* or 2.27*b* is to measure the potential with the capillary tip at decreasing distances from the electrode surface up to a distance of no less than four times the outside diameter of the tip. The *i-r* drop is then eliminated from the measurements by extrapolating the potential measurements at a given current to zero distance. This procedure minimizes problems arising from abnormal current distribution in the vicinity of the Luggin capillary and does not seriously disturb mass transport. The extrapolation procedure, however, does not readily take into account any deviations of the specific conductance within the diffusion layer. The number of measurements required is considerably increased. Furthermore, the extrapolation procedure does not lend itself to potentiostatic methods, which require that the potential of the electrode in question be maintained at fixed values relative to an external reference electrode.

The Luggin capillary may also be introduced through the rear of the electrode as shown in Fig. 2.27*d*. It has been the authors' experience that the rear Luggin capillary arrangement is preferable to placing a capillary in very close proximity to the front surface of the electrode but that problems relating to nonuniform current distribution, modifications in mass transport near the capillary, and the inclusion of *i-r* drop must still be considered. These problems can be minimized by using a very thin wall for the capillary and keeping the internal diameter as small as is practical. With the Luggin capillary introduced from the rear, it is important that the tip be fitted tightly into the electrode, flush with the electrode surface. Eisenberg, Tobias, and Wilke (28, 29) have found it particularly convenient to make rear-side Luggin capillaries of Teflon. These capillaries can be forced under pressure into the opening in the electrode to give a tight fit. The hole can be drilled in the Teflon after the insertion of the plastic and the excess plastic machined off from the solution side. For a rear-side Luggin capillary with a ratio of wall thickness to external diameter of $\frac{1}{6}$, Barnartt (4) has shown that the *i-r* drop error (ΔE) is $\Delta E = 0.28\, Id/s$, where d is the external diameter, I is the current density, and s is the specific conductance of the electrolytic solution. This expression neglects any changes in the conductance within the diffusion layer. For a rear Luggin capillary of 2-mm external diameter and a solution with a specific conductance of 0.2 mho/cm², $\Delta E = 0.28I$. Therefore, even at $I = 10^{-2}$ A/cm², several millivolts of *i-r* drop are to be anticipated.

The arrangement in Fig. 2.27*e* has been advocated by Piontelli et al. (96, 97) and minimizes problems relating to current distribution and *i-r* drop in the electrolytic solution. In instances involving appreciable concentration overpotential, however, the mass transport in the region adjacent to the tip may be substantially modified due to the effect of the tip on forced and natural convection. Luggin capillary arrangements of the type shown on Fig. 2.27*e* are most readily made by first sealing a piece of wire into the end of a glass tube at an angle of approximately 45° relative to the axis of the tube, next grinding the glass end plane, and then dissolving out the wire to leave a small capillary opening. The ground, blunt end of the glass tube must be pressed tightly against the electrode surface.

In electrochemical cells of the parallel electrode (filter press) type construction, it is often convenient to introduce the Luggin capillary through a hole in a side wall of the cell [Eisenberg, Tobias, and Wilke (55, 56)] as shown in Fig. 2.27*f*. With a nonconducting cell wall, only a small hole (e.g., 0.1 mm diameter) need be drilled in the wall and connection then can be made to the rear side by any convenient means. The hole should be positioned with the edge tangent to the plane of the electrode surface. This type of Luggin capillary arrangement has the advantage that it produces little if any disturbance of the current distribution and does not interfere with natural or forced convection to any appreciable extent. Furthermore, with large vertical electrodes such capillaries can be placed at various heights in order to establish variations in overpotential with vertical position.

The inclusion of any possible *i-r* drops in measurements with Luggin capillaries can often be established by means of the indirect or interrupter method which will be described later. It should be noted that none of the Luggin capillary arrangements exclude *i-r* drop in the electrode phase or the electrode leads and that these can be quite significant in some electrochemical systems.

The measurement of overpotential associated with porous electrodes (e.g., fuel cell electrodes) by the direct method presents major problems with respect to *i-r* drop in the solution within the pores as well as within the electrode phase. Under such circumstances it is desirable to measure the electrode potential within the porous electrode as a function of distance from the outer surface of the electrode under working conditions. One arrangement which has been used for such measurements with

Luggin capillaries is shown in Fig. 2.28*a*. The potential of the electrode in question is measured relative to reference electrode II as the position of the Luggin capillary introduced through the rear electrode is varied. For the measurements to have any significance the Luggin capillary must be pressed hard against the end of the hole drilled through the rear of the

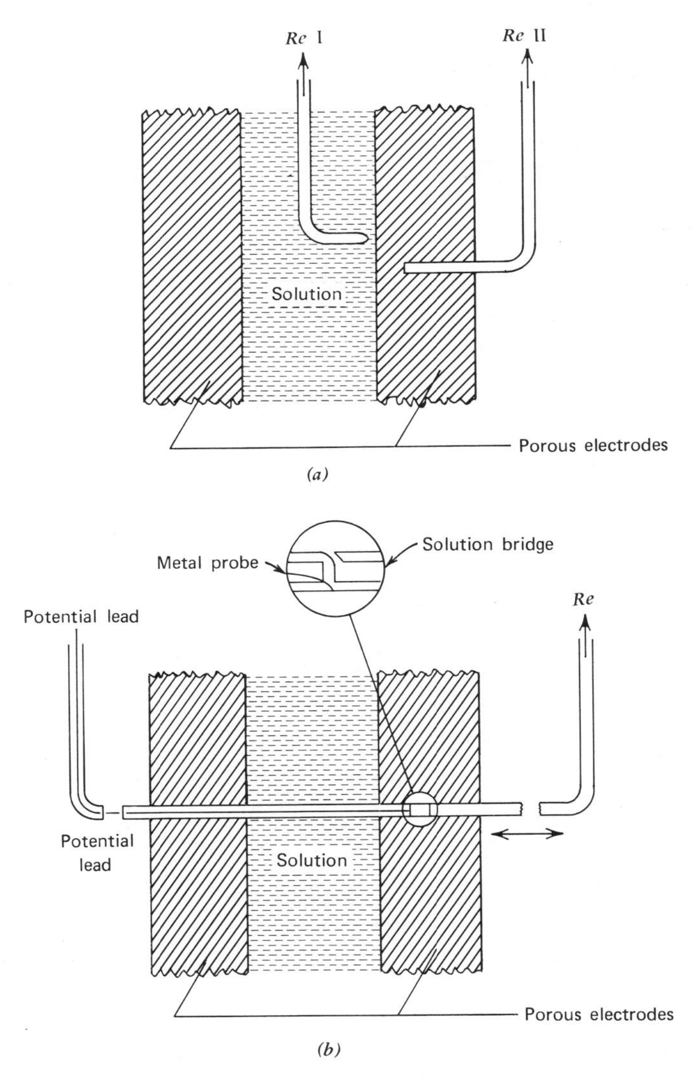

Figure 2.28 Luggin capillary arrangements for the study of porous electrodes. *Re* = reference electrode.

porous electrode. To establish the dependence on distance from the outer surface, measurements must be made either with a series of holes of different lengths sufficiently separated from each other to prevent interaction, or one hole must be progressively drilled deeper and measurements made at various depths. The overpotential associated with the outer surface can be obtained by measuring the potential of the electrode relative to reference electrode I.

The Luggin capillary arrangement in Fig. 2.28*a* disturbs the current distribution within the porous electrode. Another arrangement which reduces this complication is shown in Fig. 2.28*b*. A hole is drilled entirely through the porous electrode, perpendicular to the surface. The capillary then consists of a glass or plastic tube with a small hole in the side of the tubing. The apparent overpotential is measured as a function of distance either side of the electrode surface by moving the glass tube within the hole. It is essential that the glass tube fit tightly within the porous electrode without any space or voids between the wall of the tube and the porous electrode greater in size than the pores of the electrode. When the anode and cathode are closely spaced or the potential-distance characteristics of each are of interest, the glass tube containing the Luggin capillary can be extended through both electrodes as shown in Fig. 2.28*b*.

Neither of these Luggin capillary arrangements avoids *i-r* drop associated with the electrode phase. In some instances it may be possible to incorporate within the porous electrode inert metal probes which are insulated except at their very tips. Such a procedure is only possible if a metal is available which is highly polarized or passive with respect to any electrochemical processes at its surface in the particular electrolyte (e.g., stainless steel in oxidizing neutral and alkaline media). Such an electronic probe for sensing the potential of the electron conducting phase can be used with the arrangement shown in Fig. 2.28*b* by sealing the metal into the tube in the same perpendicular plane as the capillary hole. The probe should be flush with the outer wall of the tube and the associated lead wire carefully insulated from the electrolytic solution within the tube.

The interpretation of overpotential measurements with porous electrodes is difficult because of the interaction of such factors as *i-r* losses in the electrolyte and electrode phases, mass transport, and activation overpotential. Various authors have treated the interaction of these factors in porous electrodes [see, e.g., Ksenzhek and Stender (80, 81), Stender and Ksenzhek (115), Ksenzhek (78, 79), Newman and Tobias (90), de Levie (23), Euler (31), Euler and Muller (32), Guillou and Buvet (63), Posey (98)]. According to their treatments, the current distribution within a porous electrode may be a maximum at the front, the rear, or even both simultaneous (see Fig. 2.29), depending on the kinetics of the electrode process, the conductivities of the electrolyte and electrode phases, and the boundary conditions [see, e.g., Newman and Tobias (90)]. Whereas the overall potential-current characteristics of a porous electrode may be predicted from a detailed knowledge of the electrode kinetics, the diffusion coefficients, and the conductivities of the various phases, it is usually not possible to deduce the activation overpotential versus current density from measurements made with porous electrodes.

In measurements with porous electrodes, it is desirable to establish the concentrations of reactants and products within the electrodes at various distances from the electrode outside surface. This can be accomplished by syphoning out solution *very* slowly through the Luggin capillary within the electrode during the polarization measurements and then analyzing the solution.

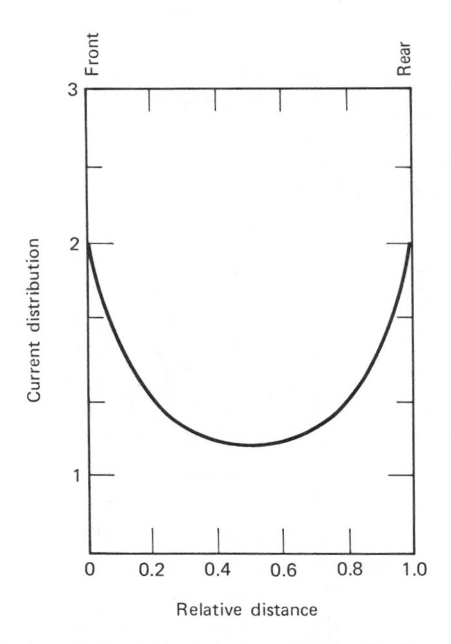

Figure 2.29 Theoretical predictions of relative current distribution versus relative distance in a porous electrode with the kinetics following the Tafel relation (see Eq. 47). For details concerning the conditions under which this plot is applicable, the reader is referred to the original paper [Newman and Tobias (90)]. Such porous electrodes as the MnO_2 cathode of the Leclanché dry cell may be expected to behave in a manner similar to that represented by this graph. (Figure used by permission of The Electrochemical Society.)

In cells with relatively large electrodes, information concerning the electrode potential (under load) as a function of position on the electrode surface can be obtained by placing the Luggin capillary at various positions along the electrode surface. In some instances this is relatively easy; for example, with vertical parallel electrodes spaced relatively far apart the Luggin capillary can be introduced from the top of the electrolyte with a scanning mechanism for varying the position in a plane parallel to the electrode surface. In other systems, cell geometry may render such scanning procedures impossible, and therefore a number of fixed position Luggin capillaries may be necessary within the cell.

Even with cells of complicated geometry, Luggin capillaries can still be introduced if sufficient ingenuity is exercised. While there undoubtedly will be errors associated with *i-r* drop and mass transport complications in the use of these Luggin capillaries, measurements will still be indicative, on at least a semiquantitative basis, of the sources of various losses and how they depend on the parameters of cell operation. The direct galvanostatic procedure with the Luggin capillary is still the only method available today which gives any reasonable indication of the overpotential as a function of position on the electrode surface for electrodes where appreciable variations in overpotential occur over the electrode surface.

The electrical equipment required for the direct galvanostatic method is probably the simplest of that for any method. A typical arrangement is shown in Fig. 2.30*a*. Even though the cell under study is such that a current will flow in the correct direction without an external voltage source, it is wise to place a power supply or battery of adequate output voltage (large compared to that of the cell) together with an adjustable resistance in series with the cell to assure adequate stability and control for the polarizing current. Both voltage-regulated and current-regulated power supplies are available commercially. The potential differences in the cell can be measured with a conventional potentiometer, a recording potentiometer (with high input impedance characteristics), or a high-impedance voltmeter such as a vacuum tube voltmeter. The voltage measuring device must not draw a current that is appreciable compared to the polarizing current or that produces any appreciable polarization or *i-r* drop in the Luggin capillary-reference electrode portion of the system. With Luggin capillaries and associated solution bridges of small diameter and large length, it is usually necessary to use a high current sensitivity galvanometer or an electrometer with the potentiometer.

A Luggin capillary also is required with steady-state potentiostatic measurements since these techniques require the potential of the electrode under study to be maintained at a given value relative to a reference electrode without interference from *i-r* drop. The potential of the elec-

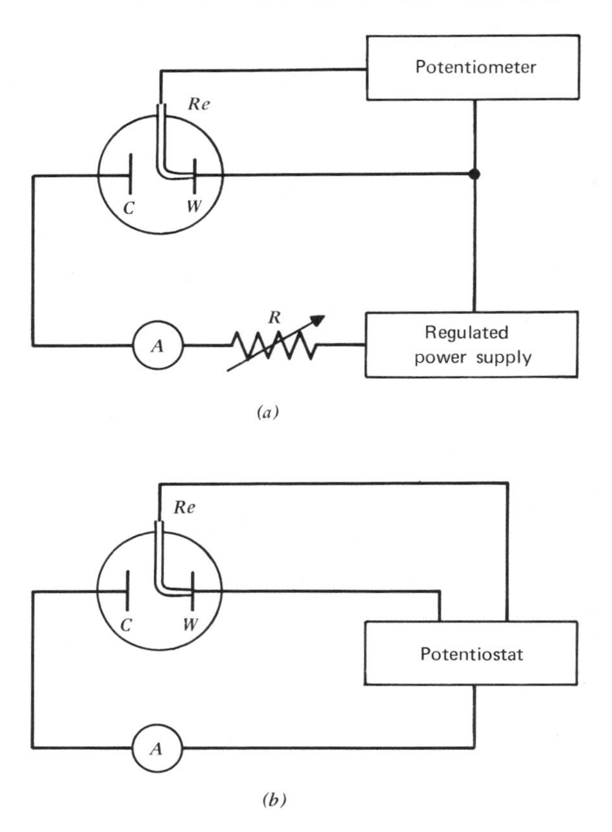

Figure 2.30 Apparatus for the direct galvanostatic and potentiostatic methods. W = working electrode, C = counterelectrode, and Re = reference. (*a*) Direct galvanostatic method; (*b*) direct potentiostatic method.

trode in question is adjusted to a particular value relative to a reference electrode with the potentiostat and the associated current at steady state measured. Potentiostats are presently available commercially with voltage regulation of ± 1 mV or better for currents from microamperes through 10^{+2} A. For steady-state measurements of the type just described, the response of the voltage-regulating circuits can be quite slow, i.e., of the order of seconds. [For nonsteady-state measurements with potential step functions, response times approaching microseconds may be needed and the design of the electronic equipment is far more critical.] Figure 2.30*b* shows a typical potentiostatic arrangement. The steady-state current at a particular voltage can be measured with a simple ammeter or recording ammeter with an accuracy of 1 percent ordinarily sufficient.

THE INDIRECT METHOD

The indirect method for measuring steady-state overpotentials can be interpreted in terms of the simplified equivalent circuit shown in Fig. 2.31. This circuit is applicable to systems in which activation overpotential is predominant. The capacitor C represents the capacitance associated with the electrical double layer at the electrode-solution interface while X is the complex impedance associated with the electrochemical reactions (Faradaic components). The component R_e is the resistance within the electrode; $(R_s)_1$ represents that fraction of the resistance within the electrolytic solution that is included in the polarized electrode-reference electrode circuit and $(R_s)_2$ that component external to this circuit. The potential difference across the capacitance C is the same the instant after interruption as the instant before interruption of the current because of the finite charge in the condenser. Similarly, the Faradaic current through X is the same immediately before and after interruption of the external current. This Faradaic current through X after interruption of the polarizing current is provided internally through the discharge of condenser C. The *i-r* drops through R_e, $(R_s)_1$, and $(R_s)_2$, however, disappear virtually instantaneously (less than 10^{-12} sec). Thus the potential measured between the polarized electrode and the reference electrode immediately after interruption of the external current is equivalent to the steady-state overpotential without the inclusion of any *i-r* drops in the bulk of the electrode, the leads, or the solution. The overpotential, however, must not decay to any appreciable extent before the potential measurement is made.

It should be noted that the Faradaic impedance X also includes a capacitance—a pseudo-capacitance—which may be quite appreciable and even much larger than the double-layer capacitance.

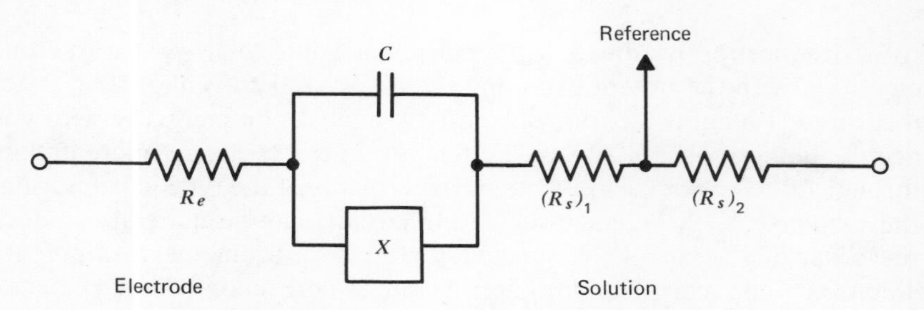

Figure 2.31 Impedance components of an electrode system. R = resistance within electrode; $(R_s)_1$ = portion of resistance in solution between working electrode and reference; $(R_s)_2$ = additional resistance in solution; X = Faradaic components of interface; and C = capacitance associated with double-layer structure.

The time dependence of the potential is shown in Fig. 2.32 with E_0 representing the steady-state polarization value measured the instant after interruption. The initial rate of the decay of the overpotential following interruption of the external current is given by the equation

$$\left(\frac{dE}{dt}\right)_0 = \frac{I}{C} \tag{51}$$

where I is the current density before interruption. In the absence of oxide films, the double-layer capacitance C is ordinarily not less than 10^{-5} f/cm^2. Thus the decay rates usually will not exceed 10^{+4} V/sec or 10^{-2} V/μsec even at current densities as high as 10^{-1} amp/cm^2 and will be proportionally lower at lower current densities. The decay rate also may be lower because of appreciable pseudo-capacitance associated with the Faradaic impedance. In systems in which concentration overpotential is predominant, the decay rate will be lower usually by several orders of magnitude. As a result, for current densities even as high as 10^{-1} amp/cm^2 an error of not more than 10 mV, and usually much less, is to be expected

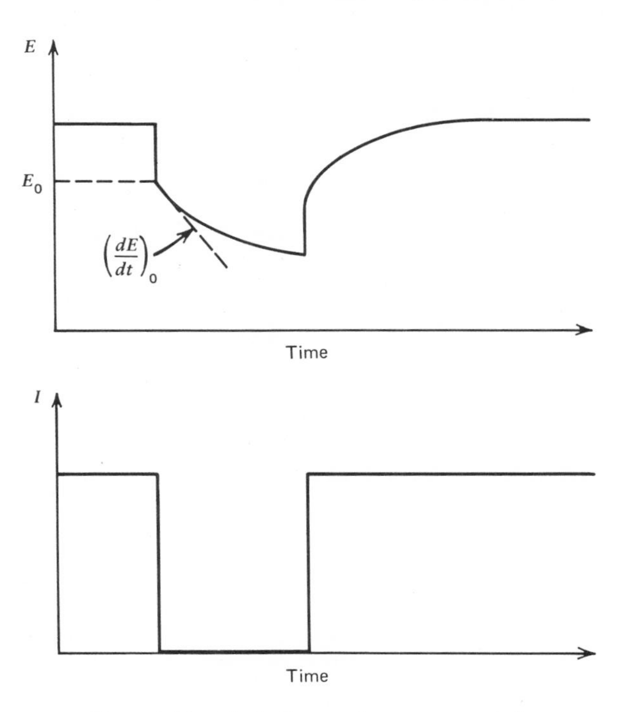

Figure 2.32 The indirect or interrupter technique.

if the potential measurements are made within 10^{-6} sec after the interruption of the polarizing current. The major exception is an electrode exhibiting passivation for which higher decay rates may be encountered.

The best arrangement, however, is to avoid the assumption that the error due to the decay of the overpotential prior to measurement is negligible and to measure the decay rate experimentally with an oscilloscope. The magnitude of any error of this type can then be evaluated and, if necessary, some correction can be made or improvements incorporated in the equipment to permit measurements in less than 10^{-6} sec. Presently available electronic equipment permits measurements within 10^{-7} sec or less after the current interruption, but specially designed polarization cells must be used for such measurements.

From Eq. (51) it is evident that the capacitance C of the ionic double layer can be evaluated from the initial rate of decay of the potential following interruption in instances where the equivalent circuit in Fig. 2.31 is adequate. The interpretation of the time dependence of the potential has been discussed by various authors [e.g., Grahame (57), Scott (108), Milner (88), Conway and Bourgault (13), Conway (12)].

In instances when both kinetic (activation) and mass transport control are substantial, the decay of the overpotential following interruption of the current can provide useful information concerning the electrochemical kinetics. The activation overpotential decays much more rapidly than the concentration gradients at the electrode surface under most circumstances. If the double layer capacity is constant and the back reaction negligible, the decay of the activation overpotential following interruption of the current is given by the equation

$$\eta = -B \log (t+\theta) + B \log \theta \tag{52}$$

where

$$\theta = \frac{BC}{i_0} \tag{52a}$$

t is the time after interruption of the current, and B corresponds to the Tafel slope (see Eqs. (47) and (48). All other symbols have the same meaning as in the discussion of Direct Methods. Equation (52) involves the assumption that the concentration of the reactants at the interface remains constant following the interruption of the current over the time period during which the decay of the activation overpotential is examined. Thus the quantity i_0 is only an apparent exchange current density, applicable for the concentrations prevailing at the electrode surface at steady state. From experimental data for the time dependence of η, the value of B and hence the transfer coefficient [α' or β' in Eq. (48)] can be determined even when the transfer coefficient cannot be obtained from steady-state measurements because of mass transport control. For times (t) short compared to θ, Eqs. (52) and (52a) reduce to Eq. (51) upon differentiation.

Figure 2.31 is not an adequate representation of the equivalent circuit for a porous electrode in which the pores are partially or completely filled with electrolytic solution. The situation is indicated schematically in a simplified manner in Fig. 2.33. The current density is not uniform over the entire electrode surface because of the resistance associated with the solution within the pores and the electrode phase itself as well as mass transport problems. Thus the Faradaic impedance X_i associated with the interfaces within the pores will differ from that on the outermost surface. The detailed analysis of interrupter measurements with porous electrodes is difficult because of the distributive nature of the interfacial impedances and series resistances. Some components of the *i-r* drops associated with R_e and R_i will be included in the potential measured immediately after the interruption of the external current because of the capacitance component C_i and C_0. These *i-r* drop effects constitute a serious deterrent to overpotential measurements for fundamental purposes on porous electrodes of the type used in various battery systems. Difficulty is encountered even in making a semiquantitative estimate of the magnitude of the error on the basis of either experimental measurements or a theoretical analysis. The direct galvanostatic method involving measurements with a Luggin capillary as a function of distance within the electrode appears better suited to studies of polarized porous electrodes than the indirect method or any of the transient methods.

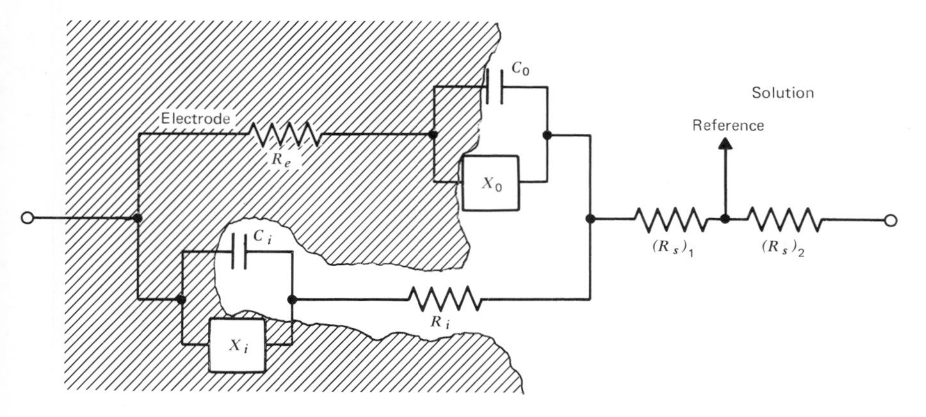

Figure 2.33 Equivalent circuit of porous electrode with solution-filled pores.

Complications also arise in the use of the indirect interrupter method with nonporous electrodes where there are *macro* variations in the overpotential and the current density distribution over the electrode surface. A good example of such a situation is a cell containing a cathode and an anode parallel to each other with part of the current being carried through the rear side of the electrodes. As with the porous electrode a part of the *i-r* drop within the solution phase will be included in the potential measured immediately after the interruption of the current. This type of error will be most pronounced in systems where the ratio of the derivative $(\partial E/\partial i)$ to the specific resistance of the solution is low compared to unity. In fundamental laboratory investigations of electrode kinetics, it is usually relatively easy to insure uniform current distribution and polarization through the use of electrode configurations such as the spherical one shown in Fig. 2.27*c*. Often this is not true for electrodes of interest in battery systems. Thus results obtained with actual battery electrodes with the indirect method may not be very accurate.

When a coating such as an oxide is present at the electrode surface, the situation becomes far more complex in measurements with the indirect method. Under some circumstances, a simplified equivalent circuit of the type shown in Fig. 2.34 may be applicable. Conduction within the oxide phase may be electronic, ionic, or both. In any event, an abnormal distribution of charge (i.e., a space-charge effect) will occur in the coating at both the metal-coating and coating-solution interfaces [see, e.g., Holmes (68), Gerischer (45), Green (60)]. These abnormal charge distributions are somewhat analogous to the electrical-double layer in the electrolytic solution and contribute a capacitive component to the impedances associated with the interfaces. In Fig. 2.34 the components C_1 and X_1 are the impedance components associated with the metal-coating interface while C_2 and X_2 represent the corresponding lumped parameters for the coating-solution interface. Thus the component C_2 reflects the capacitive components of the abnormal charge distributions in both the coating and solution phases at the interface. The component C_c represents the capacitance between the two interfaces associated with the coating as a dielectric while R_c is the parallel resistive component. It is possible to resolve the various capacitive components as indicated in Fig. 2.34 only if the effective thickness of the diffuse charged layer extending into the coating phase from each interface is sufficiently small compared to the thickness of the oxide that direct interactions between the two diffuse layers are negligible. Unfortunately, it is difficult to give a reliable criterion for an adequately thick coating since the dimensions of the diffuse charged layers in the coating depend on the charge and concentration of the charge carriers in the coating (ions, electrons, holes).

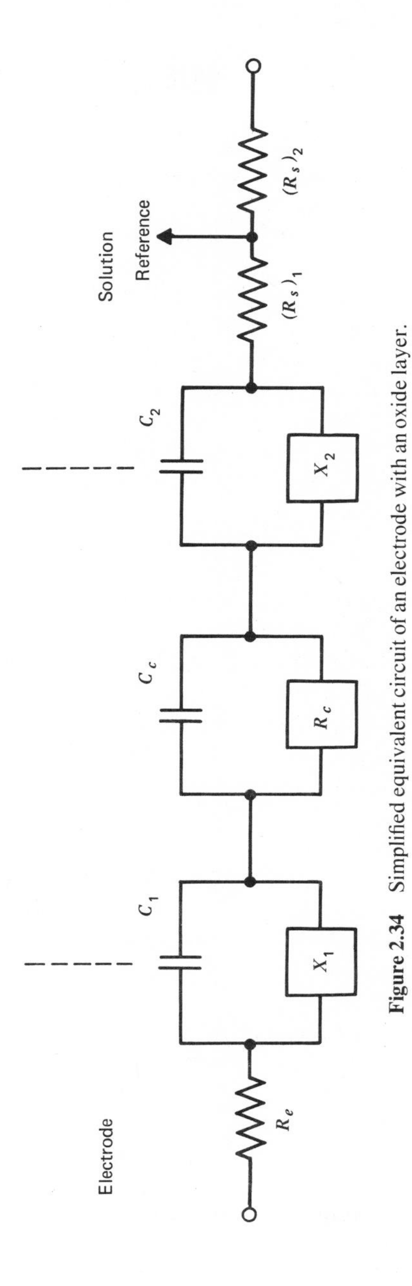

Figure 2.34 Simplified equivalent circuit of an electrode with an oxide layer.

For coatings of sufficient conductance to support reasonable current densities in battery systems, the equivalent circuit in Fig. 2.34 is probably an acceptable approximation if the coating thickness is greater than 10^{-6} cm.

The equivalent circuit in Fig. 2.34 may also be used as a crude approximation for a porous coating in which the conduction occurs primarily through electrolytic filled pores.

Upon interruption of the polarizing current in indirect measurements, the voltage corresponding to the *i-r* drop across R_c does not disappear instantaneously because of the non-Faradaic capacitance C_c. Thus *i-r* drop across the coating may be included in the overpotential measurements by the interrupter method. The rate of decay of the voltage drop across the coating will depend upon the properties of the layer. Decay rates greater than 10^4 V/sec, however, are encountered in systems with oxide layers. Interrupters that have switching transients of 10^{-6} sec are often of limited use for the study of such systems since the potential drop across R_c may decay by an unknown amount before the potential can be measured. Switching devices with switching transients as short as 10^{-8} sec are available but their use requires highly specialized electrochemical cells with careful matching of the complex impedance of the cell to the electrical equipment—a requirement not easily satisfied in most instances. Furthermore, with a relatively nonporous oxide layer of 10^{-7} cm thickness or less, it becomes difficult, if not impossible, to distinguish between activation overpotential and *i-r* drop effects within the oxide by the indicated method because the decay rates are comparable.

Periodic interruption of the current has been accomplished by most workers by means of electronic devices. Thyratrons were used for the interruption of the polarizing current as early as 1937 by Hickling (66) and more recently (1953) by Drossbach (26). Thyratron circuits, however, have definite limitations with respect to obtaining fast switching and short duration interruption of the polarizing current. Schuldiner and White (107) have used for the interruption of the current a diode with an inverse voltage pulse to cut off conduction through the diode. Their circuit is capable of switching off the current in 10^{-7} sec but restrictions in the choice of diodes (very low interelectrode capacitance requirement) appear to limit the circuit to less than 10 ma. Both triode and pentodes have been used in parallel operation in the authors' laboratory [Staicopoulos, Yeager, and Hovorka (114)] as well as by Richeson and Eisenberg (103) and Seipt (109) for currents from 10^{-6} to 5 amp with switching transients usually of the order of 10^{-6} sec but as short as 10^{-7} sec in the unit developed by Seipt.

Particularly simple devices for the switching operation are the special mechanical relays originally developed by the Western Electric Company. These relays have mercury-wetted contacts which operate in a hydrogen-pressurized environment. For low current applications (<10 ma), such relays may be operated without a protective resistive-capacitance network across their contacts, and under these circumstances they have transients for opening and closing the circuit of less than 10^{-7} to 10^{-8} sec. For currents higher than 10 ma the contacts of the relay must be protected by a capacitance-resistance network. These relays are capable of handling up to 5 amp with switching transients on closing and opening at this current of the order of 10^{-5} sec or less in duration with an adequate protection circuit in parallel with the contacts. Power transistors appear to offer some possibility for such switching applications although the transients are not likely to be much less than 10^{-5} sec and appreciable residual current may flow during the "off time" with presently available units operating at room temperatures.

Diagrams of circuits suitable for steady-state overpotential measure-

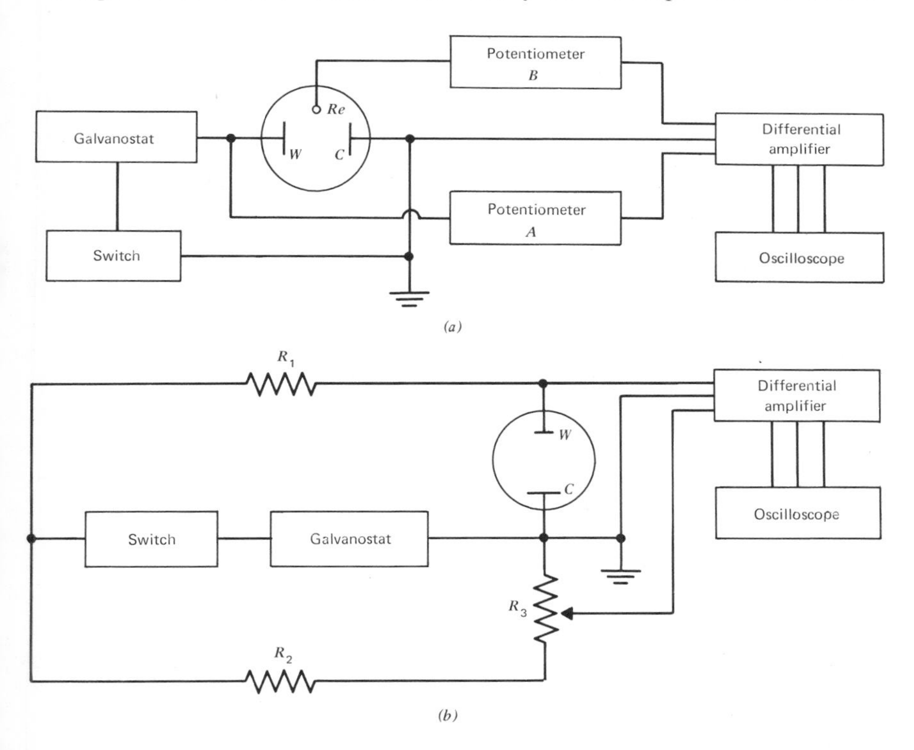

Figure 2.35 Apparatus for indirect galvanostatic (interruption) method. W = working electrode; Re = reference; and C = counterelectrode.

ments by the indirect galvanostatic method are shown in Fig. 2.35. The galvanostat is a constant source of current, which in effect may be of the same design is that used for conventional direct measurements. The potential immediately after interruption of the current as well as the time dependence of the potential for the polarized electrode relative to a reference electrode may be measured with an oscilloscope. One particularly convenient arrangement used in the laboratory of one of the authors is shown in Fig. 2.35*a*. The use of a differential amplifier makes it possible to operate with one of the electrodes (e.g., the anode) grounded and to examine the potential difference between any pair of the three electrodes: cathode-anode, anode-reference, and cathode-reference. The differential amplifier should have a band pass from d-c to at least 10^6 Hz and sensitivity sufficient to provide 5 mV/cm deflection on the oscilloscope. For measurements not requiring time resolution better than 10^{-6} sec, low impedance potentiometers may be placed in the leads to the differential amplifier as shown in Fig. 2.35*a* (potentiometer A and potentiometer B). These potentiometers may then be balanced for any particular time relative to the interruption or initiation of the current with the balancing operation very similar to that with a conventional potentiometer except that the oscilloscope is used as a time-resolved galvanometer. Potentiometers A and B are used to bring that portion of the oscilloscope trace corresponding to the instant after interruption to the zero deflection level. When properly balanced, potentiometer A will correspond to the potential difference between the cathode and anode, potentiometer B to that between the reference and the anode, and the difference between the two potentiometers to the potential difference between cathode and the reference. If potentiometers are not used in the leads to the differential amplifier, then it is necessary to calibrate directly the oscilloscope and to determine the potentials from the oscilloscope deflection at the instant of interruption.

Figure 2.35 shows an arrangement in which the counterelectrode is used as a reference electrode. This is possible when the counterelectrode does not polarize. If a differential amplifier is used with the oscilloscope as shown in Fig. 2.35, the *i-r* drop within the cell can be cancelled out through proper adjustment of resistance R.

In the past, some workers have questioned whether the direct and indirect galvanostatic methods give the same results for the steady-state polarization when applied to the same system. Little question exists today concerning agreement between the two methods when they are applied to systems for which the methods are suitable. Agreement within ± 1 mV is not unusual even for systems with high overpotential (e.g., numerically greater than 200 mV).

THE ROTATING DISK AND RING-DISK TECHNIQUES

A particularly useful technique for the study of activation overpotential when mass transport is difficult to eliminate is the rotating disk technique [see Levich (83), Riddiford (103)]. Levich has shown that for a disk electrode rotating about an axis perpendicular to the surface and through the center of the disk (Fig. 2.36), the surface concentrations of reactants and products diffusing to and from the electrode surface are uniform over the whole surface of the disk. The Levich treatment permits the calculation of the surface concentrations with considerable accuracy under most circumstances, particularly when the correction of Gregory and Riddiford (62) is included. According to Levich's treatment the surface concentration (c_e) of a reacting species is given by the equation

$$c_e = c_0\left(1 - \frac{i}{b\omega^{1/2}}\right) \tag{53}$$

and the mass transport limiting current (i_L) by the equation

$$i_L = b\omega^{1/2} \tag{54}$$

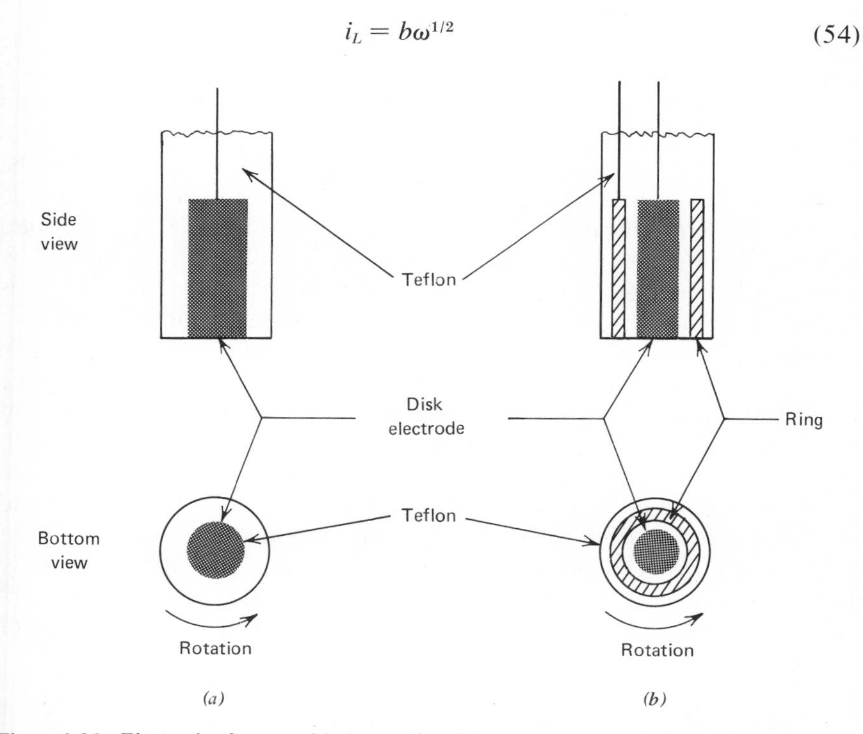

Figure 2.36 Electrodes for use with the rotating disk (*a*) and rotating disk-ring (*b*) techniques.

where ω is the rotation rate, c_0 is the bulk concentration, i is the current on the disk, and b is a constant which depends on the concentration and diffusion coefficient of the reacting species and the kinematic viscosity of the solution. Likewise for a product

$$c_e = c_0\left(1 + \frac{i}{b\omega^{1/2}}\right) \tag{55}$$

where c_e, c_0, and b now apply to a product of the electrode reaction. Thus it is possible to correct for variations in the concentrations of reactants and products at the electrode surface during the study of the kinetics of the electrode reactions. For example, for a first-order reaction for which the back reaction is negligible, Eq. (43) can be shown to become

$$i = a\rho\left(1 - \frac{i}{i_L}\right) \exp \frac{\beta nF\eta}{RT} \tag{56}$$

where a is the area of the disk electrode and i_L can be calculated by (53). This technique has been widely used, particularly for the study of the kinetics of oxygen-hydrogen fuel cell reactions [see, e.g., Yeager, Krouse, and Rao (129), Makowski, Heitz, and Yeager (85), Frumkin and Aikazjan (38), Shumilova, Zhutaeva, and Tarasevich (111)].

An important extension of the rotating disk technique involves the placing of a ring around the disk (Fig. 2.36) and the use of the ring to monitor the concentration of intermediates or products of the electrode reactions on the disk. The rotating disk-ring technique has been used by various workers (39, 111) to ascertain how much hydrogen peroxide is produced during the reduction of oxygen in alkaline solutions. The amount of peroxide reaching the ring was determined from the ring current when the ring was held at an anodic potential sufficient to oxidize all peroxide at the ring surface to O_2. The rate of peroxide production on the disk was then calculated using the treatment of Ivanov and Levich (69)[12] (1959) [see also Riddiford (113)]. In Fig. 2.37 the reduction current for O_2 on a gold-amalgamated disk is seen as a function of electrode potential and the corresponding anodic limiting ring current for the oxidation of the H_2O_2 produced on the disk. From the data in this figure and the Ivanov-Levich treatment, Frumkin and Nekrasov (39) calculated the rate constants for the reduction of the peroxide to hydroxide ions as a function of

[12]Note that the power $\frac{1}{3}$ was omitted from outside the bracketted term $[1-(3/4)(r_1/r_3)^3]$ preceding the integral in Eq. (15) of the Ivanov-Levich treatment and also Eq. (124) in Riddiford's review (103).

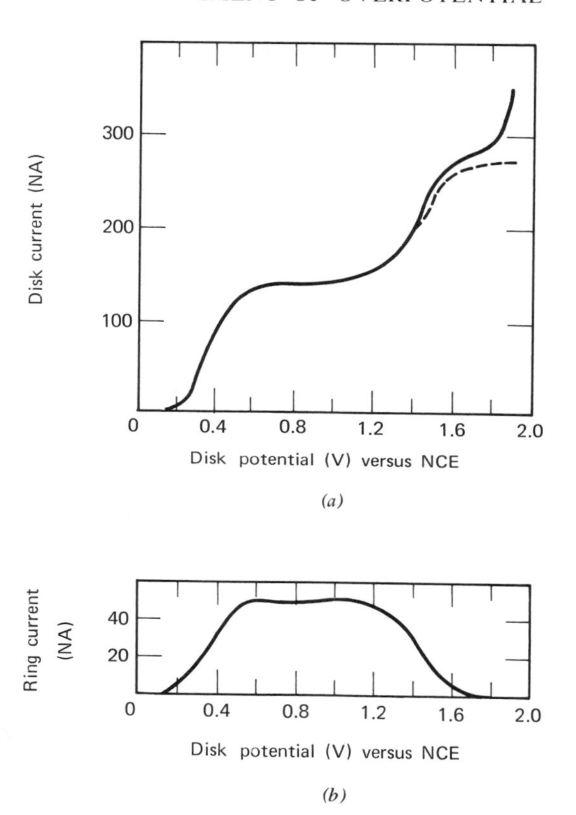

Figure 2.37 (*a*) Reduction of oxygen on an amalgamated gold disk at 2150 rpm in 0.1 *N* NaOH. (*b*) Anodic limiting current on ring for oxidation of HO_2^- produced on disk as a function of disk potential versus normal calomel electrode. Disk radius = 0.25 cm; ring internal radius = 0.275 cm; ring outside radius = 0.36 cm. The dashed line corresponds to a correction for background disk current. From (39).

disk potential. At relatively low anodic potentials, the quantitative interpretation of Fig. 2.37 indicates that the oxygen reduction proceeds principally to the peroxide state and not beyond:

$$O_2 = H_2O + 2e^- \rightleftharpoons HO_2^- + OH^- \tag{r}$$

At more anodic potentials, the oxygen reduction proceeds all the way:

$$O_2 + 2H_2O + 4e^- \rightleftharpoons 4OH^- \tag{s}$$

REFERENCES

1. Agruss, B., *J. Electrochem. Soc.*, **110**, 1097, (1963).
2. Angstadt, R., Venuto, C., and Ruetschi, P., *J. Electrochem. Soc.*, **109**, 177, (1962).
3. Barker, G., *Transactions of the Symposium on Electrode Processes*, Chap. 18 Ed. E. Yeager, J. Wiley and Sons, New York (1961). See also Barker, G., Faircloth, R., and Gardner, A., *Nature*, **181**, 247 (1958).
4. Barnartt, S., *J. Electrochem. Soc.*, **99**, 549 (1952).
5. Barnartt, S., and Forejt, D., *J. Electrochem. Soc.*, **111**, 1201 (1964),
6. Bell, G., and Huber, R., *J. Electrochem. Soc.*, **111**, 1 (1964).
7. Bewick, A., Bewick, A., Fleischmann, M., and Liler, M., *Electrochim. Acta*, **1**, 83 (1959).
8. Bichowsky, F., and Rossini, F., *The Thermochemistry of the Chemical Substances*, Reinhold Publishing Co., New York (1936).
9. Bockris, J., *Modern Aspects of Electrochemistry*, No. 1, Ed. J. Bockris, Chap. 4, Interscience Publishers, New York (1954).
10. Breiter, M., *Electrochim. Acta*, **8**, 457 (1963).
11. Breiter, M., and Volkl, W., *Z. Elektrochem.*, **58**, 899 (1954).
12. Conway, B., *Theory and Principles of Electrode Processes*, Chaps. 2 to 5, Ronald Press Co., New York (1965).
13. Conway, B., and Bourgault, P., *Trans. Faraday Soc.*, **58**, 593 (1962).
14. Conway, B., Gileadi, E., and Koxlowska, H., *J. Electrochem. Soc.*, **112**, 341 (1965).
15. deBethune, A., Licht, T., and Swendeman, N., *J. Electrochem. Soc.*, **106**, 616 (1959).
16. deBethune, A., and Loud, N., *Standard Aqueous Electrode Potentials and Temperature Coefficients at 25°C*, Clifford Hampel, Skokie, Ill. (1964).
17. Delahay, P., *New Instrumental Methods in Electrochemistry*, Interscience Publishers, New York, pp. 47–71 (1954).
18. *Id.*, *J. Phys. Chem.*, **66**, 2204 (1962).
19. *Id.*, *Double Layer and Electrode Kinetics*, Chaps. 2–6, John Wiley and Sons, New York (1965).
20. Delahay, P., and Berzins, T., *J. Am. Chem. Soc.*, **75**, 2486 (1953). See also Berzins, T., and Delahay, P., *J. Am. Chem. Soc.*, **77**, 6448 (1955).
21. Delahay, P., Senda, M., and Weis, C., *J. Phys. Chem.*, **64**, 960 (1960).
22. *Id.*, *J. Am. Chem. Soc.*, **83**, 312 (1961).
23. de Levie, R. *Electrochim. Acta*, **8**, 751 (1963).
24. Dietrick, H., Yeager, E., and Hovorka, F., *The Electrochemical Properties of Dilute Sodium Amalgam*, Tech. Report 3, Western Reserve University, Cleveland (1953). (Office of Naval Research Contract Nonr 58100.)
25. Dirkse, T., *J. Electrochem. Soc.*, **109**, 173 (1962).
26. Drossbach, P., *Z. Elektrochem.*, **57**, 548 (1953).
27. Eisenberg, M. *Advances in Electrochemistry and Electrochemical Engineering*, Vol. 2, Eds. C. Tobias and P. Delahay, Chap. 6, Interscience Publishers, New York (1962).
28. Eisenberg, M., Tobias, C., and Wilke, C., *J. Electrochem. Soc.*, **102**, 415 (1955).
29. *Id.*, *J. Electrochem. Soc.*, **103**, 356 (1956).
30. Ershler, B., *Usp. Khim.*, **21**, 237 (1952).
31. Euler, J., *Electrochim. Acta*, **8**, 409 (1963).
32. Euler, J., and Müller, K., *Electrochim. Acta*, **8**, 949 (1963).
33. *Eveready Battery Applications and Engineering Data*, Union Carbide Consumer Products Co., New York pp. 508 (~1965).
34. Fleischmann, M., and Thirsk, H., *Trans. Faraday Soc.*, **51**, 71 (1955).

35. Frumkin, A., *Z. physik. Chem.*, **164**, 121 (1933).
36. *Id.*, *J. Electrochem. Soc.*, **107**, 461 (1960).
37. *Id.*, *Advances in Electrochemistry and Electrochemical Engineering*, Vol. 4, Eds. P. Delahay and C. Tobias, Chap. 4, Interscience Publishers, New York (1965).
38. Frumkin, A., and Aikasjan, E., *Izvest. Akad. Nauk USSR*, **1959**, 202 (1959).
39. Frumkin, A., and Nekrasov, L., *Dokl. Akad. Nauk SSSR*, **126**, 115 (1959).
40. Frumkin, A., Sobol, V., and Dmitrieva, A., *J. Electroanalytical Chem.*, (in press).
41. Gerischer, H., *Z. physik. Chem.*, **198**, 286 (1951).
42. *Id.*, *Z. Elektrochem.*, **58**, 9 (1954).
43. *Id.*, *Z. Elektrochem.*, **59**, 604 (1955).
44. *Id.*, *Proceedings of the 7th Meeting of CITCE*, Butterworths Publications, London pp. 243–251 (1957).
45. *Id.*, *Advances in Electrochemistry and Electrochemical Engineering*, Vol. 1, eds. P. Delahay and C. Tobias, Chap. 4, Interscience Publishers, New York (1961).
46. Gerischer, H., and Krause, M., *Z. physik. Chem.* N. F., **10**, 264 (1957).
47. *Id.*, *Z. physik. Chem.* N. F., **14**, 184 (1958).
48. Gerischer, H., and Staubach, K., *Z. physik. Chem.* N. F., **6**, 118 (1956).
49. Gerischer, H., and Tischer, R. *Z. Elektrochem.*, **61**, 1159 (1957).
50. Gerischer, H., and Vielstich, W., *Z. physik. Chem.* N. F., **3**, 16 (1955).
51. Gierst, L. *Transactions of the Symposium on Electrode Processes*, Ed. E. Yeager, Chap. 5, John Wiley and Sons, New York (1961).
52. Glasstone, S., *Introduction to Electrochemistry*, D. van Nostrand Co., New York (1942).
53. Glasstone, S., Laidler, K., and Eyring, H., *The Theory of Rate Processes*, Chap. 10, McGraw-Hill Book Co., New York (1941).
54. Grahame, D., *Chem. Rev.*, **41**, 441 (1947).
55. *Id.*, *J. Am. Chem. Soc.*, **71**, 2975 (1949).
56. *Id.*, *J. Electrochem. Soc.*, **99**, 370C (1952).
57. *Id.*, *J. Phys. Chem.*, **57**, 257 (1953).
58. *Id.*, *J. Am. Chem. Soc.*, **76**, 4819 (1954).
59. *Id.*, *J. Am. Chem. Soc.*, **79**, 2093 (1957).
60. Green, M. *Modern Aspects of Electrochemistry*, No. 2, Ed. J. Bockris, Chap. 5, Academic Press, New York (1959).
61. Gregory D., and Riddiford, A., *J. Chem. Soc.*, **1956**, 3756 (1956).
62. *Id.*, *J. Electrochem. Soc.*, **107**, 950 (1960).
63. Guillou, M., and Buvét, R., *Electrochim. Acta*, **8**, 489 (1963).
64. Harned, H., and Owen, B., *The Physical Chemistry of Electrolytic Solutions*, Reinhold Publishing Co., New York, 3rd ed. (1958).
65. Heyrovsky, J., and Kuta, J., *Principles of Polarography*, English translation by J. Volke, Czech. Academy of Sciences, Prague, Academic Press, New York (1966).
66. Hickling, A., *Trans. Faraday Soc.*, **33**, 1540 (1937).
67. Hoar, T., *Modern Aspects of Electrochemistry*, No. 2, Ed. J. Bockris, Chap. 4, Butterworths Publications, London (1959).
68. Holmes, P., Ed., *The Electrochemistry of Semiconductors*, Academic Press, New York (1962).
69. Ivanov, Y., and Levich, V., *Dokl. Akad. Nauk SSSR*, **126**, 1029 (1959).
70. Justi, E., and Winsel, A., *Kalte Verbrennung*, Franz Steiner Verlag, Wiesbaden, Germany (1962).
71. Kirkwood, J., and Oppenheim, I., *Chemical Thermodynamics*, Chaps. 11–13, McGraw-Hill Book Co., New York (1961).
72. Kolotyrkin, J., *Trans. Faraday Soc.*, **55**, 455 (1959).

73. Kolthoff, I., and Lingane, J., *Polarography*, Vols. 1 and 2, Interscience Publishers, New York (1952).
74. Kortum, G., and Bockris, J., *Textbook of Electrochemistry*, Vol. 2, Elsevier Publishing Co., New York (1951).
75. Koutecky, J., and Levich, V., *Zhur. fiz. Khim.*, **32**, 1565 (1958).
76. Kozawa, A., and Powers, R., *J. Electrochem. Soc.*, **113**, 870 (1966).
77. Kozawa, A., and Yeager, J., *J. Electrochem. Soc.*, **112**, 959 (1965).
78. Ksenzhek, O., *Zhur. fiz. Khim.*, **36**, 243, 633 (1962).
79. *Id., Zhur. fiz. Khim.*, **37**, 2007 (1963).
80. Ksenzhek, O., and Stender, V., *Dokl. Akad. Nauk SSSR*, **106**, 487 (1956).
81. *Id., Dokl. Akad. Nauk SSSR*, **107**, 280 (1956).
82. Latimer, W., *The Oxidation States of the Elements and Their Potentials in Aqueous Solutions*, Prentice-Hall, Englewood Cliffs, N. J., 2nd Ed. (1952).
83. Levich, *Physiochemical Hydrodynamics*, English translation by Scripta Technica Inc., Prentice-Hall, Englewood Cliffs, N. J. (1962).
84. Lewis, G., Randall, M., Pitzer, K., and Brewer, L., *Thermodynamics*, McGraw-Hill Book Co., New York, 2nd ed. (1961).
85. Makowski, M., Heitz, E., and Yeager, E., *J. Electrochem. Soc.*, **113**, 204 (1966).
86. Matsuda, H., and Delahay, P., *J. Phys. Chem.*, **64**, 332 (1960).
87. *Id., J. Am. Chem. Soc.*, **82**, 1547 (1960).
88. Milner, P., *J. Electrochem. Soc.*, **107**, 343 (1960).
89. National Bureau of Standards *Handbook 71*, Washington, D. C., (1959).
90. Newman, J., and Tobias, C., *J. Electrochem. Soc.*, **109**, 1183 (1962).
91. Oldham, K., *Trans. Faraday Soc.*, **53**, 80 (1957).
92. Parsons, R., *Trans. Faraday Soc.*, **47**, 1332 (1951).
93. Parsons, R., *Modern Aspects of Electrochemistry*, No. 1, Ed. J. Bockris, Chap. 3, Academic Press, New York (1954).
94. *Id., Advances in Electrochemical Engineering*, Vol. 1, Eds. P. Delahay and C. Tobias, Chap. 1, Interscience Publishers, New York (1961).
95. Penn, R., Yeager, E., and Hovorka, F., *J. Acoust. Soc. Am.*, **31**, 1372 (1959).
96. Piontelli, R., Bertocci, U., Bianchi, G., Guerci, C., and Poli, G., *Z. Elektrochem.*, **58**, 86 (1954).
97. Piontelli, R., Bianchi, G., and Aletti, R., *Z. Elektrochem.*, **56**, 86 (1952).
98. Posey, F., *J. Electrochem. Soc.*, **111**, 1173 (1964).
99. Potter, E., *Electrochemistry*, Macmillan Co., New York (1956).
100. Pourbaix, M., *Atlas of Electrochemical Equilibria in Aqueous Solutions*, translated by J. Franklin, Pergamon Press, New York (1966).
101. Randles, J., *Discussions Faraday Soc.*, **1**, 11 (1947).
102. Richeson, W., and Eisenberg, M., *J. Electrochem. Soc.*, **107**, 642 (1960).
103. Riddiford, A., *Advances in Electrochemistry and Electrochemical Engineering*, Vol. 4, Ed. P. Delahay and C. Tobias, Interscience Publishers, New York, pp. 47–117 (1966).
104. Robinson, R., and Stokes, R., *Electrolytic Solutions*, Butterworths Publications, London, 2nd ed. (1959).
105. Sanghi, I., and Fleischmann, M., *Electrochim. Acta*, **1**, 161 (1959).
106. Schuldiner, S., *J. Electrochem. Soc.*, **106**, 891 (1959).
107. Schuldiner, S., and White, R., *J. Electrochem. Soc.*, **97**, 433 (1950).
108. Scott, W., *J. Chem. Phys.*, **23**, 1936 (1955).
109. Seipt, M., *Proceedings of the 7th Meeting of CITCE*, Butterworths Publications, London, pp. 67–85 (1957).

110. *Selected Values for the Thermodynamic Properties of Metals and Alloys*, Minerals Research Lab., Institute of Engineering Research, University of California, Berkeley (1959).
111. Shumilova, N., Zhutaeva, G., and Tarasevich, M., *Electrochim. Acta*, **11**, 967 (1966).
112. Smolders, C., *Rec. trav. chim.*, **80**, 650, 699 (1961).
113. Smolders, C. and Duyvis, E., *Rec. trav. chim.*, **80**, 635 (1961).
114. Staicopoulos, D., Yeager, E., and Hovorka, F., *J. Electrochem. Soc.*, **98**, 68 (1951).
115. Stender, V. and Ksenzhek, O., *Zhur. priklad. Khim.*, **32**, 111 (1957).
116. Stern, M., and Wissenberg, H., *J. Electrochem. Soc.*, 755 (1959).
117. Treybal, R., *Mass Transport Operations*, McGraw-Hill Book Co., New York, p. 18 (1955).
118. Vetter, K., *Z. Elektrochem.*, **55**, 274 (1951).
119. *Id.*, *Elektrochemische Kinetik*, Springerverlag, Berlin (1961).
120. *Id.*, *Transactions of the Symposium on Electrode Processes*, Ed. E. Yeager, Chap. 3, John Wiley and Sons, New York (1961).
121. *Id.*, *Z. Elektrochem.*, **66**, 577 (1962).
122. *Id.*, *J. Electrochem. Soc.*, **110**, 597 (1963).
123. Vielstich, W., and Delahay, P., *J. Am. Chem. Soc.*, **79**, 1874 (1957).
124. Vielstich, W., and Gerischer, H., *Z. phys. Chem.* N. F., **4**, 10 (1955).
125. Wales, C., *J. Electrochem. Soc.*, **109**, 1119 (1962).
126. Weissbart, J., and Ruka, R., *J. Electrochem. Soc.*, **109**, 723 (1962).
127. Yeager, E., *Transactions of the Symposium on Electrode Processes*, Ed. E. Yeager, Chap. 6, John Wiley and Sons, New York (1961).
128. Yeager, E., *Fuel Cells*, Ed. W. Mitchel. Academic Press, New York (1963).
129. Yeager, E., Krouse, P., and Rao, K., *Electrochim. Acta*, **9**, 1057 (1964).
130. Yeager, J., Yeager, E., Cels, J., and Hovorka, F., *J. Electrochem. Soc.*, **106**, 328 (1959).

APPENDIX A

LIST OF SYMBOLS

A	Tafel equation intercept [see Eq. (47)]
a	area
$a_{\pm}$	mean ionic activity
A, B, C, D, c	reacting species in silected equations
a, b, c, d	quantities of reacting species
a_i	activity of the *i*th species
$a_{\text{I}}, a_{\text{II}}$	activities of active metal in alloys I and II [see (8)]
a_2	activity of the solute
B	Tafel slope [see (47)]
b	see (53), (54)
B_a	anodic Tafel slope
B_c	cathodic Tafel slope
C	overall differential capacity
$c_{\pm}$	mean ionic concentration
C_{DL}	differential capacity of the double layer

C_G	differential capacity contribution from Gouy layer
C_H	differential capacity contribution from Helmholtz layer
c_i	concentration of *i*th component
$(c_i)_e$	concentration of *i*th component at the electrode surface
$(c_i)_H$	concentration of *i*th ion in Helmholtz plane
$(c_i)_0$	concentration of *i*th component in bulk of solution
$(c_i)_s$	concentration of *i*th ion in bulk of solution inside of the diffuse layer
d	external diameter
D_i	diffusion coefficient of *i*th component
E	potential (usually cell potential)
E°	potential corresponding to standard state = standard cell potential
E_C	rest potential
E_{ir}	*i-r* drop
E_{ox}	oxidation potential
E°_{ox}	standard oxidation potential
E_R	rational potential
E_r	thermodynamic reversible potential (usually for a cell)
E_{red}	reduction potential
E°_{red}	standard reduction potential
F	the Faraday
f_i	fugacity of *i*th component
f°_i	standard state fugacity of *i*th component
G	free energy
I	current density
i	current
i_a	internal current density in anodic direction
i_c	internal current density in cathode direction
I_C	corrosion current density
I_L	limiting current density
i_L	limiting current on disc
I_m	current through battery at maximum power
i_0	exchange current density
I_p	anodic current density for metal dissolution (see Figs. 2.23, 2.24)
I_q	cathodic current density for hydrogen gas formation (see Figs. 2.23, 2.24)
J_i	diffusion flux of *i*th component
K	a constant depending on pore structure of electrode [see (9a)]
k_s	standard rate constant

L	distance in porous electrode from electrochemical reaction zinc to the rear surface
$(m_i)_s$	molal concentration of the *i*th ion assuming complete dissociation
$(m_\pm)_s$	mean molal concentration assuming complete dissociation
n	number of electrons in over-all reaction
n_i	number of equivalents of electrons involved in electrode reaction per mole of diffusing species *i*
P_i	partial pressure of *i*th component
P_i^0	partial pressure of *i*th component at standard state
P_R	partial pressure of reacting gas at back side of porous electrode
P'_R	partial pressure of reacting gas in electrochemical reaction zone within porous electrode
P_T	total pressure
Q	heat liberated or absorbed [see (22)]
q	coulombs per mole of reaction
Q_2	heat absorbed at T_2 [see (25a)]
R	gas constant
r	resistance; change in reactions (6), (2), (2a)
s	specific conductance; amount of change [see (3b)]
t	time
T	absolute temperature
T_1	lower absolute temperature in a thermal cycle
T_2	higher absolute temperature in a thermal cycle
t_i	transference number of *i*th component
W	see (2)
w	work
W_m	maximum power of battery
x	thickness of electrolyte [see (31)]
X_I	molefraction of nonreacting gases [see (39a)]
X_R	molefraction of reacting gas [see (39a)]
Y_i	see (38a)
z_i	number of elementary charges on *i*th ion
α	cathodic transfer coefficient
α'	apparent cathodic transfer coefficient
β	anodic transfer coefficient
β'	apparent anodic transfer coefficient
γ_i	activity coefficient for *i*th component
$\gamma_\pm$	mean ionic activity coefficient
$(\gamma_i)_s$	stoichiometric activity coefficient of *i*th ion

$(\gamma_{\pm})_s$	stoichiometric mean ionic activity coefficient
ΔG	the free energy change
ΔG°	the standard free energy change
ΔH	enthalpy change
ΔH°	standard enthalpy change
ΔS	entropy change
ΔS°	standard entropy change
$\overrightarrow{\Delta U_1}\ddagger$, $\overrightarrow{\Delta U_2}\ddagger$	height of energy barrier for forward process for conditions 1 and 2 (see Fig. 2.18)
$\overleftarrow{\Delta U_1}\ddagger$, $\overleftarrow{\Delta U_2}\ddagger$	height of energy barrier for reverse process for conditions 1 and 2 (see Fig. 2.18)
$\Delta\phi_H$	potential drop across Helmholtz plane (see Fig. 2.9)
δ	effective thickness of the concentration gradient
ϵ	over-all efficiency of a cell [see (23)]
η	overpotential
η_A	activation overpotential
η_c	concentration overpotential
θ	see (52a)
κ	see (44)
μ_i	chemical potential of *i*th compound
μ_i^0	standard chemical potential of *i*th compound
ν	stoichiometric number
ρ	proportionality constant [see (56)]
σ	charge (coulombs)
Φ	coulombic efficiency [see (24)]
ϕ_m	potential of metal
ϕ_s	potential of bulk of solution
Θ	thermal efficiency [see (25)]
χ	ionic strength [see (27)]
ψ	potential drop across diffuse layer (see Fig. 2.9)
ψ_r	reversible value for potential drop across diffuse layer
ω	rotation rate

APPENDIX B

STANDARD REDUCTION POTENTIALS OF SELECTED SYSTEMS AT 25°C

Aqueous acid solutions

Electrode Reaction	E°_{red} (V)	
	L[a]	B and L[b]
$Ag^+ + e^- = Ag$	+0.7991	+0.7991
$Ag^{+2} + e^- = Ag^+$	+1.98	+1.980
$AgBr + e^- = Ag + Br^-$	+0.095	+0.0713
$AgBrO_3 + e^- = Ag + BrO_3^-$	+0.55	+0.546
$AgC_2H_3O_2 + e^- = Ag + C_2H_3O_2^-$	+0.643	+0.643
$AgCl + e^- = Ag + Cl^-$	+0.222	+0.2222
$Ag_2CrO_4 + 2e^- = 2Ag + CrO_4^{-2}$	+0.446	+0.464
$AgI + e^- = Ag + I^-$	−0.151	−0.1518
$AgIO_3 + e^- = Ag + IO_3^-$	+0.35	+0.354
$Ag_2MoO_4 + 2e^- = 2Ag + MoO_4^{-2}$	+0.49	+0.486
$AgNO_2 + e^- = Ag + NO_2^-$	+0.564	+0.564
$Ag_2SO_4 + 2e^- = 2Ag + SO_4^{-2}$	+0.653	+0.654
$Ag(S_2O_3)_2 + e^- = Ag + 2S_2O_3^{-2}$	+0.01	+0.017
$Al^{+3} + 3e^- = Al$	−1.66	−1.662
$AlF_6^{-3} + 3e^- = Al + 6F^-$	−2.07	−2.069
$Am^{+3} + 3e^- = Am$	−2.32	−2.320
$Am^{+4} + e^- = Am^{+3}$	+2.18	+2.18
$AmO_2^+ + 4H^+ + 2e^- = Am^{+3} + 2H_2O$	+1.725	+1.721
$AmO_2^{+2} + e^- = AmO_2^+$	+1.64	+1.639
$AmO_2^+ + 4H^+ + e^- = Am^{+4} + 2H_2O$	+1.26	+1.261
$AmO_2^{+2} + 4H^+ + 3e^- = Am^{+3} + 2H_2O$	+1.69	+1.694
$As^+ + e^- = As$	−2.923	–
$As + 3H^+ + 3e^- = AsH_{3(g)}$	−0.60	−0.607
$HAsO_{2(aq)} + 3H^+ + 3e^- = As + 2H_2O$	+0.247	+0.2476
$H_3AsO_4 + 2H^+ + 2e^- = HAsO_2 + 2H_2O$	+0.559	+0.560
$Au^+ + e^- = Au$	+1.68	+1.691
$Au^{+3} + 3e^- = Au$	+1.50	+1.498

[a]Data from W. Latimer, *The Oxidation States of the Elements and Their Potentials in Aqueous Solutions*, 2nd ed., Prentice Hall, Englewood Cliffs, N.Y., 1952.

[b]Data from A. de Bethune and N. Loud, *Standard Aqueous Electrode Potentials and Temperature Coefficients at 25°C*, C. Hampel, Publisher, Skokie, Ill., 1964.

AQUEOUS ACID SOLUTIONS

Electrode Reaction	E°_{red} (V) L[a]	B and L[b]
$AuBr_2 + e^- = Au + 2Br^-$	+0.96	+0.956
$AuBr_4^- + 3e^- = Au + 4Br^-$	+0.87	+0.87 (60°c)
$Au(CNS)_4^- + 3e^- = Au + 4CNS^-$	+0.66	+0.654
$AuCl_4^- + 3e^- = Au + 4Cl^-$	+1.00	+1.00
$Au(OH)_3 + 3H^+ + 3e^- = Au + 3H_2O$	+1.45	+1.45
$Ba^{+2} + 2e^- = Ba$	−2.90	−2.906
$Be^{+2} + 2e^- = Be$	−1.85	−1.847
$BiO^+ + 2H^+ + 3e^- = Bi + H_2O$	+0.32	+0.320
$Bi_2O_4 + 4H^+ + 2e^- = 2BiO^+ + 2H_2O$	+1.59	+1.593
$BiOCl + 2H^+ + 3e^- = Bi + H_2O + Cl^-$	+0.16	+0.160
$Bk^{+4} + e^- = Bk^{+3}$	+1.6	+1.6
$H_3BO_3 + 3H^+ + 3e^- = B + 3H_2O$	−0.87	−0.869
$HBrO + H^+ + e^- = \frac{1}{2}Br_{2(l)} + H_2O$	+1.59	+1.595
$Br_{2(l)} + 2e^- = 2Br^-$	+1.0652	+1.0652
$BrO_3^- + 6H^+ + 5e^- = \frac{1}{2}Br_{2(l)} + 3H_2O$	+1.52	+1.52
$C_{(graphite)} + 4H^+ + 4e^- = CH_{4(g)}$	+0.13	+0.1316
$CCl_4 + 4H^+ + 4e^- = 4Cl^- + C + 4H^+$	+1.18	+1.18
$CO_2 + 2H^+ + 2e^- = HCOOH_{(aq)}$	−0.196	−0.199
$CH_3OH_{(aq)} + 2H^+ + 2e^- = CH_4 + H_2O$	+0.586	+0.588
$C_2H_{2(g)} + 2H^+ + 2e^- = C_2H_{4(g)}$	+0.73	+0.731
$C_2H_4 + 2H^+ + 2e^- = C_2H_6$	+0.52	+0.52
$(CNS)_2 + 2e^- = 2CNS^-$	+0.77	+0.77
$Ca^{+2} + 2e^- = Ca$	−2.87	−2.866
$Cd^{+2} + 2e^- = Cd$	−0.403	−0.4029
$Ce^{+3} + 3e^- = Ce$	−2.48	−2.483
$Ce^{+4} + e^- = Ce^{+3}$	+1.61	+1.61
$(CH_3)_2SO_2 + 2H^+ + 2e^- = (CH_3)_2SO + H_2O$	+0.23	+0.23
$HCHO_{(aq)} + 2H^+ + 2e^- = CH_3OH_{(aq)}$	+0.19	—
$Cl_2 + 2e^- = 2Cl^-$	+1.3595	+1.3595
$HClO + H^+ + e^- = \frac{1}{2}Cl_2 + H_2O$	+1.63	+1.63
$HClO_2 + 2H^+ + 2e^- = HClO + H_2O$	+1.64	+1.645
$ClO_2 + H^+ + e^- = HClO_2$	+1.275	+1.275
$ClO_3^- + 3H^+ + 2e^- = HClO_2 + H_2O$	+1.21	+1.21
$ClO_4^- + 2H^+ + 2e^- = ClO_3^- + H_2O$	+1.19	+1.19
$HCNO + H^+ + e^- = \frac{1}{2}C_2N_2 + H_2O$	+0.33	+0.330

AQUEOUS ACID SOLUTIONS

Electrode Reaction	E°_{red} (V) L[a]	E°_{red} (V) B and L[b]
$Co^{+3} + e^- = Co^{+2}$	+1.82	+1.808
$Co^{+2} + 2e^- = Co$	−0.277	−0.277
$HCOOH_{(aq)} + 2H^+ + 2e^- = HCHO_{(aq)} + H_2O$	+0.056	+0.056
$Cr_2O_7^{-2} + 14H^+ + 6e^- = 2Cr^{+3} + 7H_2O$	+1.33	+1.33
$Cr^{+3} + e^- = Cr^{+2}$	−0.41	−0.408
$Cr^{+3} + 3e^- = Cr$	−0.74	−0.744
$Cu^+ + e^- = Cu$	+0.521	(+0.521)
$Cu^{+2} + e^- = Cu^+$	+0.153	+0.153
$Cu^{+2} + 2e^- = Cu$	+0.337	+0.337
$Cu^{+2} + Br^- + e^- = CuBr$	+0.640	+0.640
$Cu^{+2} + 2CN^- + e^- = Cu(CN)_2^-$	+1.12	—
$Cu^{+2} + Cl^- + e^- = CuCl$	+0.538	+0.538
$Cu^{+2} + I^- + e^- = CuI$	+0.86	+0.86
$CuBr + e^- = Cu + Br^-$	+0.033	+0.033
$CuCl + e^- = Cu + Cl^-$	+0.137	+0.137
$CuI + e^- = Cu + I^-$	−0.185	−0.1852
$Eu^{+3} + e^- = Eu^{+2}$	−0.43	−0.429
$F_2 + 2e^- = 2F^-$	+2.65	+2.87
$F_2 + 2H^+ + 2e^- = 2HF_{(aq)}$	+3.06	+3.06
$F_2O + 2H^+ + 4e^- = H_2O + 2F^-$	+2.1	+2.15
$Fe^{+2} + 2e^- = Fe$	−0.440	−0.4402
$Fe^{+3} + e^- = Fe^{+2}$	+0.771	+0.771
$Fe(CN)_6^{-3} + e^- = Fe(CN)_6^{-4}$	+0.36	+0.36
$FeO_4^{-2} + 8H^+ + 3e^- = Fe^{+3} + 4H_2O$	+1.9	+2.20
$Ga^{+3} + 3e^- = Ga$	−0.53	−0.529
$Gd^{+3} + 3e^- = Gd$	−2.40	−2.397
$GeO_2 + 4H^+ + 4e^- = Ge + 2H_2O$	−0.15	−0.15
$H^+ + e^- = H_{(g)}$	−2.10	−2.1065
$\frac{1}{2}H_2 + e^- = H^-$	−2.25	−2.25
$2H^+ + 2e^- = H_2$	0	0
$Hf^{+4} + 4e^- = Hf$	−1.70	−1.70
$Hg_2^{+2} + 2e^- = 2Hg$	+0.789	+0.788
$2Hg^{+2} + 2e^- = Hg_2^{+2}$	+0.920	+0.920
$HgBr_4^{-2} + 2e^- = Hg + 4Br^-$	+0.21	+0.223
$HgI_4^{+2} + 2e^- = Hg + 4I^-$	−0.04	−0.038

AQUEOUS ACID SOLUTIONS

Electrode Reaction	E°_{red} (V) L[a]	E°_{red} (V) B and L[b]
$I_2 + 2e^- = 2I^-$	+0.5355	+0.5355
$I_3^- + 2e^- = 3I^-$	+0.536	+0.536
$ICl_2^- + e^- = 2Cl^- + \frac{1}{2}I_2$	+1.06	+1.056
$In^{+3} + 3e^- = In$	−0.342	−0.343
$HIO + H^+ + e^- = \frac{1}{2}I_2 + H_2O$	+1.45	+1.45
$IO_3^- + 6H^+ + 5e^- = \frac{1}{2}I_2 + 3H_2O$	+1.195	+1.195
$H_5IO_6 + H^+ + 2e^- = IO_6^- + 3H_2O$	+1.6	+1.601
$IrBr_6^{-3} + e^- = IrBr_6^{-4}$	+0.99	+0.99
$IrCl_6^{-2} + e^- = IrCl_6^{-3}$	+1.017	+1.017
$IrCl_6^{-3} + 3e^- = Ir + 6Cl^-$	+0.77	+0.77
$K^+ + e^- = K$	−2.925	−2.925
$La^{+3} + 3e^- = La$	−2.52	−2.522
$Li^+ + e^- = Li$	−3.045	−3.045
$Lu^{+3} + 3e^- = Lu$	−2.25	−2.255
$Mg^{+2} + 2e^- = Mg$	−2.37	−2.363
$Mn^{+2} + 2e^- = Mn$	−1.18	−1.186
$Mn^{+3} + e^- = Mn^{+2}$	+1.51	+1.51
$MnO_2 + 4H^+ + 2e^- = Mn^{+2} + 2H_2O$	+1.23	+1.23
$MnO_4^- + e^- = MnO_4^{-2}$	+0.564	+0.564
$MnO_4^- + 4H^+ + 3e^- = MnO_2 + 2H_2O$	+1.695	+1.695
$MnO_4^- + 8H^+ + 5e^- = Mn^{+2} + 4H_2O$	+1.51	+1.51
$Mo^{+3} + 3e^- = Mo$	−0.2	−0.20
$\frac{3}{2}N_2 + H^+ + e^- = HN_{3(aq)}$	−3.09	−3.09
$N_2 + 5H^+ + 4e^- = N_2H_5^+$	−0.23	−0.23
$HN_3 + 3H^+ + 2e^- = NH_4^+ + N_2$	+1.96	+1.96
$HN_{3(aq)} + 11H^+ + 8e^- = 3NH_4^+$	+0.69	+0.695
$N_2H_5^+ + 3H^+ + 2e^- = 2NH_4^+$	+1.275	+1.275
$NH_3OH^+ + 2H^+ + 2e^- = NH_4^+ + H_2O$	+1.35	+1.35
$2NH_3OH^+ + H^+ + 2e^- = N_2H_5^+ + 2H_2O$	+1.42	+1.42
$NO_3^- + 3H^+ + 2e^- = HNO_2 + H_2O$	+0.94	+0.94
$NO_3^- + 4H^+ + 4e^- = NO + 2H_2O$	+0.96	+0.96
$2NO + 2H^+ + 2e^- = H_2N_2O_2$	+0.71	+0.712
$2NO_3^- + 4H^+ + 2e^- = N_2O_{4(g)} + 2H_2O$	+0.80	+0.803
$HNO_2 + H^+ + e^- = NO + H_2O$	+1.00	+1.00
$2HNO_2 + 4H^+ + 4e^- = H_2N_2O_2 + 2H_2O$	+0.86	+0.86

AQUEOUS ACID SOLUTIONS

Electrode Reaction	E°_{red} (V) L[a]	B and L[b]
$2HNO_{2(aq)} + 4H^+ + 4e^- = N_2O_{(g)} + 3H_2O$	+1.29	+1.29
$N_2O_4 + 2H^+ + 2e^- = 2HNO_2$	+1.07	+1.07
$N_2O_4 + 4H^+ + 4e^- = 2NO + 2H_2O$	+1.03	+1.03
$H_2N_2O_2 + 2H^+ + 2e^- = N_2 + 2H_2O$	+2.85	+2.65
$H_2N_2O_2 + 6H^+ + 4e^- = 2NH_3OH^+$	+0.496	+0.387
$Na^+ + e^- = Na$	−2.714	−2.714
$Nb^{+3} + 3e^- = Nb$	−1.1	−1.099
$Nb_2O_4 + 10H^+ + 10e^- = 2Nb + 5H_2O$	−0.65	−0.644
$Nd^{+3} + 3e^- = Nd$	−2.44	−2.431
$Ni^{+2} + 2e^- = Ni$	−0.250	−0.250
$NiO_2 + 4H^+ + 2e^- = Ni^{+2} + 2H_2O$	+1.68	+1.678
$Np^{+3} + 3e^- = Np$	−1.86	−1.856
$Np^{+4} + e^- = Np^{+3}$	+0.147	+0.147
$NpO_2^{+2} + e^- = NpO_2^+$	+1.15	+1.15
$NpO_2^+ + 4H^+ + e^- = Np^{+4} + 2H_2O$	+0.75	+0.75
$O_{(g)} + 2H^+ + 2e^- = H_2O$	+2.42	+2.422
$O_2 + H^+ + e^- = HO_2$	−0.13	−0.13
$O_2 + 2H^+ + 2e^- = H_2O_2$	+0.682	+0.6824
$O_2 + 4H^+ + 4e^- = 2H_2O_{(l)}$	+1.229	+1.229
$O_3 + 2H^+ + 2e^- = O_2 + H_2O$	+2.07	+2.07
$HO_2 + H^+ + e^- = H_2O_2$	+1.5	+1.495
$H_2O_2 + H^+ + e^- = OH + H_2O$	+0.72	+0.71
$H_2O_2 + 2H^+ + 2e^- = 2H_2O$	+1.77	+1.776
$OH + H^+ + e^- = H_2O$	+2.8	+2.85
$OsO_{4(c)} + 8H^+ + 8e^- = Os + 4H_2O$	+0.85	+0.85
$P_{(white)} + 3H^+ + 3e^- = PH_{3(g)}$	+0.06	+0.063
$H_3PO_2 + H^+ + e^- = P_{(white)}$	−0.51	−0.508
$H_3PO_3 + 2H^+ + 2e^- = H_3PO_2 + H_2O$	−0.50	−0.499
$H_3PO_4 + 2H^+ + 2e^- = H_3PO_3$	−0.276	−0.276
$Pb^{+2} + 2e^- = Pb$	−0.126	−0.126
$PbBr_2 + 2e^- = Pb + 2Br^-$	−0.280	−0.284
$PbCl_2 + 2e^- = Pb + 2Cl^-$	−0.268	−0.268
$PbI_2 + 2e^- = Pb + 2I^-$	−0.365	−0.365
$PbO_2 + 4H^+ + 2e^- = Pb^{+2} + 2H_2O$	+1.455	+1.455
$PbO_2 + SO_4^{-2} + 4H^+ + 2e^- = PbSO_4 + 2H_2O$	+1.685	+1.682

AQUEOUS ACID SOLUTIONS

Electrode Reaction	E°_{red} (V) L[a]	B and L[b]
$PbSO_4 + 2e^- = Pb + SO_4^{-2}$	−0.356	−0.3588
$Pd^{+2} + 2e^- = Pd$	+0.987	+0.987
$PdBr_4^{-2} + 2e^- = Pd + 4Br^-$	+0.6	+0.60
$PdCl_4^{-2} + 2e^- = Pd + 4Cl^-$	+0.62	+0.62
$PdCl_6^{-2} + 2e^- = PdCl_4^{-2} + 2Cl^-$	+1.288	+1.288
$PtBr_4^{-2} + 2e^- = Pt + 4Br^-$	+0.58	+0.581
$PtCl_4^{-2} + 2e^- = Pt + 4Cl^-$	+0.73	+0.73
$PtCl_6^{-2} + 2e^- = PtCl_4^{-2} + 2Cl^-$	+0.68	+0.68
$Pt(OH)_2 + 2H^+ + 2e^- = Pt + 2H_2O$	+0.98	+0.987
$PtS + 2H^+ + 2e^- = Pt + H_2S$	−0.30	−0.297
$Pu^{+3} + 3e^- = Pu$	−2.07	−2.031
$Pu^{+4} + e^- = Pu^{+3}$	+0.97	+0.97
$PuO_2^+ + 4H^+ + e^- = Pu^{+4} + 2H_2O$	+1.15	+1.15
$PuO_2^+ + e^- = PuO_2^+$	+0.93	+0.93
$PuO_2^{+2} + 4H^+ + 2e^- = Pu^{+4} + 2H_2O$	+1.04	+1.04
$Ra^{+2} + 2e^- = Ra$	−2.92	−2.916
$Rb^+ + e^- = Rb$	−2.925	−2.925
$ReO_2 + 4H^+ + 4e^- = Re + 2H_2O$	+0.252	+0.2513
$ReO_4^- + 4H^+ + 3e^- = ReO_2 + 2H_2O$	+0.51	+0.510
$ReO_4^- + 8H^+ + 7e^- = Re + 4H_2O$	+0.363	+0.362
$Rh^{+3} + 3e^- = Rh$	+0.8	+0.80
$RhCl_6^{-3} + 3e^- = Rh + 6Cl^-$	+0.44	+0.431
$RuCl_5^{-2} + 3e^- = Ru + 4Cl^-$	+0.60	+0.601
$S_{(rhombic)} + 2e^- = H_2S_{(aq)}$	+0.141	+0.142
$H_2SO_3 + 4H^+ + 4e^- = S + 3H_2O$	+0.45	+0.450
$2H_2SO_3 + H^+ + 2e^- = HS_2O_4^- + 2H_2O$	−0.08	−0.082
$2H_2SO_3 + 2H^+ + 4e^- = S_2O_3^{-2} + 3H_2O$	+0.40	+0.400
$4H_2SO_3 + 4H^+ + 6e^- = S_4O_6^{-2} + 6H_2O$	+0.51	+0.51
$S_2Cl_2 + 2e^- = 2S + 2Cl^-$	+1.23	+1.23
$SO_4^{-2} + 4H^+ + 2e^- = H_2SO_3 + H_2O$	+0.17	+0.172
$2SO_4^{-2} + 4H^+ + 2e^- = S_2O_6^{-2} + 2H_2O$	−0.22	−0.22
$S_2O_8^{-2} + 2e^- = 2SO_4^{-2}$	+2.01	+2.01
$Sb + 3H^+ + 3e^- = SbH_{3(g)}$	−0.5	−0.510
$Sb_2O_3 + 6H^+ + 6e^- = 2Sb + 3H_2O$	+0.152	+0.152
$Ab_2O_3 + 2H^+ + 2e^- = Sb_2O_4 + H_2O$	+0.48	+0.479

AQUEOUS ACID SOLUTIONS

Electrode Reaction	E°_{red} (V) L[a]	B and L[b]
$Sb_2O_3 + 6H^+ + 4e^- = 2SbO^+ + 3H_2O$	+0.581	+0.581
$Sc^{+3} + 3e^- = Sc$	−2.08	−2.077
$Se + 2H^+ + 2e^- = H_2Se_{(aq)}$	−0.40	−0.399
$H_2SeO_3 + 4H^+ + 4e^- = Se_{(grey)} + 3H_2O$	+0.74	+0.740
$SeO_4^{-2} + 4H^+ + 2e^- = H_2SeO_3 : H_2O$	+1.15	+1.15
$S_2O_6^{-2} + 4H^+ + 2e^- = 2H_2SO_3$	+0.57	+0.57
$Si + 4H^+ + 4e^- = SiH_{4(g)}$	+0.102	+0.102
$SiF_6^{-2} + 4e^- = Si + 6F^-$	−1.2	−1.24
$SiO_{2(quartz)} + 4H^+ + 4e^- = Si + 2H_2O$	−0.86	−0.857
$Sm^{+3} + 3e^- = Sm$	−2.41	−2.414
$Sn^{+2} + 2e^- = Sn_{(white)}$	−0.15	−0.136
$Sn^{+4} + 2e^- = Sn^{+2}$	+0.15	+0.15
$SnF_6^{-2} + 4e^- = Sn + 6F^-$	−0.25	−0.25
$Sr^{+2} + 2e^- = Sr$	−2.89	−2.888
$Ta_2O_5 + 10H^+ + 10e^- = 2Ta + 5H_2O$	−0.81	−0.812
$TeO_{2(c)} + 4H^+ + 4e^- = Te + 2H_2O$	+0.529	+0.529
$H_6Te_{6(c)} + 2H^+ + 2e^- = TeO_2 + 4H_2O$	+1.02	+1.02
$TeOOH^+ + 3H^+ + 4e^- = Te + 2H_2O$	+0.559	+0.559
$Te + 2H^+ + 2e^- = H_2Te_{(g)}$	−0.72	−0.718
$Th^{+4} + 4e^- = Th$	−1.90	−1.899
$Ti^{+2} + 2e^- = Ti$	−1.63	−1.628
$Ti^{+3} + e^- = Ti^{+2}$	−0.37	−0.369
$TiF_6^{-2} + 4e^- = Ti + 6F^-$	−1.19	−1.191
$TiO^{+2} + 2H^+ + e^- = Ti^{+3} + H_2O$	−0.1	+0.099
$TiO^{+2} + 2H^+ + 4e^- = Ti + H_2O$	−0.89	−0.882
$Tl^+ + e^- = Tl$	−0.3363	−0.3363
$Tl^{+3} + 2e^- = Tl^+$	+1.25	+1.25
$TlBr + e^- = Tl + Br^-$	−0.658	−0.658
$TlCl + e^- = Tl + Cl^-$	−0.557	−0.5568
$TlI + e^- = Tl + I^-$	−0.753	−0.752
$U^{+3} + 3e^- = U$	−1.80	−1.789
$U^{+4} + e^- = U^{+3}$	−0.61	−0.609
$UO_2^{+2} + e^- = UO_2^+$	+0.05	+0.05
$UO_2^{+2} + 4H^+ + 2e^- = U^{+4} + 2H_2O$	+0.334	+0.330
$UO_2^+ + 4H^+ + e^- = U^{+4} + 2H_2O$	+0.62	+0.62

AQUEOUS ACID SOLUTIONS

Electrode Reaction	E°_{red} (V) L[a]	E°_{red} (V) B and L[b]
$V^{+2} + 2e^- = V$	−1.18	−1.186
$V^{+3} + e^- = V^{+2}$	−0.255	−0.256
$VO^{+2} + 2H^+ + e^- = V^{+3} + H_2O$	+0.361	+0.359
$V(OH)_4^+ + 2H^+ + e^- = VO^{+2} + 3H_2O$	+1.00	+1.00
$V(OH)_4^+ + 4H^+ + 5e^- = V + 4H_2O$	−0.253	−0.254
$WO_{3(c)} + 6H^+ + 6e^- = W + 3H_2O$	−0.09	−0.090
$Y^{+3} + 3e^- = Y$	−2.37	−2.372
$Zn^{+2} + 2e^- = Zn$	−0.763	−0.7628
$Zr^{+4} + 4e^- = Zr$	−1.53	−1.529

AQUEOUS BASIC SOLUTIONS

Electrode Reaction	E°_{red} (V) L[a]	E°_{red} (V) B and L[b]
$AgCN + e^- = Ag + CN^-$	−0.017	−0.017
$Ag(CN)_2^- + e^- = Ag + 2CN^-$	−0.31	−0.31
$Ag_2CO_3 + 2e^- = 2Ag + CO_3^{-2}$	+0.47	+0.47
$Ag(NH_3)_2^+ + e^- = Ag + 2NH_3$	+0.373	+0.373
$Ag_2O + H_2O + 2e^- = 2Ag + 2OH^-$	+0.344	+0.345
$Ag_2O_3 + H_2O + 2e^- = 2AgO + 2OH^-$	+0.74	+0.739
$Ag_2S_{(\ )} + 2e^- = 2Ag + S^{-2}$	−0.69	−0.66
$Ag(SO_3)_2^{-3} + e^- = Ag + 2SO_3^{-2}$	+0.30	+0.295
$H_2AlO_3^- + H_2O + 3e^- = Al + 4OH^-$	−2.35	−2.33
$AsO_2^- + 2H_2O + 3e^- = As + 4OH^-$	−0.68	−0.675
$AsO_4^{-3} + 2H_2O + 2e^- = AsO_2^- + 4OH^-$	−0.67	−0.68
$H_2BO_3^- + H_2O + 3e^- = B + 4OH^-$	−1.79	−1.79
$Ba(OH)_2 \cdot 8H_2O + 2e^- = Ba + 8H_2O + 2OH^-$	−2.97	−2.99
$Be_2O_3^{-2} + 3H_2O + 4e^- = 2Be + 6OH^-$	−2.62	−2.63
$Be_2O_3 + 3H_2O + 6e^- = 2Bi + 6OH^-$	−0.44	−0.46
$BrO^- + H_2O + 2e^- = Br^- + 2OH^-$	+0.76	+0.761
$BrO_3^- + 3H_2O + 6e^- = Br^- + 6OH^-$	+0.61	+0.607
$Ca(OH)_2 + 2e^- = Ca + 2OH^-$	−3.03	−3.02

Electrode Reaction	E°_{red}(V)	
	L[a]	B and L[b]
$Cd(CN)_4^{-2} + 2e^- = Cd + 4CN^-$	−1.03	−1.028
$CdCO_3 + 2e^- = Cd + CO_3^{-2}$	−0.74	−0.74
$Cd(NH_3)_4^{+2} + 2e^- = Cd + 4NH_3$	−0.597	−0.613
$Cd(OH)_2 + 2e^- = Cd + 2OH^-$	−0.809	−0.809
$CdS + 2e^- = Cd + S$	−1.21	−1.175
$ClO^- + H_2O + 2e^- = Cl^- + 2OH^-$	+0.89	+0.89
$ClO_{2(g)} + e^- = ClO_2^-$	+1.16	+1.16
$ClO_2 + H_2O + 2e^- = ClO^- + 2OH^-$	+0.66	+0.66
$ClO_3^- + H_2O + 2e^- = ClO_2^- + 2OH^-$	+0.33	+0.33
$ClO_4^- + H_2O + 2e^- = ClO_3^- + 2OH^-$	+0.36	+0.36
$CNO^- + H_2O + 2e^- = CN^- + 2OH^-$	−0.97	−0.970
$CoCO_3 + 2e^- = Co + CO_3^{-2}$	−0.64	−0.64
$Co(OH)_2 + 2e^- = Co + 2OH^-$	−0.73	−0.73
$Co(OH)_3 + e^- = Co(OH)_2 + OH^-$	+0.17	+0.17
$Co(NH_3)_6^{+3} + e^- = Co(NH_3)_6^{+2}$	+0.1	+0.108
$CrO_2^- + 2H_2O + 3e^- = Cr + 4OH^-$	−1.2	−1.27
$CrO_4^{-2} + 4H_2O + 3e^- = Cr(OH)_3 + 5OH^-$	−0.13	−0.13
$Cr(OH)_3 + 3e^- = Cr + 3OH^-$	−1.3	−1.34
$Cu(CN)_2^- + e^- = Cu + 2CN^-$	−0.43	−0.429
$Cu(CNS) + e^- = Cu + CNS^-$	−0.27	−0.27
$Cu(NH_3)_2^+ + e^- = Cu + 2NH_3$	−0.12	−0.12
$Cu_2O + H_2O + 2e^- = 2Cu + 2OH^-$	−0.358	−0.358
$2Cu(OH_2 + 2e^- = Cu_2O + 2OH^- + H_2O$	−0.080	−0.080
$Cu_2S + 2e^- = 2Cu + S^{-2}$	−0.54	−0.89
$FeCO_3 + 2e^- = Fe + CO_3^{-2}$	−0.756	−0.756
$FeO_4^{-2} + 2H_2O + 3e^- = FeO_2^- + 4OH^-$	−0.9	—
$FeO_4^{-2} + 4H_2O + 3e^- = Fe(OH)_3 + 5OH^-$	—	+0.72
$Fe(OH)_2 + 2e^- = Fe + 2OH^-$	−0.877	−0.877
$Fe(OH)_3 + e^- = Fe(OH)_2 + OH^-$	−0.56	−0.56
$FeS_{(\ )} + 2e^- = Fe + S^{-2}$	−1.01	−0.95
$Fe_2S_3 + 2e^- = 2FeS_{(\ )} + S^{-2}$	−0.67	−0.715
$H_2GaO_3^- + H_2O + 3e^- = Ga + 4OH^-$	−1.22	−1.219
$HGeO_3^- + 2H_2O + 4e^- = Ge + 5OH^-$	−0.9	−1.03
$HfO(OH)_2 + H_2O + 4e^- = Hf + 4OH^-$	−2.50	−2.50
$Hg(CN)_4^{-2} + 2e^- = Hg + 4CN^-$	−0.37	−0.37
$HgO_{(r)} + H_2O + 2e^- = Hg + 2OH^-$	+0.098	+0.098
$HgS_{(black)} + 2e^- = Hg + S^{-2}$	−0.72	−0.69

Electrode Reaction	E°_{red} (V) L[a]	B and L[b]
$H_2O + e^- = H_{(g)} + OH^-$	−2.93	−2.9345
$2H_2O + 2e^- = H_2 + 2OH^-$	−0.828	−0.82806
$In(OH)_3 + 3e^- = In + 3OH^-$	−1.0	−1.00
$IO^- + H_2O + 2e^- = I^- + 2OH^-$	+0.49	+0.485
$IO_3^- + 3H_2O + 6e^- = I^- + 6OH^-$	+0.26	+0.26
$H_3IO_6^{-2} + 2e^- = IO_3^- + 3OH^-$	+0.7	+0.7
$Ir_2O_3 + 3H_2O + 6e^- = 2Ir + 6OH^-$	+0.1	+0.098
$La(OH)_3 + 3e^- = La + 3OH^-$	−2.90	−2.90
$Lu(OH)_3 + 3e^- = Lu + 3OH^-$	−2.72	−2.72
$Mg(OH)_2 + 2e^- = Mg + 2OH^-$	−2.69	−2.690
$MnCO_{3(ppt)} + 2e^- = Mn + CO_3^{-2}$	−1.48	−1.48
$MnO_2 + 2H_2O + 2e^- = Mn(OH)_2 + 2OH^-$	−0.05	−0.05
$MnO_4^{-2} + 2H_2O + 2e^- = MnO_2 + 4OH^-$	+0.60	+0.60
$Mn(OH)_3 + e^- = Mn(OH)_2 + OH^-$	+0.1	+0.15
$Mn(OH)_2 + 2e^- = Mn + 2OH^-$	−1.55	−1.55
$MoO_4^{-2} + 4H_2O + 6e^- = Mo + 8OH^-$	−1.05	−1.05
$N_2H_4 + 4H_2O + 2e^- = 2NH_4OH + 2OH^-$	+0.1	+0.11
$Na_2UO_4 + 4H_2O + 2e^- = U(OH)_4 + 2Na^+ + 4OH^-$	−1.61	−1.618
$2NH_2OH + 2e^- = N_2H_4 + 2OH^-$	+0.73	+0.73
$NiCO_3 + 2e^- = Ni + CO_3^{-2}$	−0.45	−0.45
$Ni(NH_3)_6^{+2} + 2e^- = Ni + 6NH_{3(aq)}$	−0.47	−0.476
$NiO_2 + 2H_2O + 2e^- = Ni(OH)_2 + 2OH^-$	+0.49	+0.490
$Ni(OH)_2 + 2e^- = Ni + 2OH^-$	−0.72	−0.72
$NiS_{(\ )} + 2e^- = Ni + S^{-2}$	−0.83	−0.830
$NO_3^- + H_2O + 2e^- = NO_2^- + 2OH^-$	+0.01	+0.01
$O_2 + e^- = O_2^-$	−0.56	−0.563
$HO_2^- + H_2O + e^- = OH_{(aq)} + 2OH^-$	−0.24	−0.245
$HO_2^- + H_2O + 2e^- = 3OH^-$	+0.88	+0.878
$O_2^- + H_2O + e^- = OH^- + HO_2^-$	+0.4	+0.413
$O_2 + 2H_2O + 4e^- = 4OH^-$	+0.401	+0.401
$O_2 + H_2O + 2e^- = HO_2^- + OH^-$	−0.076	−0.076[c]
$O_{3(g)} + H_2 + 2e^- = O_2 + 2OH^-$	+1.24	+1.24
$OH_{(g)} + e^- = OH^-$	+2.0	+2.02
$HOsO_5^- + 4H_2O + 8e^- = Os + 9OH^-$	+0.02	+0.015

[c] A value of −0.048 V has been found at 25°C by Yeager et al.(129) and −0.045 V by Bagotsky et al. [*Dokl. Akad. Nauk SSSR*, **71**, 501 (1950); *ibid.*, **85**, 599 (1952)].

Electrode Reaction	E°_{red} (V) L[a]	B and L[b]
$P_{(white)} + 3H_2O + 3e^- = PH_3 + 3OH^-$	−0.89	−0.89
$PbCO_3 + 2e^- = Pb + CO_3^{-2}$	−0.506	−0.509
$HPbO_2^- + H_2O + 2e^- = Pb + 3OH^-$	−0.54	−0.540
$PbO_2 + H_2O + 2e^- = PbO_{(r)} + 2OH^-$	+0.248	+0.247
$PbS + 2e^- = Pb + S^{-2}$	−0.95	−0.93
$Pd(OH)_2 + 2e^- = Pd + 2OH^-$	+0.07	+0.07
$H_2PO_2^- + e^- = P + 2OH^-$	−2.05	−2.05
$HPO_3^{-2} + 2H_2O + 2e^- = H_2PO_2^- + 3OH^-$	−1.57	−1.565
$Pt(OH)_2 + 2e^- = Pt + 2OH^-$	+0.15	+0.15
$PuO_2(OH)_2 + e^- = PuO_2OH + OH^-$	+0.26	−0.234
$PuO_4^{-3} + 2H_2O + 2e^- = HPO_3^{-2} + 3OH^-$	−1.12	−1.12
$Pu(OH)_3 + 3e^- = Pu + 3OH^-$	−2.42	−2.42
$Pu(OH)_4 + e^- = Pu(OH)_3 + OH^-$	−0.95	−0.963
$ReO_2 + H_2O + 4e^- = Re + 4OH^-$	−0.576	−0.577
$ReO_4^- + 2H_2O + 3e^- = ReO_2 + 4OH^-$	−0.594	−0.594
$ReO_4^- + 4H_2O + 7e^- = Re + 8OH^-$	−0.584	−0.584
$Rb_2O_3 + 3H_2O + 6e^- = 2Rh + 6OH^-$	+0.04	+0.04
$RuO_4^- + e^- = RuO_4^{-2}$	+0.60	+0.6
$S + 2e^- = S^{-2}$	−0.48	−0.447
$2SO_3^{-2} + 3H_2O + 4e^- = S_2O_3^{-2}$	−0.58	−0.571
$SO_4^{-2} + H_2O + 2e^- = SO_3^{-2} + 2OH^-$	−0.93	−0.93
$S_4O_6^{-2} + 2e^- = 2S_2O_3^{-2}$	+0.08	+0.08
$2SO_3^{-2} + 2H_2O + 2e^- = S_2O_4^{-2} + 4OH^-$	−1.12	−1.12
$SbO_2^- + 2H_2O + 3e^- = Sb + 4OH^-$	−0.66	−0.66
$Sc(OH)_3 + 3e^- = Sc + 3OH^-$	−2.6	−2.61
$Se + 2e^- = Se^{-2}$	−0.92	−0.92
$SeO_3^{-2} + 3H_2O + 4e^- = Se + 6OH^-$	−0.366	−0.366
$SeO_4^{-2} + H_2O + 2e^- = SeO_3 + 2OH^-$	+0.05	+0.05
$SiO_3^{-2} + 3H_2O + 4e^- = Si + 6OH^-$	−1.70	−1.697
$Sn(OH)_6^{-2} + 2e^- = HSnO_2^- + H_2O + 3OH^-$	−0.90	−0.93
$SnS + 2e^- = Sn + S^{-2}$	−0.94	−0.87
$HSnO_2^- + H_2O + 2e^- = Sn + 3OH^-$	−0.91	−0.909
$Sr(OH)_2 + 2e^- = Sr + 2OH^-$	—	−2.88
$Sr(OH)_2 \cdot 8H_2O + 2e^- = Sr + 2OH^- + 8H_2O$	−2.99	—
$Te + 2e^- = Te^{-2}$	−1.14	−1.143
$TeO_3^{-2} + 3H_2O + 4e^- = Te + 6OH^-$	−0.57	−0.57
$TeO_4^{-2} + H_2O + 2e^- = TeO_3^{-2} + 2OH^-$	+0.4	+0.4

Electrode Reaction	E°_{red} (V)	
	L[a]	B and L[b]
$Th(OH)_4 + 4e^- = Th + 4OH^-$	−2.48	−2.48
$Tl(OH) + e^- = Tl + OH^-$	−0.3445	−0.343
$Tl(OH)_3 + 2e^- = TlOH + 2OH^-$	−0.05	−0.05
$Tl_2S + 2e^- = 2Tl + S^{-2}$	−0.96	−0.90
$UO_2 + 2H_2O + 4e^- = U + 4OH^-$	−2.39	−2.39
$U(OH)_3 + 3e^- = U + 3OH^-$	−2.17	−2.17
$U(OH)_4 + e^- = U(OH)_3 + OH^-$	−2.2	−2.20
$HV_6O_{17}^{-3} + 16H_2O + 30e^- = 6V + 33OH^-$	−1.15	−1.154
$WO_4^{-2} + 4H_2O + 6e^- = W + 8OH^-$	−1.05	−1.05
$ZnCO_3 + 2e^- = Zn + CO_3^{-2}$	−1.06	−1.06
$Zn(CN)_4^{-2} + 2e^- = Zn + 4CN^-$	−1.26	−1.26
$Zn(NH_3)_4^{+2} + 2e^- = Zn + 4NH_{3(aq)}$	−1.03	−1.04
$ZnO_2^{-2} + 2H_2O + 2e^- = Zn + 4OH^-$	−1.216	−1.215
$Zn(OH)_2 + 2e^- = Zn + 2OH^-$	−1.245	−1.245
$ZnS_{(wurtzite)} + 2e^- = Zn + S^{-2}$	−1.44	−1.405
$H_2ZrO_3 + H_2O + 4e^- = Zr + 4OH^-$	−2.36	−2.36

APPENDIX C

THERMODYNAMIC DATA FOR VARIOUS REACTIONS AT 25°C[a]

System	Reaction	ΔH° (kcal)	ΔS° (cal/deg)	ΔG° (kcal)	Thermo-dynamics E°(V)	Open-circuit[b] E(V)
Magnesium primary battery	$Mg_{(s)} + 2MnO_{2(s)} + H_2O_{(l)} \rightarrow Mn_2O_{3(s)} + Mg(OH)_{2(s)}$	−136.4	−12.7	−132.7	2.88	2.0
Alkaline MnO_2 cell	$Zn_{(s)} + 2MnO_{2(s)} \rightarrow ZnO_{(s)} + Mn_2O_{3(s)}$	−66.9	−2.8	−66.2	1.44	1.52
	$Zn_{(s)} + 2MnO_{2(s)} - H_2Ol \rightarrow Mn_2O_{3(s)} + Zn(OH)_{2(s)}$	−68.9	−10.1	−66.0	1.43	–
Alkaline mercuric oxide	$Zn_{(s)} + HgO_{(s)}^{(red)} \rightarrow ZnO_{(s)} + Hg_{(l)}$	−61.5	+1.8	−62.1	1.35	1.35
	$Zn_{(s)} + HgO_{(s)}^{(red)} + H_2O_{(l)} \rightarrow Zn(OH)_{2(s)} + Hg_{(l)}$	−63.6	−5.5	−61.9	1.34	–
	$3Zn_{(s)} + HgSO_{4(s)} + 2HgO_{(s)}^{(red)} \rightarrow ZnSO_{4(s)} + 2ZnO_{(s)} + 3Hg_{(l)}$	−188.5	+9.5	−191.4	1.38	–
Silver chloride-magnesium reserve cell	$Mg_{(s)} + 2AgCl_{(s)} \rightarrow 2Ag_{(s)} + MgCl_{2(s)}$	−92.68	−11.9	−89.13	1.93	–
	$Mg_{(s)} + 2AgCl_{(s)} \rightarrow 2Ag_{(s)} + MgCl_{2(aq)}$	−129.74	−35.1	−119.25	2.59	1.6

[a]Thermodynamic data from W. Latimer, *The Oxidation States of the Elements and Their Potential in Aqueous Solutions*, 2nd ed., Prentice-Hall, Englewood Cliffs, N. J., 1952, unless otherwise noted. Open-circuit potentials from various sources.

[b]Open circuit voltages are listed opposite the most likely cell reaction for the early part of the discharge curve when more than one reaction is included in the table for a given battery system; these observed values are shown only for those systems for which the data are available.

System	Reaction	$\Delta H°$ (kcal)	$\Delta S°$ (cal/deg)	$\Delta G°$ (kcal)	Thermodynamics $E°$(V)	Open-circuit[b] E(V)
Cuprous chloride-magnesium reserve cell	$Mg_{(s)} + 2CuCl_{(s)} \rightarrow 2Cu_{(s)} + MgCl_{2(s)}$	−88.4	−10.8	−85.2	1.85	–
	$Mg_{(s)} + 2CuCl_{(s)} \rightarrow 2Cu_{(s)} + MgCl_{2(aq)}$	−125.5	−34.0	−115.3	2.50	1.5
Lead dioxide-cadmium reserve cell	$Cd_{(s)} + PbO_{2(s)} + 2H_2SO_{4(aq)} \rightarrow PbSO_{4(s)} + CdSO_{4(s)} + 2H_2O_{(l)}$	−77.58	+62.6	−96.3	2.09	2.1
Lead dioxide-zinc reserve cell	$Zn_{(s)} + PbO_{2(s)} + 2H_2SO_{4(aq)} \rightarrow PbSO_{4(s)} + ZnSO_{4(s)} + 2H_2O_{(l)}$	−90.1	+61.9	−108.6	2.36	–
	$Zn_{(s)} + PbO_{2(s)} + 2H_2SO_{4(aq)} \rightarrow PbSO_{4(s)} + ZnSO_{4(aq)} + 2H_2O_{(l)}$	−109.6	+10.8	−112.8	2.45	2.5
Chlorine-zinc reserve cell	$Cl_{2(g)} + ZnCl_{2(s)} \rightarrow ZnCl_{2(s)}$	−99.40	−37.3	−88.26	1.92	–
	$Cl_{2(g)} + ZnCl_{2(aq)} \rightarrow ZnCl_{2(aq)}$	−116.47	−62.3	−97.884	2.12	2.1
Zinc-iodate primary cell	$5Zn_{(s)} + 2KIO_{3(s)} + 6H_2SO_{4(aq)} \rightarrow 5ZnSO_{4(s)} + I_{2(s)} + K_2SO_{4(s)} + 6H_2O_{(l)}$	−377.6	+172.4	−428.9	1.86	–
	$3Zn_{(s)} + KIO_{3(s)} + 3H_2SO_{4(aq)}$ $3ZnSO_{4(s)} + KI_{(s)} + 3H_2O_{(l)}$	−212.7	+86.2	−238.4	1.72	–

Cupric bromide-silver cell	$CuBr_{2(s)} + Ag_{(s)} \rightarrow \frac{1}{2}Cu_2Br_{2(s)} + AgBr_{(s)}$	−15.1	+14.7	−16.4	0.71	0.74
Lead acid	$Pb_{(s)} + PbO_{2(s)} + 2H_2SO_{4(aq)} \rightarrow 2PbSO_{4(s)} + 2H_2O_{(l)}$	−75.7	+61.8	−94.1	2.04	2.1
Ionic hydride cell	$H_{2(g)} + Ca_{(s)} \rightarrow CaH_{2(s)}$	−45.1	−31.2	−35.8	0.78	–
Silver-chlorine cell	$Ag_{(s)} + \frac{1}{2}Cl_{2(g)} \rightarrow AgCl_{(s)}$	−30.4	−13.9	−26.2	1.14	–
Daniell cell	$Zn_{(s)} + CuSO_{4(s)} \rightarrow ZnSO_{4(s)} + Cu_{(s)}$	−49.9	+0.7	−50.1	1.08	1.08
Hydrogen-oxygen	$H_{2(g)} + \frac{1}{2}O_{2(g)} \rightarrow H_2O_{(l)}$	−68.3	−39.0	−56.7	1.23	0.95-1.23
	$H_{2(g)} + \frac{1}{2}O_{2(g)} \rightarrow H_2O_{(g)}$	−57.8	−10.6	−54.6	1.18	
Zinc iron cell	$Zn_{(s)} + Fe_2O_{3(s)} \rightarrow 2FeO_{(s)} + ZnO_{(s)}$	−14.1	+4.8	−15.7	0.34	–
	$Zn_{(s)} + Fe_2O_{3(s)} + 3H_2O_{(l)} \rightarrow 2Fe(OH)_{2(s)} + Zn(OH)_{2(s)}$	−23.7	−23.7	−16.6	0.36	–
Nickel-cadmium	$Cd_{(s)} + 2Ni(OH)_{3(s)} \rightarrow CdO_{(s)} + 2Ni(OH)_{2(s)} + H_2O_{(l)}$	−62.2	+16.5	−68.1	1.48	1.35
	$Cd_{(s)} + NiO_{2(s)} + 2H_2O_{(l)} \rightarrow Cd(OH)_{2(s)} + Ni(OH)_{2(s)}$	–	–	−59.9	1.30	–
	$Cd_{(s)} + NiO_{2(s)} \rightarrow CdO_{(s)} + NiO_{(s)}$	–	–	−58.0	1.26	–

System	Reaction	$\Delta H°$ (kcal)	$\Delta S°$ (cal/deg)	$\Delta G°$ (kcal)	Thermo-dynamics $E°$(V)	Open-circuit[b] E(V)
Edison cell	$Fe_{(s)} + NiO_{2(s)} + 2H_2O_{(l)} \rightarrow$	–	–	−63.0	1.37	–
	$Fe(OH)_{2(s)} + Ni(OH)_{2(s)}$	–	–	−62.6	1.36	–
	$Fe_{(s)} + NiO_{2(s)} \rightarrow FeO + NiO$					
	$Fe_{(s)} + Ni_2O_{3(s)} + 3H_2O_{(l)} \rightarrow 2Ni(OH)_{2(s)} + Fe(OH)_{2(s)}$	–	–	−62.9	1.37	–
	$Fe_{(s)} + 2Ni(OH)_{3(s)} \rightarrow FeO_{(s)} + 2Ni(OH)_{2(s)} + H_2O_{(l)}$	−65.0	+22.1	−72.7	1.58	1.6
Silver peroxide zinc alkaline cell	$2Zn_{(s)} + 2Ag_2O_{(s)} + 2H_2O_{(l)} \rightarrow 4Ag_{(s)} + 2Zn(OH)_{2(s)}$					
	$Zn_{(s)} + AgO_{(s)} + H_2O_{(l)} \rightarrow Ag_{(s)} + Zn(OH)_{2(s)}$	−155.8	−31.0	−146.6	1.59	–
	$Zn_{(s)} + AgO_{(s)} \rightarrow ZnO_{(s)} + Ag_{(s)}$	−79.2	–	−73.3	1.59	–
	$Zn_{(s)} + Ag_2O_{(s)} \rightarrow 2Ag_{(s)} + ZnO_{(s)}$	−77.2	–	−73.4	1.59	–
	$Zn_{(s)} + 2AgO_{(s)} + H_2O_{(l)} \rightarrow Ag_2O_{(s)} + Zn(OH)_{2(s)}$	−75.86	−8.1	−73.46	1.59	–
	$Zn_{(s)} + 2AgO_{(s)} \rightarrow Ag_2O_{(s)} + ZnO_{(s)}$	−80.5	–	−83.7	1.82	
		−78.48	–	−83.8	1.82	1.8
Zinc-cupric oxide alkaline cell	$Zn_{(s)} + H_2O_{(l)} + Cu_2O_{(s)} \rightarrow 2Cu_{(s)} + Zn(OH)_{2(s)}$	−45.4	−14.9	−40.9	0.89	–
	$Zn_{(s)} + Cu_2O_{(s)} \rightarrow 2Cu_{(s)} + ZnO_{(s)}$	−43.33	−7.6	−41.07	0.89	–
	$Zn_{(s)} + CuO_{(s)} + H_2O_{(l)} \rightarrow Cu_{(s)} + Zn(OH)_{2(s)}$	−48.1	−9.2	−45.5	0.99	–
	$Zn_{(s)} + CuO_{(s)} \rightarrow Cu_{(s)} + ZnO_{(s)}$	−46.1	−1.9	−45.6	0.99	–
	$Zn_{(s)} + 2CuO_{(s)} \rightarrow Cu_2O_{(s)} + ZnO_{(s)}$	−48.9	+3.8	−50.13	1.09	1.06
Air-depolarized zinc cell	$2Zn_{(s)} + O_{2(g)} + 2H_2O_{(l)} \rightarrow 2Zn(OH)_{2(s)}$	−170.4	−62.5	−151.8	1.64	1.36
	$2Zn_{(s)} + O_{2(g)} \rightarrow 2ZnO_{(s)}$	−166.3	−47.9	−152.1	1.65	–

Leclanché	$Zn_{(s)} + 2MnO_{2(s)} + 2H^+_{(aq)} \rightarrow Mn_2O_{3(s)} + Zn^{+2} + H_2O_{(l)}$	–	–	-81.8^c	1.777^c	1.55
	$Zn_{(s)} + 3Mn_2O_3 + 2H^+_{(aq)} \rightarrow 2Mn_3O_{4(s)} + Zn^{+2}_{(aq)} + H_2O_{(l)}$	–	–	-66.9^c	1.452^c	–
	$Zn_{(s)} + Mn_3O_4 + 2H_2O_{(l)} + 2H^+_{(aq)} \rightarrow 3Mn(OH)_{2(s)} + Zn^{+2}_{(aq)}$	–	–	-56.42^c	1.225^c	–
	$Zn_{(s)} + 2MnO_{2(s)} + 2NH^+_{4(aq)} \rightarrow Mn_2O_{3(s)} + Zn(NH_3)^{+2}_{2(aq)} + H_2O_{(l)}$	data not available				
	$Zn_{(s)} + 2MnO_{2(s)} + 4NH^+_{4(aq)} \rightarrow Mn_2O_{3(s)} + H_2O_{(l)} + Zn(NH_3)^{+2}_{4(aq)} + 2H^+_{(aq)}$	–	–	-44.3	0.96	–

[c]Obtained from data listed by Pourbaix (100) for zinc and manganese systems.

APPENDIX D

STOICHIOMETRIC MEAN MOLAL ACTIVITY COEFFICIENTS[a] FOR SELECTED ELECTROLYTES AT 25°C

Molality	H_2SO_4	HCl	KOH	NaOH	NH_4Cl	$ZnCl_2$
0.0005	0.639	0.928	—	—	0.924	0.77
0.01	0.544	0.904	—	0.905	0.896	0.71
0.05	0.340	0.830	0.824	0.818	0.808	0.56
0.1	0.265	0.796	0.798	0.766	0.770	0.515
0.2	0.209	0.767	0.760	0.727	0.718	0.426
0.5	0.156	0.757	0.732	0.690	0.649	0.394
1.0	0.132	0.809	0.756	0.678	0.603	0.339
2.0	0.128	1.009	0.888	0.709	0.570	0.289
3.0	0.142	1.316	1.081	0.784	0.561	0.287
4.0	0.170	1.762	1.352	0.903	0.560	0.307
5.0	0.208	2.38	1.72	1.077	0.562	0.354
6.0	0.257	3.22	2.20	1.299	0.564	0.417
7.0	0.317	4.37	2.88	1.603	0.566	0.499
8.0	0.386	5.90	3.77	2.01	—	0.607
9.0	0.467	7.94	4.86	2.55	—	0.737
10	0.559	10.44	6.22	3.23	—	0.898
11	0.643	13.51	8.10	4.10	—	—
12	0.742	17.25	10.5	5.19	—	1.294
13	0.830	21.8	13.2	6.50	—	—
14	0.967	27.3	15.8	8.04	—	1.73
15	1.093	34.1	19.6	9.74	—	—
16	1.234	42.4	24.6	11.58	—	2.18
17	1.387	—	—	13.47	—	—
18	—	—	—	15.41	—	2.63
19	—	—	—	17.38	—	—
20	—	—	—	19.33	—	3.06
22	—	—	—	23.1	—	3.46
25	—	—	—	28.0	—	—

[a]All values from R. Parsons, *Handbook of Electrochemical Constants*, Academic Press, New York, 1959.

3

Primary Cells with Caustic Alkali Electrolyte

ERWIN A. SCHUMACHER

On the basis of the electrolyte used, a survey of commercially used primary batteries reveals a number of advantages for the aqueous sodium or potassium hydroxide systems. Among the common acids, nitric, for example, serves as cathodic reactant as well as electrolyte, but it attacks zinc and is itself reduced to noxious nitrogen oxides. Hydrochloric acid is somewhat volatile and can be used only in moderate concentrations; moreover, it reacts with cathodic reactants such as manganese dioxide or copper oxide. The usefulness of sulfuric acid is limited by the relatively low solubility of the anodic reaction product, zinc sulfate. The acids themselves are bulky and inconvenient to transport, in contrast with sodium or potassium hydroxide, which are available in dry form and easy to handle.

Salts, such as the ammonium and zinc chlorides of the Leclanché system, have fared better because of the great demand for portable dry cells, to which they are adapted. However, dry-cell construction introduces complicating factors that make it difficult to maintain uniform discharge voltage, such as concentration and pH changes at operating anodes and cathodes, and the time required to re-establish equilibrium (recuperation) by diffusion.

The caustic alkalis present a generally more favorable picture. Their solutions are stable and they can be used over a wide temperature range. Their wasteful attack on zinc is slight, especially when the latter is amalgamated, and they are compatible with conventional cathodic reactants such as the oxides of copper, manganese, mercury, and silver. Zinc solubility is high and, for caustic concentrations at which the anodic end-product is ZnO, the electrolyte becomes virtually nonvariant. The cells not only have high electrical output per unit of weight or volume but also show a more nearly horizontal voltage-discharge curve than their acid or salt counterparts. Caustic alkali electrolyte is used in the only wet cells of commercial importance as well as in the most advanced fuel cells, and its comparatively recent introduction into "dry" cells has served further to increase its importance in primary battery technology.

Though there are numerous references to alkaline systems with liquid electrolyte in the early technical literature on the primary battery, the first commercially successful cell of this type, the zinc:copper oxide couple developed by Lalande and Chaperon (1), did not appear until the 1880s. First produced in Europe, shortly thereafter modified by Edison (2) and subsequently by others in the United States (see Chap. 4), it still finds applications, e.g., in railway signal or track circuits, where high capacity, low cost, and well-maintained operating voltage are required. It superseded the Daniell cell and, indeed, was the only wet primary battery in large-scale production as late as the 1920s.

The first active competition encountered by the Lalande cell came from the long-foreshadowed development (3, 4) of special carbon cathodes (5), capable of utilizing atmospheric oxygen, which could be substituted for the copper oxide electrodes of the conventional wet cell. With the advantage of a more stable system and almost 100 percent higher operating voltage, the air-depolarized batteries (Chap. 8) have displaced the Lalande cells in many railway installations and have expanded their usefulness to applications where the copper oxide units have proved inadequate.

The alkaline zinc:silver oxide couple (Chap. 6), notable for its excellent heavy-drain characteristics and high ampere- and watt-hour capacity per unit of weight or volume, was developed (6) during World War II to meet military requirements. The system is somewhat unstable, and considerable care is required, particularly in the fabrication of dry types, to prevent migration of dissolved silver to the anode and resultant attack on the zinc. Wet cells are generally fabricated in reserve form, to be activated by addition of water or electrolyte just before use.

Since wet cells generally do not enter the broad consumer markets, the development of dry-type units, long delayed because of technical difficulties, was a further step in the commercial progress of the alkaline battery. The copper oxide system proved to be too unstable for long-time use and too low in operating voltage to gain acceptance. Small air-depolarized cells were marketed for use in hearing-aid sets (7) but with continued miniaturization of equipment these have been replaced by systems more adaptable to rapidly changing dimensional requirements.

As size reduction of electronic devices proceeded and power requirements were reduced, material costs of battery components became progressively less important, and barriers to the large-scale utilization of relatively expensive cathodic reactants, such as the oxides of mercury (Chap. 5) and silver (Chap. 6) were removed. Similarly, cadmium and indium have been considered as anodes despite their obvious cost handicap. The resulting inroad into the field once exclusively occupied by the Leclanché dry battery has been countered, in part, by the development of alkaline $Zn:MnO_2$ dry cells (Chap. 7), which, in addition to excellent heavy-drain and improved low-temperature performance, show promise also as secondary units.

ANODES

The metals that have been given most consideration as anodes for alkaline cells are discussed in this chapter. Their theoretical potentials and their ampere-minute equivalents are shown in Table 3.1. Despite attractive technical features possessed by some of these metals, the present

TABLE 3. 1. PRINCIPAL ANODE METALS FOR PRIMARY ALKALINE CELLS

Metal	Reaction	Potential (8) (E_b°)	Theoretical (amp-min/g)
Cadmium	$Cd + 2(OH)^- = Cd(OH)_2 + 2e^-$	−0.809	28.9
Indium	$In + 3(OH)^- = In(OH)_3 + 3e^-$	−1.0	42.0
Zinc	$Zn + 4(OH)^- = Zn(OH)_4 + 2e^-$	−1.216	49.2
	$Zn + 2(OH)^- = Zn(OH)_2 + 2e^-$	−1.245	49.2
Aluminum	$Al + 4(OH)^- = H_2AlO_3^- + H_2O + 3e^-$	−2.35	178.9
Sodium	$Na + 4(OH)^- = NaOH + 2e^-$	−2.71	69.9

dominant position of zinc does not appear to be subject to any major change in the immediately foreseeable future.

ZINC

In the conventional alkaline wet cells, zinc anodes are used in massive or plate form. Alkaline dry cells, on the other hand, employ amalgamated powdered zinc, either compacted or mixed with gelling materials to form semirigid electrodes. The porous structure of the latter, associated with large surface areas, is of particular advantage in cells for which high current density and maximum ampere- and watt-hour capacity per unit of weight and volume are desired. Zinc of high purity, e.g., 99.85–99.90 percent, is customary for both cell types, with the addition, in the case of the powdered metal, of approximately 0.04–0.06 percent of lead for greater resistance to corrosion.

Amalgamation is necessary for reduction of wasteful corrosion. It increases the normally high hydrogen overvoltage of zinc and provides a means for obtaining a uniform equipotential surface.[1] During discharge

[1]Unamalgamated zinc monocrystals may show as much as 20 millivolts difference between the base and side faces (9). Amalgamation should remove this local corrosion-couple effect.

the mercury concentrates mainly on the surface, thereby increasing protection effectiveness.

Mercury may be added by direct application to the electrode surface, by deposition from mercury salt solutions, or by incorporation with molten zinc. For granular dry-cell anodes, the mercury is simply mixed with metal powder. The cast zinc plates of conventional alkaline wet cells contain 0.5–2.0 percent mercury as compared with the 8–15 percent content of the powdered electrodes.

Distribution of the mercury when incorporated at room temperature proceeds primarily by grain-boundary diffusion. The spangled appearance of rolled zinc sheets, observed for a limited range of mercury concentration, is caused by lifting of the surface crystals as the mercury moves into crystal boundaries. Exposure to moderate temperatures, e.g., 60–100°C for short periods, accelerates spatial distribution within the grains (10), and reduces the brittleness characteristic of room-temperature amalgamation.

Great care is exercised in the manufacture of alkaline batteries to avoid contaminants that accelerate corrosion. With partly soluble depolarizers, such as silver oxide, the electrode separators serve to minimize diffusion of the dissolved species and resultant attack on the anode. However, with attention to zinc quality, to amalgamation, and to absence of metallic impurities stemming from other system components, zinc efficiency in the wet cells may be as high as 95 percent during a year of normal usage. As further protection against corrosion and gas generation in the dry type cells, substantial amounts of zinc oxide are added initially to the electrolyte. With these precautions cells may be sealed, although gas-releasing means are generally provided as protection against inadvertent pressure build-up (11).

Potential measurements referred to the normal hydrogen electrode in conventional battery electrolytes at room temperature show:

1.32 V for amalgamated zinc (1% Hg) in 6*N* zinc-free NaOH (12).

1.317 V for amalgamated zinc (10% Hg) in 7.7 *N*, zincate-saturated KOH (13).

Variation in mercury concentration in the range below 2 percent may change potentials by 10–15 mV .

Aluminum

Aluminum is a soft metal, ductile and malleable at room temperature, and adapted to rolling at 100–150°C. In air at room temperature it readily develops a thin, adherent, substantially impervious film of aluminum oxide. Despite its high theoretical potential and electrochemical equivalent, light weight, and low cost, aluminum has not yet achieved entry into the alkaline battery field. In most aqueous environments it either acquires passivity due to its oxide film or is attacked as in the basic solutions which remove the oxide surface.

For pure aluminum anodes[2] in one molar caustic electrolyte, coupled with manganese dioxide-carbon cathodes (14), an initial operating potential of about 1.24 V, dropping quickly on continuous, moderate load (3 ma/cm^2) to 0.85–0.90 V, and leveling off at about 0.75 V has been reported. Capacity to 0.72 V cutoff (26 hr) was given as 50 amp-hr (45 watt-hr)/kg total cell weight.

Recent exploratory efforts to utilize aluminum anodes in alkaline electrolyte (15) were concerned with reserve-type cells with either carbon or porous nickel, air-depolarized cathodes. The 10.5 M potassium hydroxide electrolyte, modified by additions of zinc oxide, aluminum hydroxide, or alkyl dimethylbenzyl ammonium salts to minimize anode corrosion, was periodically renewed. Open circuit voltages of 1.65 to 1.7 V and operating potentials of about 1.3 V per cell at current densities of the order of 15–20 ma/cm^2 were observed with anodes of pure aluminum or with Al-Mg, Al-Zn, and Al-Hg alloys. In all instances, however, it was found necessary to exclude air from contact with the anodes during idle periods to avoid excessive corrosion.

The large volume of aluminum hydroxide precipitate formed during service appears to be an obstacle to achievement of good ampere-hour output from small size cells. Larger batteries with three-molar potassium hydroxide electrolyte and provision for continuous removal of precipitated aluminum hydroxide have been proposed to obtain high kilowatt-hour output per unit of weight and volume at moderate to high current densities.

Although these observations modify earlier concepts regarding the instability of aluminum anodes in strongly basic electrolytes, the practical utilization of this metal in conventional alkaline batteries remains an unsolved problem. For reserve-type constructions offering moderate energy densities for short-term applications, aluminum remains of interest in view of its high ampere-hour equivalent.

Indium

Indium is a silvery white metal, softer than lead and quite ductile. Stable in dry air at room temperature, it oxidizes rapidly when heated to form indium trioxide. Experimental evidence points to trivalent indium

[2]Analysis not stated: heavy metals, e.g., iron, were reported to be detrimental to stability.

hydroxide as the anodic product in alkaline solution (16). Hydrogen overvoltage is high—as for mercury—and less local-action corrosion is observed than with either cadmium or zinc. Pure indium polarizes excessively in alkaline electrolytes at current densities above 2.3 ma/cm^2.

Mechanical rigidity is improved by alloying. Bismuth in low concentrations is preferred, because the resulting anodes will maintain current densities up to 30 ma/cm^2 with little polarization. Bismuth addition offers the further advantage of minimizing the harmful effects of impurities normally found in commercial grades of indium.

Combination of indium-bismuth anodes with mercuric oxide cathodes and potassium hydroxide electrolyte produces cells of excellently maintained open-circuit voltage (1.15–1.16 V). Experimental cells with over-all volume of 1.3 cm^3, and 1.9 cm^2 of active electrode area, gave up to 70 percent of rated capacity at room temperature at a discharge rate of 5.3 ma/cm^2. Operability down to −18°C (0°F) was indicated for such cells at 3.3 ma/cm^2. The preferred electrolyte is 35–40 percent KOH containing $\frac{1}{2}$–1 percent tartrates, glucose, or sucrose additive to promote formation of a gelatinous $In(OH)_3 \cdot H_2O$ precipitate.

Such cells have been proposed for use in battery-operated wrist watches. Their stability and good temperature characteristics suggest applications to still other uses requiring constant voltage, e.g., precision instruments or secondary reference standards.

Cadmium

Cadmium is quite malleable and ductile at normal temperatures. Tarnishing occurs at room temperature in moist air; with temperature elevation, the oxide is readily formed. Exposure to carbon dioxide-containing atmosphere will produce dense, adherent carbonate films. Cadmium is embrittled by simple heating to 80°C, or by addition of small amounts of zinc (17). In view of the toxicity of cadmium and its compounds, special precautions are necessary to avoid hazards to personnel handling these materials (18).

Cadmium anodes in strongly alkaline electrolyte form oxide and hydroxide coatings (19) characterized by porosity, conductivity, and low solubility. Cadmium resists passivation even at high rates of discharge or at low temperatures. It is more stable than zinc in basic electrolyte, less sensitive to contaminants, and together with its high hydrogen overvoltage lends itself to sealed cell construction.

Experimental cadmium: mercuric oxide: potassium hydroxide cells (20) with 3.8 amp-hr capacity (0.95 ocv) operated at room temperature at substantially full capacity on 0.1 amp drain (25 ma/cm^2) with voltage maintained at 0.87 V. At 0.5 amp load, the operating voltage held between 0.84 and 0.80 V for approximately 60 percent of rated capacity. After 4 months shelf at 74°C (165°F), similar cells retained 93 percent of their original capacity.

Though well adapted to the construction of primary cells offering excellent shelf stability and good voltage maintenance, cadmium is presently too costly to be competitive with zinc. Where the cost can be spread, as in reversible systems, cadmium has become an important ingredient for the manufacture of storage batteries.

Sodium

This metal is theoretically attractive in view of its high electrode potential, but its extreme reactivity makes it technically impractical with aqueous electrolytes. However, when dissolved in a large excess of mercury, sodium is sufficiently stable to permit its use for the construction of high-output cells (21) (discussed in greater detail in Chap. 9). An amalgam containing approximately 0.55–0.60 percent by weight of sodium is fluid at room temperature. Combination of this with an active carbon-oxygen electrode in 5–6 N sodium hydroxide electrolyte gives an open circuit potential of approximately 2.05 V and operating voltages ranging from 1.6–1.8 V at current densities of the order of 75–100 ma/cm. In 13–14 N sodium hydroxide electrolyte, the amalgam is sufficiently stable for use as a secondary reference electrode for 1 to 2 months.

Cells utilizing sodium amalgam anodes with metal oxide cathodes also have been proposed (22), but commercialization has not been achieved. The amalgam is prepared in the form of solid tablets containing 23 percent sodium, one or two tablets being added periodically to a pool of mercury in the cell. With copper oxide such cells (ocv, 1.8) will operate at 1.1–1.3 V under moderate load (about 5 ma/cm^2); with monovalent silver oxide (ocv, 2.35) operating potential is 2.0–2.1 V; with oxygen cathodes the cell potential on 5 ma/cm^2 drain will range from 1.75 to 1.80 V.

CATHODES

A summary of cathodically reducible materials, other than organic, having potential or actual utility for alkaline primary cells is given in Table 3.2. Of the many cathodic materials considered (23), only the oxides of mercury, silver, manganese, and copper or the electrodes using oxygen have found their way into commercial cells. Iron oxide, a low-cost material, suffers from low potential and significant solubility in alkaline solution. Lead dioxide presents an even greater solubility problem. Nickelic oxide, costly and not particularly well suited for the

TABLE 3.2. PRINCIPAL CATHODE MATERIALS FOR ALKALINE CELLS

Material	Reaction	Potential[a] ($E_b°$)	Theoretical Ampere-Minutes per gram
Cupric oxide	$CuO + 2H_2O + 2e = Cu_2O + 2OH^-$	−0.159	40.6
Cuprous oxide	$Cu_2O + H_2O + 2e = Cu + 2OH^-$	−0.357	20.3
Iron oxide	$Fe_2O_3 + H_2O + 2e = 2FeO + 2OH^-$	−0.56	20.2
Lead oxide	$PbO_2 + H_2O + 2e^- = PbO + 2OH^-$	+0.247	13.4
Manganese dioxide	$MnO_2 + H_2O + e^- = \frac{1}{2} Mn_2O_3 \cdot H_2O + OH^-$	+0.188[b]	18.5
Mercuric oxide	$HgO + H_2O + 2e^- = Hg + 2OH^-$	+0.098	14.8
Nickel oxide	$NiO_2 + H_2O + 2e^- = Ni_2O_3 + 2OH^-$	+0.49	19.5
Silver peroxide	$2AgO + H_2O + 2e^- = Ag_2O + 2OH^-$	+0.570	26.0
Silver monoxide	$Ag_2O + H_2O + 2e^- = Ag + 2OH^-$	+0.344	13.9
Oxygen	$O_2 + 2H_2O + 4e^- = 4OH^-$	+0.40	201.4
Oxygen (carbon)	$O_2 + H_2O + 2e = HO_2^- + OH^-$	−0.076	100.7

[a]W. C. Latimer (8), except as otherwise noted.

[b]Values for MnO_2 estimated from data of Brenet and Monsard (24); measurements on high grade commercial oxides are generally more cathodic than the calculated value.

voltage stability required of primary cells, has found application largely in secondary batteries. Organic depolarizers compatible with alkaline electrolytes, e.g., the nitro compounds of benzene, toluene, naphthalene (25), have very high ampere-hour equivalents (some in excess of 100 amp-min/g), but they have not yet found commercial acceptance.

MERCURIC OXIDE

This depolarizer operates at substantially constant voltage since only one reduction step is involved. Its solubility in alkaline electrolytes is quite low and any mercury deposited on the zinc simply augments the amalgamation normally employed to stabilize the anode. The cells, as described in Chap. 5, display long shelf life and provide a high energy output per unit volume. High cost of mercury and its compounds and a moderately poor weight-ratio of available oxygen to total oxide are deterrents to wider use. For miniature batteries, in which fabrication costs tend to overshadow material expense, the mercury cells compete effectively with other systems.

Silver oxide

Silver oxide is finding increasing use in batteries where miniaturization and high volumetric efficiency are the major requisites (cf. Chap. 6), as in hearing-aid devices. The effect of oxide solubility on anode corrosion is controlled chiefly by use of interelectrode barrier-diaphragms which minimize the transfer of dissolved silver. For applications where high power density is a prime requirement, and long periods of inactive shelf are involved, the reserve type alkaline silver oxide: zinc primary cell is an outstanding source of portable energy.

Copper oxide

The black cupric form has been in commercial use since the 1880s as the cathode member of the Lalande cell (cf. Chap. 4). The principal application for this system has been in railway and similar signaling service, where long periods of maintenance-free operation are required. The cells are made in reserve form, for assembly and activation by the consumer. Their low operating voltage (0.6–0.7 V) is a definite drawback to wider use. Copper oxide solubility in the alkaline electrolyte is sufficient to accelerate zinc corrosion, making closed-circuit operation desirable for full realization of built-in capacity.

Manganese dioxide

A relatively recent addition to the list of important cathode materials for alkaline primary cells (cf. Chap. 7), manganese dioxide, is used in dry cells with amalgamated, powdered zinc anodes, and generally with potassium hydroxide electrolyte. The depolarizer is characterized by good heavy-duty capability, operability over a wide range of temperature, and stability during storage. Dry alkaline cells made with it will deliver rated ampere-hour capacity even on heavy continuous discharge rates at which their Leclanché counterparts become inefficient. The voltage of the manganese dioxide drops off under load, resembling in this respect its behavior in the Leclanché system. The cell voltage changes are related primarily to cathode operation since the losses due to anodic and cathodic *p*H changes, characteristic of the conventional dry cell, are not encountered with strong alkaline electrolyte.

Oxygen

Using active carbon, the oxygen cathode has long been in commercial use in alkaline air-depolarized cells. Reduction of oxygen at such electrodes (cf. Chap. 8) proceeds through a peroxyl step, its potential being several tenths volt below the thermodynamically calculated value for oxygen. In combination with zinc in alkaline solution, it produces a high energy output per unit of cell volume at well-maintained voltage. Because of the stability of the carbon cathode, cells are equally effective for intermittent and continuous discharge. When supplied with oxygen, these cathodes support current densities in the range of 100–150 ma/cm^2; with air at room temperature the useful upper limit is about 25–30 ma/cm^2. Recent developments in cathode construction, involving thin layers of carbon supported on iron or nickel screens, permit major reduction in volume. Cathodes of this type are finding use in fuel cells (cf. Chap. 9) through their ability to handle high currents with oxygen and substantially higher currents with air than the electrodes employed in commercial primary cells.

ELECTROLYTES

A number of factors enter into the selection of the alkali species for the electrolyte of the alkaline type cells with zinc anodes. Both sodium and potassium hydroxides are readily available as high-grade bulk chemicals either as solids or in concentrated solutions containing a minimum of metallic impurities. They are highly soluble and can be used over a wide temperature range. Freezing point curves for the two materials are shown in Fig. 3.1.

The alkali electrolytes are distinguished by high conductivities over the normally used range of concentrations, thus contributing to the low impedance of the alkaline cells. Specific conductance data for both NaOH and KOH, including the depressant effect of added zincate ions, are presented in Fig. 3.2. Though variation exists in the conductance data obtained (26) by different workers, the values on which the curves are based are considered to be reliable. The depressant effect on conductance with increasing zinc hydroxide additions to either sodium or potassium hydroxide solutions may be explained by the disappearance of mobile hydroxyl ions and formation of alkali zincate.

Sodium hydroxide is cheaper than potassium hydroxide on a bulk basis, with further cost saving accruing when equimolar solutions are considered. It is generally used in wet cells, where relatively large volumes of electrolyte are required, and very low temperatures are not encountered. The usual concentration range of 5.5–6.25 *N* for commercial cells is a compromise of conductivity and low temperature tolerance.

Caustic potash electrolyte has a lower freezing temperature, −65°C (−85°F) for the 7.2 *N* (eutectic) solution as compared with −28°C (−18°F) for 6.25 *N* (eutectic) sodium hydroxide solution, and thus is

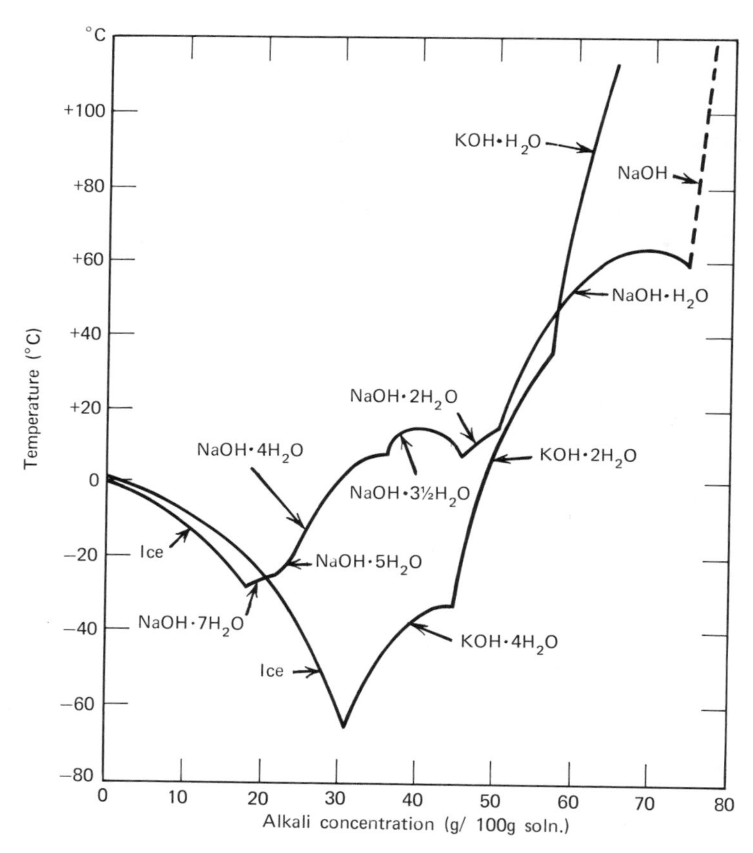

Figure 3.1 Freezing points of NaOH and KOH solutions. Derived from data in International Critical Tables, Vol. IV, 235; (1928), and material, by permission, from Hooker Chemical's "Nialk Caustic Potash" (1955) and Pittsburgh Plate Glass Industries, "Caustic Soda" (1969).

more suitable where low temperature operation is required. Its conductivity is a further advantage. The potash electrolytes employed in wet cells range from about 6 to 7.5 *N*, again reflecting the desire to combine best conductivity and freezing point values. In dry-type alkaline cells, they are used at a concentration level of 9–10 *N* to achieve the ZnO precipitation necessary for operation of this type of cell.

As with other electrochemical systems, much of the work on the improvement of alkaline cells has been directed toward increasing the ampere-hour capacity per unit weight or volume, particularly the latter. In this effort, the volume of electrolyte required for operation becomes the limiting factor since incremental additions of high-density anodic

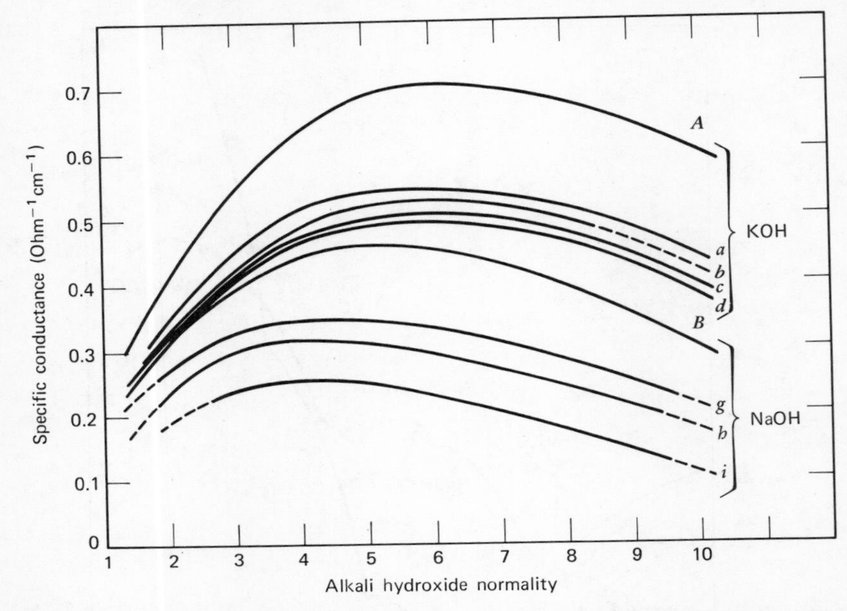

Figure 3.2 Conductivity curves: NaOH-ZnO and KOH-ZnO solutions at 30°C [data from (26)]. (*A*) KOH only:

(*a*) 1.0 mole ZnO : 4.33 mole KOH
(*b*) 1.0 mole ZnO : 3.71 mole KOH
(*c*) 1.0 mole ZnO : 3.37 mole KOH
(*d*) 1.0 mole ZnO : 3.00 mole KOH

(*B*) NaOH only:

(*g*) 1.0 mole ZnO : 4.05 mole NaOH
(*h*) 1.0 mole ZnO : 3.03 mole NaOH
(*i*) 1.0 mole ZnO : 1.76 mole NaOH

or cathodic constituents have comparatively small effect on cell size. Thus the ampere-hour capacity of commercial alkaline primary cells is a function of the solubility of the complexes formed by anodic solution of zinc in sodium and potassium hydroxide electrolyte and of the physical form of the solid by-products crystallizing from these solutions when saturation is reached.

Solubility relationships in NaOH solutions

The problems of zinc-complex solubility and nature of the ionic species obtained in alkali solutions have received extensive attention. Experimental data for the existence of colloidal zinc hydroxide, $Zn(OH)_3^-$, and $Zn(OH)_6^{-2}$ have been reported (27), but the most recent work (28) in this area, based on Raman spectra, nuclear resonance studies, and electrochemical measurements, leaves little doubt that $Zn(OH)_4^{-2}$ is the most prevalent or perhaps the only form of zinc ion present over a broad range of alkali and zinc concentration.

The numerous solubility studies found in the literature (29) cover different temperatures and experimental conditions, making quantitative comparison difficult. Reproducibility of data is satisfactory only when precautions are observed in the preparation of the initial solid phases, in the procedures used for solution preparation[3] and in allowance of sufficient time for attainment of true solubility equilibria. Table 3.3

TABLE 3.3. THE SYSTEM $Zn(OH)_2$-ZnO-NaOH-H_2O[a]

NaOH Normality	Zinc Concentration as Moles ZnO/liter at 20°C						Terminal Solid Phase
	5 days	20 days	60 days	100 days	160 days	200 days	
4.15	0.80	0.80	0.80	0.65	0.45	0.45	ZnO
5.80	1.50	1.50	1.10	0.80	0.65	0.65	ZnO
7.30	2.40	1.65	1.20	1.15	1.10	1.10	ZnO
8.80	3.25	2.30	1.65	1.60	1.55	1.55	ZnO
10.10	2.80	–	–	2.05	–	1.95	ZnO

[a]Estimated from data by Scholder and Hendrich (29) for solutions initially saturated at 20°C with freshly made zinc hydroxide. For zincate solutions (approximately 6 *N* NaOH) prepared anodically in the author's laboratory a full year has occasionally been required to reach equilibrium.

summarizes studies with sodium hydroxide solutions by Scholder and Hendrich (29), who found it essential in most instances to carry on equilibration at 20°C for many months in silvered flasks. At low alkali concentration (up to 3 *N* NaOH) or for short equilibration periods (1–3 weeks) in stronger solutions, the final solid phase is rhombic zinc hydroxide. With higher alkali concentration (4–10 *N* NaOH) and increasing time of contact between solutions and solid, a gradual phase shift to zinc oxide is observed.

In commercial wet cells (approximately 6 *N* NaOH) discharged to exhaustion, the normal solid end-product is crystalline zinc hydroxide, and only occasionally is zinc oxide found as a minor component. The discrepancy between battery experience and the Scholder, Hendrich data may have its origin in stabilizing components present in commercial alkali, e.g., silicates, and/or the greater resistance to phase change by the large, well-formed zinc hydroxide crystals found in operating cells.

[3]Contamination of test solutions, e.g., with silicates leachable from glassware, will interfere with equilibration.

Further solubility measurements by Scholder and Hendrich on zinc hydroxide in sodium hydroxide solutions up to 17 *N* disclose hydrated [$NaZn(OH)_3 \cdot 3H_2O$], and anhydrous [$NaZn(OH)_3$] monosodium zincates as solid phases.

The solubility diagram shown in Fig. 3.3 illustrates the equilibria for the concentration range applicable to commercial cells. Freshly prepared zinc hydroxide (curve *AC*) shows distinctly higher solubility than partially dehydrated or aged hydroxide (curve *AD*), or completely water-free zinc oxide (curve *AG*). Curve *AB* has been inserted to indicate the higher values[4] observed during discharge of alkaline wet batteries. Vertical line *aa'* illustrates the equilibrium change noted on occasion in conventional cells, e.g., in 6 *N* NaOH, when the usually observed rhombic zinc hydroxide phase is followed by deposition of less soluble zinc oxide, with a corresponding drop in zinc concentration in the electrolyte to line *AF*.

[4]Study of the zincate complexes present in these "supersaturated" electrolytes apparently was not included in the recent work on complex identification (28) and it may be that species other than $Zn(OH)_4^{-2}$ can form in battery electrolytes during anodic dissolution of zinc.

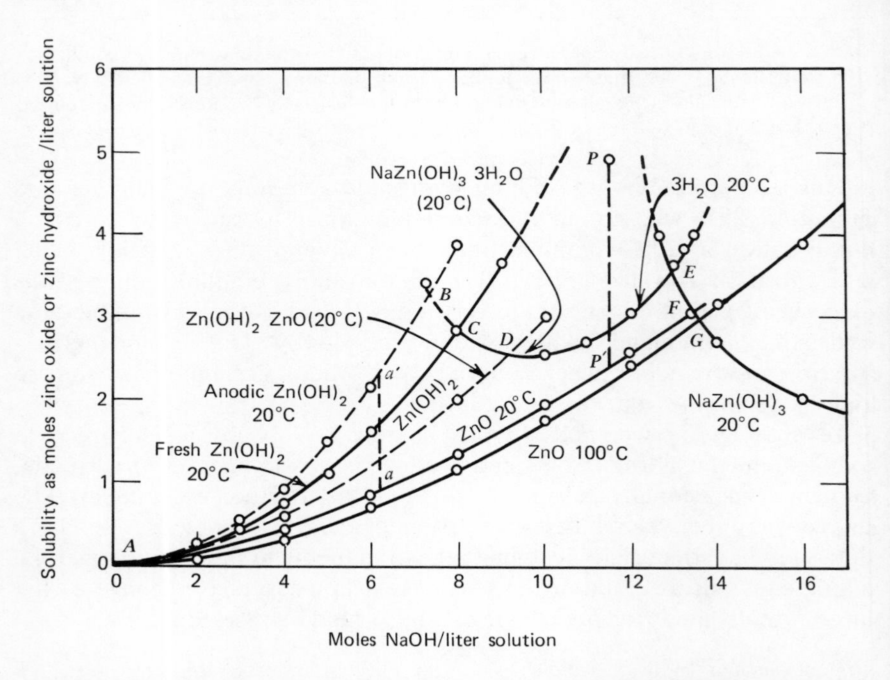

Figure 3.3 Solubility of $Zn(OH)_2$ and ZnO in NaOH solutions. **Derived from data in** (29).

An area of metastability bounded by curve *CE* is indicated for zinc hydroxide dissolved in sodium hydroxide of about 8–13 *N*. The extrapolated maximum, in excess of 5 moles, is in contrast with the normally attainable maximum of approximately 3 moles/liter in commercial wet cells. As ZnO separates from the unstable zincate solutions, the zinc concentration drops to the curves obtained for ZnO-NaOH solutions. Other solid phases encountered in the metastable range are reported as zinc hydroxide and hydrated monosodium zincate, $NaZn(OH)_3 \cdot 3 H_2O$, the latter as needlelike, silky crystals (30) which convert to ZnO as indicated by vertical line *pp'*. Solutions above 8 *N* concentration show rapid conversion of the solid phases to zinc oxide. This undoubtedly is a major factor in the differentiation of wet cells utilizing 5.5–6.25 *N* sodium hydroxide electrolyte, and the "dry" cells containing from 8–12 *N* alkaline solutions.

The solubility of zinc oxide is below that of $Zn(OH)_2$ over the range of 1–14 *N* alkali and is unaccompanied by metastable areas or formation of solid phases other than oxide. At approximately 13.4 *N* NaOH, the curve for ZnO intersects that of the anhydrous $NaZn(OH)_3$. Zinc oxide is slightly less soluble at 100°C than at 20°C. Partially dehydrated or aged zinc hydroxide (curve *AD*) is noticeably less soluble than the freshly precipitated hydroxide, but more so than calcined zinc oxide.

Commercial Practice. Commercial cells with approximately 6 *N* NaOH electrolyte are generally operated under conditions producing a slow approach to saturation with zincate and a corresponding slow growth of initially fine zinc hydroxide into large rhombic or octahedral crystals. These do not convert into zinc oxide in cells, though in the final stage of discharge, and after the electrode surfaces are substantially covered with an impervious layer of zinc hydroxide crystals, some zinc oxide may precipitate from the electrolyte. Discharge of experimental wet cells containing higher NaOH concentration, e.g., 9 *N*, produces an amorphous blue zinc oxide, nonadherent to the electrodes, permitting longer operation and commensurately greater ampere-hour capacity per unit volume of electrolyte. However, these higher concentrations of sodium hydroxide are not compatible with freezing point requirements for practical applications of the wet cells (cf. Fig. 3.1), and further depress conductivity (cf. Fig. 3.2).

Miniature "dry" cells with approximately 9 *N* NaOH electrolyte yield only zinc oxide as the solid phase. The electrolyte is therefore substantially nonvariant and, at temperatures above 15°C, the ampere-hour output per milliliter of electrolyte is from 8–16 times as great as that normally obtained from standard wet-cell electrolyte.

Solubility relationships in KOH solutions

Potassium hydroxide solutions do not dissolve as much zinc hydroxide or zinc oxide as sodium hydroxide solutions of equivalent strength (31), (Fig. 3.4), and as in the case of the latter, zinc oxide is the less soluble material. Anodic dissolution of zinc electrodes in KOH electrolytes yields variably higher concentrations of zincate solutions (32) than obtained by conventional solubility measurements on precipitated $Zn(OH)_2$. Contrary to the behavior of zinc hydroxide in NaOH solutions, no breaks are reported in the solubility curves for KOH solutions up to at least 13–14 *N*. With zinc oxide solute Dirkse (31) observed a sharp drop in solubility of about 14.5 *N* KOH but found only zinc oxide as the solid phase. Potassium zincate crystals ($KHZnO_2$) are observed only when strongly alkaline solutions (13–15 *N* KOH) are cooled to 0°C; these crystals are redissolved by careful heating (32).

As in the case of wet cells with 6 *N* NaOH electrolyte, zinc hydroxide crystals are formed at room temperature in operating cells containing potassium hydroxide of equivalent strength. Such cells yield approximately the same ampere-hour capacity as their counterparts with sodium

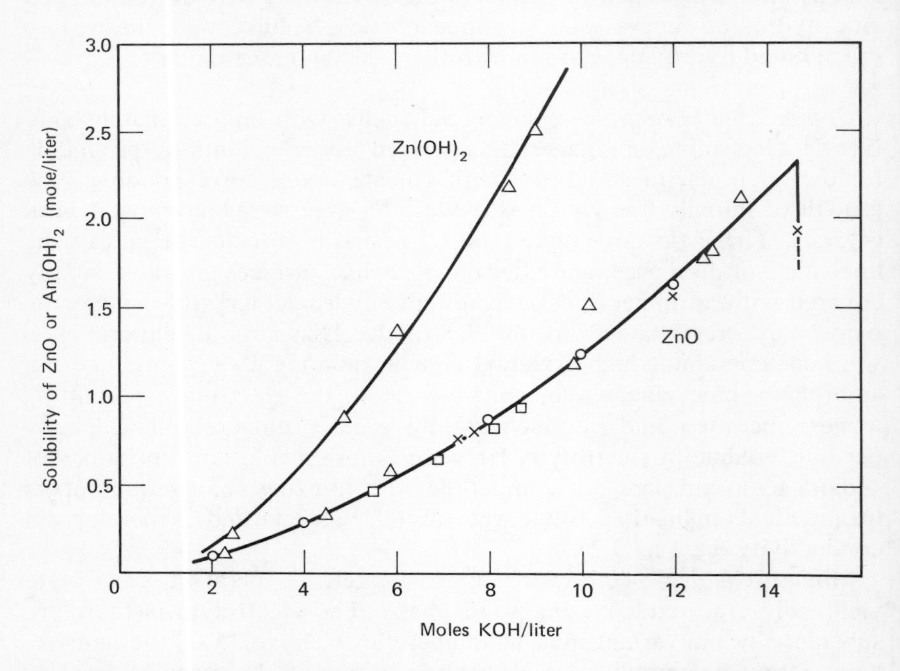

Figure 3.4. Solubility of $Zn(OH)_2$ and ZnO in KOH solutions at 25°C. × = Dirkse; □ = Baker, Trachtenberg; △ = Sochevanov; ○ = Hildreth (unpublished data). From (31).

hydroxide but offer somewhat higher voltages because of better electrolyte conductivity, and permit operation at lower temperature. At higher potassium hydroxide concentration, e.g., above 8 N, amorphous blue zinc oxide is produced with a large rise in ampere-hour output per milliliter of electrolyte. Such electrolyte is commonly utilized in "dry" type alkaline primary cells.

Anodic Films

Polarization measurements with zinc anodes in sodium and potassium hydroxide solutions and study of the development of zinc hydroxide and zinc oxide films on zinc surfaces (33) indicate that hydroxide or oxide films[5] form initially at the grain boundaries of the surface zinc crystals, as a result of the greater rate of anodic solutions at these points. The film then spreads across the electrode until the surface is substantially covered. Adherent blue coatings of zinc oxide are observed at low alkali concentrations (0.1–1.0 N) and low zincate concentrations in the bulk electrolyte (34), white zinc hydroxide in the intermediate range, and soft blue oxide at high alkali concentrations (8–12 N) and high zincate ion content.

Amorphous hydroxide as well as several crystalline zinc hydrates have been described (35). Those designated as α, β, γ, and δ forms—listed in order of increasing stability—are obtained by controlled addition of dilute alkalis (0.1–0.5 N NaOH) to dilute solutions of zinc salts. By variation in the precipitation conditions that α and β crystals convert to the γ type, which have the form of needlelike crystals or sometimes of small spheroidal particles. The ∂ modification appears as small platelets or double pyramids when strong zinc solution is treated with a 0.35 N NaOH; aging in the solution produces some ϵ crystals. The gamma hydroxide is observed on occasion as needles or prisms in cells operating below 0°C. Small rhombic crystals (ϵ) are obtained by addition of 1.0 N alkali to supersaturated zincate solution, and large rhomboids (ϵ) are obtained in the normal operation of commercial wet cells with sodium or potassium hydroxide electrolytes.

Deposits of blue oxide are observed on amalgamated sheet-zinc anodes in 9 N sodium hydroxide, which, through stress development, separate from the metal to permit continued discharge. These deposits are porous, shrink when dried, and lose color when exposed to air. Oxide release proceeds rapidly when amalgamated zinc particles are formed into anodes.

The blue color of the zinc oxide, whether produced by simple anodic polarization or cell discharge, does not appear to be related to impurity factors in the zinc but is believed to stem from excess zinc atoms in the

[5]The form of the solid phase is controlled to a large extent by current density and concentration of alkaline electrolyte; mixtures of oxide and hydroxide are also observed.

oxide lattice. Even when prepared to prevent mechanical occlusion of metallic zinc, the blue oxide is a semiconductor developing potentials[6] in alkaline solution approaching, in some instances, the value for zinc. The concept of semiconductivity is supported by others, e.g., Hayward and Trapnell (36).[7] The wet blue oxides formed in "dry" cells are sufficiently conducting to require care in the placement and selection of the separator member to prevent shorting across the anode-cathode electrolyte gap.

CAPACITY EXTENSION

Capacity Extenders

Anode : electrolyte reactions during discharge of cells at typical current densities, e.g., 1–10 ma/cm² of anode surface, may be visualized as proceeding in the following general manner on the basis of the solubility relationships outlined. Thermodynamic and kinetic data for the transition from the solutions to solid phases (Eqs. (4) and (8)) are not available (37).

1. Zinc hydroxide end product (wet cells, 5.5–7 *N* electrolyte):

$$Zn = Zn^{+2} + 2e^- \tag{1}$$
$$Zn^{+2} + 4OH^- = Zn(OH)_4^{-2} \tag{2}$$
$$Zn(OH)_4^{-2} + 2Na^+ \rightarrow Na_2Zn(OH)_4 \tag{3}$$
$$Na_2Zn(OH)_4 \rightarrow Zn(OH)_2 + 2NaOH \tag{4}$$

2. Zinc oxide end product ("dry" type cells, 8–12 *N* electrolyte):

$$Zn = Zn^{+2} + 2e^- \tag{5}$$
$$Zn^{+2} + 4OH^- = Zn(OH)_4^{-2} \tag{6}$$
$$Zn(OH)_4^{-2} + 2Na^+ \rightarrow NaZn(OH)_3 + NaOH \tag{7}$$
$$NaZn(OH)_3 \rightarrow ZnO + NaOH + H_2O \tag{8}$$

The most effective method of increasing ampere-hour capacity, as already noted, is the use of electrolyte in the range of 8–12 *N* alkali to promote the formation of zinc oxide as insoluble end product. This expedient, unfortunately, is not applicable to general-purpose wet cells in which sodium hydroxide is employed, because of limitations of permissible operating temperatures and conductivity, and it has serious economic disadvantage for large wet cells utilizing potassium hydroxide electrolyte. However, two other methods are available to the battery technologist for

[6]Unpublished studies.

[7]According to these authors semiconductor zinc oxide is characterized by excess zinc present as interstitial metal atoms: "in the oxide . . . the equilibrium between the interstitial atoms and ions permits appreciable liberation of electrons and hence appreciable conductivity."

substantial enhancement of output. Incorporation of soluble complexing agents such as alkali silicates will delay precipitation of the anodic reaction products well beyond their normal appearance. Such additives, in concentrations of the order of 25–50 g/l of electrolyte, are used principally in the alkaline wet cells of the railway signal cell types.

Even more effective is the addition of solid reactants such as calcium hydroxide to the cell, which serves to convert the soluble zinc complex into insoluble calcium zincate (38) with simultaneous regeneration of the electrolyte:

$$2Na_2Zn(OH)_4 + Ca(OH)_2 + 2H_2O \rightarrow CaZn_2O_3 \cdot 5H_2O + 4NaOH \quad (9)$$

Thus 1 g of solid hydrated lime can remove about 1.76 g of zinc from solution, equivalent to 1.45 amp-hr of service. This expedient is described more fully in Chap. 8, devoted to the alkaline oxygen-zinc cell.

Extension of electrolyte effectiveness by means of complexing agents, electrolyte regenerants, or high-alkali concentrations to promote zinc oxide precipitation is illustrated in Table 3.4.

TABLE 3.4 VOLUMETRIC EFFECTIVENESS OF ALKALINE ELECTROLYTES

Electrolyte System		End Product	Electrolyte Requirement per Ampere-hour
Conventional electrolyte	(5–7 N NaOH)	$Zn(OH)_2$	8 ml[a]
Conventional electrolyte plus sodium silicate	(5–7 N NaOH)	$Zn(OH)_2$	5–5.5 ml
Conventional electrolyte plus $Ca(OH)_2$	(5–7 N NaOH)	$Ca \cdot Zn_2O_3 \cdot 5H_2O$	1.5–2.5 ml
Strong electrolyte	(8–12 N NaOH)	ZnO	0.5–1 ml

[a]Based on wet cell practice with the zinc:cathode assembly mounted 4–5 in. above the bottom of the battery jar, permitting gravity separation of the heavier zincate solution, and crystallization of some zinc hydroxide on bottom and side walls before the electrodes become coated and inoperative.

The technical innovations made in the alkaline primary cell field in relatively recent times have brought a considerable number of important new cells into being, and improvements in the systems previously available have opened up a new era of portable battery operated devices. It is highly probable that this impetus will stimulate still further system development.

REFERENCES

1. de Lalande, F., and Chaperon, G., French Pat. 143,644 (1881); U.S. Pat. 247,110 (1883); *Compt. rend.*, **97**, 164 (1883); **112**, 1243 (1891).
2. Anon., *Elektrotech. Z.*, **11**, 377 (1890); **13**, 476 (1892); **21**, 205 (1900).
3. Walker, W., and Wilkins, F. H., U.S. Pats. 524,229; 524,291 (1894).
4. Nyberg, H. D., *Z. Elektrochem.*, **30**, 549 (1924).
5. Heise, G. W., and Schumacher, E. A., *Trans. Electrochem. Soc.*, **62**, 383 (1932); Schumacher, E. A., Hamister, V. C., and Heise, G. W., U.S. Pat. 2,010,608 (1935).
6. Denison, I. A., *Trans. Electrochem. Soc.*, **90**, 387 (1946); White, J. C., Pierce, R. T., and Dirkse, T. P., *ibid.*, **90**, 467 (1946).
7. Schumacher, E. A., and Heise, G. W., *J. Electrochem. Soc.*, **99**, 191C (1952).
8. Latimer, W. M., *Oxidation Potentials*, Prentice-Hall, Englewood Cliffs, N.J. (1952).
9. Akhmetov, N. S., and Vozdvizhensky, G. W., *J. Appl. Chem., USSSR*, **29**, 1297 (1956).
10. Pleteneva, N. A., and Fedoseeva, N. P., *Dokl. Akad. Nauk., SSSR*, **151**, 384 (1963).
11. Booe, J. M., *J. Electrochem. Soc.*, **99**, 197C (1952).
12. Heise, G. W., Schumacher, E. A., and Fisher, C. R., *Trans. Electrochem. Soc.*, **92**, 241 (1947); also (7).
13. Friedman, M., and McCauley, C. F., *Trans. Electrochem. Soc.*, **92**, 195 (1947).
14. Brull, L., *Ricera Scientifica*, **1**, 525 (1937).
15. Zaromb, S., *J. Electrochem. Soc.*, **109**, 1125 (1962); Zaromb, S., and Foust, R. A., *ibid.*, **109**, 1191 (1962); Bockstie, L., Trevethan, D., and Zaromb, S., *ibid.*, **110**, 267 (1963); Kuz'mina, A. V., and Kuz'min, L. L., *Protective Metallic and Oxide Coatings, Corrosion and Electrochemistry*, N. P. Fedot'ev, Ed., p. 330. Orig. Russian version published by Izdatel'stovo Nauka, Moscow/Leningrad, 1965. (Translated version by Israel Program for Scientific Translations, Ltd., 1968.)
16. Boswell, T. L., U.S. Pat. 2,683,184 (1954); *J. Electrochem. Soc.*, **105**, 239 (1955).
17. Brady, G. S., *Material Handbook*, 7th ed., McGraw-Hill Book Co., New York (1951).
18. Sax, N. I., *Handbook of Dangerous Materials*, Reinhold Pub. Corp., New York (1951).
19. Huber, K., *J. Electrochem. Soc.*, **100**, 376 (1953).
20. Klein, M. G., and Eisenberg, M., *J. Electrochem. Tech.*, **3**, 58 (1965).
21. Yeager, E. B., *J. Electrochem. Soc.*, **108**, 67C (1961); Schumacher, E. A., et al., AD 301978 (April 1958); Miller, K. D., *Chem. Eng. Prog.*, **57**, 140 (1961).
22. Walz, H. P., *Radiotechnik*, p. 713 (Dec. 1949), Brit. Pat. 640,496 (1950); U.S. Pat. 2,646,458 (1953).
23. Glickman, R., and Morehouse, C. K., *J. Electrochem. Soc.*, **104**, 589 (1957); Morehouse, C. K., Glicksman, R., and Lozier, G. S., *Proc. Inst. Rad. Eng.*, **46**, 1462 (1958).
24. Pourbaix, M., *Atlas d'Equilibres Electrochemiques*, Gauthier-Villars, Paris (1963).
25. Bauer, W. C., U.S. Pat. 1,134,093 (1915).
26. *International Critical Tables*, **6**, 254 (1929); Gmelin, *Handbuch anorg. Chemie*, **2**, 229 (1927); Mehta, S. M., and Kabadi, M. B., *J. Ind. Chem. Soc.*, **16**, 223 (1939); *id.*, *J. Univ. Bombay*, **18A**, 39 (1949); Darken, L. S., and Meier, H. F., *J. Am. Chem. Soc.*, **64**, 621 (1942); Manvelyan, M. G., Keanoyan, T. V., Eganyan, A. G., and Kocharyan, A. M., *Izvest. Akad. Nauk. Armyan SSSR*, **9**, 3 (1956); Cooper, J. E., and Fleischer, A., *Alkaline Silver Oxide-Zinc Batteries*, sponsored by A. F. Aeropropulsion Laboratory, Dayton (1964); Dyson, W. H., Schreir, L. A., Sholette, W. P., and Salkind, A. J., *J. Electrochem. Soc.*, **115**, 556 (1968).
27. Scholder, R., and Weber, H., *Z. anorg. Chem.*, **215**, 355 (1933); Despande, V. V., and Kabadi, M. B., *J. Univ. Bombay*, **20A**, 28 (1951).
28. Dirkse, T. P., *J. Electrochem. Soc.*, **101**, 328 (1954); Newman, G. H., and Blomgren,

G. E., *J. Chem. Phys.*, **43**, 2744 (1965); Fordyce, J. S., and Baum, R. L., *J. Chem. Phys.*, **43**, 843 (1965).

29. Klein, O., *Z. anorg. Chem.*, **74**, 157 (1912); Muller, E., *Z. Elektrochem.*, **33**, 134 (1927); Dietrich, H. G., and Johnson, J., *J. Am. Chem. Soc.*, **49**, 1419 (1927); Fricke, R., *Z. anorg. Chem.*, **172**, 234 (1928); Scholder, R., and Weber, H., **215**, 355 (1933); Scholder, R., and Hendrich, G., *ibid.*, **241**, 76 (1939); Iofa, Z., Mirlina, S., and Moiseyeva, N., *Zhur. Priklad. Khimii*, **22**, 983 (1949).
30. Fricke, R., and Hume, H., *Z. anorg. Chem.*, **172**, 242 (1928); Feitknecht, W., *Helv. Chim. Acta*, **13**, 314 (1930); *ibid.*, **32**, 2294 (1949); Martus, M. L., *Trans. Electrochem. Soc.*, **53**, 183 (1928); Iofa et al., (29); Scholder and Hendrich (29).
31. Sochevanov, V. G., *J. Gen. Chem. USSR*, **20**, 1119 (1952); Dirkse, T. P., *J. Electrochem. Soc.* **106**, 154 (1959); Baker, C. T., and Trachtenberg, I., *ibid.*, **114**, 1045 (1967).
32. Mallory Co., P. R., ATI 192786 (Dec. 1948); Iofa et al. (29).
33. Fricke(30); Feitknecht(30); and Iofa et al.(29); Dirkse, T. P., Dewit, D., and Shoemaker, R., *J. Electrochem. Soc.*, **115**, 442 (1968).
34. Jirsa, F., and Loris, K., *Z. phys. Chem.*, **113**, 235 (1924); Hedges, E. S., *J. Chem. Soc.*, **1533**, 2580 (1926); *ibid.*, 969 (1928); Vernon, W. H., and Stroud, E. G., *Nature*, **142**, 477 (1938); Huber, K., *Helv. chim. Acta*, **26**, 1037 (1947); *id.*, *J. Electrochem. Soc.*, **100**, 376 (1953); Iofa et al. (29); Landsberg, R., *Z. phys. Chem.*, **206**, 291 (1957); Bartlet, H., and Landsberg, R., *ibid.*, **222**, 217 (1963); Hampson, N. A., Tarbox, M. J., Lilley, J. T., and Farr, J., *J. Electrochem. Soc.*, **110**, 174C (1963).
35. Newberry, E., *J. Chem. Soc.*, **109**, 1066 (1916); Hedges (34); Huber, K., *Z. Elektrochem.*, **48**, 26 (1942).
36. Hayward, D. O., and Trapnell, B. M., *Chemisorption*. Butterworth, Washington, D.C., p. 259 (1964).
37. Sochevanov, V. G., (31); Scholder and Hendrich (29).
38. Heise, G. W., U.S. Pat. 1,835,867-8 (1931); 1,864,652 (1932).

4

The Alkaline Copper Oxide:Zinc Cell

ERWIN A. SCHUMACHER

Despite much early work on solid cathodic reactants (1), it was not until 1881 that Lalande and Chaperon (2) devised the first practical element employing granular cupric oxide electrodes, zinc anodes, and aqueous caustic electrolyte. This was followed by various attempts to commercialize the system—among them the Cupron cell (3) in Germany and the Neotherm cell (4) in England both using agglomerated copper oxide cathodes which could be regenerated by thermal reoxidation after discharge.

Investigation of the system (5) by Edison in the United States led to innovations in construction, particularly as regards agglomerated cathodes, which met with considerable success. In 1889 a plant was established to produce batteries in a wide range of sizes, 15–900 amp-hr, for the rapidly developing markets in railway signaling, telephony, telegraphy, operation of small motors, electrotherapy, etc., previously monopolized by the gravity (Daniell) and wet Leclanché cells. In 1895, the basic structural patents of the Waterbury cell were granted for annular molded CuO cathodes and annular zinc anodes (6)[1]. By 1910 many of the present-day features of the cell had been introduced (7), (8), including metal frames to support the molded cathodes, heat- and alkali-resistant glass battery jars (9), silicate additions to the electrolyte to improve capacity, copper-coated cathodes, and complete factory assembly for ease of installation by the user. Cells employing granular copper oxide were in use as late as 1960, but are no longer of commercial importance.

Despite a relatively low operating voltage, nominally 0.6–0.7 V, the copper oxide:zinc cell with molded cathodes finds many applications where capacity, flat discharge curve, and stability are dominant service requirements. It is currently employed for track circuits, centralized traffic control, switchlamp lighting, crossing signals, approach lighting,

[1]Manufacture of the Waterbury cell was discontinued about 1960.

semaphore signals, and for kindred uses, either as the direct source of electrical energy or as standby in the event of main-line power failure.

COMMERCIAL CELLS

Commercial railway cells range in capacity from 75–1000 amp-hr, with the 500 and 1000 amp-hr units representing the bulk of the cells manufactured. They are generally available in cylindrical, barrel-shaped, or rectangular glass jars, although a few are made with steel containers. According to the nature of the service application, the cells may have from three electrodes per unit assembly, i.e., one cathode and two anodes for minimum current requirement, up to nine or more electrodes where the duty cycle is severe (10). Table 4.1 lists some typical 500 amp-hr commercial cells and the maximum continuous-discharge rates at 20°C compatible with the indicated capacity for each type.

TABLE 4. 1. CURRENT RATINGS OF TYPICAL 500 AMP-HR ALKALINE COPPER OXIDE CELLS

Type of Service	Electrodes[a] per Cell	Permissible Continuous Discharge Rates (amp at 20°C to 0.50 V)
Light	3	1.75
Medium	5	2.25
Heavy	9	6.50

[a]Each electrode assembly has one anode plate more than the number of cathode elements.

Although several distinctly different cell constructions were available up to recent years, only the flat-plate types, such as the five-electrode assembly shown in Fig. 4. 1, are still in use. These utilize agglomerated copper oxide electrodes in approximately rectangular form, mounted between dimensionally matching zinc anodes. The oxide plates are fastened in steel or coppered steel frames, which provide mechanical support and the means for establishing electrical connection to the external positive terminal. Porcelain insulators, fastened to the cathodes, permit mounting of the zinc anodes in rigid parallel alignment with the cathodes, and ensure mechanically strong units completely assembled at the factory. The weights of copper oxide and zinc are in balanced relationship, sufficient to yield an ampere-hour output in moderate excess of the battery rating.

In practice, the cathode-anode assemblies are shipped to the user together with watertight containers of granular or flake alkali, and

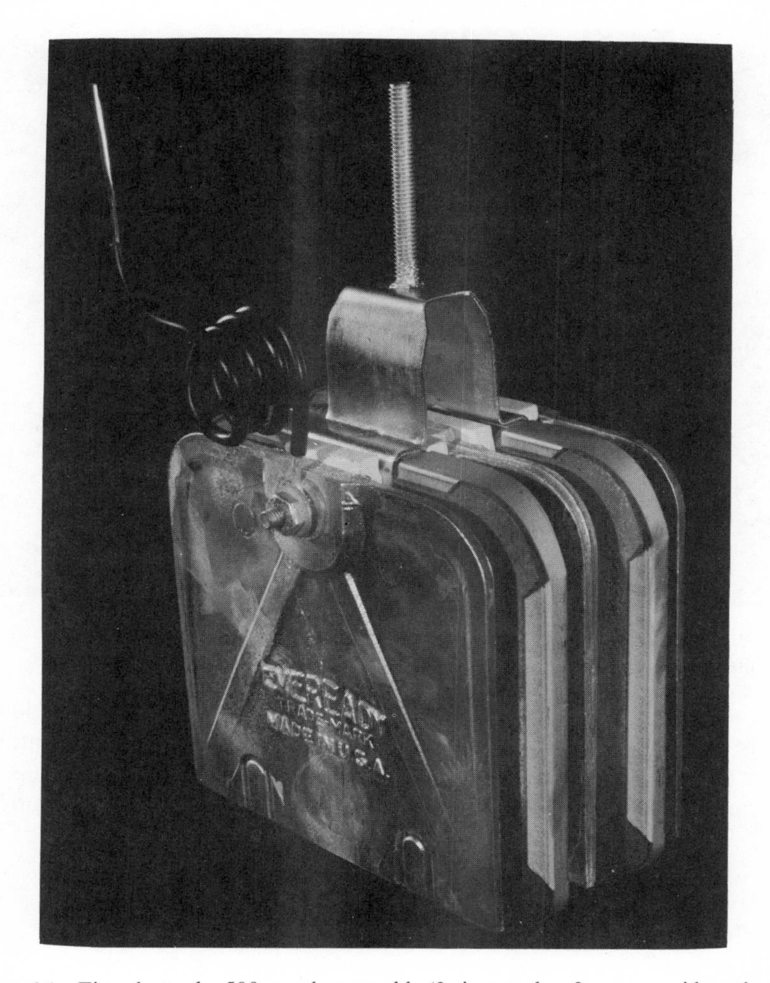

Figure 4.1 Five electrode; 500 amp-hr assembly (3 zinc anodes; 2 copper oxide cathodes).

separate bottles of oil. Figure 4.2 shows a typical three-electrode cell with components for field assembly. The user removes the spent electrolyte and assembly of the exhausted cells[2] from the battery jars, prepares fresh solution, and fastens the new electrode assemblies to the porcelain cover.

[2]The spent oxide: zinc assemblies are customarily returned for credit toward fresh units, the manufacturer reclaiming the copper and blending with virgin oxide for reuse in the fabrication of new plates. Because of the ready oxidizability in air of the copper in the spent units, accompanied by heat generation, special precautions must be taken in the handling of the exhausted assemblies.

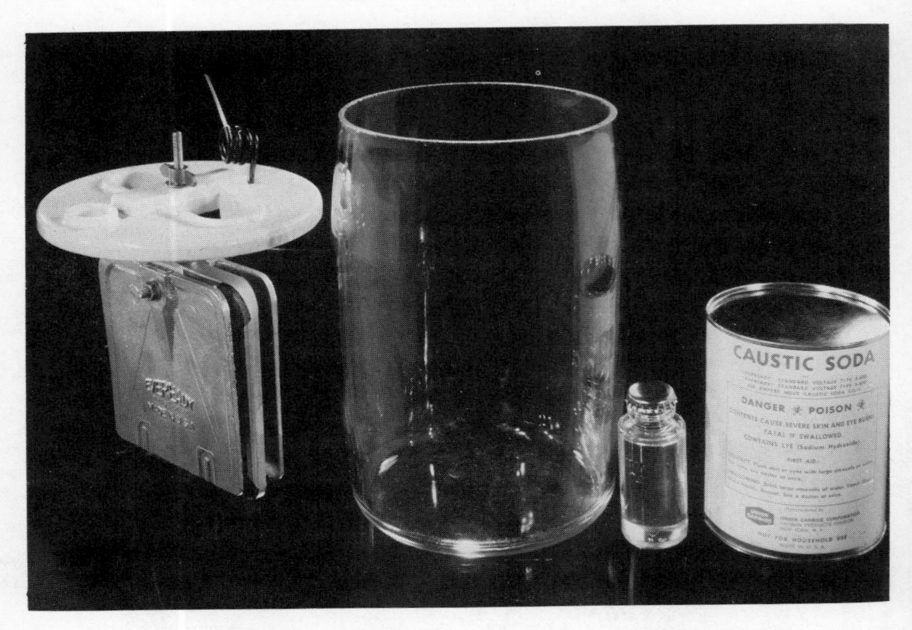

Figure 4.2 Three-electrode; 500 amp-hr assembly with battery jar, granular sodium hydroxide, and oil.

The electrolyte increases in density during discharge, (see Table 4.2) zincate-rich solution accumulating in the bottom of the cell container where the precipitation of end product, $Zn(OH)_2$, is first noted as the cell approaches exhaustion. The electrode assembly is therefore mounted

TABLE 4.2. CHANGE IN ELECTROLYTE CONCENTRATION DURING CELL DISCHARGE[a]
(500 amp-hr CuO cell at 21°C on 1 amp continuous drain)

	Specific Gravity of Electrolyte		
Ampere-Hours Discharge	One Inch Below Solution Surface	One Inch From Bottom of Jar	Midway Top and Bottom
0	1.215	1.215	1.215
100	1.221	1.245	1.239
200	1.229	1.273	1.271
300	1.266	1.304	1.304
400	1.264	1.333	1.332
500	1.256	1.364	1.360
550	1.268	1.373	1.373

[a]From (8).

in the jar as high as is practicable to delay crystallization of end product on the active surfaces (8).

The alkali hydroxide used in the copper oxide cell should be of high purity (low heavy-metal content) and the electrolyte (5.5–6.5 N) prepared with it preferably is made with water[3] of potable quality. Except where cell operation at low temperatures requires potassium hydroxide electrolyte, caustic soda is employed to reduce cost. The alkali is usually furnished in the form of flakes, cubes, or coarse particles, to avoid the dusting associated with powdered material. Where powdered material is employed, the dusting can be controlled by mixing oil with alkali before packaging for shipment.[4]

Oil is furnished with all commercial cells to provide a seal protecting the electrolyte from adsorption of atmospheric carbon dioxide and evaporation of water and to prevent electrolyte creepage on the exposed portions of the battery jar and assembly. It is most commonly supplied in separate glass containers, but its use in electrolyte-soluble capsules, packaged with dry alkali, has been proposed (11).

Granular copper oxide

Though no longer manufactured, cells with cathodes of granular oxide (first disclosed by Lalande and Chaperon) for a long time played a significant role in the development of the copper oxide-zinc caustic system. In the simplest construction, the oxide, placed loosely on a tinned iron plate, was at the bottom of the cell (12) and an annular zinc anode at the top. Such cells, characterized by high internal resistance, were particularly suited for low drain, "closed circuit" use. In their final form, the electrodes were mounted vertically from the cover with two anodes used per cathode. The cathode was a rectangular, perforated tinned-iron basket packed with granular cupric oxide. A layer of paper (13) prevented finely divided oxide from sifting through the perforations and by fixing soluble copper, reduced corrosive attack on the zinc anodes. With the paper-lined basket construction, electrical contact with the copper oxide was made by means of a centrally mounted iron vane introduced as cathode collector. This arrangement (14) permitted oxide reduction to start at the center of the electrode when discharge

[3]The solid caustic must be added to the water with vigorous stirring until completely dissolved. Heat of solution of caustic is high, and without adequate agitation, a solid layer of slowly soluble sodium hydroxide monohydrate, $NaOH \cdot H_2O$ may form at the bottom of the jar.

[4]Solution of the alkali releases the oil, which then rises to the electrolyte surface. Though convenient, this method has the drawback that when the electrode assembly is lowered into the electrolyte, the copper oxide may adsorb sufficient oil to affect its cathodic behavior.

was initiated, proceeding toward the outer layers with gradual decrease in the distance between the anodes and the effective cathode reaction sites. The internal resistance of the cell dropped as discharge progressed, resulting in a flatter discharge curve and better maintenance of operating voltage.

DISCHARGE CHARACTERISTICS

Tests simulating field service cover a wide range of discharge frequency and rates, but since the output of the cells is not greatly influenced by the degree of intermittency, measurements of ampere-hour capacity and operating voltage are conveniently carried out as continuous-drain tests. Performance of a medium-duty 500 amp-hr cell, using sodium hydroxide electrolyte, with three anodes and two cathodes is shown in Fig. 4.3 for several discharge rates at room temperature.

The voltage curves, even for moderately light load (1 ma/cm^2), show no inflection due to stepwise reduction of the depolarizer, e.g.,

$$CuO \rightarrow Cu_2O \quad \text{and} \quad Cu_2O \rightarrow Cu \tag{1}$$

the observed operating potentials being more nearly related to the Cu_2O/Cu couple than the CuO/Cu_2O electrode. Lowering of voltage with progressive discharge is largely due to rise in internal resistance as zincate ion concentration rises and is more pronounced, as expected,

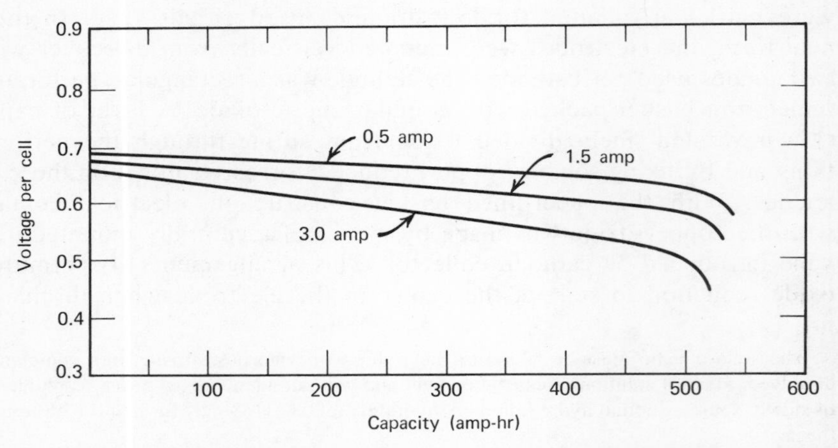

Figure 4.3 Continuous discharge characteristics of 5-electrode; 500 amp-hr copper oxide cell at 21°C; electrolyte 6 *N* NaOH [current density at 3.0 amp is approximately 6.6 ma/cm^2 (6.1 amp/ft^2)].

with NaOH than with KOH electrolyte (cf. Chap. 3). The rapid drop in potential with approaching exhaustion reflects saturation of the electrolyte with zincate and deposition of impervious reaction products on the anode surfaces. Because the cells are typically constructed as balanced systems, very little service is available after the normal endpoint is established, and replacement of anodes or cathodes will not significantly influence ampere-hour output.

Operating potential drops with temperature, and below about −18°C (0°F), cells with sodium hydroxide electrolyte will not deliver rated capacity at useful voltage levels even at moderate discharge rates (see Fig. 4.4). For railway signaling applications, the minimum safe operating temperature is about −4°C (25°F). Where lower ambient temperatures (down to −18°C) are encountered, batteries are kept in insulated boxes set into the ground to a depth of 20–30 in. For extremely low temperatures battery wells are employed, extending perhaps 7 ft below ground level and wide enough to permit entry by the signal maintainer. Potassium hydroxide electrolyte, which offers a much lower freezing point and better conductivity, can be used to advantage for operating at low temperatures (−40°C for experimental cells) but has found no significant acceptance in commercial cells.

Table 4.3 briefly summarizes the relationship between maximum permissible current drain on continuous or intermittent discharge and cell temperature from −4 to 32°C (25 to 90°F) for representative copper oxide-zinc-caustic soda cells (15). The differences in permissible current drain tend to diminish with rising temperature.

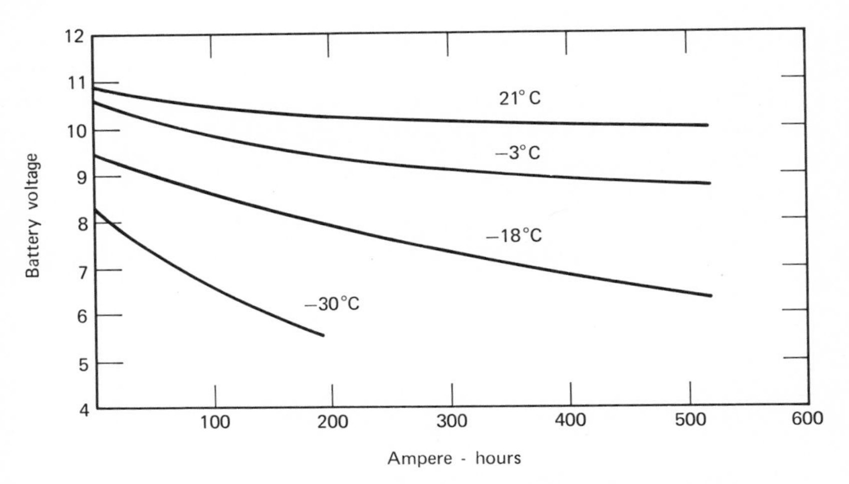

Figure 4.4 Effect of temperature on color-lighting, signal operation test.

TABLE 4.3. EFFECT OF TEMPERATURE ON MAXIMUM CURRENT FOR RATED AMPERE-HOUR OUTPUT

Cell Type	A.A.R.[b] Class	Type of Discharge[f]	Maximum Drain (Amperes) to Rated Capacity at Various Solution[a] Temperatures (°F)					
			25°	30°	40°	50°	60°	70–90°
3–500	2[c]	Continuous	0.6	0.7	0.8	1.1	1.3	1.75
		Intermittent	1.0	1.1	1.4	1.75	2.0	2.2
5–500	3[d]	Continuous	0.8	0.9	1.2	1.5	1.9	2.25
		Intermittent	1.3	1.4	1.7	2.0	2.2	2.60
5–1000	5[e]	Continuous	1.0	1.2	1.8	2.2	2.8	3.5
		Intermittent	2.5	2.7	3.1	3.7	4.0	4.5

[a]Nominally 6 *N* sodium hydroxide.
[b]Assoc. Amer. Railroads Class designation.
[c]Class 2. 3-electrode, 500 amp-hr cells.
[d]Class 3. 5-electrode, 500 amp-hr cells.
[e]Class 5. 5-electrode, 1000 amp-hr cells.
[f]Intermittent test: e.g., $\frac{1}{2}$-min drain every $\frac{1}{2}$ hr.

GENERAL CONSIDERATIONS

The over-all reactions of the alkaline copper oxide-zinc system are fairly straightforward. The cathode undergoes reduction as previously indicated:

$$CuO \rightarrow Cu_2O \rightarrow Cu \tag{2}$$

However, the freshly formed copper reacts quickly with the cupric oxide to form a thin layer of cuprous oxide, the active interface becoming essentially a mixture of CuO and Cu_2O. As discharge proceeds, the outermost layers are reddish in color, a sharply defined line marking the junction with the residual black oxide. Only a thin layer of black oxide remains when the cell reaches exhaustion.

The anode reactions in the electrolyte concentration range (5.5–6.5 *N*) customarily employed for the commercial copper oxide cells are pictured as follows:

$$Zn \rightarrow Zn^{+2} + 2e \tag{3}$$

$$Zn^{+2} + 4(OH)^{-2} \rightarrow Zn(OH)_4^{-2} \tag{4}$$

and

$$2Na^+ + Zn(OH)_4^{-2} \rightarrow Na_2Zn(OH)_4 \qquad (5)^5$$

[5]This form is preferred over the earlier Na_2ZnO_2 in view of the evidence of recent years for the $Zn(OH)_4^{-2}$ ion (16).

Accordingly, the operation of the cell may be described as

$$CuO + Zn + 2NaOH + H_2O \rightarrow Cu + Na_2Zn(OH)_4 \qquad (6)$$

Initial open circuit voltage of the cell is approximately 1.16–1.18 V dropping to 0.96–0.98 V during discharge. The higher voltage of the zinc-cupric oxide couple, e.g., 1.20–1.24 V, is seldom observed in practice since the commercial cathode, made with coppered surfaces and supporting frames, quickly undergoes reduction at the active electrode:electrolyte interface when cells are activated.

The initially high resistivity of the cathode is gradually lowered through formation of metallic copper and cuprous oxide as discharge progresses. The electrolyte, on the other hand, increases in resistivity as the zincate ion concentration rises (see Chap. 3) contributing materially to the overall drop in operating voltage with use.

ELECTROLYTE

The common forms of the commercial 500 amp-hr copper oxide-alkali-zinc cells employ sodium hydroxide electrolyte and require, on the average, slightly in excess of 4 liters of 6 *N* electrolyte, equivalent to about 8 ml/amp-hr of rated capacity. This concentration represents the best compromise of high conductivity and minimum freezing point and for this reason is preferred over more concentrated solutions even though the latter can dissolve more zinc at room temperature to provide higher ampere-hour capacity.

Substitution of the more expensive potassium hydroxide electrolyte is indicated where excessively low temperatures are encountered or where underground installations are too costly. By virtue of its better conductivity, potash electrolyte also contributes to higher operating voltages. To compensate for the lower solubility of zincates in this electrolyte, higher concentrations, e.g., 6.5–7.0 *N*, should be used than for caustic soda.

Upon saturation of the electrolyte with zincate ions, crystals of octahedral zinc hydroxide [$Zn(OH)_2$] begin to deposit on the inner surfaces of the cell container and eventually on the electrodes in adherent, impervious form, to terminate cell operation. At low temperatures, e.g., 0°C, the zinc hydroxide crystals may be replaced by long, needlelike crystals of alkali zincate, but these will redissolve at room temperature followed by zinc hydroxide precipitation. Zinc oxide is not normally encountered as an end product but can be found at higher alkali concentrations or with excessively high current drains.

The requirement for periodic visual inspection of cell conditions imposed by railway safety practice makes the glass battery container a necessary part of the assembly and to a considerable extent has made it difficult to introduce capacity-increasing means which interfere with visibility. Silicate additives to stabilize the zincate solutions can be employed in moderate amounts (25–50 g/liter of electrolyte) but are likely to induce cloudy solutions toward end of life. Electrolyte regeneration by means of the lime:zincate reaction (Chap. 3) has been slow to find acceptance, but is currently receiving attention because of its effective employment in other alkaline cells.

ANODE

The zinc anodes employed in the present alkaline copper oxide cells are of high purity (99.8–99.9 percent) cast as plates with 0.5–2.5 percent mercury content, and in fresh 6*N* sodium hydroxide electrolyte have a potential of −1.32 V referred to the standard hydrogen electrode (17). Sufficient excess metal is provided to maintain mechanical strength and active surface area throughout the entire use period, thus ensuring delivery of rated capacity at recommended current drains. For ordinary service conditions this may amount to perhaps 25–50 percent of the theoretical requirement. Copper oxide is somewhat soluble in the electrolyte, creating local corrosion couples during intermittent discharge, which increases the requirement for excess zinc. On continuous load, migration of soluble copper complexes from the cathodes is retarded, and thus minimizes the excess zinc corrosion problem.

The iron washers, nuts, and bolts used in direct contact with the anodes in commercial cells would normally accelerate zinc corrosion; however, when copper plated and amalgamated they acquire a sufficiently high hydrogen overvoltage to minimize attack. Similarly, the copper wire connector is insulated up to the external negative terminal to prevent corrosion acceleration by contact with air, and is also amalgamated at the point of contact with the zinc.

Anodes intended for railway signal cells are provided with several panels of reduced thickness located either at the lower edge or in the body of the plate (18). During the late stages of discharge, these thinner sections drop out to warn the signal maintainer of the need for cell renewal. For a progressive indication of delivered capacity, the individual panels may be varied in thickness.

CATHODE

The copper oxide used in the manufacture of electrodes is derived chiefly from rolling mill operations, turnings, and washed material salvaged from spent batteries. It is roasted to convert all metal and cuprous oxide to the cupric state and then is ground to the desired fineness. As in the earliest forms of the Lalande unit and the annular type, the cathode plates of present-day cells are formed by agglomeration of powdered copper oxide with binders, e.g., caustic soda or silicate. A mixture of oxide, caustic alkali, and copper sulfate produces a particularly strong agglomerate (19). The mixtures are molded in presses, and then baked at temperatures of 650–950°C (1200–1750°F) to complete the bonding action.

Because of the high electrical resistance of cupric oxide (about 2×10^6 ohm-cm), it is common practice to heat the baked plates in a reducing atmosphere to produce a superficial conducting layer of metallic copper (20). An alternative procedure is to spray or otherwise deposit zinc or aluminum on the cathode surfaces to effect reduction when the electrodes are immersed in electrolyte. Zinc wires fastened between anodes and cathodes to create internal short circuits of regulated duration have also been used to give the desired copper oxide reduction (8).

Despite the apparent simplicity of the overall cell reactions (Eq. 6), the cathodic reduction process is not a simple one. Though charge:discharge measurements (21) show the two-step reduction expected for the reactions

$$2CuO + H_2O + 2e = Cu_2O + 2(OH)^- \tag{7}$$

$$Cu_2O + H_2O + 2e = 2Cu + 2(OH)^- \tag{8}$$

the commercial primary cells discharged at normal rates (0.5–6.0 ma/cm^2) yield reproducible, well sustained voltage curves without inflections, but well below the potentials expected for the cupric oxide values shown in Table 4. 4. Using a value of -1.32 V for zinc in zinc-free 6*N* NaOH (17) gives initial open circuit voltages of 0.96 V for the Cu_2O/Cu-Zn couple, and 1.20–1.24 V for the Zn-CuO/Cu_2O or $Cu(OH)_2/Cu_2O$ couples, respectively. The latter are somewhat higher than the 1.16–1.18 V observed for commercial cells.

TABLE 4.4. POTENTIALS OF COPPER OXIDE IN ALKALINE SOLUTION

Electrode	Potential versus Hydrogen Electrode
Cu_2O/Cu	-0.358 V[a]
$Cu(OH)_2/Cu_2O$	-0.08 V[a]
CuO/Cu_2O	-0.12 V[b]

[a]From (22).
[b]From (23).

Because cupric oxide reacts with metallic copper to form the monovalent oxide,

$$CuO + Cu \rightarrow Cu_2O \tag{7}$$

which is less readily polarized than the divalent oxide, the operating voltage of the commercial cell represents a mixed potential controlled primarily by the cuprous component. The two reduction steps presumably proceed simultaneously in the operation of commercial cathodes, with the chemical reaction supplying the cuprous oxide at the active interface. Used by itself, the Cu_2O operates at a slightly higher potential than observed for cells with CuO (24) but it is not employed commercially since for equal oxide weights, ampere-hour capacity is less for cuprous oxide, and equal capacity would seriously increase cost.

Potentials of the order of 0.82–0.86 V are observed when commercial cells are discharged at very low drains, e.g., 0.1–0.2 ma/cm^2, supporting the long-held belief that cupric oxide is inherently a "poor" depolarizer. However, high potentials may be obtained on normal drains following an initial period of overload or short circuit, these conditions apparently creating a temporary network of copper to bring a larger area of cupric oxide into effective use. With time the reaction between copper and copper oxide produces cuprous oxide at the active interfaces, thereby establishing the lower mixed potential on continued discharge.

Since the manner of cupric oxide preparation can influence its stability in alkaline solution, it is not unlikely that voltage peaking occurrences may be associated with formation of cuprate ions which act as the primary depolarizer constituent until exhausted or renewed by standing. Calcining of copper oxide as commonly practiced in the United States reduces oxide solubility, though minor high voltage plateaus may be observed when cells are started in hot electrolyte.

High-voltage plateaus up to 0.88 V at customary battery loads have also been reported for cupric oxide-zinc cells when subjected to intermittent discharge (25). Gradually increasing the open circuit periods, from 1 up to 7 hr, lengthens the peak voltage region (see Table 4. 5), as does temperature elevation during discharge or rest. The peaks are gradually lost after prolonged stand periods (1–2 months) and after extended discharge.

Hydration of both oxides leading to cuprite and cuprate ions is suggested as an important factor in the electrode discharge. Thus in the steady-state cathodic process in the CuO electrode, some reduction through an ionic mechanism may be expected in addition to the usual solid-phase reduction. The latter process becomes controlling unless conditions are established (rest and/or temperature) which permit hydration to proceed

TABLE 4.5. VOLTAGE PEAKING IN CuO:NaOH: Zn CELLS VERSUS OPEN CIRCUIT TIME

Open Circuit Intervals (hours)	Hours on Load		
	Above 0.875 V	Above 0.80 V	Above 0.665 V
1	3	10	12
3	30	65	190
5	35	175	200
7	40	340	350

Data from (25).
NOTE: Cells discharged at 2 ma/cm^2 in 5.5 *N* NaOH at 20°C with atmospheric oxygen excluded.

at rates commensurate with the cathodic reduction. With sufficient hydration of the divalent oxide, formation of hydrated monovalent oxide is accelerated, thereby promoting "high" voltage operation. A schematic representation (Fig. 4. 5) of the interfaces between copper and the active components is given by Flerov (25).

The electrical resistivity of conventional copper oxide cathode plates is a significant factor in the low-voltage operation of commercial cells. Thus when granular oxide is pressed into retaining metal screens, it will support current densities well in excess of the ratings for standard signal cells. Combination of such cathodes with powdered zinc electrodes, for example, yields cell voltages of the order of 0.6–0.7 V (IR included) at 70 ma/cm^2 in 7 *N* KOH at room temperature.

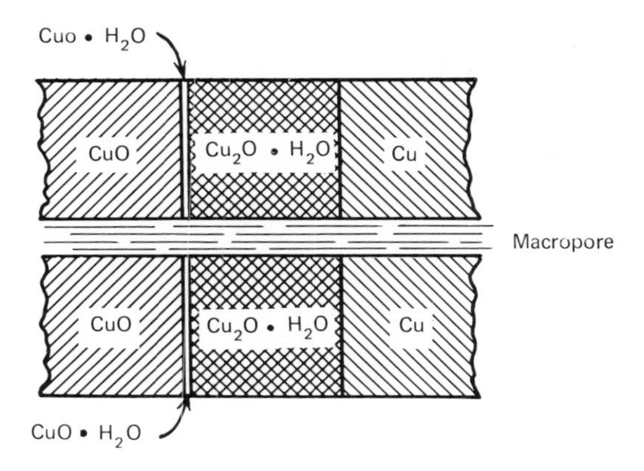

Figure 4.5 Schematic representation of interphases in a CuO-Cu_2O-Cu electrode.

HIGH VOLTAGE COPPER OXIDE

Addition of elemental sulfur or sulfur compounds (26) to the copper oxide cathode to enhance its potential was practiced[6] for many years. Voltage increments up to 0.2 V during the life of the cell were obtained with additions of $\frac{1}{2}$–2 percent sulfur to the oxide by exposure to hydrogen sulfide. Excellent results have also been reported by Flerov (25) for concentrations as low as 0.15 percent, this later work ascribing the beneficial effect of the small amount of sulfur to improvement in the conductivity of the cuprous oxide, thereby accelerating the reaction between copper and hydrated cupric oxide which favors rapid reformation of the cuprous hydroxide layer consumed during discharge. An alternative mechanism[7] is based on the assumption that sulfur retards the reaction between cupric oxide and copper by formation of copper sulfide and thus promotes reduction of the higher voltage divalent oxide.

Efforts to improve the performance of the sulfured copper oxide cathode for use in high-energy batteries (27) indicate that as with plain CuO, greater current-carrying ability may be obtained by molding a mixture of fine copper oxide and sulfur into a metal grid. Addition of sulfur as a binder (up to 33 percent by weight) to the copper oxide is claimed to raise cell voltage up to 0.95 V at 33 ma/cm^2 drain, although more recent work by others (23) indicates that large amounts of sulfur should depress the potential of the copper oxide (see Table 4.6). Mixtures of oxide, sulfur, and copper powder applied to metal supports apparently offer no advantage over the simple oxide: sulfur blends.

Copper peroxides have been reported (28) and their preparation confirmed by the author[8] with evidence of distinctly different x-ray patterns

TABLE 4.6. EFFECT OF SULFUR ON CuO POTENTIAL[a]

Electrode	Voltage versus HgO/Hg
CuO (no sulfur)	−0.22 V
CuO + 5% S	−0.075 V
CuO + 17% S	−0.177 V
CuO + 29% S	−0.20 V
Sulfur (with graphite)	−0.35 V
Cu_2S/Cu	−0.64 V[b]

[a]Measurements in 30 percent KOH at room temperature from (23).
[b]From (22).

[6]Primarily for operation of low-drain track circuits ("Columbia" High Voltage Cell).
[7]Unpublished studies at the Union Carbide Research Laboratories.
[8]Treatment of $Cu(OH)_2$ at 0°C with hydrogen peroxide.

than obtained with normal cupric or cuprous oxides. The treated cupric hydroxide material contains a relatively small fraction of peroxidized component, about 5 percent, which decomposes slowly in contact with 6 *N* NaOH electrolyte to form a blue cuprate solution. Open circuit voltage at 25°C measured against zinc in 6 *N* NaOH is about 1.31 V. Discharged in a nitrogen atmosphere using NaOH electrolyte, at current densities up to 40 ma/cm^2 (graphite addition used for conductivity), the peroxidized material was found to be 0.35–0.42 V more cathodic than normal cupric oxide: graphite electrodes. Cell voltages at 10 ma/cm^2 were 1.0 V and 0.58 V, respectively for the peroxidized oxide and the cupric oxide.

The zinc: copper oxide alkaline system is handicapped by a low operating voltage, and copper oxide, though relatively efficient in terms of its ampere-hour equivalent (1.48 g/amp-hr) and total utilization of combined oxygen cannot be considered an inexpensive cathodic reactant. However, because of its applicability to rugged construction, its flat discharge curve, and its ability to operate for long periods of time with minimum maintenance over a wide range of service conditions, the cell remains an important commercial unit.

REFERENCES

1. de la Rive, A., *Arch. de l'Élec.* **3**, 112 (1843).
2. de Lalande, F., and Chaperon, G., French Pat. 143,644 (1881); German Pat. 22,702 (1882); U.S. Pat. 274,110 (1883); *Compt. rend.* **97**, 164 (1883); **112**, 1243 (1891).
3. Lorenz, R., *Z. Elektrochem.*, **4**, 305 (1898).
4. Allmand, A. J., and Ellingham, H. J., *Principles of Applied Electrochemistry,* Longmans, Green & Co., New York, p. 211 (1924).
5. Anon., "The Edison Lalande Cell", *Elektrotech. Z.*, **11**, 377 (1890); **13**, 476 (1892).
6. Gordon, R. W., U.S. Pat. 542,049 (1895).
7. Garrity, P. A., *Proc. Sig. Div. Am. Railway Assn.*, **16**, 428 (1919); Spangler, J. M., *ibid.*, **16**, 567 (1919).
8. Martus, M. L., *Trans. Electrochem. Soc.*, **53**, 183 (1928); *ibid.*, **68**, 151 (1935); Edison T. A. Industries, *Packaged Power 1889–1964*, Bloomfield, N.J.
9. Assoc. Am. Railroads, *Signal Section Manual,* Part 9, Spec. 113–39 (1939).
10. *Id., Signal Section Manual,* Art. 6, Spec. 87–39 (1939).
11. Martus, M. L., Becker, E. H., and Schoenmehl, C. B., U.S. Pat. 1,532,252 (1925).
12. Marshall, E. L., *Trans. Electrochem. Soc.*, **25**, 467 (1918); Smith, E. C., and Marshall, E. L., U.S. Pat. 1,017,064 (1912).
13. Heise, G. W., U.S. Pat. 1,563,980 (1925).
14. Wolfe, F. J., U.S. Pat. 1,564,741 (1925).
15. Am. Railway Signalling Principles and Practice, Chap. V, "Batteries" (1948).
16. Fordyce, J. S., and Baum, R. L., *J. Phys. Chem.*, **43**, 843 (1965); Newman, G. H., and Blomgren, G. E., *ibid*, p. 2744.
17. Heise, G. W., Schumacher, E. A., and Fisher, C. R., *Trans. Electrochem. Soc.*, **92**, 173 (1947).

18. Eddy, A. I., U.S. Pat. 2,189,463 (1940); Heise et al. (17).
19. Edison, T. A., U.S. Pat. 1,207,382 (1916); U.S. Pat. 1,386,095 (1921).
20. Eddy, A. I., U.S. Pat. 2,157,072 (1939).
21. Johnson, W. M., *Trans. Electrochem. Soc.*, **1**, 187 (1902).
22. Latimer, W. C., *Oxidation Potentials,* Prentice-Hall, Englewood Cliffs, N.J. (1952).
23. Amlie, R. F., Hohner, H. N., and Ruetschi, P., *J. Electrochem. Soc.*, **112**, 1073 (1965).
24. Benner, R. C., and French, H. F., U.S. Pat. 1,415,860 (1922).
25. Flerov, V. N., *J. Appl. Chem., USSR,* **30**, 1677 (1957); *Russ. J. Phys. Chem.,* **37**, 934 (1963).
26. Erwin, R. W., U.S. Pats. 1,282,057 (1918); 1,295,459 (1919); Benner, R. C., and French, H. F., U.S. Pats. 1,255,283 (1918); 1,316,761 (1919), 1,415,860 (1922); Armstrong, G. W., U.S. Pat. 1,624,460 (1927).
27. Catalytic Research Corp., AD 238818, (April 1960).
28. Mellor, J. W., *Comprehensive Treatise on Inorganic and Theoretical Chemistry,* Longmans, Green & Co., London, Vol. III, p. 147, (1923).

5

The Mercuric Oxide : Zinc Cell

SAMUEL RUBEN

Although the effectiveness of mercuric oxide as cathodic reactant with alkaline electrolyte has long been recognized, early efforts (1), (2), (3), (4) to utilize this material in conventional wet cells led to no practical result. However, with the comparatively recent (5), (6), (7) successful development of the Zn : KOH : HgO dry cell, mercuric oxide has become one of our most important commercial cathode materials.

The RM[1] zinc : mercuric oxide dry cell with alkaline electrolyte was developed during the early part of World War II. It was produced on a large scale for military purposes (the "Handy Talkie" and "Walky Talky" radio transceivers) and proved generally useful in applications demanding batteries of high capacity per unit volume and of good heavy-drain characteristics. It has also found general acceptance in nonmilitary uses; improvements in design and reduction in size along with miniaturization of electronic equipment, particularly through the advent of transistorized devices, has made it an important commercial battery, available in a wide variety of shapes and sizes (8).

THE CELL SYSTEM

The cell system

$$Zn(Hg) : KOH\,(ZnO)_{x(aq)} : HgO$$

is basically simple, comprising amalgamated zinc, a 40 percent caustic potash solution saturated with zinc oxide, and mercuric oxide. The

[1] "RM," a contraction of the names of Samuel Ruben, the inventor, and P. R. Mallory & Co., Inc., the battery manufacturer, is the official designation given the cell by the U.S. Signal Corps.

calculated potential of the zinc electrode in cell electrolyte for the electrode reaction

$$Zn + 4\,OH^- \rightarrow ZnO_2^{-2} + 2\,H_2O + 2\,e^-$$

is −1.332 V; that for the mercuric oxide

$$HgO + H_2O + 2\,e^- \rightarrow Hg + 2\,OH^-$$

is +0.012. The deviation from the standard values (−1.216 and +0.098 respectively is due to the high caustic concentration and the presence of dissolved zinc oxide (9). The algebraic sum of the electrode potentials (1.344) is in satisfactory agreement with the open-circuit voltage of the commercial RM cell [1.35 (7)].

When the cell is in service, further addition of zinc to the saturated electrolyte through the anodic reaction causes direct precipitation of zinc oxide[2]

$$ZnO_2^{-2} + H_2O \rightarrow ZnO + 2\,OH^-$$

The over-all cell reaction is therefore simply

$$Zn + HgO \rightarrow ZnO + Hg + 2\ \text{Faradays}$$

Since the electrolyte is, in effect, nonvariant during discharge, only a minimal quantity (about 1.3 g/amp-hr of cell capacity) is required. This, together with the high density of HgO, makes possible a very compact cell, of high coulombic capacity per unit of volume. A number of other factors sharply differentiate the RM cells from conventional wet batteries and have contributed to its successful development; these are noted in the following discussion of cell design and construction.

DESCRIPTION OF CELLS

The conventional RM cell is manufactured in either cylindrical or flat form, shown in Figs. 5.1 and 5.2, respectively. Despite structural differences, both use essentially the same type of component materials.

The anode is a compressed unit of amalgamated zinc powder, providing the porous structure and large active area requisite for effective utilization.

[2]A detailed discussion of zinc solubility and the insoluble complexes formed at different caustic concentrations is given in Chap. 3.

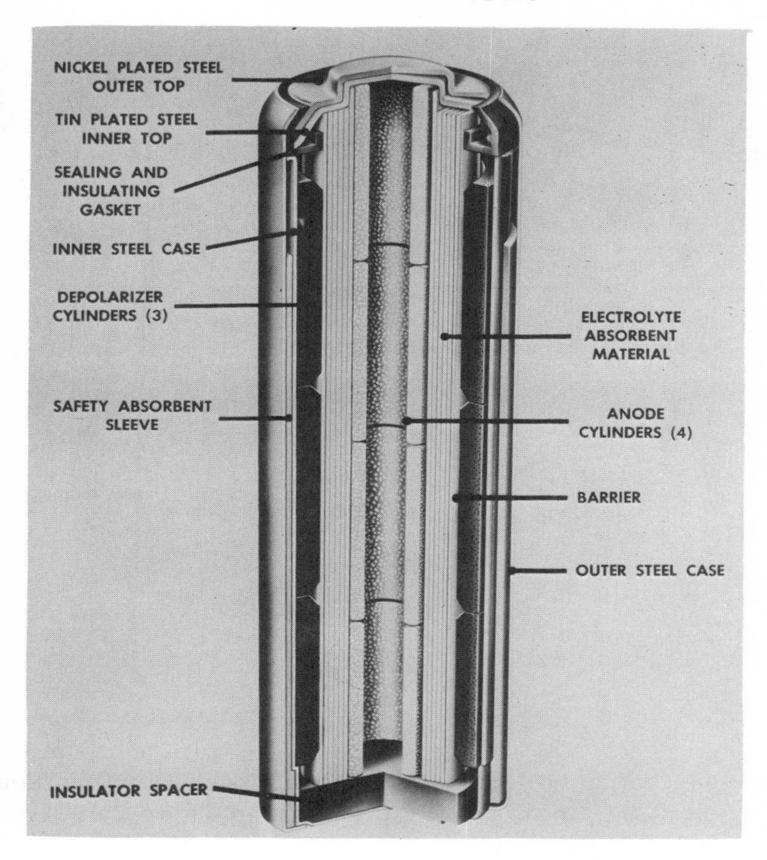

Figure 5.1 Pressed-powder cylindrical cell.

The high-density pressed cathode member is a mixture of mercuric oxide with 5–10 percent of high purity micronized graphite, which increases conductivity and, during discharge, minimizes the coalescence of droplets of mercury formed by cathodic reduction of the oxide.

The cathode is made of greater coulombic capacity than the anode to ensure complete utilization of zinc during service discharge. This is a safeguard against evolution of hydrogen and development of pressure due to anodization of residual zinc in an exhausted cell allowed to remain connected to the load circuit. For applications in which initial voltage is not critical, the excess capacity may be supplied by additions of manganese dioxide (which raise open circuit voltage to 1.4).

The separators or spacers between anode and cathode consist of absorbent, e.g., felted cellulose to immobilize the electrolyte, and a

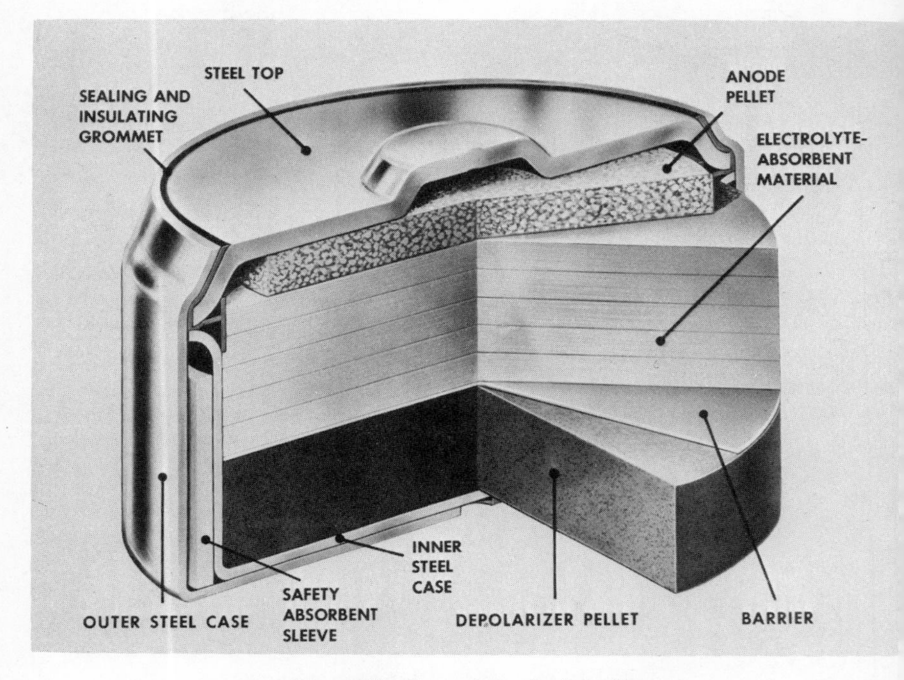

Figure 5.2 Pressed powder flat cell.

barrier layer next to the cathode, of parchment or microporous plastic to prevent internal short circuit and cell discharge by accumulation of bridging materials such as free mercury, graphite, or certain forms of semiconductive zinc oxide.

For special-purpose cells, the spacer may be replaced by a gelatinized electrolyte, made by the addition of 4 percent of sodium carboxymethyl cellulose to hot electrolyte (40 percent KOH solution, saturated with ZnO and heated to 120°C (248°F). The mixture is poured into the cathode container and allowed to set on the cathode surface, which previously has been coated with a more oxidation-resistant barrier film of alkalized polyvinyl alcohol.

Contact of the cathode member to its steel container offers no problem. Tinned steel is used at the anode: in contact with the amalgamated zinc, this steel acquires an amalgam coating which effectively prevents attack on zinc (and hydrogen evolution) due to local couples.

An outer steel can holds anode and cathode containers tightly against a plastic grommet. Properly constructed cells should show no gas pressure but, since imbalance of cell components or contact of cathodic impurities with the zinc anode may cause hydrogen evolution, this construction provides an automatic vent (path of gas shown by arrows in Fig. 5.3). Specifically, excessive pressure causes temporary separation of anode and cathode containers, allowing gas to escape through the sidewall space, which contains a ring of paper to absorb any small quantities of electrolyte that might be expelled with the gas.

"Wound-anode" cell

Anode design is not limited to compressed or pelleted powdered zinc. In the "wound anode" cell (Fig. 5.4) the zinc electrode is in the form of a flat spiral of corrugated metal and an absorbent paper spacer-strip. The

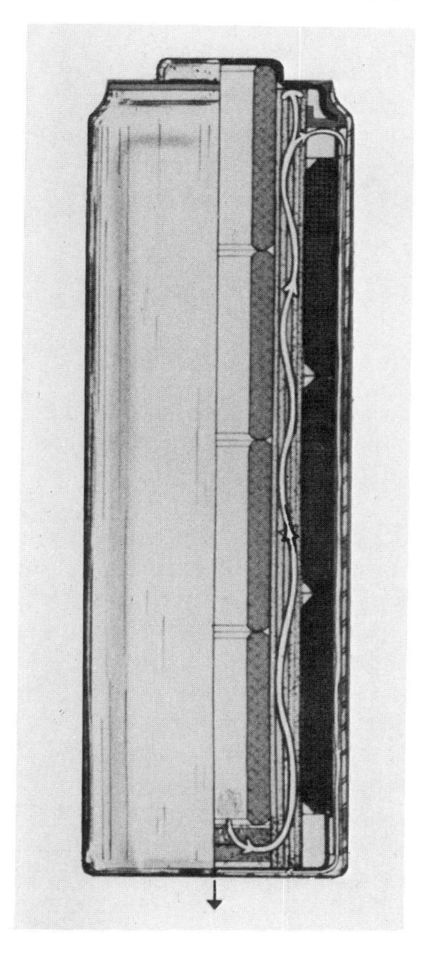

Figure 5.3 Automatic venting of mercury cells.

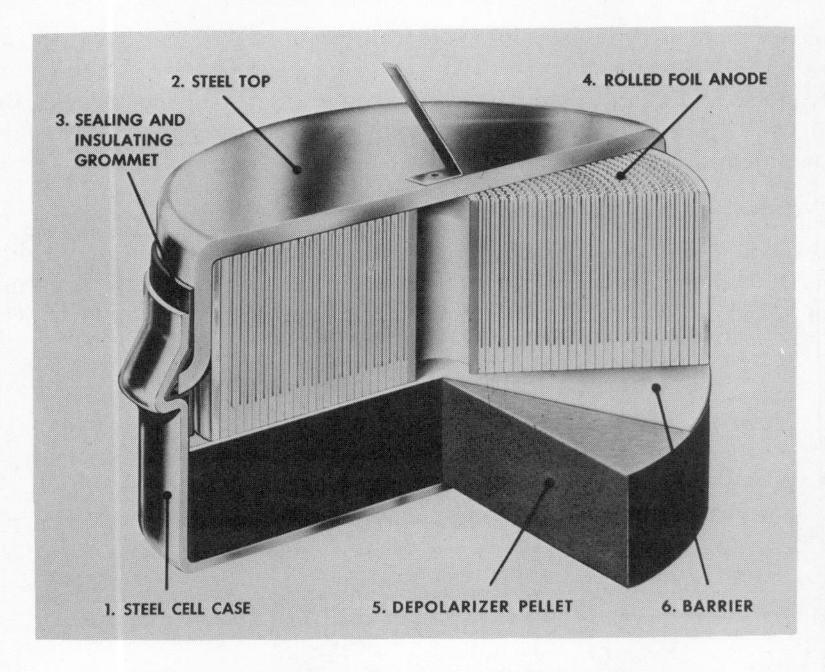

Figure 5.4 Wound-anode pressed powder cathode cell.

anode is wound in an offset manner, zinc protruding at the top for good terminal connection, paper at the bottom, in contact with the barrier layer. Zinc is corrugated to provide ample space for the zinc oxide formed during discharge, and to maintain adequate electrolytic paths despite the accumulation of reaction product. The difficulty of attempting to roll amalgamated zinc can be avoided with this construction since mercury, (10 percent by weight), applied to the exposed end of the zinc spiral, rapidly and uniformly distributes itself over the zinc.

"CROWN" CELL

In this design (cf. Fig. 5.5), found useful for miniature cells and particularly well adapted to series connection, the depolarizing mix of mercuric oxide and graphite or mercuric oxide, manganese dioxide, and graphite is pressed into a steel can which is welded to a steel positive terminal cap. Porous felt-type cellulose serves as barrier and electrolyte-absorbent spacer between anode and cathode. Amalgamated powdered zinc, pressed into a plastic ring, is in contact with the tinplate negative terminal cap to form the porous anode. The cell is sealed by crimping the caps containing the active elements over the shoulders of the plastic ring.

Figure 5.5 Cells and batteries of crown cap construction.

For normal service at temperatures above 15°C (59°F), caustic soda solution containing dissolved zinc oxide may be used as electrolyte; for more severe conditions the potassium hydroxide formula is required.

SERVICE CHARACTERISTICS

In contrast with the conventional Leclanché dry battery, the RM depends on a substantially nonvariant system, with no change during discharge, either in electrolyte composition or *p*H. Since reduction takes place as a one-step reaction without noticeable effect of intermediate mercurous oxide formation, the drop in cathode potential is almost negligible, and even the accumulation of solid zinc oxide causes only slight rise in internal resistance. The net result of these factors is a well-sustained voltage curve even on continuous discharge (Fig. 5.6).

Further analysis of the typical discharge curve (Fig. 5.7) shows an equilibrium period (T_2) represents practically 97 percent of the cell life when the current drain is relatively low, e.g., for an RM 1 cell at 1 ma drain. Voltage is constant (at 98 percent of no-load potential) within less than $\pm\frac{1}{2}$ percent during this period. At higher levels of drain, such as 20 ma, the potential will vary by ± 2 percent from 90 percent of no-load level,

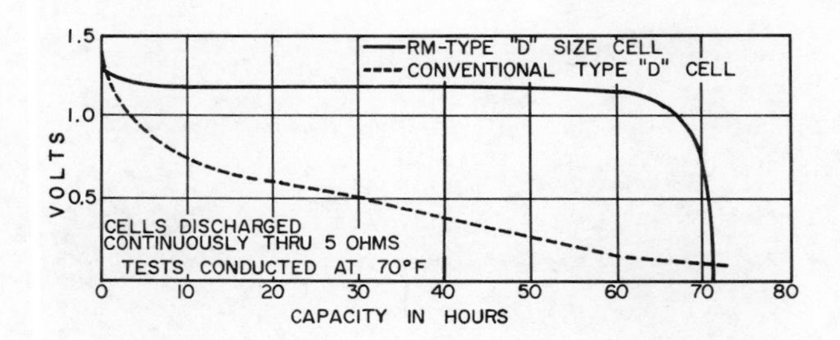

Figure 5.6 Comparison of RM and Leclanché dry cells.

and the equilibrium period is approximately 67 percent of cell life. Larger cells, or groups of smaller cells in parallel, would show better equilibrium period characteristics for equivalent rates of drain. Since the current density per unit of electrode area determines the equilibrium period, it is obvious that by proper battery design the voltage may be kept as constant as desired to meet the demands of any particular circuit.

The rated current discharge of the pressed powder anode is about 1.55 amp-dm² (0.1 amp/in.²) of cathode area. This, however, is somewhat dependent upon the density of the pressed element and the particle size of the cathode and anode powders. The relation of the current drain to the service life and capacity for an RM 1 cell is shown in Fig. 5.8. Since the ampere-hour capacity (1 amp-hr) is relatively constant for the range of

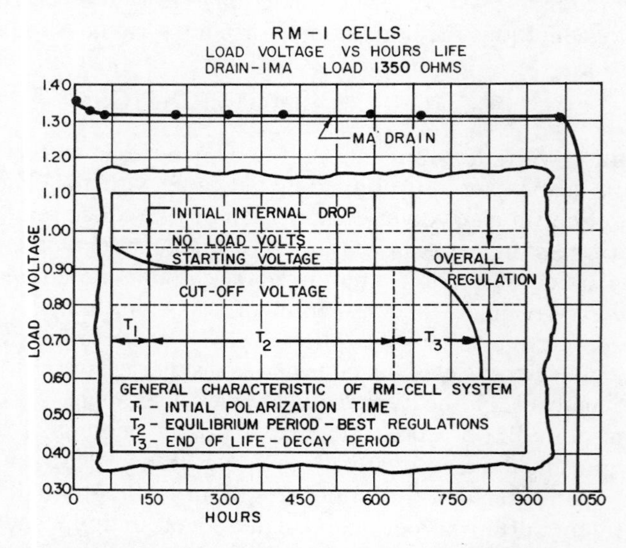

Figure 5.7 General characteristics of zinc-mercuric oxide cells.

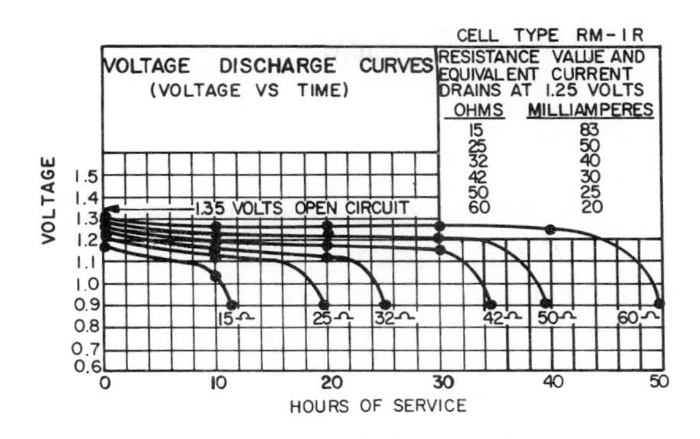

Figure 5.8 Discharge characteristics of RM1R cell at various drain rates.

moderate discharge rates shown, the watt-hour output is roughly proportional to the operating voltage.

Effect of temperature

The operating temperature range of zinc-mercuric oxide cells depends in part upon the specific anode design and size of cell for a given rate of discharge. For most commercial applications, such as hearing aids, the range is indicated in Fig. 5.9. Maximum service life is reached at approximately 70°F (21°C), after which output remains fairly constant, except at low drains, with rise in temperature up to 160°F (71°C).

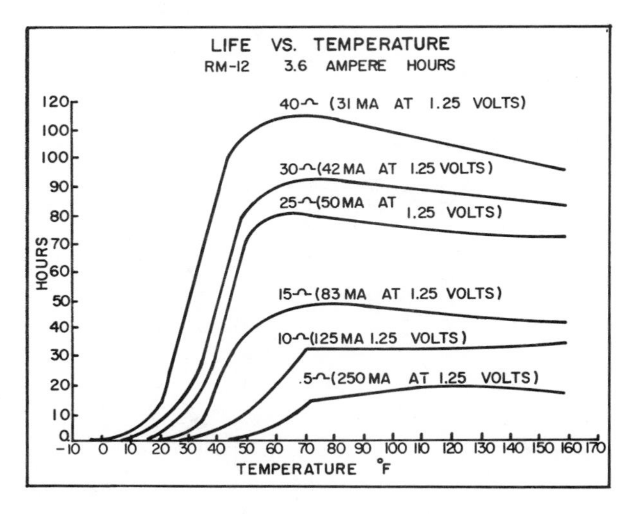

Figure 5.9 Operating temperature range of commercial pressed-powder cells.

The relation of cell voltage to temperature at low current drains is shown in Fig. 5.10. This curve illustrates the typical performance of RM 12 R cylindrical zinc-mercuric oxide cells over the temperature range of −54°C (−65.2°F) to 60°C (140°F) for current drains encountered in many applications. Voltage depression is slight at low temperatures when drains are 100 μamp or less in the larger cell types, or when intermittent drains of this order are used in smaller cells.

Low-temperature operation is an important consideration for such military applications as air-sea rescue and Sonobuoy units. These devices require a compact power source of high reliability that can operate over a wide temperature range with a storage requirement of a minimum of five years under wide limits of ambient temperatures. For these and other low-temperature applications, a cell of wound-anode construction may be used, with deepened corrugations to provide large contact area, ample space for zinc oxide, and adequate electrolyte path. The discharge characteristics of such a unit are shown in Fig. 5.11.

Internal resistance

Because of the porous structure of the solid zinc oxide formed during service, there is no substantial rise in internal resistance due to the build-up of reaction products (until the anode is completely consumed.) This is graphically shown by the ac resistance values of a typical cell noted in Fig. 5.12.

Shelf life

The excellent shelf quality of RM cells is shown in Figs. 5.13 and 5.14. After a year of storage at room temperature (70°F) it will be noted, open

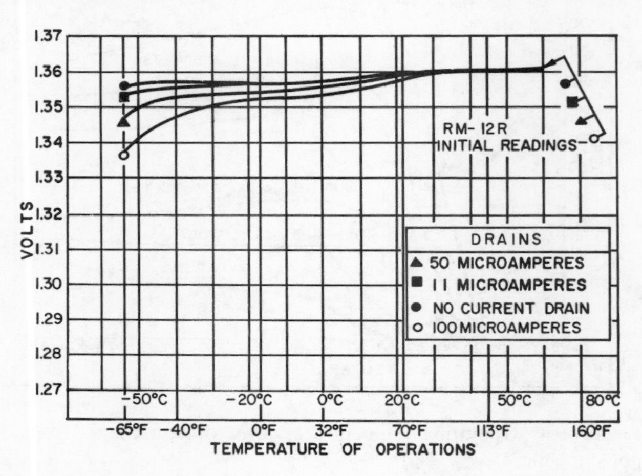

Figure 5.10 Voltage variation with temperature of pressed-powder cells.

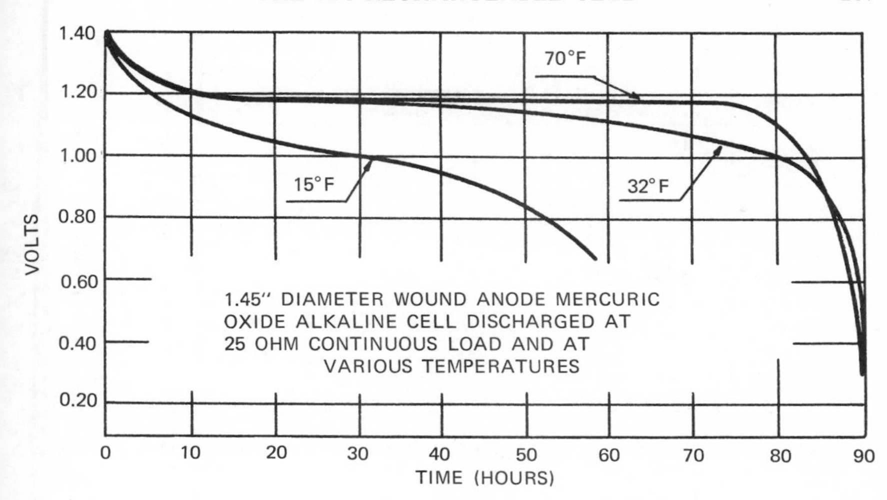

Figure 5.11 Discharge characteristics of RM 1450 at various temperatures.

circuit voltage of typical cells dropped only 0.006 V (0.011 in 5 years) and 97 percent of the original service capacity was maintained. On longer storage periods, mercury cells on low-drain tests showed a 3 percent loss in capacity in 1 year, and only 9 percent in 3 years.

THE RM RECHARGEABLE CELL

Though the standard RM cell can be discharged and recharged a limited number of times, certain modifications in design and composition are requisite for a safe and efficient secondary unit (10).

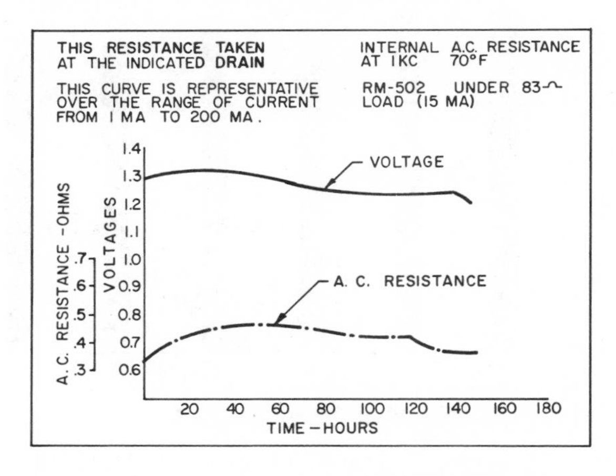

Figure 5.12 A-c resistance of cell during discharge.

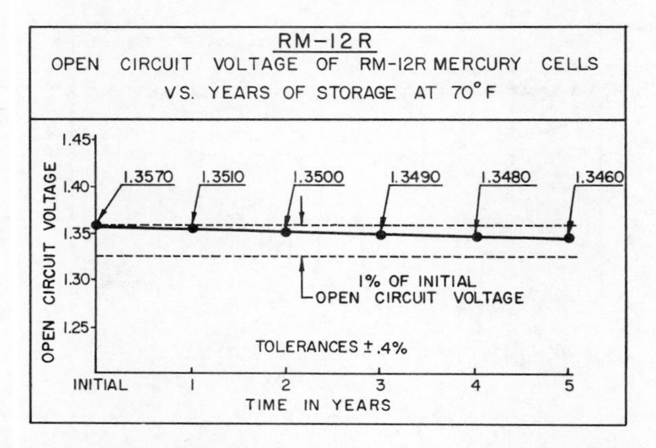

Figure 5.13 Open-circuit voltage of RM 12 cells versus years of storage at 21°C (70°F).

Cell construction

Cell structure, essentially similar to that of the wound-anode primary unit, is shown in Fig. 5.15. The cathode is a mixture of 80 percent mercuric oxide and 20 percent silver powder pressed into a disk and consolidated under high pressure into the cathode container. A steel screen, placed on the bottom of the cathode container, is embedded into the cathode upon consolidation, thus maintaining continuous electrical contact to all sections of the cathode despite volume changes in the cathode during the charging and discharging cycles.

The silver powder serves two functions other than affording electronic

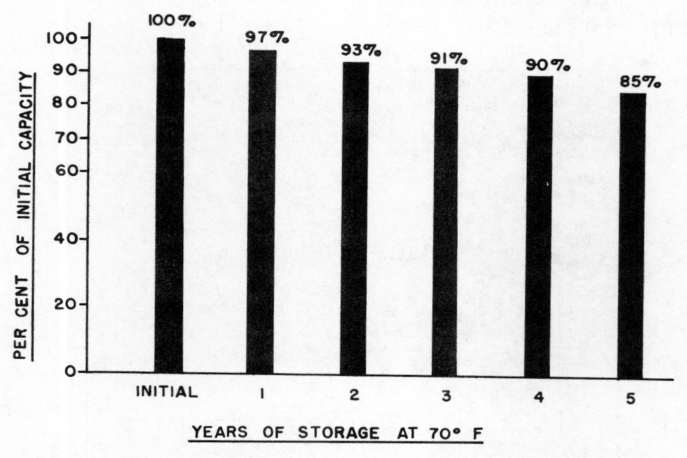

Figure 5.14 Average available capacity for typical mercury cell after storage.

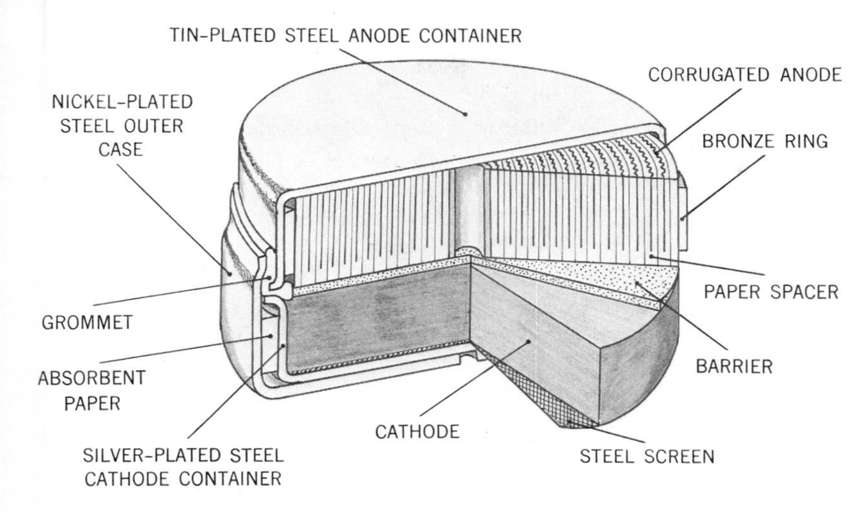

Figure 5.15 Modified structure of wound anode zinc-mercuric oxide cell.

conduction to the cathode. When the mercuric oxide is electrochemically reduced to metallic mercury, the silver amalgamates with it and prevents the formation of mobile mercury globules, thus serving to keep the mercury uniformly distributed in the cathode and promoting the efficient reconversion of mercury to mercuric oxide on charge. During charge, some of the silver near the surface of the cathode is converted to silver oxide, thereby creating a voltage gradient between the surface and interior of the cathode. This gradient permits the reconversion of much thicker cathode sections than would be possible with mercuric oxide alone. In addition, the silver provides a rapid rise in voltage when the mercury has been completely reconverted to mercuric oxide. This rapid rise in potential is adequate to operate a cutoff relay or to reduce the current to a near-zero floating value on a fixed charging voltage.

The corrugated zinc of the primary cell is replaced by a corrugated zinc-plated 85/15 copper zinc alloy strip. This alloy backing provides a constant area for the deposition of zinc during charge and helps to maintain a constant distance between anode and cathode. The corrugation depth is sufficient to allow adequate space for zinc oxide produced on discharge and maintain adequate electrolytic paths.

The anode, like that of the primary cell anode, is wound in an offset manner so that the absorbent paper between the zinc plate strip protrudes at one end and the spacer width of paper on the other. When wound it is retained in rolled form by means of a bronze sleeve instead of a styrene sleeve. It is amalgamated by application of 10 percent of its metal weight with mercury added to the exposed metal end after absorption of electrolyte into its paper spacer.

The barrier material should be resistant to chemical attack, and microporous, like sintered magnesium oxide, synpor, a microporous polyvinyl chloride, or polypor, a nylon cloth coated with a microporous resin. The construction shown in Fig. 5.15 utilizes a sintered magnesium oxide disk in direct contact with the cathode.

Rechargeable cells have narrower operating limits than primary unit and are subject to modifications in structure and size for specific service applications. The data in Fig. 5.16 were obtained over a period of about 1 year for a cell, as described, designed for equipment requiring 35 ma drain on a 12-hour cycle.

GENERAL CONSIDERATIONS

The use of the zinc-mercuric oxide cell has assumed an important position in the battery industry. Its application has been to both military and commercial portable communication equipment where a compact source of high energy per unit volume is specified and where long storage ability is required. Industrial uses are numerous and varied, many

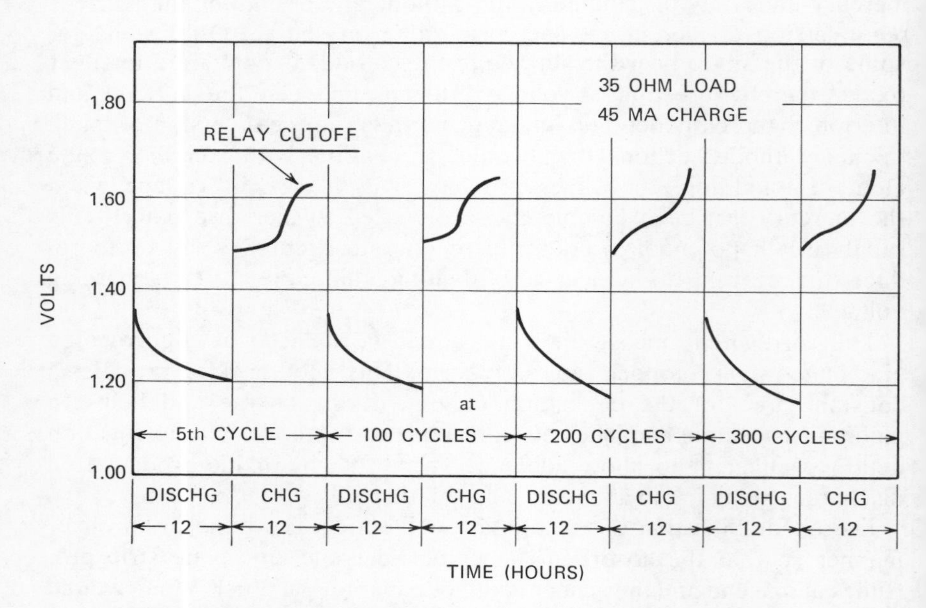

Figure 5.16 Cyclic performance of rechargeable zinc-mercuric oxide cells.

involving special packaging of multicell combinations. The selection of mercury cells for these applications is usually based on one or more of the following characteristics

1. Long storage life.
2. Good voltage regulation.
3. High power per unit volume.
4. Ability to withstand high temperatures.
5. Constant no-load or light-load voltage.

The steel-can structure allows the zinc-mercuric oxide cells to operate under conditions of high stress where their ability to withstand pressures up to 5000 lb/in.^{2} or high vacuum can be of value. Applications of this type have been in meteorological balloons, oceanographic instruments, missile and satellite electric power sources. The compact construction afforded by the high-pressure consolidation of its elements in their respective containers and immobilized electrolyte with constant pressure contact affords reliable operation under the stresses of impact, vibration, or acceleration.

Voltage reference batteries

The constant and reproducible emf of the zinc/mercuric oxide cells allows them to be used as secondary standards of potential. When used as reference cells, they are not as sensitive to mishandling as are the more expensive standard cells. Cells that are accidentally short-circuited on

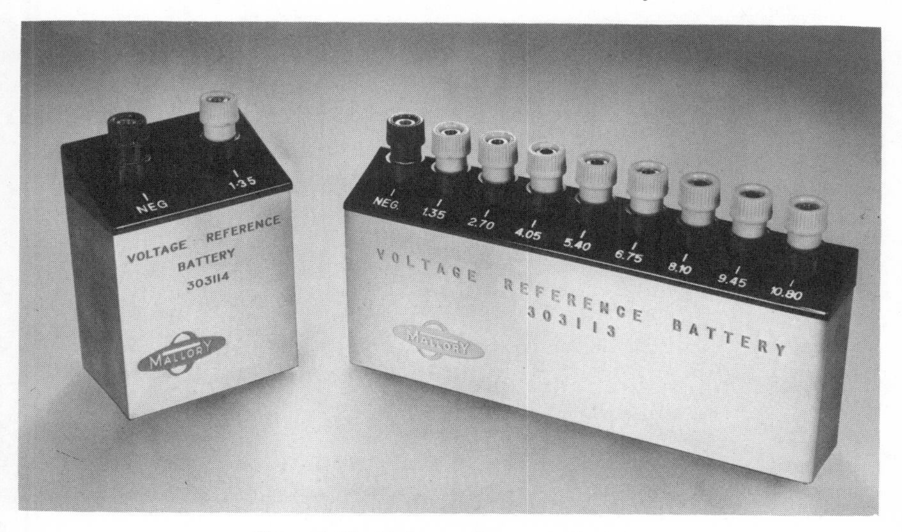

Figure 5.17 Voltage reference batteries.

measuring equipment will not be permanently affected and will return to their rated voltage.

The voltage-reference battery is manufactured in two forms, as illustrated in Fig. 5.17. One is a single cell, and the other a battery of eight voltage-stabilized mercury cells connected in series and assembled in a rugged container having a voltage tap from each cell. This provides reference voltages ranging from that of a single cell (1.35 V) to the total of eight cells (10.80 V).

The voltage stability of the battery is maintained within ½ percent over a period of 5 years in normal use. Rugged construction makes possible use in applications where batteries may be subjected to shock or vibration, or operation in varying positions. The reference battery is quite useful for virtually all applications where standard cell precision is not required. Its voltage range to 10.80 V and the capability of delivering small amounts of current without harm contribute to its broad utility. Although useful in certain instruments, the major application is for general laboratory uses such as the calibration of oscilloscopes and other instruments and potential measurement.

REFERENCES

1. Clarke, C. L., U.S. Pat. 298,175 (1884).
2. Morrison, W., Ger. Pat. 273,143 (1910).
3. Bronsted, J. N., Ger. Pats. 290,748 (1915); 72,147 (1916); Brit. Pat. 16,474 (1915); U.S. Pat. 1,219,074 (1917).
4. Ludwigsen, Ger. Pat. 296,398 (1916).
5. Ruben, S., U.S. Pat. 2,422,045 (1947).
6. *Id.,* "Balanced Alkaline Dry Cells, *Trans. Electrochem. Soc.,* **92**, 183–93 (1947).
7. Friedman, M., and McCauley, C. E., "The Ruben Cell, A New Alkaline Primary Dry Cell Battery," *Trans. Electrochem. Soc.,* **92**, 195–215 (1947).
8. Service Data, Mallory Battery Co., Tarrytown, N.Y.
9. Latimer, W. M., *Oxidation Potentials.,* 2nd ed., Prentice-Hall, Englewood Cliffs, N.J. (1953).
10. Ruben, S., U.S. Pat. 2,571,616 (1951).
11. Herbert, W. S., *J. Electrochem. Soc.,* **99**, 190–1 C (1952).

BIBLIOGRAPHY

Booe, J. M., "The Alkaline Mercuric Oxide Cell," *J. Electrochem. Soc.,* **99**, 209–14 C (1952).

Bro, P., and Epstein, J., *Electrochim. Acta,* **10**, 471-3 (1965).

Clune, R. R., Proc. 14*th* Ann. Power Sources Conference, U.S.A. Signal R. & D. Lab., Fort Monmouth, N.J. (1960).

Id., Proc., Advances in batt. tech. Symposium, *Electrochem. Soc.,* pp. 195–217 (1965).

Dalfonso, J. L., *10th* Ann. Power Sources Conference, U.S. Signal R. & D. Lab., pp. 11–14 (1956).

Garrett, A. G., *Batteries of Today,* Research Press, Dayton, Ohio, pp. 84–7 (1957).

Hallows, R. W., "Dry Battery Development. The R. M. Primary Cell," *Wireless World* (May 1948).

Ko, W. H., and Neuman, M. R., Implant biotelemetry and micro-electronics," *Science,* **156**, 351–60 (1967).

Potter, E. C., *Electrochemistry,* The Macmillan Co., New York, pp. 375–7 (1956).

Przbyta, F., and Kelly, F. J., "Structure of mercuric oxide cathodes," *J. Electrochem. Soc.,* **114**, 71 C (1967).

Ruben, S., Int. Comm. for Electrochem. Thermodynamics and Kinetics (CITCE), 17*th* mtg. (1966).

Vinal, G. W., *Primary Batteries,* John Wiley and Sons, New York, pp. 304–21 (1950).

Walkley, A., *J. Electrochem. Soc.*, **99**, 209–14 C (1952).

Yosizawa, S., Takehara, Z., and Namba, Y., *J. Electrochem. Soc. Japan,* **30**, 185–92 (1962).

6

The Silver Oxide: Zinc System

THEDFORD P. DIRKSE

The silver oxide: zinc system is unique as a power source because it has a watt-hour output per unit of weight or volume greater than that of most primary batteries and because it can deliver this output at a high current rate and at a high, fairly constant voltage level. Zinc is an outstanding anode and silver oxide is preeminent among cathodes. The combination of these with an effective and highly conductive electrolyte, KOH, is responsible for the exceptionally good performance characteristics obtained.

The open circuit voltage of the zinc-silver oxide battery may be 1.85 or 1.60 V, depending on the oxidation state of the cathode; in either case, the discharge voltage is around 1.5 V for the major part of the discharge at normal drain.

Despite the limit on the use of this system imposed by the high cost of silver and its compounds, several important areas of usefulness have been developed commercially. The larger wet cells, in addition to their use as storage batteries, have become important in military and aerospace applications, where cost is a secondary consideration. Since material costs become progressively less important as size of the unit is reduced, miniature dry cells have found wide acceptance as power sources for hearing aids, electric watches, and the like.

A wet battery of this type was proposed by Clarke (8) as early as 1883, though this and subsequent disclosures (26, 33) in the patent literature covered the construction only of secondary units. Apparently little or no use was made of these disclosures, probably because of cost considerations and the poor cycle-life of the system. Although silver oxide is only slightly soluble, the diffusion of dissolved silver to the anode can result in self-discharge and poor charging performance. In 1941 André (2) suggested the use of a cellophane separator to overcome these difficulties. His work brought about a revival of interest in the zinc-silver oxide-alkali system as a secondary battery.

The combination of cadmium anode with silver oxide cathode and KOH electrolyte is used primarily as a secondary battery. The energy density figure is not as high as for the silver-zinc system and the discharge voltage per cell is about 0.4 V lower. However, it has longer cycle life than the silver-zinc system and is used in portable electric appliances and in orbiting scientific satellites.

During World War II, as the result of some preliminary work at the National Bureau of Standards (11), the U.S. military services became interested in the zinc: silver oxide system as a primary battery, stimulating the development of the present day commercial units.

THEORETICAL CONSIDERATIONS

The zinc: silver oxide battery has a flat discharge voltage curve (Fig. 6.1), since the electrodes are not highly polarized and the zinc complexes formed (discussed in detail in Chap. 3) are highly soluble in KOH electrolyte. Indeed, the amount of dissolved zincate, $Zn(OH)_4^{-2}$, reaches a value considerably above that corresponding to a solution of KOH saturated with ZnO (12). Thus, except for very long discharges, the precipitation of zinc as ZnO is insignificant and does not interfere with the discharge process. At higher current rates, however, ZnO or $Zn(OH)_2$ may form faster than it can dissolve in the KOH, and so interfere with the electrode process.

The silver oxide electrode may exhibit a two-level discharge voltage if large quantities of AgO are present (cf. curve *a*, Fig. 6.1). This is due to reactions 1 and 2:

$$2AgO + H_2O + 2e^- \rightarrow Ag_2O + 2OH^- \quad (1)$$

$$Ag_2O + H_2O + 2e^- \rightarrow 2Ag + 2OH^- \quad (2)$$

The potentials of these two reactions differ by about 0.25 V. The electrical capacity per gram of silver is twice as great in AgO as in Ag_2O. The length of the higher voltage level depends on such factors as temperature, discharge current rate, and length of stand before discharge. The discharge voltage drop to the lower level occurs while there is still a significant amount of AgO in the electrode (39) and does not mark the completion of reaction (1) and the onset of reaction (2).

Though the presence of Ag_2O can be detected on a discharging electrode shortly after discharge begins, it is questionable whether a large amount of Ag_2O is present on the electrode (assuming the electrode was AgO to begin with) at any time during discharge. Rather, it appears that

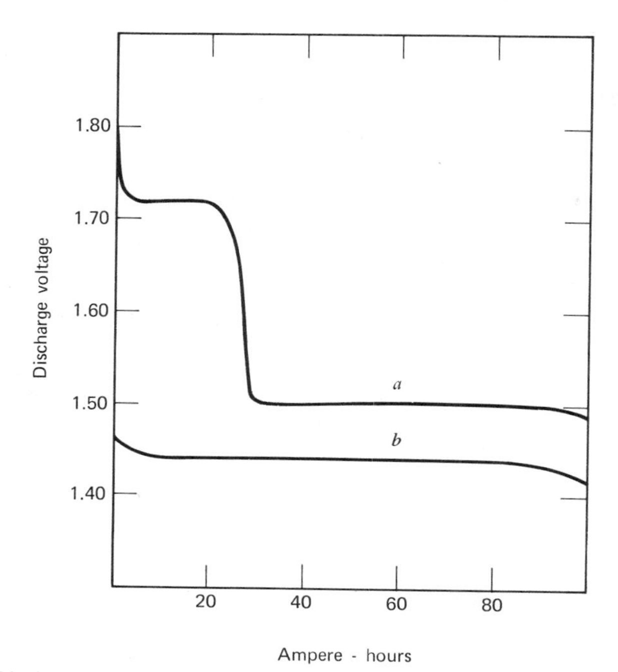

Figure 6.1 Discharge voltage of a 100 amp-hr silver-zinc battery at room temperature. The discharge current is 10 amp for curve *a* and 100 amp for curve *b*.

reactions 1 and 2 proceed simultaneously during discharge without a large accumulation of Ag_2O. However, sufficient Ag_2O does form to give a good x-ray diffraction pattern (41).

Two mechanisms have been suggested for the discharge of the AgO electrode. According to the first the AgO is reduced to Ag_2O or to Ag at the surface (39, 7).* As the surface concentration of AgO decreases, the discharge current density on it increases to the point where the process is polarized to the potential associated with reaction 2. During this process, the Ag produced reacts with undischarged AgO to form Ag_2O, which is further discharged. However, the rate of reaction of Ag with AgO in the presence of KOH has not yet been established.

The second mechanism assumes that the discharge process involves the migration of O^{-2} ions through the silver (ion) lattice (16). These ions are removed as OH^- ions by reaction with water. The potential of the electrode then is determined by the concentration of O^{-2} ions in the silver electrode at the interface. An O^{-2} ion gradient then exists in the electrode.

*This may proceed through a dissolution-precipitation reaction (cf. Miller, B., *J. Electrochem Soc.*, **117**, 491 (1970).

X-ray diffraction evidence, however, indicates the presence of discrete particles of Ag_2O and AgO, though of a poorly crystalline variety (41).

In any event, the result is that the electrode during discharge consists largely of Ag and AgO. Both of these have a low electrical resistance, whereas Ag_2O has a high resistance (30). Hence, because there is no large accumulation of Ag_2O in the electrode during discharge, the electrode has a low electrical resistance throughout the discharge process.

The two-level discharge voltage is often objectionable and may necessitate the use of voltage regulators in the equipment. However, the upper level can be eliminated with no appreciable loss of capacity by a pretreatment of the AgO before discharge. Since the potential of the electrode is set to a large extent by or at the electrode-electrolyte interface, only the surface layer need be treated to reduce AgO to Ag_2O or Ag, to bring the voltage down to the lower level (Fig. 6.1*b*). At the same time the capacity of the electrode will essentially be that of AgO. Fischbach (21) has suggested the use of reducing agents such as photographic developers, but the partial reduction is more often brought about by a carefully controlled thermal treatment.

Much of the uncertainty about the cathode mechanism is due to an incomplete knowledge of the nature of AgO. It does react very slowly with the electrolyte to form Ag_2O and liberate oxygen (1), but it appears that only a thin film of Ag_2O is formed in this way. AgO also reacts with Ag in the presence of the electrolyte to form Ag_2O (14), but the parameters of this reaction have not yet been established.

Attempts have been made to determine the standard potential of reaction 1 and the results obtained indicate a value of about 0.60 V versus the standard hydrogen electrode (3, 4, 18). However, it is difficult to get stable, reproducible results. Because of this uncertainty some reservations must also be placed on other thermodynamic data determined by emf measurements. Yet about the only thermodynamic data available, those of Bonk and Garrett (4), were determined in just this way.

The thermodynamic characteristics of Ag_2O are better known. The standard potential of reaction 2 is 0.34 V on the standard hydrogen electrode scale (24, 17). Other thermodynamic values have been determined by Hamer and Craig (24).

Ag_2O is soluble in the electrolyte to the extent of about 10^{-4} molar. The dissolved material may diffuse to the zinc electrode and react with it to deposit Ag on the zinc (13).

$$Zn + 2Ag(OH)_2^- \rightarrow Zn(OH)_4^{-2} + 2Ag \qquad (3)$$

giving rise to local cell action. Eventually the metallic deposits may form short circuits between the electrode plates, a difficulty that can be minimized by a proper selection of separators and tight packing of the plates.

STRUCTURAL DETAILS

Wet cells

Figure 6.2 shows some of the structural details of a silver-zinc battery. The positive plate is essentially AgO with a surface treatment so that it discharges at the lower voltage level (curve *b*, Fig. 6.1). This electrode may be fabricated in several ways. AgO is commercially available and can be pressed onto a silver screen grid. The plates may also be made by pressing and sintering Ag powder on a silver grid, followed by electrolytic oxidation in a KOH solution at a low current rate (1–5 ma/cm^2) to form AgO. The Ag powder plaques may also be obtained by pasting the grid with Ag_2O and then reducing this thermally to Ag at temperatures of 425–550°C. With overheating the surface particles begin to fuse, lowering the surface area and tending to reduce the capacity of the electrode. If the sintering temperature is too low, a poor bond may be formed between the silver particles and the grid. The efficiency of utilization of silver oxide is rather high but it varies with the length of time of sintering (44). On plates formed by electrolytically oxidizing sintered silver, the capacity delivered at the 10-min rate is about 80 percent of theoretical, assuming that the silver has been completely oxidized to AgO.

Fischbach and Almerini (22) have suggested a process for making the silver oxide plates by mixing Ag_2O with a solution of a high polymer. This mixture is pasted on the grid, dried, thermally reduced, and finally anodized to AgO.

The zinc electrode, or negative plate, may also be produced in several ways. Sheet zinc can be used, but ordinarily an electrode of larger surface area is required, such as that obtained by electroplating a zinc sponge on a grid from an alkaline plating bath. Another method consists of pasting ZnO on a grid and then reducing this electrolytically to zinc in a KOH solution. To reduce the reaction on stand between zinc and the electrolyte, the formed electrode may be amalgamated by treatment with a solution of a mercuric salt or by incorporating HgO in the pasted ZnO electrode.

The electrolyte is usually a 30–40 percent KOH solution, concentrations near the eutectic (31 percent KOH, freezing point about −62°C) ordinarily being preferred for low-temperature work. Since such solutions are rather viscous, especially at lower temperatures, it is customary to provide a system for warming the electrolyte before the cell is activated. Prewarming has the added advantage of eliminating the low open-circuit or operating voltage which prevails at the beginning of discharge when

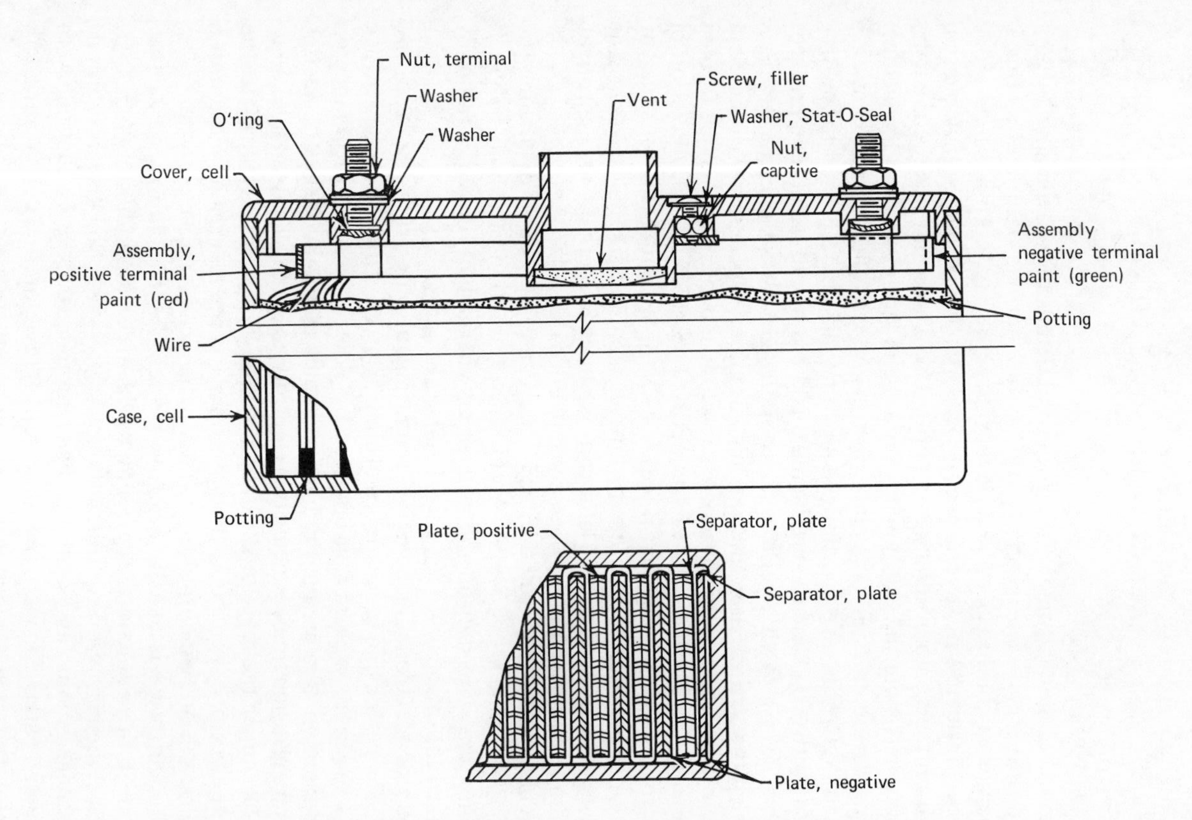

Figure 6.2 Structural details of a silver-zinc primary battery. Courtesy of The Eagle-Picher Co.

the batteries have been in cold storage. Wetting agents such as cresol may be added to the electrolyte to decrease the time necessary for activation.

Many of the structural details of modern silver-zinc batteries are proprietary information. This is especially true of the separators with which the positive and negative electrodes are wrapped to slow down the self-discharge processes and to insulate the closely packed plates from each other.

Separators should prevent the diffusion of the zincate ions and the dissolved silver oxide, and many different substances such as nylon, dynel, and cellulosics have been suggested for this purpose. This requirement is especially important for batteries that are to be recharged, rather than for units that are to be discharged but once, immediately after activation. Ideally, separators should have a low electrical resistance and should not interact chemically with the electrolyte, with any dissolved oxides, or with the active materials on the electrode, but no membrane as yet meets all of these requirements. Ion-exchange membranes have also been suggested (36), but these are mainly for use in secondary batteries. A survey of the effectiveness of various separator materials has been made by Lander and Rhyne (29).

The separators appear to prevent transport of the dissolved silver oxide by reacting with the latter and fixing the silver in the separator. This is analogous to the action of the cellulosic separator used in some forms of the copper oxide cell (43).

Dry cells

Miniature zinc-silver oxide dry cells, currently ranging from about 0.3–0.5 cc in size, 1.2–2.5 g in weight, and from 60–150 ma-hr in discharge capacity, have found wide commercial acceptance for use in hearing aids, electric watches, microminiature lamps, and the like. A typical construction is shown in Fig 6.3. This cell is of the "button"type, comprising an anode of powdered amalgamated zinc admixed with gelatinized KOH electrolyte and separated from the silver oxide cathode by an absorbent layer of cellulosic material. In some applications, e.g., operation of electric watches, where service conditions are uniform, NaOH may be substituted for KOH. Where the relatively rapid loss of power as a cell approaches exhaustion is undesirable, as in hearing aid service, manganese dioxide may be substituted in part for silver oxide to obtain a more gradual reduction of the working voltage.

The discharge curves for a small dry cell given in Fig. 6.4 show a capacity of slightly over 1000 ma-hr, the equivalent of 40–65 hr of hearing aid service, or 1 year's operation of an electric watch.

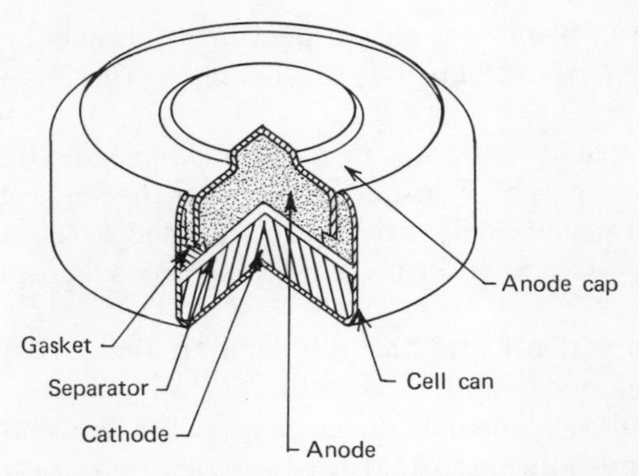

Figure 6.3 Dry type, zinc:silver oxide cell. Courtesy of Consumer Products Division, Union Carbide Corp.

Unlike the wet types, the dry cells have excellent shelf characteristics, some commercial brands showing service maintenance as high as 90 percent after storage periods of 1 year at 70°F.

Experimental models of a dry-tape type battery have been described (45). The active materials are applied to a tape which is fed continuously or intermittently to a set of current collectors that activate the system, allowing the electrochemical reactions to take place. The advantages claimed for this concept are: (*a*) reduction of concentration and activation polarization through continuous feed of reactant and removal of products and the continuous feed of fresh electrode surface; and (*b*) activation of the components occurring only as they are needed, thus permitting unlimited storage of the unactivated portion of the tape. A disadvantage is the necessity of auxiliary equipment to run the tape and

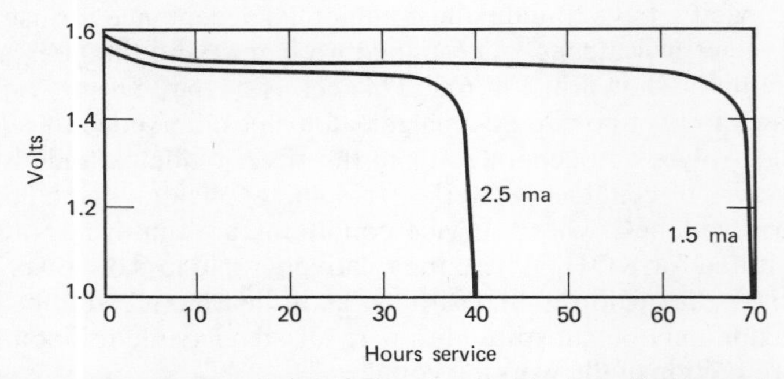

Figure 6.4 Operating characteristics of zinc:silver oxide dry cell at room temperature. Courtesy of Consumer Products Division, Union Carbide Corp. Cell data: diameter, 1.16 cm. (0.455 in.); height, 0.42 cm. (0.165 in.); weight, 1.7 g. (0.06 oz.); service capacity, 1.00 mah.

the expenditure of energy to drive the motors. Cathodic utilization of 90 percent at 1 amp/in.^{2} has been reported.

GENERAL CONSIDERATIONS

One of the main advantages of the zinc-silver oxide battery is its theoretically high watt-hour output per unit weight and volume. Considering only the weight of active materials, the system should theoretically deliver 0.42 watt-hr/g (193 watt-hr/lb) and 3.1 watt-hr/cc (51 watt-hr/in.^{3}). These figures do not include the weight of grids, separators, battery case, electrolyte, activating mechanism or *i-r* drop and consequently, are limiting values that are unattainable practically.

Present commercial batteries will deliver up to 120 watt-hr/lb. (0.265 watt-hr/g and 11.4 watt-hr/in.^{3} under ideal conditions. Generally, however, the values are around 30 percent of these figures and in certain applications they may be even lower, decreasing with reduction in cell size and with increase in discharge current rate. The corresponding values for silver-zinc storage batteries are still lower. In such batteries more separation is needed to prevent the self-discharge reactions over a period of time which includes stand as well as cycling. This also allows for more deterioration of the separators by reaction with the silver oxides in the presence of the electrolyte.

One of the factors working to decrease the energy output per unit weight and volume is deterioration during the period of idleness after activation, which often makes the safe "wet stand time" a matter of 10 hours or less. Thicker and more efficient separator materials add to the internal resistance of the battery and reduce the discharge current rate available but make it possible to achieve wet-storage times of the order of months.

The intrinsically poor shelf life of wet cells is customarily overcome by filling the batteries with electrolyte only at the time they are to be used. This is especially true for batteries that are to be discharged at a high current rate. The electrolyte may be stored in a container attached to the battery case (cf Fig. 6.5) to be added either manually or automatically shortly before the battery is to be discharged. Various devices such as a moving piston or an expanding bladder may be used for activation, and it is possible to fill a battery in as little as 0.2 sec, although times up to a few seconds are more common. In general, to achieve the shorter activation time, less membrane protection must be used, which means shorter wet storage life. Activation devices add to the weight and volume of the battery and thus reduce the power output available per unit weight and volume. However, such devices also enable the battery to be stored dry

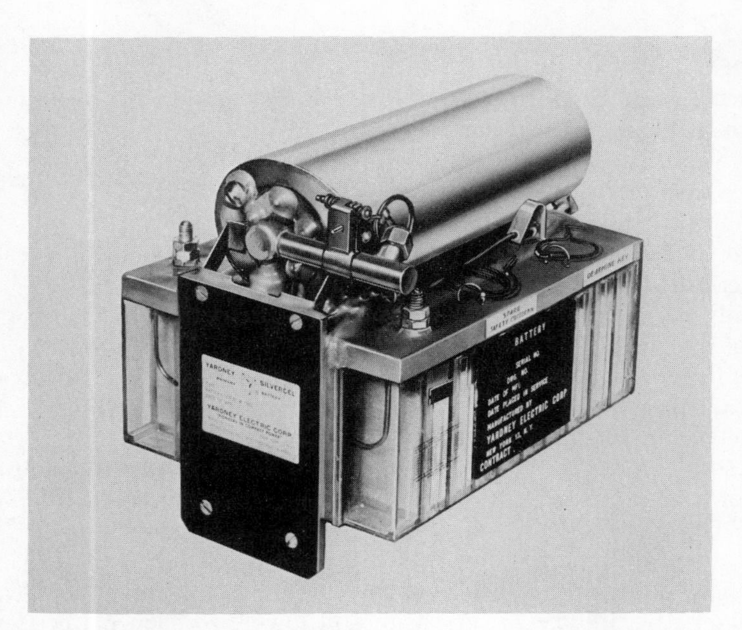

Figure 6.5 Automatically activated silver-zinc primary battery. Discharge characteristics: 135 amp for 12 min at 24–27 V. Size: 500 in.^{3}; weight: 20 lb. Courtesy Yardney Electric Corp.

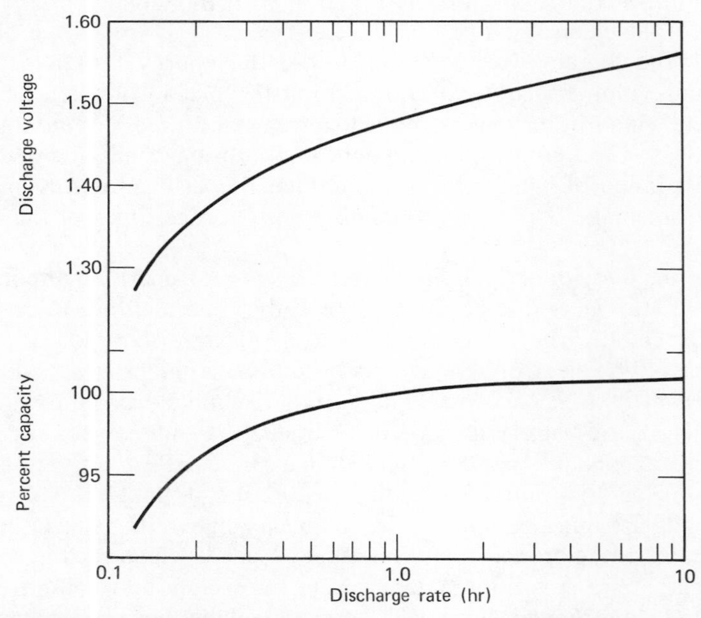

Figure 6.6 Variation of capacity and discharge voltage of a 100 amp-hr silver-zinc battery with discharge current rate at room temperature.

for long periods provided the temperature is not high enough to decompose the silver oxides. Temperatures up to 80°C can be tolerated.

There are several features that make the silver-zinc battery useful for many specialized applications. A rather constant voltage is obtained on discharge, particularly if the cathode has been treated to remove the higher voltage level due to reaction 3, (Fig. 6.1*b*). This flat voltage curve is maintained even at high discharge current rates though, of course, at lower levels (Fig. 6.6). Starting voltage may be low at the beginning of discharge, but as the discharge progresses, the cell warms up and the voltage rises and soon levels out. Both the capacity and operating voltage decrease at temperatures below 15°C (Fig. 6.7), but a reasonably flat discharge curve is maintained.

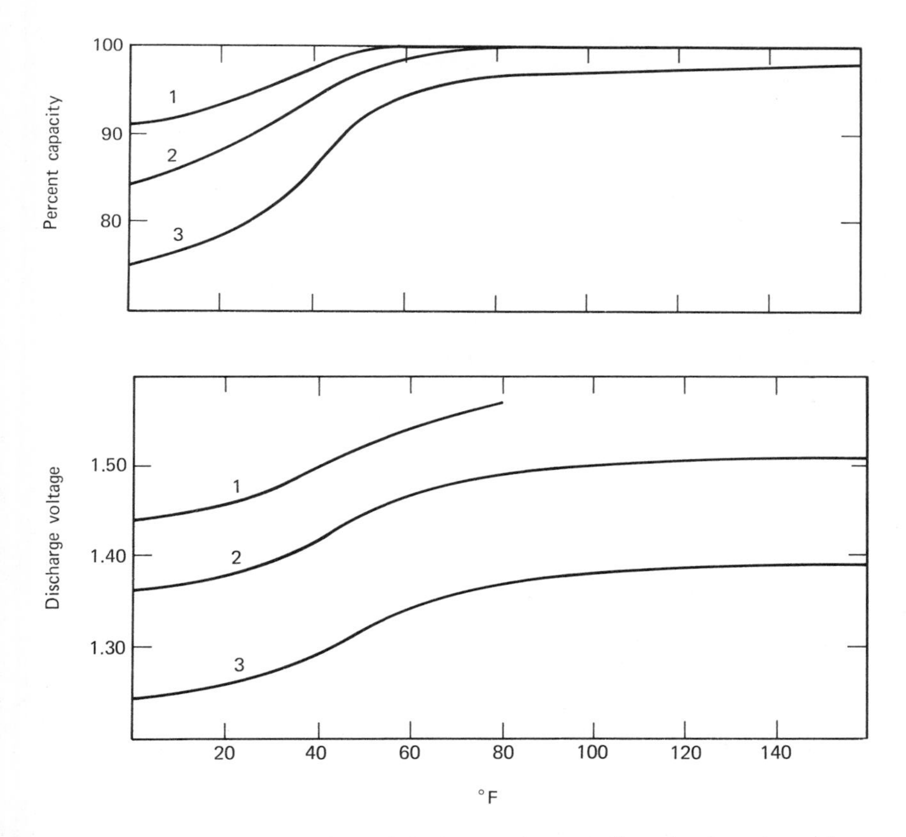

Figure 6.7 Variation of capacity and discharge voltage of silver-zinc batteries with temperature. Discharge rates for a 100 amp-hr battery: 10 amp for curve 1; 100 amp for curve 2; 500 amp for curve 3.

Batteries are available which will discharge at current drains equal to 30 times the rated ampere-hour capacity, or up to 0.5 amp/cm^2. The limiting factor here appears to be the heat generated within the cell at these high rates.

There is, however, also a limitation on the use of the silver-zinc battery. Because of its short shelf life, its capacity generally must be withdrawn in one discharge. Unless there is considerable separator protection it is not practicable to partially discharge a battery and allow it to stand for some time before withdrawing the rest of the built-in capacity.

Silver-zinc batteries are ideally suited for use in missiles and other devices where a single discharge is needed, where there is a premium on space and weight, and where a high current drain is desired. Because of progress and changes in this area, sizes of silver-zinc batteries have not been standardized to the extent they have been for other systems. Wet cell models range from single cells weighing a few ounces and having a capacity of the order of 1 amp-hr to larger batteries weighing over 2000 lb. Some of these units contain two sections, each having different capacities and different discharge voltages. Depending on the discharge current rate, the energy density figure may be as high as 100 watt-hr/lb.

ACKNOWLEDGMENTS

The Cook Battery Co., the Yardney Electric Corp., and the Eagle-Picher Co. have supplied information on operational characteristics of the wet cells, and the Consumer Products Division of Union Carbide Corp. the data for dry cells as used in this chapter.

REFERENCES

1. Amlie, R. F., and Ruetschi, P., *J. Electrochem. Soc.*, **108**, 813 (1961).
2. André, H., *Bull. soc. Franc. elec.*, (6) **1**, 132 (1941).
3. Bedwell, M. E., "Thermodynamics of the Zinc-Silver Oxide Cell" Report No. Chem. 515, Royal Aircraft Establishment, Farnborough, England Feb. 1958.
4. Bonk, J. F., and Garrett, A. B., *J. Electrochem. Soc.*, **106**, 612 (1959).
5. Booe, J. M., *ibid.*, **99**, 197 C (1952).
6. Butler, E. A., "Discharge Behavior of the AgO-Ag Electrode," Technical Report No. 32-535, Jet Propulsion Laboratory, Pasadena, Calif. (22 December 1963).
7. Cahan, B. D., Ockerman, J. B., Amlie, R. F., and Ruetschi, P., *J. Electrochem. Soc.*, **107**, 725 (1960).
8. Clarke, C. L., Brit. Pat. 1932 (1883).
9. Cohen, G. L, and Atkinson, G., *Inorg. Chem.*, 3, 1741 (1964).
10. Cooper, J. E., and Fleischer, A., "Battery Separator Screening Methods" AF Aero Propulsion Lab., Wright-Patterson Air Force Base, Ohio (1964).
11. Denison, I. A., *J. Electrochem. Soc.*, **90**, 387 (1946).

12. Dirkse, T. P., *ibid.*, **102**, 497 (1955).
13. Dirkse, T. P., and De Haan, F., *ibid.*, **105**, 311 (1958).
14. Dirkse T. P., and Wiers, B., *ibid.*, **106**, 284 (1959).
15. Dirkse, T. P., *ibid.*, **106**, 453; 920 (1959).
16. *Id.*, *ibid.*, **107**, 859 (1960).
17. *Id.*, *J. Chem. and Eng. Data*, **6**, 538 (1961).
18. *Id.*, *J. Electrochem. Soc.*, **109**, 173 (1962).
19. Dirkse, T. P., and Vander Lugt, L. A., *ibid*, **111**, 629 (1964).
20. Eidensohn, S., *ibid.*, **99**, 252 C (1952).
21. Fischbach, A., U.S. Pat. 2,739,179 (1957).
22. Fischbach, A., and Almerini, A. L., U.S. Pat. 2,811,572 (1957).
23. Fischbach, A., and Hochberg, F., U.S. Pat. 2,640,864 (1953).
24. Hamer, W., and Craig, N., *J. Electrochem. Soc.*, **104**, 206 (1957).
25. Hawkins, B. R., "Investigations Leading to the Development of a Primary Zinc-Silver Oxide Battery of Improved Performance Characteristics," Final Report on Contract NAS-8-5493 of Eagle-Picher Co., Joplin, Mo. (31 July 1964).
26. Jungner, E. W., Germ. Pat. 110,210 (1900); U.S. Pats. 670,024 (1901); 692,298 (1902).
27. Kinoshita, K., *Bull. Chem. Soc. Japan,* **12**, 164 (1937).
28. *Id.*, *ibid.*, **12**, 366 (1937).
29. Lander, J. J., and Rhyne, J. W., "Silver Oxide-Zinc Battery Program", Technical Report on Contract No. AF 33(600)-41600, Delco-Remy, Anderson, Ind. (12 Jan. 1961).
30. Le Blanc, M., and Sachse, H. *Physik. Z.*, **32**, 887 (1931).
31. McMillan, J. A., *J. Inorg. Nucl. Chem.*, **13**, 28 (1960).
32. Mendelsohn, M., and Horowitz, C., U.S. Pat. 2,872,362 (1959).
33. Morrison, W. U.S. Pat. 975,980, 975,981, 976,092 (1910).
34. Moulton, J. D., and Enters, R. F., U.S. Pat. 2,615,930 (1952).
35. Scatturin, V., Bellon, P., and Salkind, A., *Ricerca sci.*, **30**, 1034 (1960).
36. Shair, R. C., Bruins, P. F., and Gregor, H. P., *Ind. Eng. Chem.*, **48**, 381 (1956).
37. Solomon, F., U.S. Pat. 2,818,462 (1957).
38. Vinal, G. W., *Primary Batteries*, John Wiley and Sons, New York (1950).
39. Wales, C. P., and Burbank, J., *J. Electrochem. Soc.*, **106**, 920 (1959).
40. Wales, C. P., *ibid.*, **108**, 395 (1961).
41. Wales, C. P., and Burbank, J., *ibid*, **112**, 13 (1965).
42. White, J. C., Pierce, R. T., and Dirkse, T. P., *ibid.*, **90**, 467 (1946).
43. Wolfe, F. J., U.S. Pat. 1,564,741 (1925).
44. Wylie, G. M., "Investigation of AgO Primary Batteries," Final Report on Signal Corps Contract No. DA-36-039-SC-78319, Electric Storage Battery Co., Raleigh, North Carolina. DDC AD-267 953 (31 August 1961).
45. *Chem. and Eng. News,* p. 42 (Feb. 24, 1964).

7

The Alkaline Manganese Dioxide:Zinc System

N. Corey Cahoon and Harry W. Holland

A wet cell with zinc anode, manganese dioxide:carbon cathode, and caustic hydroxide electrolyte was reported first in the 1880s (1). A dry counterpart was described in 1912 (3), but despite these and subsequent disclosures, particularly in the patent literature (1), (2), (3), (4), (5), (6), (7), (8) as compiled by Herbert (9), no commercial wet cell has been developed, and it was not until 1949 that a successful dry unit was placed on the market.

The dry type, to which the present discussion is limited, has several attractive features. It is capable of well-sustained operating voltage at high current drains; it is substantially better than the conventional Leclanché dry cell at low temperatures; it has long shelf life. The system is reversible, within limits, and can be used as a low-cost, efficient storage battery.

GENERAL CONSIDERATIONS

The cell system is basically simple, comprising, in general terms, an amalgamated zinc anode, an electrolyte of strong KOH solution containing dissolved zinc, and a compacted cathode of manganese dioxide and carbon. Many of the parameters involved are common to alkaline batteries and have been discussed in previous chapters, but a substantial amount of research and development was required to produce an acceptable commercial unit.

The open-circuit voltage of the alkaline cell is 1.54. In simplified form, the over-all reaction, on discharge, may be expressed as

$$2Zn + 3MnO_2 \rightarrow Mn_3O_4 + 2ZnO \qquad (1)$$

As the equation suggests, the electrolyte is substantially nonvariant, zinc oxide being formed as solid reaction product.

The success of the cell has been dependent partly on the proper selection and combination of battery constituents and partly on structural features.

ANODE

Anode configuration, as in the other alkaline dry cells, is determined in large measure by the type of service for which the finished cell is intended. The solid ZnO reaction product accumulates, usually in adherent form, at the anode surface, and flat zinc sheet is therefore generally suitable only for relatively light service. To overcome this difficulty, various expedients to increase anode surface have been described, notably the use of multiple layers of perforated or corrugated zinc foil or, preferably, of compacted zinc powder (10). The suspension of powdered zinc in gelled electrolyte (11) (in which the metallic particles remain in electrical contact), commonly used in commercial practice, is less sensitive to passivation than the other configurations because of the larger amounts of electrolyte located near the active anode surfaces.

ELECTROLYTE

The electrolyte commonly used is a strong aqueous solution of potassium hydroxide in the concentration range (e.g., 30 percent KOH) favoring the precipitation of solid ZnO when saturation occurs during cell discharge (cf. Chap. 3). Because of higher conductivity and lower freezing point, KOH is preferred to NaOH in commercial cells. The solution of ZnO or $Zn(OH)_2$ in the electrolyte, a practice of long standing in the battery art (12), reduces corrosion of zinc during storage and is generally favored for rechargeable cells. According to Ruben (13), the zinc content may be 10–20 percent that of KOH.

The electrolyte is immobilized with gelling agent, commonly sodium carboxymethylcellulose, which may be added in small amounts to the electrolyte solution. The higher quantities produce a very stiff gel, which discourages movement of liquid but also reduces conductivity. Reduced conductivity becomes an increasingly serious problem with lowering of temperature.

Where absorbent separators are used, the chosen material must be resistant to chemical attack by the electrolyte and the strongly oxidizing environment of the cathode. The severity of attack increases with temperature, and the materials used must be selected with the anticipated service conditions in mind. For general-purpose cells, regenerated cellulose "nonwoven fabric" has proved satisfactory, but some units use a blend of cellulose and vinyl chloride-acetate fibers. Nylon and dynel fibers are also suitable.

CATHODE

The cathode is an electrolyte-wet, compressed mixture of carbon or graphite and MnO_2, the latter preferably the electrolytic, predominantly gamma-oxide[1]. Mix ratio may vary with the type of service for which a cell is designed; higher proportions of carbon, e.g., reduction in MnO_2:C ratio from 5:1 to 4:1 favors low-temperature operation by reducing internal resistance.

THE CATHODE REACTION

Manganese dioxide is used as the cathode reactant or depolarizer in both the alkaline MnO_2:Zinc and the Leclanché dry cell. The alkaline electrolyte containing an aqueous solution NaOH or KOH represents such a different chemical environment from the $ZnCl_2$-NH_4Cl-H_2O of the Leclanché system that little similarity in reaction mechanism exists.

The mechanism of the cathode reaction in the alkaline MnO_2-Zinc system has been studied by a number of workers and their results are presented in an interesting series of papers. For many years efforts to utilize MnO_2 as the cathode in an alkaline cell failed to furnish any reasonable output of electrical energy. Herbert (9) provided the first description of the behavior of MnO_2 as a cathode in the crown type cell which appeared on the market in 1949 where he identified the reaction product as Mn_2O_3. This cell was essentially a low-current unit, usually applied in series connected stacks of cells. Some years later, studies at the Union Carbide Consumer Products Research Laboratories by Cahoon and Korver (14) provided the basis for the commercial development of practical batteries. In contrast to the earlier type of unit this new cell was capable of carrying unusually heavy currents. In support of this, they noted three steps for the cathode reaction of electrolytic MnO_2 in experimental cells:

1. The simultaneous reduction of MnO_2 to $Mn(OH)_2$ and an intermediate oxide tentatively identified as Mn_4O_7 through the following reactions:

$$4MnO_2 + 2K^+ + 2e^- + H_2O \rightarrow Mn_4O_7 + 2KOH \qquad (2)$$

$$MnO_2 + 2K^+ + 2e^- + 2H_2O \rightarrow Mn(OH)_2 + 2KOH \qquad (3)$$

2. The simultaneous reduction of Mn_4O_7 to $Mn(OH)_2$ and Mn_3O_4:

$$3Mn_4O_7 + 10K^+ + 10e^- + 5H_2O \rightarrow 4Mn_3O_4 + 10KOH \qquad (4)$$

$$Mn_4O_7 + 6K^+ + 6e^- + 7H_2O \rightarrow 4Mn(OH)_2 + 6KOH \qquad (5)$$

[1]Several allotropic varieties of MnO_2 are currently used by the battery industry. Gamma, produced by the electrolytic process, is the most important synthetic MnO_2 used commercially. Beta MnO_2 may be prepared in a variety of ways, one of which is described by Kozawa and Powers (27). Other varieties, e.g. alpha MnO_2 is only referred to briefly in this chapter. However all three allotropic forms occur in nature.

3. The final stage, in which Mn_3O_4 is reduced to $Mn(OH)_2$:

$$Mn_3O_4 + 2K^+ + 2e^- + 4H_2O \rightarrow 3Mn(OH)_2 + 2KOH \quad (6)$$

Step 1 was associated with a drop in working voltage from an initial level of 1.46 to about 1.20; step 2 with a drop from 1.20 to about 0.95; and step 3 with that part of the discharge below 0.95 V. Both x-ray diffraction and analytical evidence for the well known product, Mn_3O_4, were presented. $Mn(OH)_2$ was determined by its solubility in a chloride electrolyte. The evidence for the intermediate Mn_4O_7 is based on the assumption that only a single insoluble cathode product was formed and the finding that during the early stages of discharge the original MnO_2 was reduced at a rate of 0.112 g MnO_2/amp-min, as compared with 0.108 g MnO_2/amp-min calculated from Eq. (2). These data were obtained in small experimental cells made without much more electrolyte than could be absorbed by the electrodes and separator. The resulting x-ray data are shown in Fig. 7.1.

Bell and Huber (15) examined the cathode reaction in beaker type cells where a much greater amount of electrolyte was present than is possible in a commercial unit. Furthermore, the use of low current drains obtained from loads of 100 ohms/cell, equivalent to 0.5 ma/g MnO_2, gave ample opportunity for concentration equilibrium to be maintained throughout the discharge. These authors relied on the x-ray diffraction technique to establish the identification of the products of the cathode reaction. Figure 7.2 shows the discharge data obtained on γ-MnO_2, of electrolytic origin from a Japanese source, as well as the times at which cells were taken off test for x-ray examination.

They also found three stages in the discharge with electrolytic MnO_2

1. MnO_2-$MnO_{1.7}$ — Homogeneous phase reduction producing α-MnOOH (7)
2. $MnO_{1.7} - MnO_{1.47}$ — Heterogeneous system producing γ-Mn_2O_3/Mn_3O_4 (8)
3. Below $MnO_{1.47}$ — Heterogeneous system producing $Mn(OH)_2$ (9)

In their analysis they applied Vetter's (15a) thermodynamic treatment. According to this the homogeneous reaction is shown when the potential varies with the composition, in this case the X value in MnO_x, whereas the heterogeneous reaction, indicating two phases, is shown by a potential that remains constant over a range of compositions. These criteria are valid only when the rate of solution of the oxide or hydroxide is so low that equilibrium exists in the system. Even though their experimental cells were discharged at a low current of about 0.5 ma/g MnO_2, Bell and Huber (15) point out that their $MnO_{1.7}$ composition closely approximates

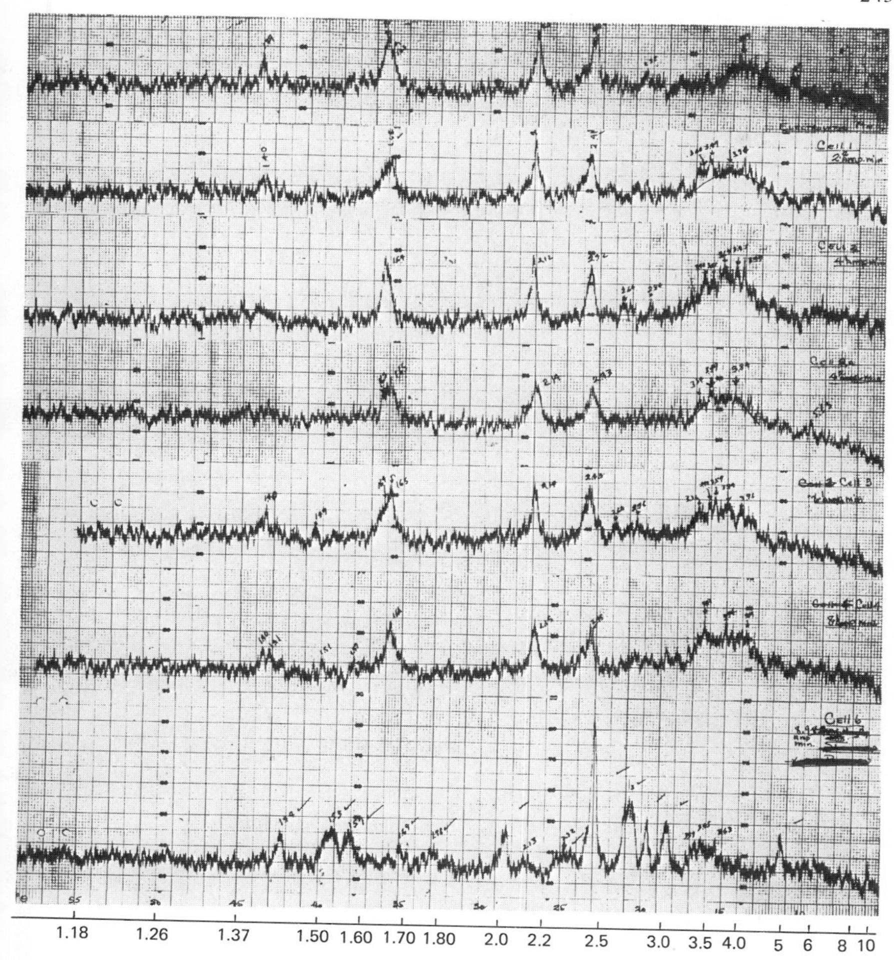

Figure 7.1 X-ray diffraction patterns of insoluble cathode products at various levels of discharge. Upper scale shows diffraction angle in °θ. Lower scale shows d values in Ångstroms.

the Mn_4O_7 tentatively proposed by Cahoon and Korver in their work. They also suggest that the higher current density used by Cahoon and Korver, 60 ma/g MnO_2, probably was responsible for the $Mn(OH)_2$ found throughout the discharge.

Bell and Huber also examined the cathode reaction in cells made with a Caucasian natural MnO_2 which was predominately β-MnO_2. They found a similar stepwise discharge to that reported above for electrolytic MnO_2 but containing five steps with products similar to those reported above.

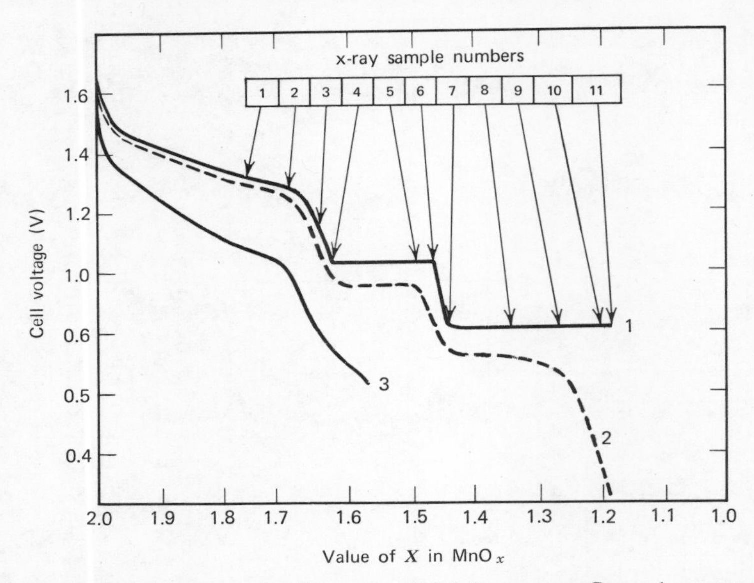

Figure 7.2 Discharge curves of cells made with MnO_2 gamma. Curve 1 = open circuit voltage after recuperation; Curve 2 = closed circuit voltage on 100 ohms 100 hours per week; Curve 3 = closed circuit voltage on 5 ohms continuous for comparison; Arrows show when cathode samples were taken for X-ray tests.

These authors examined the products formed when a synthetic α MnO_2 was used as cathode reactant. Their results were summarized thus:

1. Reduction in homogeneous phase to $MnO_{1.83}$ characterized by a rapidly falling potential and an unchanging crystal structure.
2. Reduction of the oxide $MnO_{1.83}$ proceeds directly to $MnO_{1.5}$ and a heterogeneous system is developed characterized by a constant potential and structures typical of α-MnO_2, γ-Mn_2O_3/Mn_3O_4 and γ-MnOOH.
3. Reduction of the oxide $MnO_{1.5}$ proceeds directly to $Mn(OH)_2$ and a second heterogeneous system develops characterized by a lower, constant potential and structures typical of γ-Mn_2O_3/Mn_3O_4, γ-MnOOH and $Mn(OH)_2$.

The work of Kozawa and a number of different collaborators has introduced several techniques to the field with interesting results. Thus Kozawa and Yeager (16) applied the electrode structure used earlier by Vosburgh et al. (17, 18, 19, 20) in Leclanché cell studies. This consisted of a short length of spectroscopic graphite rod on which a thin layer of MnO_2, about 10 mg, had been deposited electrolytically. Such electrodes were discharged in a variety of concentrations of aqueous KOH electrolyte which did not contain zinc, in contrast to the work of Bell and Huber as well as that of Cahoon and Korver where zinc was present.

Kozawa and Yeager used a beaker-type cell with a large amount of electrolyte and currents from 0.11 to 33 ma/g MnO_2 electrode. Under the lightest drain conditions in 9 *M* KOH a two-step discharge was found (shown in Fig. 7.3);

1. Homogeneous system: MnO_2 to $MnO_{1.5}$ forming α-MnOOH
2. Heterogeneous system: $MnO_{1.5}$ to $MnO_{1.0}$ forming $Mn(OH)_2$

This finding corresponds closely to steps 1 and 2 found by Bell and Huber and both authors observed a reddish violet color in the electrolyte or separator near the end of the first step. They believe the color is evidence for an Mn^{+3} ion or complex. Kozawa and Yeager propose an electron-proton[2] mechanism for step 1 in which

$$MnO_2 + H^+ + e^- \rightarrow MnOOH \tag{10}$$

or

$$MnO_2 + H_2O + e^- \rightarrow MnOOH + OH^- \tag{11}$$

The second step represents the reduction of the trivalent oxyhydroxide MnOOH to $Mn(OH)_2$ as proposed by earlier workers. They also found

[2] An electron-proton mechanism has been proposed previously by Coleman (21), Vosburgh (22), and Scott (23), for cathode reduction in the Leclanché cell.

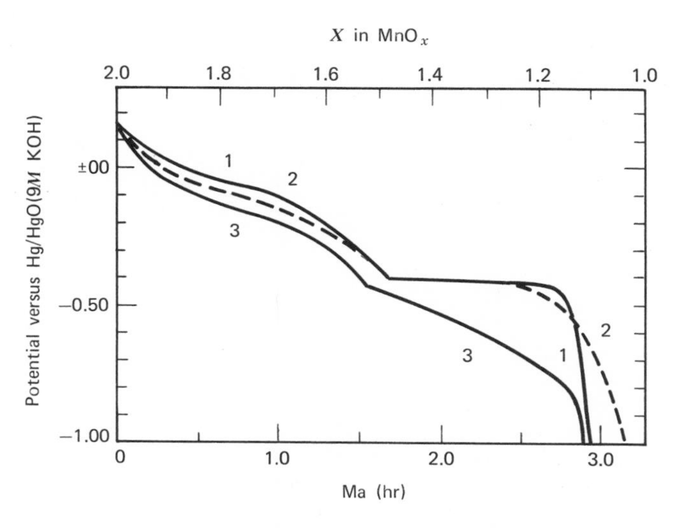

Figure 7.3 Discharge curves at constant current in 9*M* KOH at 23°C. The x value in MnO_x was calculated from the amount of electricity drawn. Curve 1, at 0.11 ma/electrode; curve 2, at 0.33 ma/electrode; curve 3, at 3.0 ma/electrode.

that the effectiveness of the second step was dependent on KOH concentration and current density, and proposed that it operated through a mechanism involving the solution and electrochemical reduction of the three-valent manganese oxide MnOOH thus:

$$\alpha\text{-MnOOH} \rightleftarrows \text{Mn III in solution} \tag{12}$$

$$\text{Mn III} + e^- \rightarrow \text{Mn II} \tag{13}$$

$$\text{Mn II in soln} \rightleftarrows \text{Mn(OH)}_2 \text{ solid} \tag{14}$$

Kozawa and Powers (24) have extended the study of the potential developed by the cathode in KOH electrolytes as it is discharged by the homogeneous phase reaction in the range $MnO_{2.0}-MnO_{1.50}$. They have developed the following equation relating the potential of the oxide to the degree of reduction:

$$E = E^\circ - \frac{RT}{F} \ln \frac{[\text{Mn}^{+3}]_{\text{solid}}}{[\text{Mn}^{+4}]_{\text{solid}}} \tag{15}$$

corresponding to the electron-proton reduction reaction previously given, as diagramatically presented in Fig. 7.4. A good correlation exists between measured and calculated values for the range MnO_2 to about

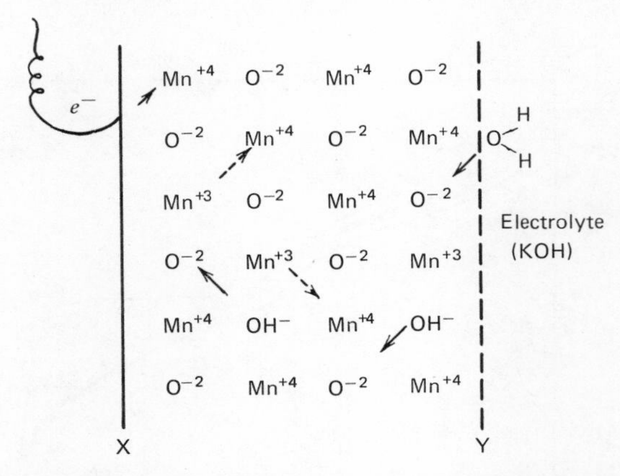

Figure 7.4 Schematic presentation of the solid phase (Mn^{+4}—Mn^{+3}—O^{-2}—OH^-) during the discharge of MnO_2. The arrows show the directions of movement: solid line = proton movement; dashed line = electron movement; X, MnO_2 – electronic conductor interface; Y, MnO_2 – solution interface.

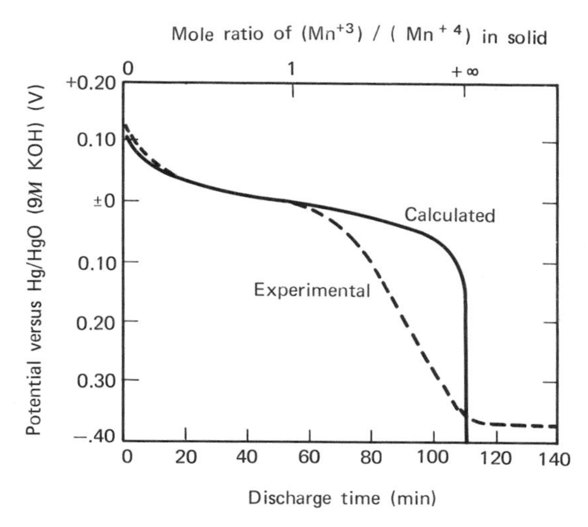

Figure 7.5 Comparison of potential versus log $(Mn^{+3})/(Mn^{+4})$ relation in $9M$ KOH at 23°C.

$MnO_{1.7}$, as shown in Fig. 7.5. Confirming Bell and Huber's findings, these authors also observed a lattice dilation of the α-MnO_2 structure during the discharge in this composition range. Brenet (25) has been given the credit for first suggesting the possibility of the lattice dilation without additional evidence for a second separate manganese oxide phase in his studies of the MnO_2 cathode in the Leclanché system. However, the evidence presented by Brenet did not include analytical confirmation of the x-ray diffraction patterns.

Kozawa and Powers show that the presence of zinc in solutions of $9\,M$ KOH does not interfere with the first stage but reduces the effectiveness of the second stage of reduction.

A paper by Boden, Venuto, Wisler, and Wylie (26) describes the results of studying the cathode discharge products by a combination of x-ray diffraction and analytical methods. They used electrolytic MnO_2 and graphite formed into a compressed pellet suspended on a platinum lead-wire in a vial of 37 percent KOH electrolyte. Their cell was discharged at an initial current of 14 ma/g MnO_2. Figure 7.6 shows representative values of cathode potential and of the i-r loss at the cathode obtained in an experimental run. The discharge proceeded in two well defined steps:

1. From $MnO_{1.92}$ to $MnO_{1.52}$ the lattice became dilated without the development of a new manganese oxide phase as shown by the x-ray diffraction. This step agrees with the findings of Bell and Huber and Cahoon and Korver.

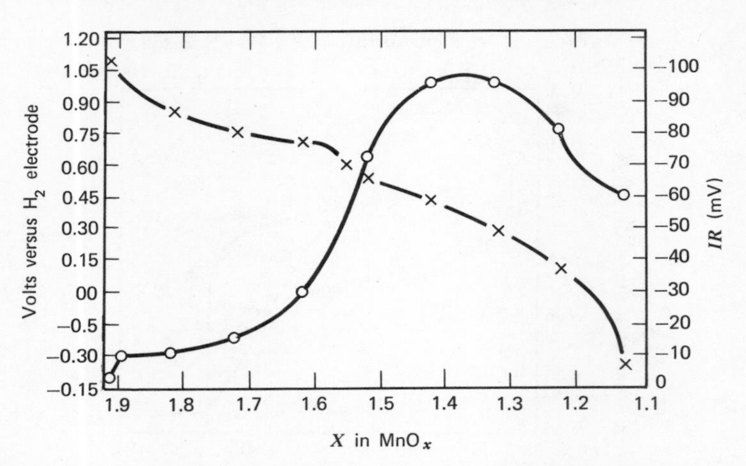

Figure 7.6 Discharge curve of a manganese dioxide electrode in 7*M* KOH; C.D., 5 ma/cm². × points show the potential change as given on the left ordinate; ○ points show the development of IR in the cathode on the right ordinate.

2. (*a*) From $MnO_{1.52}$ to $MnO_{1.44}$, Mn_3O_4 appears as a product again in agreement with Cahoon and Korver data. (*b*) At or about $MnO_{1.34}$, $Mn(OH)_2$ develops as the final product. [We have taken the liberty of dividing the second step into two parts since two separately identified products are reported.]

Kozawa and Powers (27), in a second paper, investigated the differences between γ-MnO_2 and β-MnO_2 as cathodes in KOH solution. They recognized that many previous comparisons of these materials in Leclanché cells and at least one in alkaline cells showed the γ-MnO_2 greatly superior to the β-MnO_2. For this comparison β-MnO_2 was made by heating γ-MnO_2 (electrolytic) sample at 400°C for 10 days and identified as β MnO_2I. Some reduction occurred and the sample was treated with dilute H_2SO_4, washed thoroughly, and dried. This product was identified as β MnO_2 II. Measurements of the pore spectra by mercury intrusion and B.E.T. techniques showed the samples of the original γ-MnO_2 and the final β-MnO_2 had substantially the same pore structure. The surface area of the original γ-MnO_2, 39.7 m²/g, decreased to 17.4 m²/g for the β-MnO_2 during the preparation. Although this sample of β-MnO_2 is not the equivalent of the γ-MnO_2 in surface area, as had been hoped, it is probably closer than the earlier sample of natural Caucasian β-MnO_2 used by Bell and Huber. In 9*M* and 1*M* KOH the γ-MnO_2 gave substantially the same curves reported earlier for electrolytic MnO_2 plated on carbon rods and discharged in the same electrolyte. The β-MnO_2 sample gave 100–200 mV lower potential values than the γ-MnO_2 in the discharge curve from MnO_2 to $MnO_{1.5}$.

These authors compared the observed cell voltage for the γ-MnO_2 and β-MnO_2 samples. Whereas the γ-MnO_2 provides satisfactory agreement between observed and calculated potential values, the β-MnO_2 data provide a different relationship from that given by the γ-MnO_2. Figure 7.7 shows a representative chart taken from their paper.

Kozawa and Powers (28) extended their studies by measuring polarization at the MnO_2 cathode in KOH and KOD electrolytes under a variety of current drains. They defined polarization as the difference between the i-r-free CCV and the OCV at any particular stage in the discharge. These values were taken periodically throughout the course of the test. This method of examining cell potential behavior emphasizes differences which might appear less pronounced had the total polarization been presented in the usual way. A part of their data is shown in Fig. 7.8.

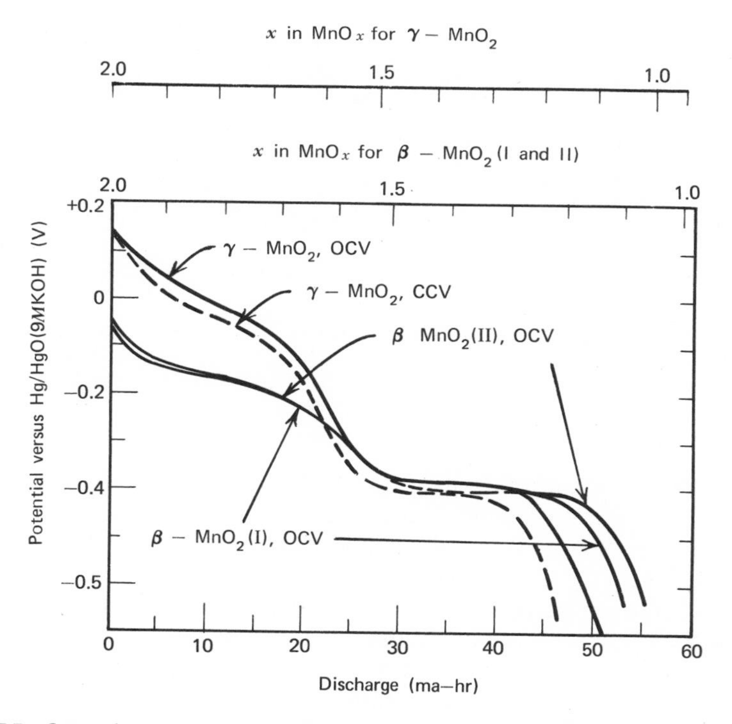

Figure 7.7 Open-circuit potential at various stages of the discharge of γ-MnO_2 and β-MnO_2 in 9*M* KOH. MnO_2 sample (100 mg) mixed with graphite and coke was discharged at 0.35 ma for 17 hr; then the circuit was opened and the final open-circuit voltage (OCV) was obtained after sufficient recovery time. The closed circuit voltage (CCV) includes the *IR* drop, which was 10–15 mV. β MnO_2I was prepared by heating a sample of electrolytic gamma MnO_2 at 400°C for 10 days. β MnO_2II was prepared by leaching a part of the β MnO_2I product in 0.1 M H_2SO_4 at 30°–40°C for 12 hours, to remove any residual Mn^{2+}, followed by washing and drying.

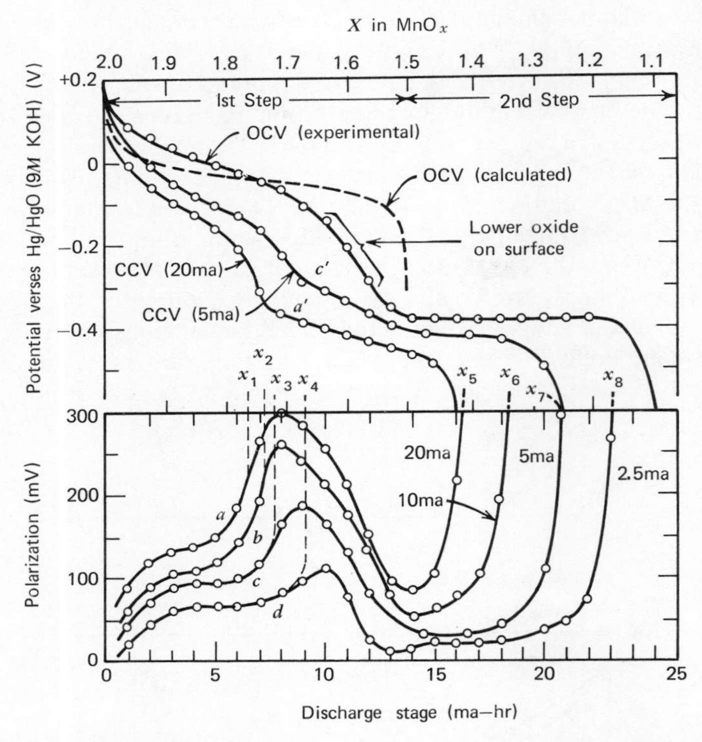

Figure 7.8 Constant current polarization of γ-MnO_2. The currents shown on the curves are for a single cell containing 50 mg γ-MnO_2 sample. The OCV (calc.) was calculated as described previously. Curves *a* and *a′* are polarization and CCV, respectively, at 20-ma discharge. Curves *c* and *c′* are those at 5-ma discharges.

As a result these authors show a maximum in polarization developing at about $MnO_{1.7}$ with a minimum at about $MnO_{1.5}$ for a series of discharges made with γ-MnO_2. Under the heaviest drain condition of 20 ma, equivalent to 400 ma/g MnO_2, the curve begins to approach the shape of the three-step curves of Bell and Huber and Cahoon and Korver. Similar tests with β-MnO_2 do not show the maximum as clearly as in the former case but the minimum is present as before. The development of the maximum in the γ-MnO_2 polarization curves is attributed to the development of a new, lower surface oxide, which accounts for at least part of the divergence between the observed and calculated values of open-circuit potential. They showed that the polarization values were considerably greater, 10–30 percent in KOD solutions, than in KOH in accordance with expectations based on the dimensions of the OD^-

and OH^- ions. Tafel relationship with slopes of 130 mv/decade provide a basis for stating that the rate determining step in reaction

$$MnO_2 + H_2O + e^- \rightarrow MnOOH + OH^- \tag{16}$$

has a value of about $\frac{1}{2}$ in the equation

$$\eta = \frac{RT}{\alpha nF}\left(\ln\frac{i_0}{i}\right) = a - b\log i \tag{17}$$

Table 7.1 shows examples of exchange current values in the early part of the discharge with γ-MnO_2.

TABLE 7.1. EXCHANGE CURRENTS

	Stages of the Discharge					
	1 ma-hr		2 ma-hr		3 ma-hr	
Electrolyte	ma/g	ma/m$^{2(a)}$	ma/g	ma/m$^{2(a)}$	ma/g	ma/m$^{2(a)}$
In 9*M* KOH	30.0	0.71	20.4	0.51	17.6	0.45
In 9*M* KOD	22.0	0.59	14.8	0.38	10.6	0.27

From (28).
[a]Based on the BET surface area 39.7 m^2/g.

Boden, Venuto, Wisler, and Wylie, in a second paper (29) reviewed the discharge of the MnO_2 cathode in preparation for their discussion of the processes that occur during the charge cycle. Their findings during the discharge step are in agreement with their earlier results. They found the system to be reversible if the reduction of the MnO_2 does not proceed to the 0.6 c* stage at which Mn_3O_4 develops. Figure 7.9 shows the discharge was limited to 0.89 c. This would approximately correspond to a reduction to $MnO_{1.78}$. Their x-ray data show that under these favorable conditions the γ-MnO_2 is produced during charging. When the discharge produced Mn_3O_4, the system was not as reversible since Mn_3O_4 is apparently not reoxidized to a higher oxide during charge.

Kang and Liang (30) have investigated the reversibility of the MnO_2 electrode with special emphasis on the oxidation process during charging.

*The abbreviation "c" is often used to indicate the amount of charge in papers dealing with rechargeable cells. Here the terms "0.6 c" and "0.89 c" indicate that the cells have been discharged from the initial state of 1.0 c to 0.6 c or 40%. Similarly at 0.89 c cells have delivered 11% of their theoretical capacity. The authors (2a) based their calculation of theoretical capacity on the cathodic reduction of Mn^{4+} to Mn^{2+}.

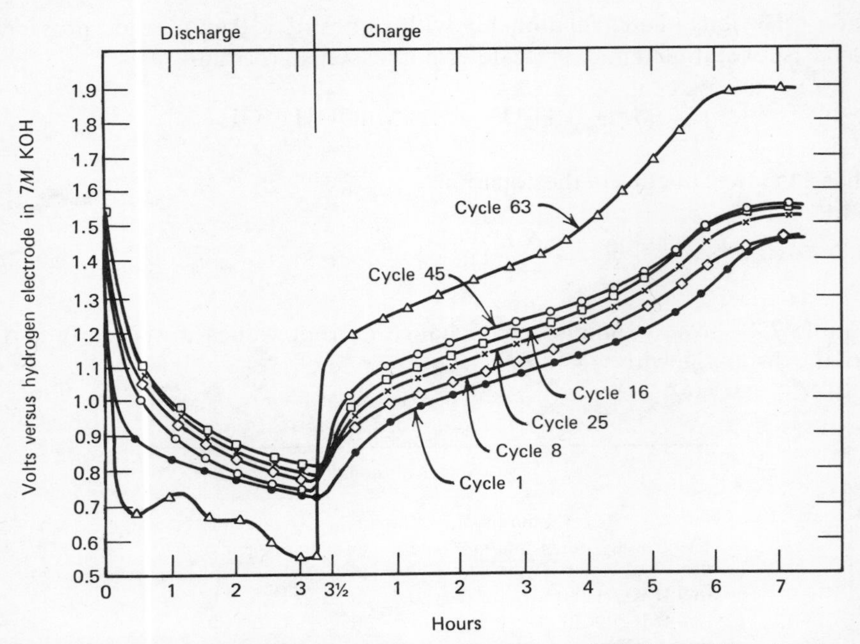

Figure 7.9 Alkaline MnO_2 electrode cycle test. Discharge to 0.89 C at a C.D. = 5 ma/cm²; charge to 1.0 C + 1% over chg., at a C.D. = 2.5 ma/cm².

They examined the electrodes by chemical and x-ray diffraction methods and subsequent cathodic reduction. They reported that the efficiency of the electrochemical oxidation of the electrode decreases as the KOH concentration increases from 1 *M* to 10 *M*. In 1 *M* KOH, MnO_2 is reduced to a species tentatively identified as an active form of Mn (III)-oxide which can be efficiently reoxidized to MnO_2. In 10 *M* KOH, however, MnO_2 is reduced first to Mn(III) oxide and subsequently to Mn(II) oxide. This manganese (II) oxide can be recharged only to an inactive form of manganese (III) oxide, which is not reoxidized efficiently to MnO_2. The obvious conclusion to this complex situation is that the reaction mechanism during the charge process still requires further elucidation.

In reviewing the various conditions used by the several experimenters whose studies have been described it becomes clear that two major variables are present. The first is the current density used during the discharge which has varied from 0.11 to over 60 ma/g MnO_2. The second is the volume of electrolyte used in the experimental cell, which has also varied considerably. It is difficult to assess the effect of the volume of the electrolyte present in an experimental cell especially when no precise relation to the cathode by-product has been suggested or established.

A summary of the reaction steps described by the various authors, limited to the results obtained when γ-MnO_2 is used as the starting material, shows that all workers who used γ-MnO_2 as cathode reactant discharged their experimental cells at currents of 0.5 ma/g or higher and utilized the x-ray diffraction technique to identify the cathode products found that Mn_3O_4 or γ-Mn_2O_3 was produced. Some authors have emphasized that the similarity in x-ray patterns of γ-Mn_2O_3 and Mn_3O_4 make it very difficult to differentiate between these compounds. Since it appears unlikely that two such oxides of manganese could be expected to have such similar patterns, we suggest that the bulk of the evidence is for the formation of Mn_3O_4. Those workers who have utilized a combination of coulombic and potential methods have indicated that α-MnOOH is produced in the early part and near the midpoint of the discharge. Indeed there is no direct evidence of this type to suggest the formation of Mn_3O_4. However, on the very lightest current drains it may well be that no Mn_3O_4 is produced.

The final step, the reduction to $Mn(OH)_2$ as end product, is generally accepted by all workers. However, it is the first step in the discharge where some disagreement exists, mainly as a result of lack of research tools to identify the products. It could be that the Mn_4O_7 of Cahoon and Korver (14), the $MnO_{1.7}$ of Bell and Huber (15), the $MnO_{1.75}$ of Kozawa and Powers (24), and the $MnO_{1.82}$ of Boden (26) are all substantially the same rather indefinite, amorphous material. Only future effort with new methods can resolve this situation. At the present time it appears that over a wide range of current densities the following steps may occur during the discharge:

1. The reduction of $MnO_{1.92}$ to $MnO_{1.7} \rightarrow \begin{cases}\text{lattice dilation and/or} \\ \text{amorphous product}\end{cases}$
2. The reduction of $MnO_{1.7}$ to $MnO_{1.5} \rightarrow \alpha$-MnOOH
3. The reduction of $MnO_{1.5}$ to $MnO_{1.33} \rightarrow Mn_3O_4$
4. The final reduction to $MnO_{1.0} \rightarrow Mn(OH)_2$

COMMERCIAL BATTERIES

"Crown" cells, product of Ray-O-Vac Co., Madison, Wisconsin

First to appear on the market were batteries of series-stacked, relatively small, flat cells, of 1.25 to 2.25 cm. diameter, initially designed for radio "B" circuits but, as market demand shifted, also adapted to other uses. They owe their name to their terminal caps, patterned after bottle closures. Their construction is shown in Fig. 7.10 (9).

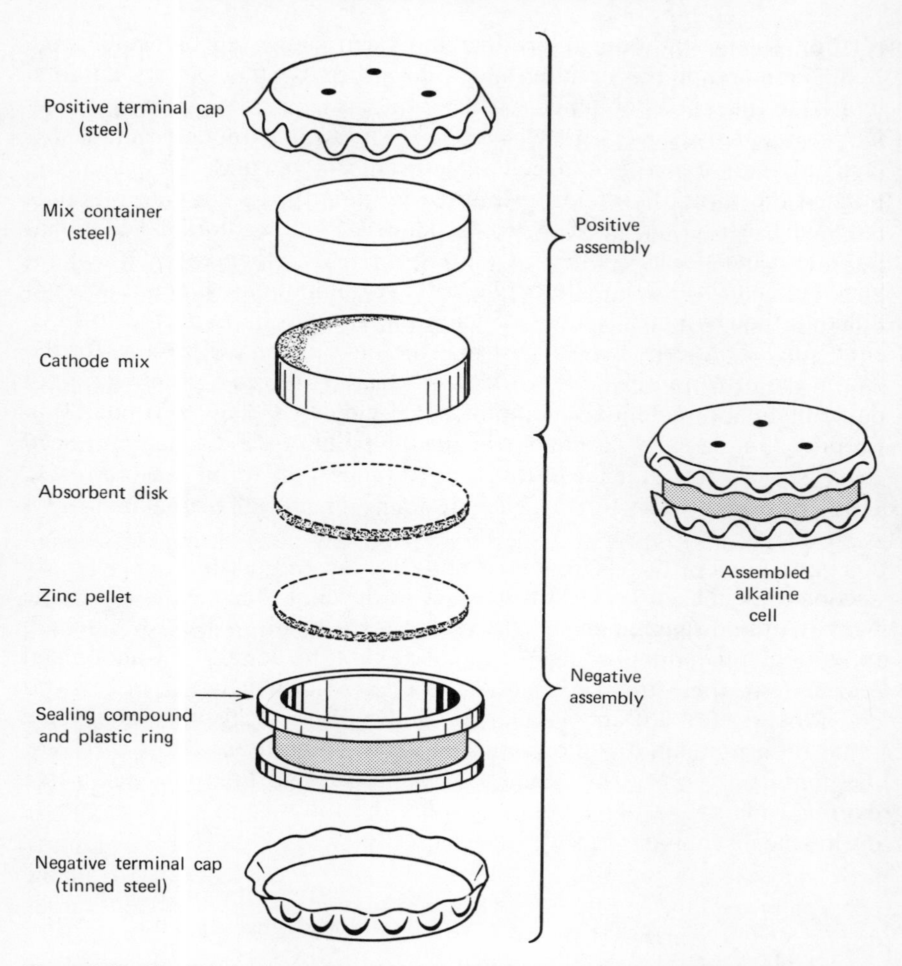

Figure 7.10 Exploded view of Ray-O-Vac alkaline cell. (Courtesy of Electric Storage Battery Corp.)

The cathode assembly includes a small steel can, into which a pellet or disk of MnO_2:C mix is molded, and a steel terminal cap. Good electrical contact between can and cap is assured by spot welding.

The anode of powdered amalgamated zinc, molded under pressure to form a pellet of high surface area, is contained in a polymethacrylate ring, the latter coated at both edges with an elastomeric sealing compound. An absorbent disk of felted cellulose fiber is placed next to the zinc. The negative terminal cap is of tinned steel. Anode, cathode, and separator with the necessary electrolyte are combined, and crimping of terminal caps completes the assembly. Sufficient pressure is applied to force the sealing compound tightly against both terminal caps and to form a gasket which prevents leakage of electrolyte.

Since the cell terminals completely cover opposite faces, the crown cells are well suited to series connection by simply stacking, one above another. Usually, a thermoplastic sleeve is shrunk around a stack to confine the cells and to exert the longitudinal force needed to maintain intercell electrical contacts. Batteries up to a practical limit of about 20 cells are built in this manner. Longer stacks can be made with different confining members, or several stacks can be connected with wires.

Performance of the 900C-size crown cell is summarized in Fig. 7.11. These data resulted from constant-resistance discharge at 70°F 6 hours/day, 5 days/week. This intermittent program simulates that of a portable radio.

Cylindrical cells

Alkaline units similar in shape and sizes to the conventional Leclanché dry cells have also appeared on the market. Figures 7.12 and 7.13 are cross-sectional views of two of these products.

The cathode member, a mix of γ-MnO_2 and carbon or graphite, is molded as a hollow cylindrical layer against a steel can, which serves as cell container and current collector. A thin separator of nonwoven vinyl-cellulose fabric lines the inner surface of the cathode.

The centrally located anode is a hollow cylinder of powdered, amalgamated zinc compacted with a high concentration of sodium carboxymethylcellulose. The remaining interior space, as indicated in Fig. 7.12, is filled with a 30 percent solution of KOH gelled with sodium CMC, thus providing a large volume of electrolyte in the immediately available position requisite for the anodic reaction under conditions of continuous discharge and high current drain.

The "Eveready" cells are closed by crimping a nylon seal (31) between the cathode can and a steel inner cell bottom disk. This construction results in a very tightly closed cell which effectively prevents the creepage of caustic electrolyte out of the unit. The nylon sealing disk has molded in its surface a kidney-shaped frangible diaphragm (32) which, along with a special washer (33), provides a fail-safe pressure release should internal pressures develop inside the unit from abuse, indiscriminate charging, or other causes. The space between the active cell volume and the exterior cover is designed to accommodate expanding gas and liquid should the diaphragm rupture.

For cells to be used as rechargeable units some changes in structural details are needed. Since zinc plated during discharge may be deposited in dendritic form, thus causing short-circuits, a thicker separator is

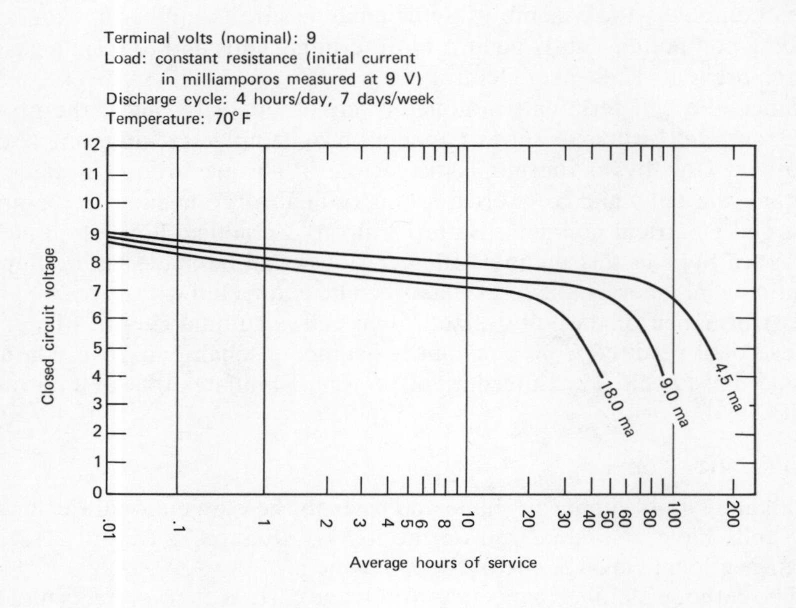

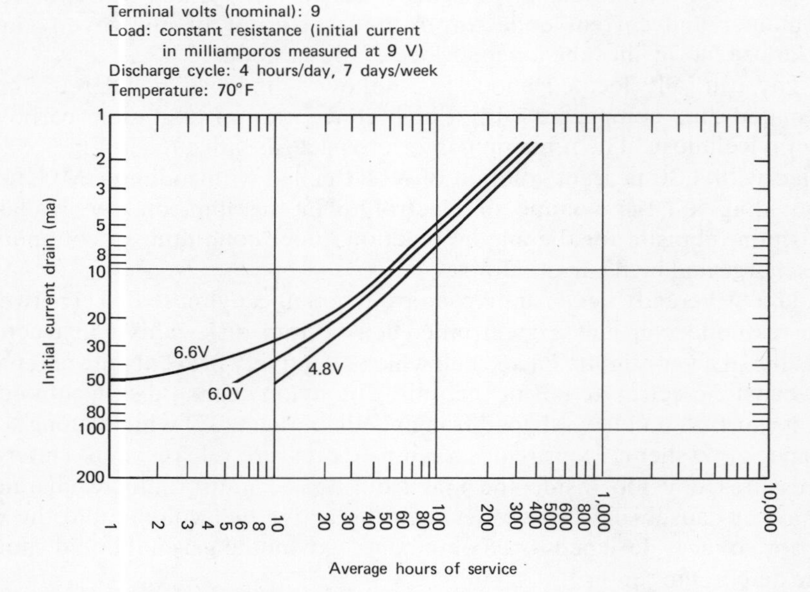

Figure 7.11 Service data for a battery of crown type cells. (From Engineering Manual Ray-O-Vac Corp.)

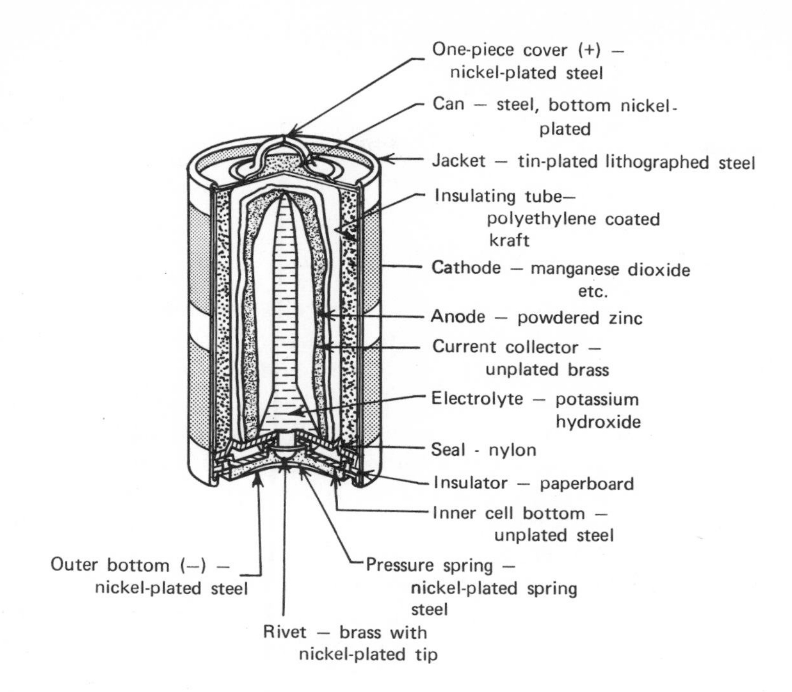

Figure 7.12 Cutaway of alkaline cell (primary type). (Courtesy of Union Carbide Corp.)

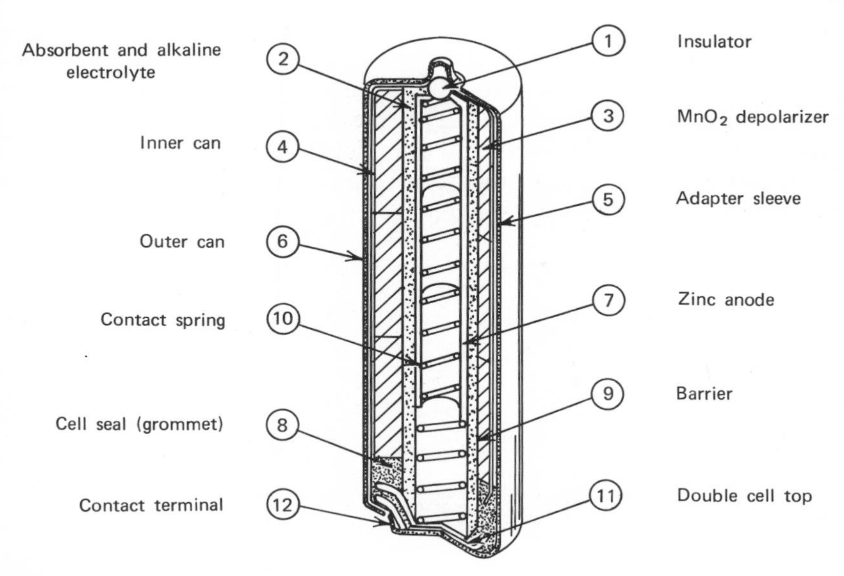

Figure 7.13 The construction of an alkaline – MnO_2 cell, courtesy of Mallory Battery Company, a Division of P.R. Mallory & Co., Inc.

advisable. The addition of a fibrous binder (chopped "Dynel" monofilament) to the cathode mix helps to maintain the integrity of the cathode under the mechanical stresses, due to changes in volume accompanying the charge and discharge cycle.

PERFORMANCE

One of the outstanding features of the cell is its high output under continuous heavy load. Its superiority in this respect over the conventional Leclanché unit is shown in Fig. 7.14 and Table 7.2. Figure 7.15 shows a comparison of alkaline and Leclanché dry cells on continuous service at 70°F (34). The advantage of the alkaline cell decreases as the severity of service decreases, and on very light and/or intermittent drain the alkaline cell tends to lose its economic advantage.

Cells can be used at temperatures as low as −20°C, at which point they are capable of more than twice the output of special purpose (low-temperature) Leclanché batteries on standard tests as shown in Table 7.3. Service maintenance over long periods of storage is better than that of the Leclanché counterpart. The alkaline MnO_2-zinc system is manufactured in a range of sizes as shown in Table 7.4

Since the electrolyte of the alkaline system is substantially nonvariant, the ampere-hour capacity is largely dependent on the quantity of cathodic reactant that can be packed into a cell. Mercuric oxide, with a density about 2.3 times that of electrolytic manganese dioxide, thus offers an advantage over the latter material. In cost, however, the situation is reversed, HgO being the more expensive by a factor of about 20. For small units, e.g., the "crown" cells previously described, material cost is of minor consequence; for the larger, e.g., C and D sizes, the mercuric oxide cell is at an economic disadvantage.

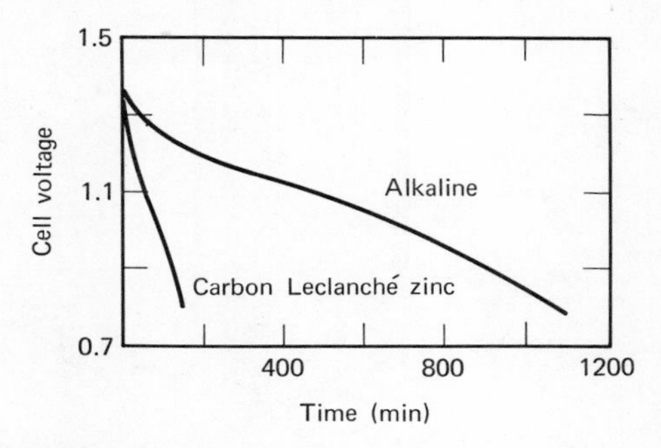

Figure 7.14 Discharge curves on alkaline and Lechanché D-size cells on a constant resistance test using a 500-ma starting lead. (Courtesy of Union Carbide Corp.)

TABLE 7.2. REPRESENTATIVE DISCHARGE DATA ON "D" SIZE ALKALINE CELLS

"EVEREADY" NO. E95
Estimated hours Service at 70°F

Schedule	Starting Drains (milliamperes)	Load (ohms)	Estimated Hours to Cut-off Voltage							
			0.6 V	0.65 V	0.7 V	0.8 V	0.9 V	1.0 V	1.1 V	1.2 V
24 hrs/day (or less)	10	150			1350	1150	1050	910	780	700
	30	50			450	400	350	315	265	210
	50	30			270	240	210	185	160	120
	150	10			80	72	66	59	47	30
	200	7.5			57	52	48	43	32	17
	250	6			44	40	37	32	22	11
	300	5			35	32	29	25	17	7.5
	350	4			29	27	24	20	12.5	5.2
	375	3.75			27	24.5	22	18.5	11	4.5
	600	2.4			15	13.5	11.7	8.5	3.5	
	637	2.25			14	12.5	10.7	7.5	3	
	1350	1.0			5.1	4.3	2.9	1.0	0.3	
	1500	0.85			4.4	3.5	2.1	0.8	0.2	
	2400	0.5			2.0	1.4	0.4	0.1		
	3000	0.35			1.2	0.7	0.1			
	4000	0.25			0.5	.15				
4 min/hr, 8 hr/day, 16 hr rest, (Light Industrial Flashlight Test)	637	2.25		17.5						
4 min every 15 min, 8 hr/day, 16 hr rest, (Heavy-Industrial Flashlight Test)	350	4					25.5		15	
8 amp (starting) for 8 sec–2.2 amp (starting) for next 52 sec – this cycle repeated continuously to 0.6 V cut off (Electronic Photoflash Test)			160 flashes							

Courtesy of Union Carbide Corp.

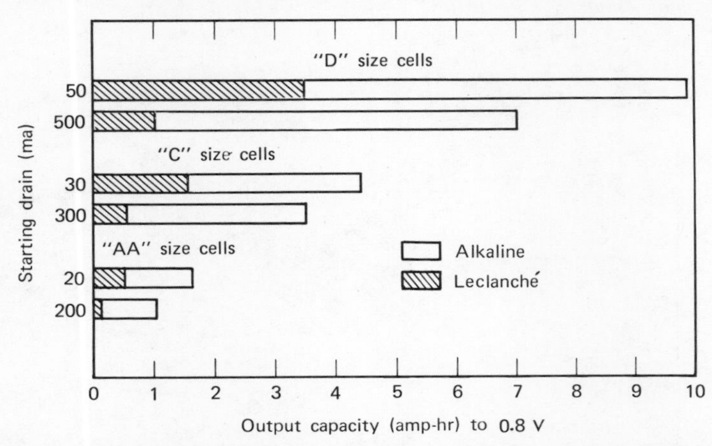

Figure 7.15 Alkaline MnO_2–Zinc versus Leclanché cells at 70°F on continuous tests. (Courtesy of Union Carbide Corp.)

Though not as efficient a storage battery as the nickel: cadmium unit, the alkaline manganese system is much less expensive and in many applications provides current at lower cost. However, certain precautions are required to ensure satisfactory operation.

It is recommended, for example, that not more than about 25 percent of primary capacity or 2.0 amp-hr for a D-size cell be drained at a single discharge and that, to avoid irreversible reaction, operating voltage should not be allowed to fall below 0.9. Charging may normally amount to 120 percent of the ampere-hour discharge, though up to 150 percent can be used without injury to the cell. Voltage-limited tapered-current chargers give greater cycle life than constant current devices. Typical results are shown in Fig. 7.16.

TABLE 7.3. ALKALINE VERSUS LECLANCHÉ LOW TEMPERATURE DUTY

Continuous 0.5 Ampere Starting Drain (Service to 0.8 V)

	Service		Capacity	
Temperature	Alkaline E-95	Leclanché 950	Alkaline E-95	Leclanché 950
70°F	1,200 min	154 min	6.93 aH	0.89 aH
32°F	835	100	4.45	0.58
0°F	275	50	1.39	0.26
−20°F	162	9	.82	0.04

Courtesy of Union Carbide Corp.

TABLE 7.4. PHYSICAL CHARACTERISTICS OF "EVEREADY" ALKALINE PRIMARY BATTERIES

"Eveready" Battery Numbers	Voltage	Number and size of Cells	Maximum Dimensions (inches)				Terminals	Weight
			Diameter	Length	Width	Height		
EPX825	1.5	1-825	0.905			0.228	Flat	0.026 oz
E90	1.5	1-"N"	0.470			1.130	Flat	0.35 oz
E90E	1.5	1-"N"	0.470			1.130	Flat	0.35 oz
E89	1.5	1-"$\frac{1}{2}$AA"	0.563			1.087	Flat	0.4 oz
E92	1.5	1-"AAA"	0.410			1.745	Flat	0.4 oz
E91	1.5	1-"AA"	$\frac{9}{16}$			$1\frac{31}{32}$	Flat	0.75 oz
E94	1.5	1-"$\frac{1}{2}$D"	$1\frac{11}{32}$			$1\frac{13}{64}$	Flat	2.1 oz
E93	1.5	1-"C"	$1\frac{1}{32}$			$1\frac{15}{16}$	Flat	2.2 oz
E95	1.5	1-"D"	$1\frac{11}{32}$			$2\frac{13}{32}$	Flat	4.5 oz
532	3.0	2-"1"	0.664			1.670	Snap	0.8 oz
529	3.0	2-"$\frac{1}{2}$AA"	$\frac{9}{16}$			2	Flat-Negative Recessed	0.75 oz
523	4.5	3-"1"	0.662			1.965	Flat	1.17 oz
531	4.5	3-"1"	0.662			2.290	Snap	1.25 oz
520	6.0	4-"G"		$5\frac{17}{32}$	$4\frac{21}{32}$	$2\frac{21}{32}$	Insulated Knurls	2 lb, 8 oz

Courtesy of Union Carbide Corp.

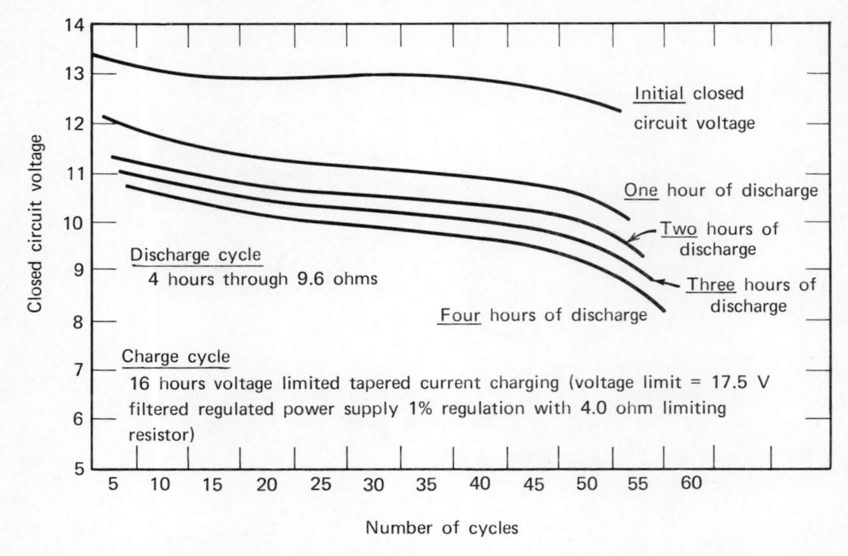

Figure 7.16 Changes in battery performance with every fifth cycle (561 Battery on 1.25 amp rate for 4 hr.) (Courtesy of Union Carbide Corp.)

REFERENCES

1. Leuchs, Ger. Pat. 24,552 (1882).
2. Yai, U.S. Pat. 746,227 (1903).
3. Achenbach, Ger. Pat. 265,590 (1912).
4. *Id.*, Ger. Pat. 279,911 (1913).
5. Heil, U.S. Pat 1,195,677 (1916).
6. Böcker and Eichoff, Ger. Pat. 343,336 (1918).
7. Malichatka, Ger. Pat. 456,423 (1926).
8. Wonder, Fr. Pat. 793,617 (1936).
9. Herbert, W. S., *J. Electrochem. Soc.*, **99**, 190C (1952).
10. Ruben, S., U.S. Pat. 2,576,266 (1951).
11. King, B. H., U.S. Pat. 2,593,893 (1952).
12. Treadwell, A., *The Storage Battery*, The Macmillan Co., New York pp. 103–105 (1906).
13. Ruben, S., U.S. Pat. 2,422,045 (1947).
14. Cahoon, N. C., and Korver, M. P., *J. Electrochem. Soc.*, **106**, 745 (1959).
15. Bell, G. S., and Huber, R., J., *Electrochem. Soc.*, **111**, 1–6 (1964); *Electrochem. Acta*, **10**, 509–12 (1965).
15a. Vetter, K. S., *Z. Electrochem.*, **55**, 274 (1951); **66**, 577 (1962).
16. Kozawa, A., and Yeager, J. F., *J. Electrochem. Soc.*, **112**, 959 (1965).
17. Chreitzberg, A. M., Jr., Allenson, D. R., and Vosburgh, W. C., *J. Electrochem. Soc.*, **102**, 557 (1955).
18. Vosburgh, W. C., Pribble, M. J., Kozawa, A., and Sams, A., *J. Electrochem. Soc.*, **105**, 1 (1958).

19. Vosburgh, W. C., and Pao-Soong Lou, *J. Electrochem. Soc.*, **108**, 485 (1961).
20. Kozawa, A., and Vosburgh, W. C., *J. Electrochem. Soc.*, **105**, 59 (1958).
21. Coleman, J. J., *Trans. Electrochem. Soc.*, **90**, 545 (1946).
22. Vosburgh, W. C., *J. Electrochem. Soc.*, **106**, 839 (1959).
23. Scott, A. B., *J. Electrochem. Soc.*, **107**, 941 (1960).
24. Kozawa, A., and Powers, R. A., *J. Electrochem. Soc.*, **113**, 870 (1966).
25. Brenet, J. P., 8th C. I. T. C. E., Madrid, Butterworths Scientific Pubs., 7.4, 394 (1956).
26. Boden, D., Venuto, C, J., Wisler, D., and Wylie, R. B., *J. Electrochem. Soc.*, **114**, 415 (1967).
27. Kozawa, A., and Powers, R. A., *Electrochem. Tech.*, **5**, 535 (1967).
28. *Id., Electrochem. Soc.,* **115**, 122 (1968).
29. Boden, D., Venuto, C. J., Wisler, D., and Wylie, R. B., *J. Electrochem. Soc.,* **115**, 333 (1968).
30. Kang, H. Y., and Liang, C. C., *J. Electrochem. Soc.*, **115**, 6 (1968) .
31. Carmichael, R., Spanur, F., and Klun, H., U.S. Pat. 3,069,489 (1962).
32. Carmichael, R., and Vulpio, W. A., U.S. Pat. 3,218,197 (1965).
33. Spanur, F., U.S. Pat. 3,314,824 (1967).
34. Winger, J., 18th Ann. Proc. Power Sources Conf. (1964).

8

Zinc: Oxygen Cells with Alkaline Electrolyte

Erwin A. Schumacher

The search for an oxygen electrode (1) has been carried on sporadically since early in the nineteenth century, beginning with observations of the improved performance of Volta piles exposed to air or oxygen, and including some serious efforts to market "air-depolarized" cells. For the alkaline systems, however, it was not until the 1920s that independent investigations in Europe (2) and the United States (3) paved the way to commercial success.

In its simplest form, the oxygen unit may be considered a Lalande cell in which the copper oxide electrode has been replaced by a porous carbon cathode adapted to utilize atmospheric oxygen. Its chief advantage, other than the unlimited supply of no-cost cathodic reactant, is voltage and watt-hour capacity roughly twice those of its Lalande counterpart. For many applications it is one of the most efficient of contemporary galvanic systems (4).

Developed originally for railway use and for operation of radio sets in areas where power lines had not yet penetrated, the modern battery, available in a wide range of sizes, is adapted to many new fields of service requiring large capacity, ability to maintain moderately large current output at fairly constant voltage, and long stand-life after activation. It is suited, therefore, for mine, highway, off-shore and on-shore markers, buoys, railway signaling installations, switch lamps, telephone, telegraph, and radio operation, and similar services. Most commonly employed as the primary source of electrical energy, it also finds application as a stand-by unit to furnish current in the event of powerline failure. Rechargeability for heavy duty applications is receiving increasing attention.

WET CELLS

Large wet cells are constructed individually to meet specific service requirements. Early types (3, 4), factory-assembled in plastic cases, were made as one- or two-cell units with open circuit voltages of about 1.45 V/cell. Their current output was largely dependent on the carbon electrode, a molded block cathode, 4.5 in. × 4.75 in. × 1.25 in., by way of example, delivering 0.65 amp continuously at room temperature, equivalent to a current density of 3.1 amp/ft^2 (3.3 ma/cm^2) of immersed active surface (30 in^2) Other cells intended for higher current output embody cathode designs which shorten the path from air to active cathode surface by providing ventilating holes (5) or hollow electrodes. With these features the current density can be doubled or tripled.

Railway signal cells

Although service requirements cover a wide range of applications—semaphore, color lighting, searchlight-approach signals or switch lights among them—and may vary with traffic density, the bulk of such service can be handled with two cell sizes having nominal ratings of 500 and 1000 amp-hr.

The 500 amp-hr cell shown in Fig. 8.1 is designed for field assembly in a heat-resistant glass jar of approximately 4-liter capacity (6). Its component parts, a zinc:carbon electrode assembly, canned granular caustic soda or potash, and oil, are separately packed and shipped dry, thus insuring complete freedom from deterioration during storage. Without alteration of the carbon electrode, the three-electrode assembly is

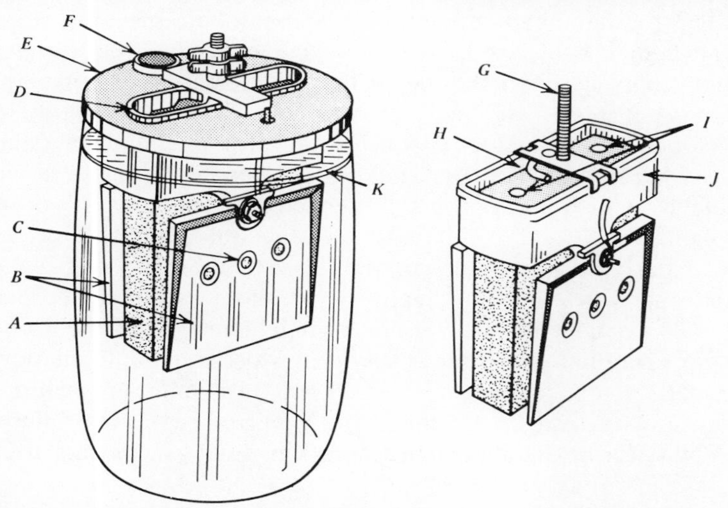

Figure 8.1 Air-depolarized cell, railway signal type. Right: cell renewal unit; left: assembled cell. A—carbon cathode; B—zinc anodes; C—exhaustion indicator panels; D—breather opening; E—porcelain cover; F—filler opening for oil; G—movable suspension bolt; H—flexible cable cathode connection; I—carbon ventilating openings; J—plastic collar; K—oil layer.

converted to 1000 amp-hr capacity by using a larger battery jar with commensurate increase in the weights of the caustic alkali and anode components.

A plastic or metal collar on the emergent portion of the cathode guards the carbon against direct contact with the oil layer. Ventilating holes are incorporated in the carbon to provide the access of air necessary to sustain heavy discharge. Connecting the holes in "V" fashion within the body of the electrode improves ventilation and uniformity of current distribution, thus minimizing uneven anode consumption (7).

The amalgamated zinc anodes are made thicker at the top to compensate for the higher current densities that may occur in this region. It is customary to fabricate the anodes with service-indicating panels (thinner than the adjacent plate sections) to provide visual evidence of approaching exhaustion. These panels may be located in the body of the anode or at the lower edges (8).

A wide range of current requirements is met in railway service, varying from perhaps 50 ma on continuous discharge to a high, as in semaphore operation, of 30-sec discharges at 3 amp repeated at $\frac{1}{2}$-hr intervals. Many cells are utilized in multiunit batteries, e.g., up to nine or ten in series, and, for extra-heavy-duty service, in series-parallel arrangement. For track circuit, relay, and similar low-voltage operation, cells are sometimes paralleled in groups of two or three to minimize renewal frequency on heavy traffic lines.

Some of the properties—freezing points, conductivity, and zinc solubility data for sodium hydroxide and potassium hydroxide solutions—are given in Table 8.1 (cf. also Chap. 3). At room temperature, a signal

TABLE 8.1. SOME PROPERTIES OF ALKALI HYDROXIDE ELECTROLYTES

Alkali Normality	Specific Conductance[a] (mho/cm at 30°C)		Freezing Point[a] (°C)		Zinc Solubility[b] (mole ZnO/liter) (20°C)	
	NaOH	KOH	NaOH	KOH	NaOH	KOH
3	0.395	0.540	−12°	−15°	0.35	0.25
4	0.445	0.635	−18°	−24°	0.65	0.40
5	0.460	0.690	−26°	−33°	1.00	0.60
6	0.453	0.705	−27°	−47°	1.35	0.75
7	0.430	0.695	−23°	−65°	1.75	1.0
8	0.395	0.675	−16°	−50°	2.20	1.3
9	0.353	0.645	−7°	−42°	2.80	1.65

[a]Values for zinc-free solution.

[b]Measurements made on freshly precipitated $Zn(OH)_2$; results expressed as mole ZnO/liter.

cell with sodium hydroxide electrolyte should be capable of operating at 2.0 amp on continuous discharge, and up to 3.0 amp intermittently (Fig. 8.2). For temperatures down to about 0°C (32°F), caustic soda electrolyte is generally satisfactory, but high continuous discharge below this temperature point can lead to cathode deterioration due to peroxide accumulation at the working electrode surface. For lower temperatures, potassium hydroxide, in which the alkali peroxide is more soluble than in sodium hydroxide solutions, permits continuous operation at 1.0 amp, or 2.5–3.0 amp intermittently (Fig. 8.3) even at −18°C (0°F). Mixed electrolytes have been proposed to combine the economy of the sodium hydroxide with the better conductivity and low-temperature tolerance afforded by potassium hydroxide (9).

Plastic-Bonded Cathodes. Another type of cell (see Fig. 8.4), marketed for a number of years, employed cathodes of active granular carbon, bonded with plastic and molded to perforated iron plates (10). Such electrodes were thinner, mechanically stronger, and more uniform in activity than block cathodes, permitting higher drain and giving more

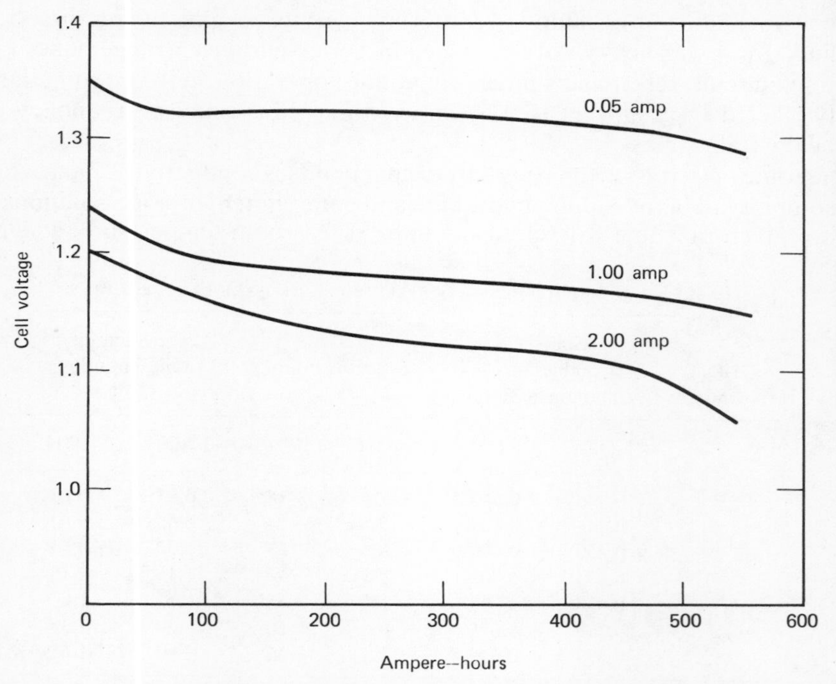

Figure 8.2 Continuous discharge characteristics of AC-500 air-depolarized railway cell at 24°C with 6.1 *N* NaOH electrolyte.

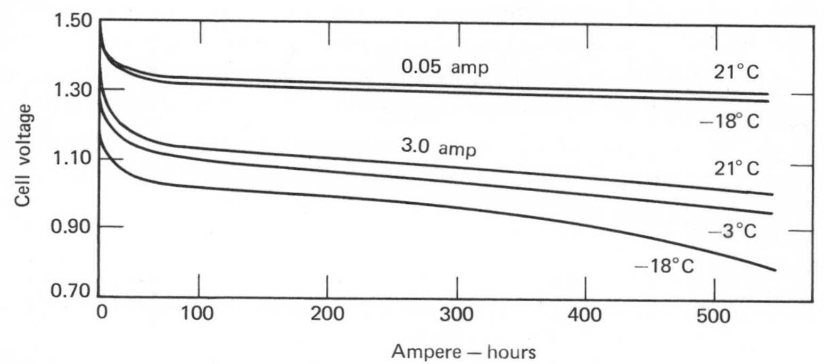

Figure 8.3 Air-depolarized signal cell on signal operating test (3.0 amp for 30 sec every $\frac{1}{2}$ hour (lower curves) and 0.05 amp (upper curves) for remainder of each interval (EVEREADY AC-500).

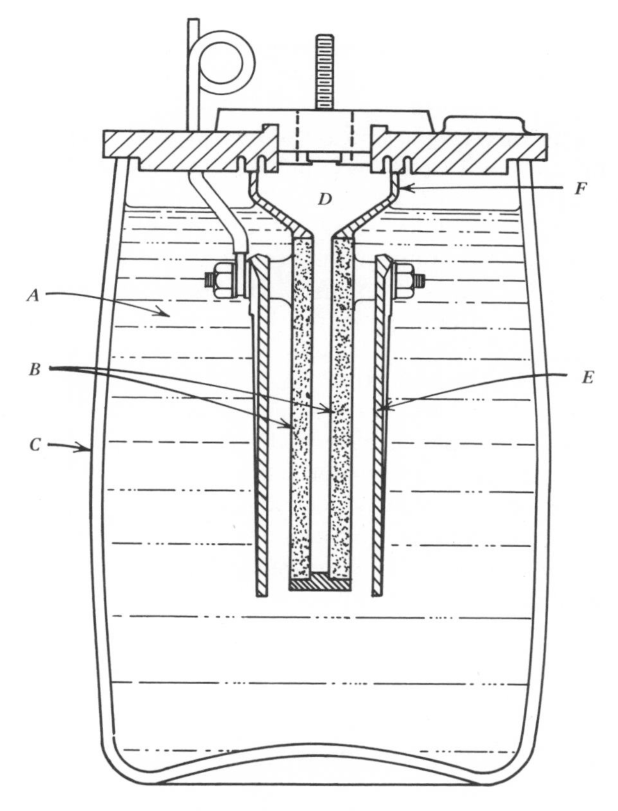

Figure 8.4 Railway signal cell with hollow cathode. *A*: electrolyte; *B*: carbon panels on perforater metal; *C*: glass jar; *D*: air chamber; *E*: zinc anodes; *F*: plastic frame.

uniform distribution of current as a result of the metal reinforcement. The cathode panels, mounted in pairs to furnish an internal air space, were combined by a plastic injection-molding operation to produce a unitary structure having means for mounting the zinc anodes in precise parallel alignment. As in the case of the conventional air-depolarized signal cells, these carbon: zinc assemblies were sold with cans of granular caustic for electrolyte preparation.

General-Purpose Cells

Units in this category are marketed as fully assembled and sealed reserve cells of the water-activated type, ranging in capacity from 300 to 3000 amp-hr. Originally developed for d-c radio sets, they currently find their principal uses in Coast Guard buoy service and other marine signalling installations, in railway switch-lamp lighting, telephone, and general instrument operation.

The development of a general-purpose battery, adapted to handling and shipping without damage, free from deterioration during storage, and simple to prepare for service, presents a number of problems. Carbon electrodes are usually fragile and should be protected against breakage. Caustic alkali solution is corrosive, and its preparation outside the battery cannot be entrusted to unskilled users; the solid solute must therefore be kept inside the cell container. This in turn means that without vigorous stirring—a difficult operation in the restricted space available—a solid layer of hydrated caustic, e.g., $NaOH \cdot H_2O$, slow to dissolve, may form at the bottom of the container, increasing the time required for solution equilibration and, in extreme cases, permitting penetration of dilute electrolyte into the carbon electrode. Finally, as with NaOH, the heat of solution may be sufficient to raise the electrolyte temperature to 85–100°C, at which a conventional plastic case might soften and the carbon electrode would be less resistant to electrolyte penetration. These and other difficulties have substantially been avoided in cells currently available.

Commercial general-purpose batteries, either as single or multicell units, are assembled in plastic cases. Figure 8.5 shows a unit[1] of two cells in series, the one on the right before activation, the one on the left after activation. The unactivated cell is fully sealed with a removable strip (*I*) of cellophane or equivalent, covering the emergent carbon electrode (*C*). The amalgamated zinc anodes (*B*) are held in grooves in the walls of the battery case. Sealing material (*E*) in the cathode well, and an iron rod engaging the bottom of the cathode, serve to support and locate this electrode.

[1]This battery, a product of Union Carbide Corporation, was first marketed around 1930 (3) and subsequently modified (4) as shown in Fig. 8.5.

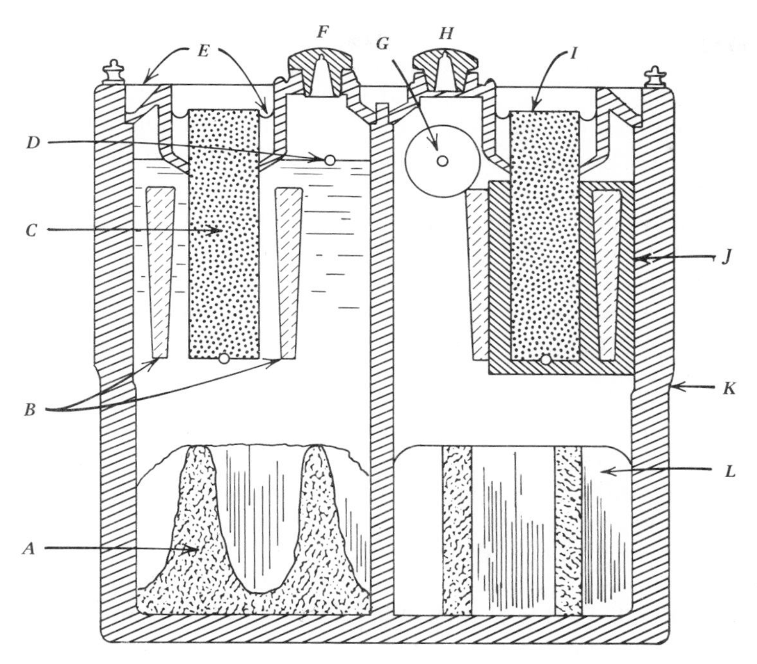

Figure 8.5 General purpose air-depolarized battery, cross section view of activated (left) and nonactivated (right) cells. *A*: lime cake after activation; *B*: zinc anode; *C*: carbon cathode; *D*: electrolyte-level indicator; *E*: top seal; *F*: filler plug; *G*: cast hydrated alkali with lime; *H*: filler plug with frangible seal; *I*: removable seal over carbon; *J*: cast hydrated alkali cylinder with lime addition; *K*: composition case; *L*: lime : cellulose cakes.

Use of caustic as the hydrate (11) instead of an anhydrous form offers a number of advantages. $NaOH \cdot H_2O$ has a melting point of only 65°C and can be readily cast (*J*) directly around the cell assembly, thereby protecting it from breakage during shipment and handling. It dissolves rapidly and, being positioned in the upper portion of the cell, provides a satisfactory solution without the need for stirring. Its heat of solution is low, so that the initial electrolyte temperature is kept at safe levels, e.g., 40–45°C.

Hydrated lime and cellulose floc, in the form of x-shaped blocks (*L*), are added to regenerate exhausted electrolyte (12) (as described in the section on Cell Components). Activation is accomplished by breaking the frangible diaphragm (*H*) in the cover, completely filling the cell with tap water, and removing the seal from the carbon. Solution proceeds rapidly, no stirring being required, and a cell is normally ready for service within 15 min. To insure complete solution uniformity, a cylinder (*G*) of hydrated caustic [to which hydrated lime has been added (13) to slow down the rate of solution] is cast around a supporting metal rod (*D*). Since there is some shrinkage in electrolyte volume as solution proceeds, the supporting rod serves also as a level indicator, visible through the filler opening. The lime-cellulose cakes partially disintegrate, forming a reactant mass (*A*) more readily penetrated by exhausted electrolyte than powdered lime alone.

The general-purpose cell has excellent shelf life before activation and can be kept on activated stand for relatively long periods of time. From 80–90 percent of rated capacity can be obtained on discharges covering 2–3 years of use. Continuous discharge characteristics are illustrated in Fig. 8.6 for a representative commercial two-cell air depolarized battery.

In the Edison general-purpose cell (see Fig. 8.7) a mixture of lime pellets (14) and sodium hydroxide granules substantially fills the electrolyte cavity. On activation the alkali dissolves, leaving the lime as a permeable reaction mass. Cells are usually manufactured either as single units rated at 2400 to 3300 amp-hr or as two-cell batteries with nominal ratings of 1000 amp-hr.

A general-purpose cell without the lime feature, manufactured by LeCarbone (15), has a nominal 2000 amp-hr rating at 1.0 amp continuous discharge (Fig. 8.8). It uses a cylindrical carbon cathode with an annular, cast zinc anode electrically coupled with a second anode member of

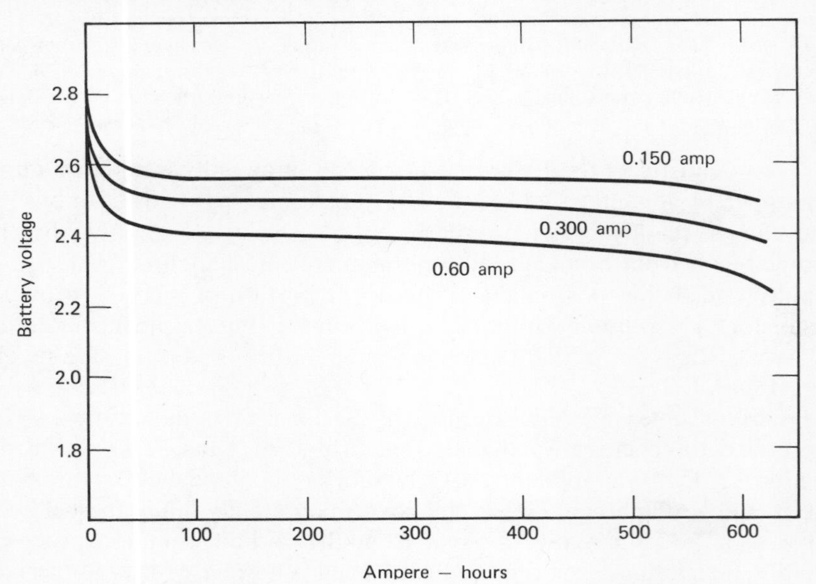

Figure 8.6 T-2600 general purpose air-depolarized battery on continuous discharge at room temperature (6.2 *N* NaOH electrolyte).

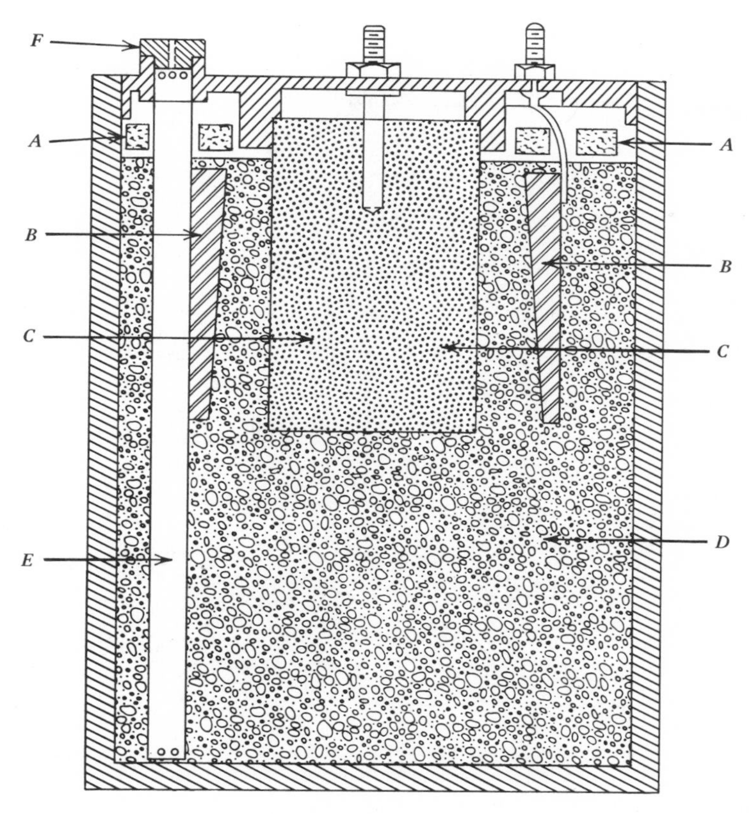

Figure 8.7 Edison air-depolarized cell with pelleted lime and granular caustic soda. *A*: cast caustic sticks; *B*: cylindrical zinc anode; *C*: carbon cathode; *D*: mixture of pelleted lime and granular caustic soda; *E*: filler tube for water; *F*: fillet tube cap; *G*: cover.

coiled sheet zinc resting on the bottom of the battery case. The space between the coils is filled with glass fibers and a layer of the same material covers the top of the spiral. The electrolyte cavity contains granular caustic alkali dissolved by water addition when the cell is activated.

During discharge the electrolyte at the bottom of the cell is soon enriched with zincate ions due to diffusion restriction induced by the glass fiber in and around the zinc spiral. The concentration cell thus formed between principal (*E*) and auxiliary (*J*) anode elements causes a wasteful continuous small flow of current intended to minimize formation of adherent zinc hydroxide crystals. Augmenting the sacrificial current function, alkali concentration differentials are promoted by the dual anode construction bringing the bottom electrolyte strata into the higher

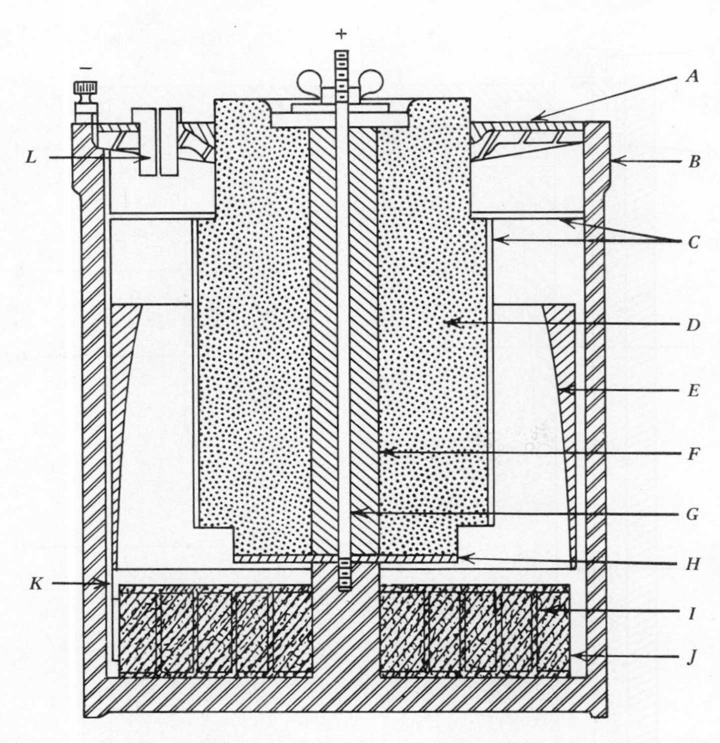

Figure 8.8 LeCarbone air-depolarized cell with multi-element anode. *A*: cover; *B*: case; *C*: glass fiber mat; *D*: cylindrical carbon cathode; *E*: zinc cylinder; *F*: seal; *G*: cathode assembly bolt; *H*: metal contact disk for cathode; *I*: glass fiber; *J*: zinc spiral; *K*: metal connector for anode elements; *L*: filler-hole cap.

electrolyte concentration region associated with formation of amorphous, nonadherent zinc oxide (cf. Chap. 3). Ampere-hour output per unit of weight or volume is somehwat less than obtained with most lime-containing cells.

Table 8.2 summarizes battery dimensions and ampere-hour output in terms of unit volumes and weights of representative general-purpose cells. The most efficient of these batteries in terms of current output, watt-hours per pound of weight and watt-hours per cubic inch are 2-S-J-1, 2P-30, and Tr-3300 units. Figure 8.9 pictures a typical high-ampere-hour capacity, two-cell battery equipped with carrying handle and ventilated covers for the air electrodes. The cells in this battery may be parallel-connected internally to provide 3000 amp-hr at 1.25 V (cf. 2P-30 listed in Table 8.2).

TABLE 8.2. PHYSICAL AND ELECTROCHEMICAL PARAMETERS OF GENERAL PURPOSE AIR DEPOLARIZED CELLS

	T-2600[a]	CG-212[a]	CG-124[a]	2P-30[a]	2-S-J-1[b]	TR-3300[b]	608A[c]
Number of cells	2	2	1	1	2	1	1
Nominal voltage	2.4	2.4	1.2	1.2	2.4	1.2	1.2
Nominal capacity (amp-hr)	600	1,200	2,400	3,000	1,000	3,300	2,000
Cont. disch. rating (amp)	0.660	0.600	1.32	2.50	0.50	1.00	1.00
Overall dimensions[d]							
Length (in.)	9.59	10.55	10.55	10.55	8.31	8.00	8.50
Width (in.)	6.41	7.28	7.28	7.28	7.38	8.00	8.50
Height (in.)	9.84	10.00	10.00	10.00	8.63	12.50	10.00
Total volume (in.^{3})	605	770	770	770	530	800	720
Wt. activated (lbs)	30	41.5	41.5	38	29	42	32
Amp-hr/lb (activated)[e]	40	58	58	79	69	79	63
Amp-hr/in.^{3}	2.0	3.1	3.1	3.9	3.8	3.5	2.8
Milliamps/in.^{3}	2.1	1.6	1.6	1.3	1.9	1.3	1.4
Watt-hr/lb	48.0	69.4	69.4	94.8	82.8	94.3	76.8
Watt-hr/in.^{3}	2.4	3.75	3.75	4.68	4.54	4.20	3.3

[a]Union Carbide batteries.
[b]Edison batteries.
[c]LeCarbone battery.
[d]Case dimensions – terminals not included.
[e]Per cell.

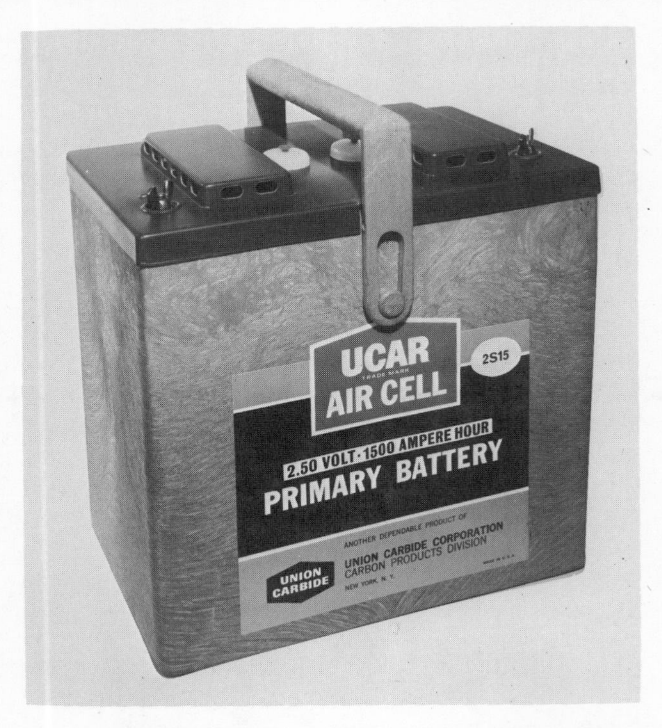

Figure 8.9 Typical high-capacity air-depolarized battery. Dimensions as for 2 P-30, Table 8.2.

At normal ambient temperatures, maximum continuous-discharge rates up to 1.25 amp/ma, single cell are indicated. Lower discharge rates, 250 to 500 ma, are recommended for some cells when full capacity is essential. If operation is conducted at reduced temperature, current output should be reduced to prevent electrode damage. Under normal operation, cell voltages are maintained in the approximate range of 1.20–0.90 V until capacity has been achieved. The discharge curves (see Fig. 8.6) show that the air-depolarized cells have markedly stable operating voltages. On the lighter drains the rate of change is of the order of 1-3 × 10^{-6} V/hr. Cells intended for low-drain service requiring very constant voltage may be kept on constant light discharge, 5–10 ma, to avoid the changes otherwise observed during the period immediately following initiation of service.

Because the general-purpose cells are constructed so that even with transparent battery cases it is difficult to inspect the anodes for state of exhaustion, one manufacturer provides a small reusable ampere-hour meter, based on movement of a mercury column between two electrodes, which gives a continuous measure of delivered capacity. Another method of estimating exhaustion in cells without lime regenerant is based on the change in electrolyte resistance with use. This requires corrections for temperature and adjustment when NaOH and KOH are interchanged.

DRY CELLS

The dry alkaline air-depolarized cells have not met with the same success as the wet versions, and commercialization has been restricted to miniature units designed for use with hearing-aid devices. The first development of this nature (Fig. 8.10), a cell approximately 2.0 × 1.1 × 0.6 in. in size, was made with two plastic-bonded carbon cathodes formed on perforated metal supports, and injection molded to make them an integral part of the external plastic container(16). The centrally located, amalgamated zinc anode, immersed in zinc oxide-containing caustic soda electrolyte thickened with starch, was spaced from the cathode surfaces by liners of cellulosic material. The plastic container heat sealed against the negative terminal provided a liquid-tight closure. The cells were completely assembled at the factory with protective tapes over the breather apertures, removable when the cells were put into use. This unit was capable of operating for 100–110 hr on a 55 ma load (4 ma/cm^2) at discharge voltages of the order of 1.15–1.0 V at room temperature. Discharge below 15°C was not recommended.

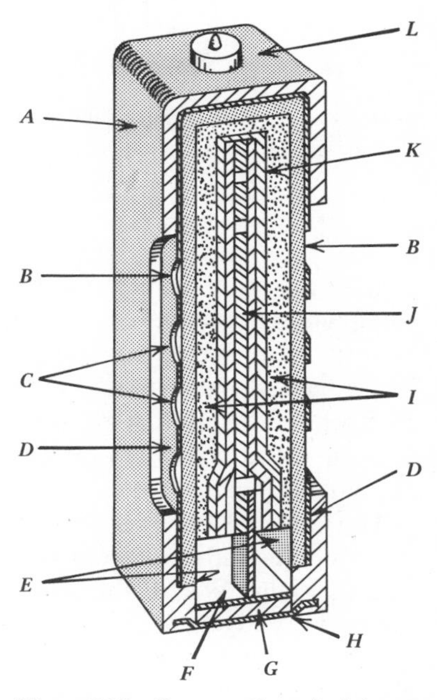

Figure 8.10 Cross section of miniature dry air-depolarized cell. *A*: plastic case; *B*: carbon cathode; *C*: ventilating apertures; *D*: metal support for carbon; *E*: carbon; *F*: air space; *G*: inner seal; *H*: negative terminal; *I*: gelled electrolyte; *J*: zinc anode; *K*: paper liner; *L*: positive terminal.

A smaller, cylindrical version of the "dry" hearing aid cell is shown in Fig. 8.11. This is approximately 0.65 in. in diameter and 0.62 in. high, yielding 1.2–1.4 amp-hr at 20–30 ma discharge. The active carbon powder used in this cell was mixed with thermoplastic binder, and formed

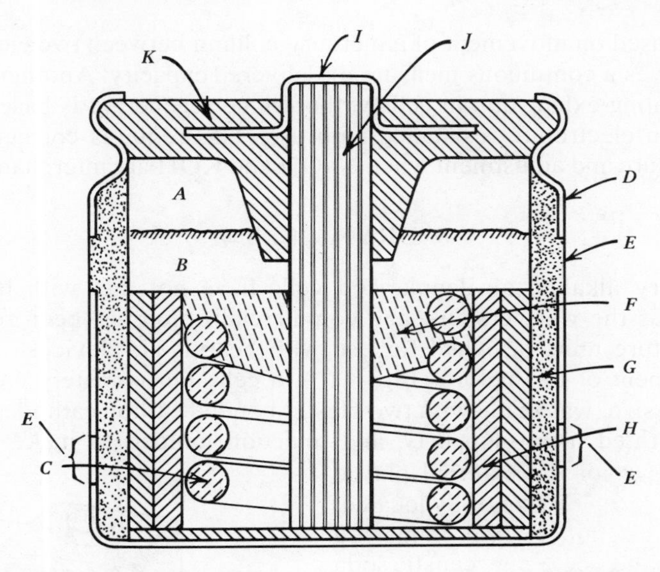

Figure 8.11 Cross section of miniature air cell. *A*: air space; *B*: gelled electrolyte; *C*: coiled zinc anode; *D*: steel container; *E*: ventilating aperture; *F*: zinc bridge to negative collector; *G*: carbon cathode; *H*: paper liner; *I*: negative terminal; *J*: negative collector post; *K*: seal.

against the inner wall of the perforated metal container. A helical coil of amalgamated zinc wire served as the active anode with connection to the external negative terminal effected by a compressed zinc powder bridge. Starch-gelled, 9*N* NaOH containing dissolved zinc oxide constituted the electrolyte. Instead of packaging with a plastic tape over the breather openings, the cell was housed in an air-tight polyethylene capsule until put into service.

The internal changes observed in these small units during discharge were significantly different from those observed in the conventional wet alkaline cells. Where the latter are characterized by the precipitation of impervious layers of well crystallized white octahedral zinc hydroxide, the small units invariably contained layered deposits of porous, amorphous blue zinc oxide (cf. Chap. 3), continuously separating from the zinc surface until the entire electrolyte cavity was filled. Thus the over-all change in the "dry" cells corresponded to the utilization of atmospheric oxygen at the cathode, and production of stoichiometrically equivalent weights of zinc oxide at the anode:

$$2\,Zn + O_2 \rightarrow 2\,ZnO \tag{1}$$

The electrolyte showed no significant change in composition during discharge since zinc oxide was a component of the fresh solution and lime reactant was not used. Nevertheless, the dry-type, air-depolarized cells gave greater output in terms of electrolyte usage per ampere-hour than the larger wet cells. Values as low as 0.6 to 0.7 cc/amp-hr were observed with this construction.

An innovation, described in the European literature(17), proposes construction of an air-depolarized flashlight cell with an internal, hollow cylinder of carbon as the cathode member, mounted in the unit so as to receive air at both end openings. The container, which serves as the negative terminal, is a steel cylinder closed at both ends with flexible gaskets fitted tightly between the cathode and external can as shown in Fig. 8.12. A paste of amalgamated zinc flakes combined with alkaline electrode thickened with carboxymethylcellulose is placed between the cathode and outer container. A sorbent paper separator located next to the cathode prevents direct contact with zinc particles.

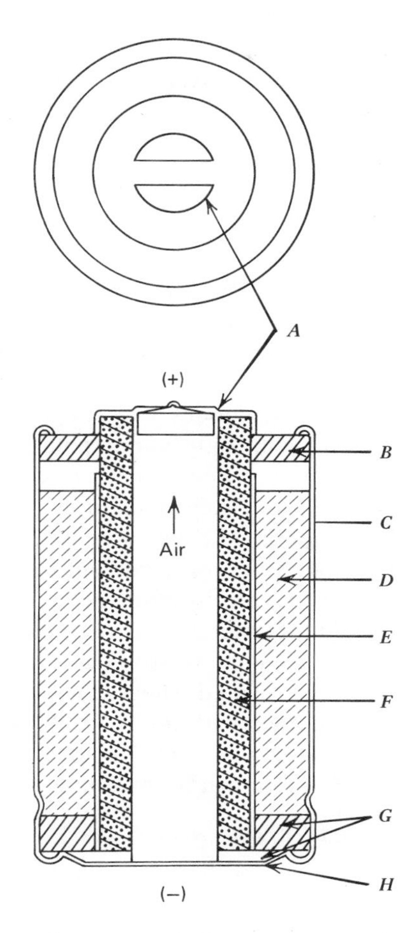

Figure 8.12 Air-depolarized alkaline flashlight cell. *A*: steel cap with ventilation openings; *B*: rubber seal; *C*: steel can; *D*: amalgamated flake zinc in KOH gel; *E*: separator; *F*: carbon tube electrode; *G*: rubber seal; *H*: steel bottom with opening.

In a modified version, an enlarged, gas-permeable carbon cylinder is employed in the dual role of cathode and cell container. Exceptionally high capacity, in terms of normal Leclanché cell performance, is indicated, output on moderately heavy continuous load (150–200 ma) running as high as 15 amp-hr for a "D" size cell. There is relatively little difference between intermittent and continuous discharge, until intermittent usage becomes so infrequent that desiccation, carbonation, and other aging factors detract from operability. As with other dry cell constructions, utilization of built-in capacity generally is satisfactory over a 2–3 month use period.

The anodic reaction closely resembles that observed in the small, domestic, dry air-depolarized cells, zinc oxide being the sole by-product. The large anode and cathode areas made possible by this construction undoubtedly contribute to the well maintained voltages.

The cathode is described as a fine-grained carbon, treated with metal salts, such as cobalt and aluminum nitrates (17), and ignited to deposit spinel-type catalysts in the pores of the electrode. This treatment increases the porosity of the carbon and simultaneously improves its operating effectiveness. Wetproofing is applied to retard electrolyte penetration.

CELL COMPONENTS

Anode

The present commercial alkaline air-depolarized cells rely exclusively on zinc anodes. These are generally cast of high-purity metal, 99.9 percent or better, containing from $\frac{1}{2}$ to 2 percent mercury added as outlined in Chap. 3.

The cast anodes will support current densities of the order of 35–45 ma/cm^2 with only minor polarization, but higher rates of discharge lead to premature filming with zinc hydroxide and zinc oxide as well as early cell failure. Amalgamated granular, flaked, or powdered zinc supported in immobilized electrolyte or compacted into porous electrodes makes it possible to increase current output at good efficiency. Further extension of area by electrolytic reduction of zinc oxide, hydroxide, or carbonate pasted into metal grids permits current densities of the order of 200–300 ma/cm^2. When electrolytic reduction of the zinc compounds is conducted with the pasted sheets held under mild pressure, 0.054–0.25 kg/cm^2 (18), anodes with up to 85–90 percent porosity can be obtained, capable of sustaining current densities as high as 3000–3500 ma/cm^2 in 32 percent KOH solution at room temperature.

Electrolyte (additives)

The finding that hydrated lime will remove dissolved zinc from caustic solution through the formation of an insoluble calcium zincate (19), thus in effect regenerating exhausted electrolyte, has made possible a substantial reduction in the weight and volume of the alkaline battery. Though the efficacy of the lime reaction may vary with operating conditions and the mole ratio of zinc to lime in the calcium zincate formed may range from 1:1 to as high as 2.5:1, the general reaction may be expressed as

$$2\,Na_2\,Zn(OH)_4 + Ca(OH)_2 + 2\,H_2O \rightarrow CaZn_2O_3 \cdot 5\,H_2O \rightarrow 4\,NaOH \qquad (2)$$

X-ray examination confirms the existence of the solid reaction product as a crystalline calcium zincate [*d*-spacing (Å) for the principal lines: 3.13, 5.02, 2.88, 4.13], and not as a mixture of lime with zinc oxide or hydroxide.

On the basis of Eq. (2), 1 g $Ca(OH)_2$ can remove from solution an amount of zinc equivalent to 1.45 amp-hr of service or, in terms of volume, 1 g $Ca(OH)_2$ can replace about 10–12 cc of electrolyte. In commercial practice, this means that the electrolyte requirement, normally 7–9 cc amp-hr, can be reduced to 3 cc or less by the lime reaction.

Lime may be introduced as a powder or as pellets, or briquets which disintegrate in electrolyte. When lime is used in substantial amounts, special precautions may be necessary, such as the addition of extenders such as cellulose floc (20) to ensure an electrolyte-permeable reaction mass.

Reactions similar to the formation of calcium zincate have been observed with barium and strontium hydroxide[2] in the operation of air-depolarized cells employing 6*N* sodium hydroxide. The barium reaction products had compositions ranging from $BaZn_2O_3 \cdot 4H_2O$ to $BaZn_2O_3 \cdot 6H_2O$.[3] The strontium zincate product was characterized by inclusion of zinc oxide and zinc hydroxide precipitates and was not quantitatively identified.

Additives such as sodium or potassium silicates stabilize the electrolyte by complexing the zincate ions, and thereby permit achievement of higher zincate concentrations than suggested by solubility measurements on conventional electrolytes (cf. Chap. 3). Termination of cell operation is marked by deposition of the usual adherent zinc hydroxide crystals on the working faces of the electrodes. Although not as effective as the lime: zincate regeneration reaction, significant capacity increase is afforded by the silicates. They are of particular value maintaining maximum visibility in the electrolyte as an aid in judging state of cell exhaustion. From 25 to 50 g of the additive per liter of solution can increase capacity by as much as 50 percent. Thus a cell normally rated at 500 amp-hr can deliver up to 750 amp-hr.

[2]Unpublished studies.

[3]Scholder and Weber (21), working with solutions of ZnO in 20–25 percent NaOH solutions, were unsuccessful in producing calcium zincate by their experimental procedures but reported the formation of compounds of the type: (*a*) $BaZn(OH)_4 \cdot 5H_2O$, or $BaZnO_2 \cdot 7H_2O$, crystallizing between 0°–−10°C; and (*b*) $BaZn(OH)_4 \cdot H_2O$, or $BaZnO_2 \cdot 3H_2O$, crystallizing at 20°C.

CARBON CATHODES

Carbon electrodes of a variety of configurations have been made, ranging from simple molded blocks and cylinders to various modifications of "ventilated" electrodes (5). In some cell types, the carbon electrode has been used as the cell wall. Electrode thickness has varied from 0.035 to 0.10 in. in small cells, up to 3 in. in large units; the air-travel distance to the electrolyte : carbon interface has ranged from 0.035 to 7 in. or more.

Many materials and processes have been disclosed for the formation or carbon cathodes utilizing atmospheric oxygen. Recognition of the advantages of well developed surfaces has led to the adoption of adsorbent-type carbons derived from peat, sawdust, charcoal, industrial wastes, blacks, etc., used separately or in combination, and bonded with sugars, molasses, tars, pitches, and other carbonizable binders (22). Mechanical strength and conductivity are imparted by calcining teatment. Electrodes of active granular carbon agglomerated with resinous binders supported on, or bonded to, porous or perforated metal (23) can be made thinner and stronger and with more uniform current distribution, than conventional, block-type cathodes.

With the added impetus provided by fuel cell investigations, some general criteria have been established which serve as guides for the formulation of operable, stable carbon electrodes. The electrodes for commercial air-depolarized cells should have sufficient porosity, combined with pore size control, to give the necessary gaseous transport to the reaction sites without allowing electrolyte to penetrate into adventitious gross pores. The preferred upper limit of desirable macropore size is 20–50 μ with a spectrum of 0.05–50 μ. However, porosity and pore size are not the only factors determining operating quality and, as shown in Fig. 8.13, there is a wide divergence among block carbon electrodes made by different battery manufacturers. A porosity range of 50–70 percent is indicated for block electrodes operating on air (24), the larger values preferred when the electrode is intended for higher current density of the order of 10 ma/cm^2. To some extent the porosity limitation can be reduced with thin electrodes and with elevation in operating temperature. The response to rise in temperature of very thin, plastic-bonded active carbon molded on porous metal plates, with a breathing area equal to that of the electrochemically functioning surface, is shown in Table 8.3.

Desired activity and surface development are obtained by exposing granular, or molded carbon to air, steam, flue gas, or their mixtures at temperatures of 800–1000°C, the time of treatment depending largely on the temperature selected for processing. Activating chemicals added to the raw materials or to the preformed electrodes may be in the form of salts or oxides of manganese, cobalt, silver, etc. (25), which leave

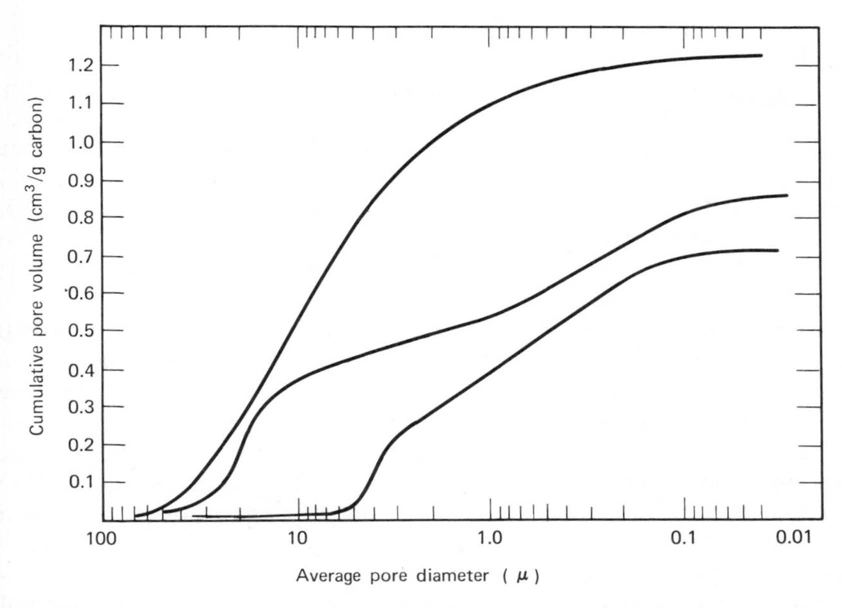

Figure 8.13 Macropore distribution of typical block cathodes used in commercial air-depolarized cells.

residues capable of accelerating decomposition of cathodically generated peroxide. Total surface areas of the order of 200–500 m^2/g have been reported for processed electrodes (26). From 200 to 350 m^2/g represents a useful range for carbon employed in commercial air-depolarized cells.

TABLE 8.3. INFLUENCE OF TEMPERATURE ON THIN AIR/CARBON CATHODE
(0.035-in. electrodes, 9 N KOH electrolyte)

Current Density (ma/cm^2)	Cathode : Zinc Potential (i-r free) (V)			
	−5°C	25°C	50°C	70°C
2	–	1.33	1.36	1.385
5	1.32	1.32	1.35	1.375
15	1.28	1.30	1.33	1.365
25	1.24	1.275	1.30	1.35
50	–	1.20	1.265	1.325
100	–	–	1.20	1.29
200	–	–	–	1.185

NOTE: Potentials measured after 175 to 500 hours of continuous operation; data from Clark, Darland, and Kordesch (23).

Electrodes processed at 800–1000°C are fairly repellent to caustic alkali electrolyte. Indeed, for relatively short discharge periods or with high caustic concentrations, carbon electrodes show excellent operating characteristics without further treatment. However, for conventional cells, where long periods of service are required, carbon electrodes are impregnated with waxes, oils, greases, rubber, or other elastomers in solvents which can subsequently be removed by evaporation (27). Properly applied wetproofing treatment does not materially reduce permeability or surface area, and has only minor influence on depolarization effectiveness in commercial cells.

Simple comparison of the effectiveness of the many forms of carbon cathodes is difficult in view of the empiricism which attended the earlier development of the air-depolarized cathodes. Evaluation of depolarizing or oxidizing properties in terms of physical and chemical properties of the carbon include measurement of temperature changes during peroxide decomposition under controlled conditions (28), as indicated in Table 8.4. Useful for routine comparison of closely related carbon preparations, it is necessary in the use of this test to guard against fortuitous contamination or additives which themselves may be decomposition agents. Variation in thermal treatment, and in particle sizing, can be equally detrimental to test significance. Furthermore, measurements of this type meet only one of the several requirements for electrode effectiveness. Discharge tests on the finished product and expression of effectiveness in terms of the potential: current density relationship still offer the most useful evaluation procedure since these integrate the many factors involved in the formulation of a suitable electrode with its ultimate performance over many months of use in batteries.

Although limited in its capabilities as an oxygen electrode when the commercial block-type carbon cathode ($1\frac{1}{4}$ in. thick) is simply exposed to

TABLE 8.4. INDICATED RELATIONSHIP BETWEEN PEROXIDE DECOMPOSITION AND CELL SERVICE

	Peroxide Decomposition	Discharge Test		
Carbon	Temperature Rise in 3 Minutes	Hours Service to: 1.00 volt	0.90 volt	0.80 volt
A	8.8°C	2.00	2.8	> 48
B	1.5	0.25	1.8	14
C	0.7	< 0.10	0.7	3

PROCEDURE: 0.2 g of sized carbon is mixed with 10 ml distilled water plus 25 ml of 3 percent hydrogen peroxide. In 3 min the temperature rise should be no less than 1°C when the test is started at 20°C. [From Wisfield (28).]

the atmosphere, current output can be increased measurably by circulation of air over the breathing surface of the carbon. If air is blown through the electrode into the electrolyte, current densities can be raised several fold. With oxygen instead of air, current densities of 150–250 ma/cm^2 are attainable (29). Use of plastic-bonded carbon panels ($\frac{1}{8}$ to $\frac{1}{4}$ in.) on perforated metal supports with silver catalyst applied to the active carbon surfaces permits operation in oxygen for many months at 50–100 ma/cm^2 (30). Performance of these electrodes is only moderately improved by use of pressurized oxygen. Active baked-carbon electrodes ($\frac{3}{16}$ in. thick) treated with platinum catalyst permit operation in oxygen at up to 350–400 ma/cm^2 for significant periods of time with relatively small polarization losses (see Fig. 8.14).

The several stages of oxygen utilization by the carbon cathode are pictured as follows:

1. Absorption of atmospheric oxygen at the active sites.
2. Transfer of loosely sorbed gas to the active interface region primarily by bulk diffusion through gas filled macropores.

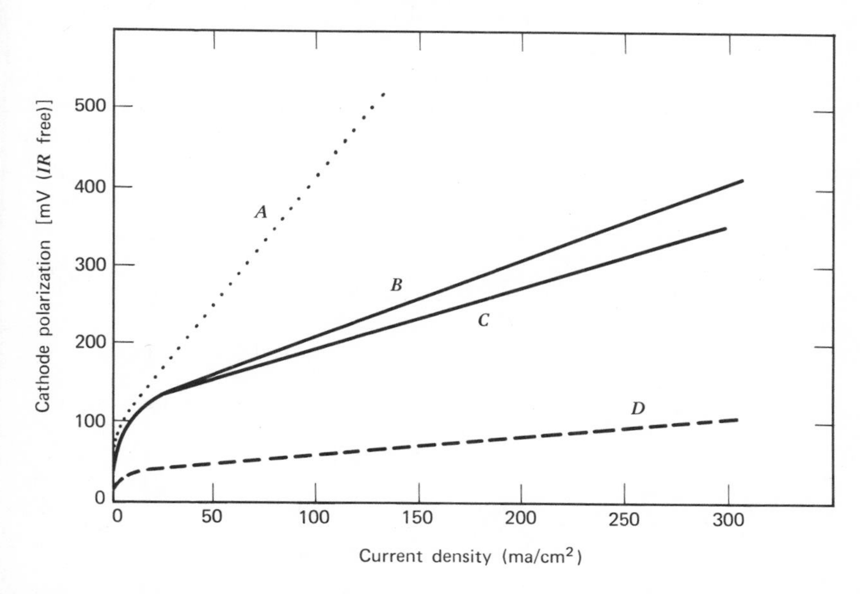

Figure 8.14 Air cell cathode operation with oxygen. Room-temperature tests in 6 *N* NaOH. Plastic-bonded carbon on perforated steel support. Curve *A*: commercial form; $\frac{3}{8}$ in. thick; oxygen at atmospheric pressure. Curve *B*: Like *A*; silver catalyst added; improved conductivity; oxygen at atmospheric pressure. Curve *C*: same as *B*; operation with oxygen at 7 atm pressure. Curve *D*: fuel cell carbon cathode; $\frac{3}{16}$ in. thick; rhodium/iridium catalyst; oxygen at atmospheric pressure.

3. Movement of oxygen from the macropores through the micropore structure to the electrochemically active reaction sites at the electrolyte: electrode interface. (Surface migration of adsorbed oxygen molecules is considered a less important factor for practical porous carbon electrodes.)

4. Cathodic reduction of the oxygen to peroxyl ion at the active interface.

5. Removal of peroxyl ions from the surface chiefly by decomposition, with minor contributions resulting from electrolytic transport and diffusion into the bulk solution.

It has long been recognized (31) that the activation and thermal treatment of active carbon is associated with the development of surface oxides, in amounts ranging from perhaps ½ to 1 percent of oxygen by weight following treatment at elevated temperature, e.g., 1000°C, to as much as 20 percent with treatment at 150–300°C (32). At the same time, acid or base adsorption is also affected. As indicated in Table 8.5, measurements on sucrose carbon calcined in nitrogen (33) for several hours at 950°C prior to activating treatments show that lower base adsorption, preferable for alkaline cell cathodes, is obtained at elevated temperatures.

Very large heats of adsorption are observed in the initial stages of chemisorbed oxygen complex formation at ambient temperature following evacuation at 1000°C, e.g., 70,000–90,000 cal mole (34), with subsequent sorption increments dropping rapidly to normal physical adsorption values of 5000–6000 cal/mole. The chemisorbed oxides are located at

TABLE 8.5. EFFECT OF THERMAL TREATMENT ON ACID: BASE SORPTION BY ACTIVE CARBON

Thermal Processing						Acid: Base Adsorption[a]	
Pretreatment			Final Treatment			(Milli Equiv. Adsorbed per g C)	
Gas	Temp (°C)	Hours	Gas	Temp (°C)	Hours	HCl	NaOH
–	–	–	O_2	150	10.0	0.102	1.048
–	–	–	O_2	400	7.5	0.029	0.395
O_2	1,000	7.5	O_2	400	37.5	0.002	0.462
CO_2	1,000	5.0	O_2	400	40.0	0.000	0.462
–	–	–	O_2	1,000	7.5	0.172	0.000
–	–	–	CO_2	1,000	5.0	0.281	0.000

[a]One gram samples of calcined sucrose carbon were shaken intermittently at room temperature with 50.0 ml of 0.01 N HCl or NaOH, then filtered and 25.0 ml portions titrated for residual acid and base. [From (33).]

active centers, rather than as continuously dispersed films of highly sorbed gas, and are removable only at elevated temperatures, when they are released as carbon monoxide and dioxide. Additional amounts of oxygen may be held in a labile state, at or near the active sites, probably in proportion to the abundance of the latter, and replaceable by diffusion in operating electrodes from the ambient atmosphere.

Though the extent to which the chemisorbed carbon oxides might serve as primary depolarizers is not clear,[4] the most logical current explanation is that the physically sorbed air in the carbon electrode serves as the major source of oxygen for the cathodic reaction in commercial cells.

By using either air or oxygen over a moderately wide range of alkali concentration, the potential of the carbon:oxygen electrode is associated specifically with the formation of peroxyl ions and is reversible with respect to the latter (35). Therefore in place of the reaction postulated for the standard oxygen electrode

$$O_2 + 4H^+ + 4e \rightleftarrows 2H_2O \qquad E^\circ = 1.23 \tag{3}$$

the reaction involving the active carbon cathode may be pictured as follows:

$$O_2\,(\text{adsorbed on the active surface}) + 2e \rightleftarrows O_2^{-2}\,(36)^5 \tag{4}$$

and

$$O_2^{-2} + H_2O \rightleftarrows OH^- + HO_2^- \tag{5}$$

with decomposition

$$HO_2^- \rightarrow \tfrac{1}{2}O_2 + OH^- \tag{6}$$

or

$$O_2\,(\text{adsorbed}) + H_2O + 2e \rightleftarrows OH^- + HO_2 \tag{7}$$

[4]It may be questioned if chemisorption, which takes place only on a clean carbon surface (after treatment at 1000°C in vacuum), will be a significant process at normal temperatures in a functioning electrode. No operating advantages are noted for electrodes containing from 5 to 10 percent of fixed oxides (by low temperature oxidation) over electrodes with perhaps less than 1 percent of oxygen fixed at 1000°–1100°C.

[5]Formation of negatively charged oxygen atoms or molecules have been proposed as the specific mechanism leading to peroxyl ions. This point of view is receiving increasing support.

from which

$$E^\circ = E - 0.0295 \log \frac{pO_2 \cdot H_2O}{OH^- \cdot HO_2^-} \tag{8}$$

$$E^\circ = -0.076 \tag{37}$$[6]

Measurements at Case Western Reserve University by E. Yeager and students (39) show that with active carbon, or graphite electrodes, the standard reduction potential for the O_2/HO_2^- couple in alkaline solution is -0.048 V. This is in good agreement with a value of -0.046 V obtained by other workers (40) for the reduction of oxygen on mercury in alkaline solution. The original measurements by Berl (35) with active carbon electrodes in strongly alkaline solution and without junction-potential correction yielded a value of -0.042 V. Other polarographic studies (41), using alkaline solutions buffered with borate gave a value of -0.063 ± 0.005 V. The weight of the experimental evidence suggests that a value of perhaps -0.05 V is a reasonable compromise, having greater validity than the long-used Latimer data.

Because of the very low peroxyl concentrations, 10^{-10}–10^{-11} mole/liter normally existing at the surface of active carbon cathodes, the potential is considerably more cathodic than that calculated from Eq. (8). Voltages obtained for commercial carbons in 6 N sodium hydroxide solutions (conventional electrolyte for large wet cells) may vary as much as 100 mV depending on prior carbon processing and whether or not metal oxides are present. Representative electrode potentials are of the order of 0.15–0.20 V less cathodic than the theoretical oxygen potential; in combination with zinc anodes, the open-circuit voltages of the cells are generally in the range of 1.42–1.47 V.

In accordance with Eq. (8), the oxygen electrode potential varies with alkali concentration (42) as shown in Fig. 8.15. The slope of the O_2/HO_2^- curve is about 30 to 32 mV/pH unit, which is in reasonably good agreement with the value of 29 mV postulated for the two electron process. Nonlinearity increases with rising alkali concentration, reflecting the rapid change in activity coefficient in these solutions.

No cathodic reduction of the peroxyl ion is noted at active carbons in alkaline electrolyte (35). However, in the usual alkaline air-depolarized cells, chemical decomposition of peroxide at the active electrode surface is essentially instantaneous and is complete when operation is conducted at normal temperature and light-to-moderate rates of discharge. (Peroxide

[6]This value was calculated by Latimer from thermodynamic data published by Lewis and Randall in 1923 (38).

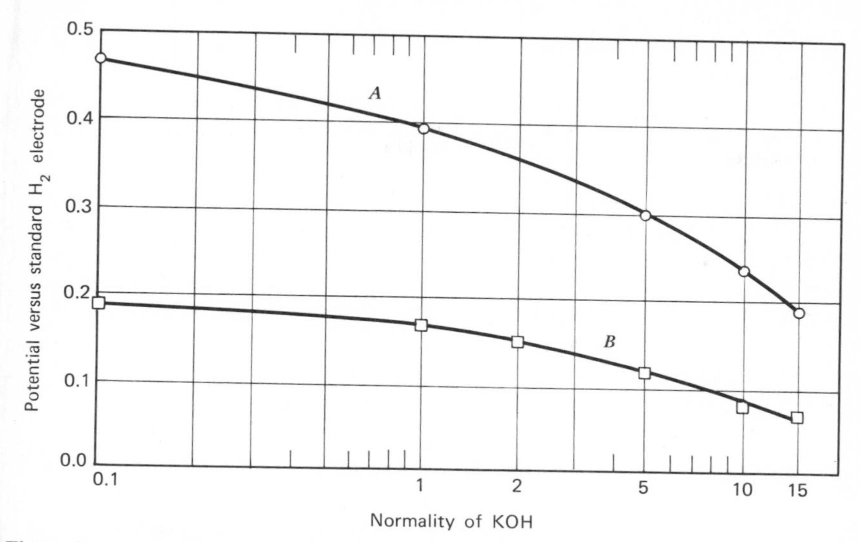

Figure 8.15 Alkali concentration versus potential of oxygen at active carbon electrodes at 20°C. Curve *A*: theoretical: $O_2 + 2H_2O + 4e \rightarrow 4OH^-$. Curve *B*: measured: $O_2 + H_2O + 2e \rightarrow HO_2^- + OH^-$.

is more stable at low temperature, e.g., 0°C, and can lead to cathode deterioration unless peroxide-decomposing catalysts are present in the electrode.) The oxygen thus liberated is sorbed at the solution-electrode interface and used electrochemically in lieu of equivalent atmospheric oxygen. The cathodic process, therefore, is considered equivalent to a four-electron reaction, even though the direct reduction of oxygen involves only two electrons.

THE ZINC:OXYGEN CELL AS A HEAVY DRAIN UNIT

The period since the 1920s has seen the emergence of the alkaline zinc:air-cathode system as a commercial primary cell as dependable as its zinc:copper oxide prototype, but with twice its operating voltage. With enhanced current output made possible through the use of improved cathodes, more effective zinc anodes (18), and electrolyte adapted to promote the zinc oxide reaction, a broader range of industrial utility may be anticipated for the zinc:air-depolarized cell.

Perhaps of even greater impact on battery technology has been the concomitant development of porous carbon electrodes, capable of operating on oxygen at very high current densities, which have contributed importantly to the progress of recent years toward technically practical continuous-feed primary and fuel cells. By combining the new carbon: oxygen electrodes with high area anodes, a new family of primary cells is foreseen, offering outstanding possibilities for packaged power applications where high energy densities and stability are critical requirements.

Stemming from developments in the fuel cell field, electrodes made of such porous metal as silver or noble-metal catalyzed nickel have been proposed for use in alkaline electrolyte with oxygen and zinc anodes. Such cathodes have considerable potential value for special high energy density applications where higher cost or relatively poorer standlife are not dominant considerations. The carbon:oxygen electrode used without noble metal catalyst operates at slightly lower voltage but is more effective at room temperature and is less sensitive to fluctuation in gas supply pressure. It has the further advantage of good stability for standby or intermittent use.

Secondary cell characteristics are also receiving attention in view of the progress in other areas on rechargeable zinc anodes (cf. Chap. 6, Silver Oxide:Zinc Systems), and the very considerable attractiveness of a system in which air or oxygen can be substituted for metal-oxide cathodes. Since the carbon electrode will oxidize destructively when subjected to anodic charging currents, a third electrode of perforated or porous metal such as iron or nickel is proposed to receive current during the charging cycles (43).

REFERENCES

1. Haldane, H., and Nicholson's, J., **4**, 313 (1800); Davy, H., *Phil. Mag.*, **11**, 325 (1800); Walker, W., and Wilkins, F. H., U.S. Pat. 524,229 and 524,391 (1894); Jungner, E. W., U.S. Pat. 1,370,119 (1921); Nyberg, H. D., *Z. Electrochem.*, **30**, 549 (1924).
2. Oppenheim, R., U.S. Pats. 1,544,030; 1,545,801; 1,552,871 (1925); 1,599,061 (1926); Ziegenberg, R., *Electrotech Z.*, **54**, 131 (1933).
3. Heise, G. W., and Schumacher, E. A., *Trans. Electrochem. Soc.*, **62**, 383 (1932); Heise, G. W., U.S. Pats. 1,835,867-8 (1931); 1,863,791; 1,865,652 (1932); 1,924,314 (1933); Heise, G. W., *Trans. Electrochem. Soc.*, **93**, 9P (1948).
4. Heise, G. W., Schumacher, E. A., and Fisher, C. R., *J. Electrochem. Soc.*, **92**, 173 (1947); Schumacher, E. A., Heise, G. W., *ibid.*, **99**, 191C (1952); Schumacher, E. A., Proc. 10th Annual Battery R & D Conf., Ft. Monmouth (1956); Anderson, E. T., *Signalman's J.*, pp. 16–18 (June 1959).
5. Nyberg (1); Fournier, E., French Pat. 636,664 (1928); Oppenheim, R., U.S. Pat. 2,097,077 (1936); Thompson, H. G., U.S. Pat. 2,044,923 (1936); Schumacher, Heise (4); Schumacher, E. A., Heise, G. W., U.S. Pat. 2,017,280 (1939); Heise, G. W., Schumacher, E. A., U.S. Pat. 2,207,734; 2,207,763 (1940).
6. Heise, Schumacher, and Fisher (4); Schumacher and Heise (4).
7. Fisher, C. R., Heise, G. W., and Schumacher, E. A., U.S. Pat. 2,572,918 (1951).
8. Eddy, A. I., U.S. Pat. 2,189,463 (1940); Heise, Schumacher, and Fisher (4); Schumacher and Heise (4).
9. Schumacher, E. A., and Heise, G. W., U.S. Pat. 2,863,789 (1958).

10. Fisher, C. R., Southworth, J., Schumacher, E. A., and Bennett, R. J., U.S. Pat. 2,724,010 (1955).
11. Heise, G. W., U.S. Pat. 1,924,314 (1938).
12. Heise, G. W., and Schumacher, E. A., U.S. Pat. 2,180,995 (1939).
13. Heise, G. W., and Schumacher, E. A., U.S. Pat. 2,077,562 (1937).
14. Dunham, L. S., and Jegge, E. O., U.S. Pat. 2,450,472 (1948).
15. Oppenheim, R., U.S. Pat. 2,118,712 (1938); 2,146,348 (1939).
16. Marsal, P. A., and Fox, R. P., U.S. Pat. 2,597,116 (1952); Brooks, P. S., and Schumacher, E. A., U.S. Pat. 2,597,117-8 (1952); Schumacher, E. A., and Bennett, R. J., U.S. Pat. 2,597,119 (1952).
17. Kordesch, K., and Marko, A., *Radiotechnik*, p. 407 (1950).
18. Shepherd, C. M., and Langelan, H. C., *J. Electrochem. Soc.*, **109**, 657-661 (1962); *ibid.*, **114**, 8 (1967).
19. Heise, G. W., U.S. Pat. 1,835,867-8 (1931); 1,864,652 (1932); Heise and Schumacher (3).
20. Heise, G. W., and Schumacher, E. A., U.S. Pat. 2,180,955 (1939); Dunham, L. S., and Jegge, E. O., U.S. Pat. 2,450,472 (1948).
21. Scholder, R., and Weber, H., *Z. Anorg. Allgem. Chem.*, **215**, 355 (1933).
22. Walker, Wilkins; Jungner; Nyberg (1); Jungner, E. W., U.S. Pat. 1,370,119 (1921); Schumacher, E. A., Hamister, V. C., and Heise, G. W., U.S. Pat. 2,010,608 (1935); Spolek pro Chemikou, Brit. Pat. 644,990 (1950).
23. Berl, E., U.S. Pat. 2,275,281 (1942); Schumacher and Heise (4), Schumacher (4); Winkler, G., and Reinhardt, O., U.S. Pat. 2,641,623 (1953); Fisher, Southworth, Schumacher, and Bennett (10); Clark, M. B., Darland, W. G., and Kordesch, K. V., Proc. 18th Annual Power Sources Conf., Ft. Monmouth (1964).
24. Heise and Schumacher (3).
25. Spolek pro Chemikou (22); Watanabe, J., and Shiramoto, T., *J. Electrochem. Soc., Japan*, **19**, 274 (1951) thru C. A. **46**, 1891 (1952), C. A. **47**, 52 (1953); Marko, A., and Kordesch, K., U.S. Pat. 2,615,932 (1952); 2,669,598 (1954).
26. Parker, W. E., and Marek, R. W., Proc. 16th Annual Power Sources Conf., Ft. Monmouth (1962).
27. Jungner, E. W., Brit. Pat. 145,018 (1920); Heise and Schumacher (3); Ziegenberg (2).
28. Wisfeld, W., U.S. Pat. 2,190,817 (1940); Carbo-Norit Union, Ger. Pat. 675,601 (1939); 700,907 (1940); Bratzler, K., *Z. Elektrochem.*, **54**, 81 (1950); Watanabe and Shiramoto (25).
29. Heise, G. W., *Trans. Electrochem. Soc.*, **75**, 119 (1939); Janes, M., *Trans. Electrochem. Soc.*, **77**, 411 (1940).
30. Schumacher, E. A., Drengler, G. R., Clark, M. B., and Uline, L. J., "Continuous Feed Sodium Amalgam Oxygen Primary Battery," *A.D.* **301**, 978 (1958); Yeager, E., "The Sodium Amalgam Oxygen Continuous Feed Cell," *A.D.* **248**, 803 (1960).
31. Rhead, T. F., and Wheeler, R. W., *J. Chem. Soc.*, **101**, 846 (1912); *ibid.*, **103**, 1210 (1913); Kruyt, H. R., and deKadt, G. S., *Kolloid Z.*, **47**, 33 (1929); Schoenfeld, F. K., *Ind. Eng. Chem.*, **27**, 375 (1935); Strickland-Constable, R. F., *Trans. Faraday Soc.*, **34**, 1074 (1939); Emmett, P. H., *Chem. Rev.*, **43**, 69 (1949); Smith, R. N., *Quarterly Rev. (London)*, **13**, 287 (1959); Boehm, H. P., and Diehl, E., *Z. Elektrochem.*, **66**, 642 (1962).
32. Emmett (31); Weller, S., and Young, T. F., *J. Am. Chem. Soc.*, **70**, 4155 (1948).
33. Bartell, F. E., and Lloyd, L. E., *J. Am. Chem. Soc.*, **60**, 2120 (1938).
34. Blench, E. A., and Garner, W. E., *J. Chem. Soc.*, **125**, 1288 (1924); Garner, W. E., *Nature*, **114**, 932 (1924); Keyes, F. G., and Marshall, M. J., *J. Am. Chem. Soc.*, **49**, 156 (1927); Marshall, M. J., and Bramston-Cook, H. E., *J. Am. Chem. Soc.*, **51**, 2019 (1929).

35. Berl, W. G., *Trans. Electrochem. Soc.*, **83**, 253 (1943); Weisz, R. W., and Jaffe, S. S., *ibid.*, **93**, 128 (1948).
36. Abel, E., *Monatshefte Chem.*, **82**, 39 (1951).
37. Latimer, W. M., *Oxidation Potentials*, 2nd ed., Prentice-Hall, Englewood Cliffs, N.J. (1952).
38. Lewis, G. N., and Randall, M., *Thermodynamics*, McGraw-Hill Book Co., New York, p. 487 (1923).
39. Yeager, E., Krouse, P., and Rao, K., *AD*, **409**, 121 (1963).
40. Yoblokova I., and Bagotsky, V., *Dokl. Akad. Nauk SSSR*, **85**, 599 (1952).
41. Kern, D. M., *J. Am. Chem. Soc.*, **76**, 4208 (1954).
42. Kordesch, K. V., *Fuel Cells*, G. J. Young, Ed., Reinhold Pub. Corp., New York, p. 15 (1960).
43. *Id.*, Australian Pat. 176,889 (1953); Gumucia, R. S., U.S. Pat. 3,219,486 (1965).

9

Fuel and Continuous-Feed Cells

RALPH ROBERTS

The past three decades have been marked by a renewed interest in the fuel cell as a power source. The concept of an efficient converter for the direct production of electricity from commercially available fuels has been taken up anew as a scientific challenge, and much progress has been made. In this chapter, the status of both the basic science and the technology will be reviewed, although no attempt will be made to cover the literature in its entirety nor to include details of the large variety of cells that have been investigated.

The fuel cell may be defined as an isothermal electrochemical power source in which the anodic and cathodic reactants are continuously fed into the cell, with the reaction products removed as formed to maintain an essentially invariant system. The most advanced systems at the present time are based on the use of hydrogen and oxygen or air with aqueous electrolytes. A second type, the indirect or redox fuel cell, is an electrochemical system in which the fuel and oxidant are consumed in the chemical regeneration of the cell reactants, rather than directly in the cell. The last converter to be discussed, the closed-cycle system, is one in which the cell products are processed externally to the cell, by means of heat or other energy source, to reform the reactants. These are refed to the cell and the cycle repeated.

HISTORY

The first attempt to obtain electricity directly from a fuel has been ascribed to Sir Humphrey Davy (1), who in 1802 reported a cell with a carbon anode and aqueous nitric acid as cathodic reactant. Twenty such cells in series were required to decompose water. The first hydrogen-oxygen cell was that of William R. Grove in 1839 (2). It consisted of two platinum strips immersed in acidified water, the emergent portion of one exposed to hydrogen, the other to oxygen. (The hydrogen and oxygen were obtained by electrolysis of water with a voltaic pile.) A strong galvanometer deflection was obtained. In 1842 Grove (3) described a similar, multicell battery, with platinized platinum electrodes, which was able to "effect the decomposition of water by means of its composition."

Numerous attempts to utilize carbon in aqueous electrolytes met with indifferent success. Among the first to attempt the development of a carbon-consuming cell using a fused-salt electrolyte (1855) was A. C. Becquerel (4). His cell consisted of a carbon and a platinum electrode immersed in a fused nitrate. As pointed out by Baur and Tobler (5), this was actually a nitrate-nitrite concentration cell resulting from the reduction of the nitrate ion by carbon.

The growth of thermodynamics increased the understanding of the relationship between electrochemical and heat energy and led Ostwald (6) in 1894 to conclude that the way to solve that most important of all technical problems—the providing of inexpensive energy—must be found in electrochemistry. He further stated that if a galvanic cell directly producing electrical energy from carbon and the oxygen of the air, in a quantity which satisfactorily approximates the theoretical value, were discovered, then a technical revolution would be near.

Even before this conclusion of Ostwald, investigators had started the search for fused salt cells in which carbon reacts electrochemically with oxygen. One of the most persistent workers was Baur (7), who devoted a lifetime of effort to this problem. [He and Tobler (5) have reviewed this and other fuel cell work through 1933.] Nevertheless, up to the time of World War II no fuel cell of technological interest had been developed. Since then significant advances have been made in both the basic electrochemistry of systems and their technology and, by the 1960s, specialized applications of the fuel cell had been achieved and broader ones were being sought.

THERMODYNAMIC CONSIDERATIONS

It has been shown in Chapter 2 that the reversible potential of cells can be calculated if the free energy change for the system is known. The potential for the hydrogen : oxygen cell is given by the relationship

$$E = E^\circ_{H_2} - E^\circ_{O_2} - \frac{RT}{2F} \ln \frac{(a_{H_2O})}{(P_{H_2})(P_{O_2})^{1/2}} \tag{1}$$

where E = electromotive force of cell
E° = standard electrode potential

R = Boltzman constant
T = absolute temperature (°K)
F = value of the faraday
a_{H_2O} = activity of water
P_{H_2,O_2} = pressure, in atmospheres, of reactant gases

In cells with fixed water activity, i.e., constant electrolyte concentration, the electromotive force is independent of product formation and dependent on the pressure of the reacting gases.

In the case of oxygen reacting at a noncatalytic carbon electrode in an alkaline electrolyte, the reversible reaction has been shown by Berl (8) to be

$$O_2 + H_2O + 2e^- = HO_2^- + OH^- \tag{2}$$

i.e., with formation of peroxyl, rather than

$$\tfrac{1}{2}O_2 + H_2O + 2e^- = 2OH^- \tag{3}$$

simple reduction to hydroxyl ion. The higher oxidation potential of reaction 3 makes it more desirable; however, the kinetic factors, which will be discussed later, determine the actual reaction taking place, and in alkaline electrolytes this is reaction 2.

The single electrode potential for reaction 2 is given by

$$E = E^\circ - \frac{RT}{2F} \ln \frac{(a_{HO_2^-})(a_{OH^-})}{(P_{O_2})(a_{H_2O})} \tag{4}$$

This emphasizes the desirability of keeping the peroxyl ion concentration low and the oxygen pressure high.

The reversible electrode reactions for carbon monoxide, hydrocarbons, and other carbon-containing fuels have not been defined experimentally and the electrode potentials must be calculated from thermochemical quantities. Using such data, Broers (9) has computed the electromotive force for selected carbon-containing fuels as a function of temperature, as shown in Fig. 9.1. These curves also reveal the importance of entropy considerations in the thermodynamics of cells with gaseous reactants and products. The relationship between the electromotive force, for the standard state, as related to the entropy, is shown in the following:

$$-nE^\circ F = \Delta G_0 = \Delta H_0 - T\Delta S_0 \tag{5}$$

where n = number of electrons transferred
F = value of the faraday

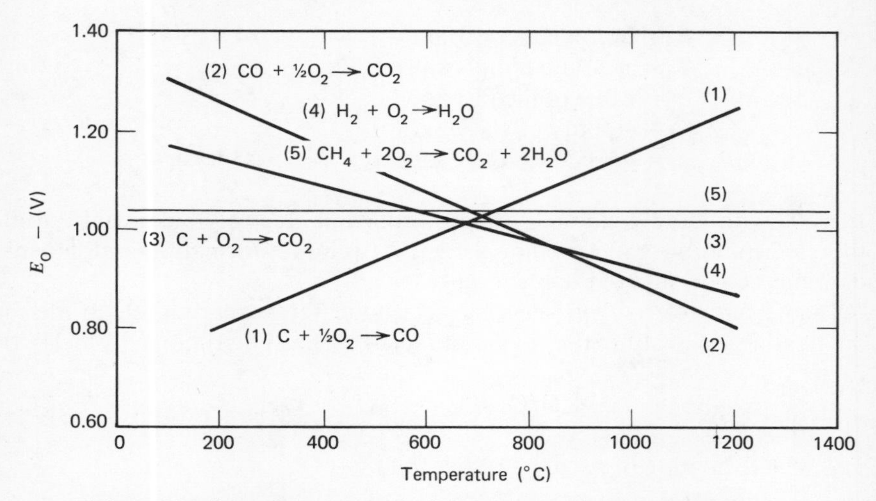

Figure 9.1 Dependence of standard electromotive force on temperature.

$\Delta G_0 =$ standard free energy change
$\Delta H_0 =$ standard heat of reaction
$\Delta S_0 =$ standard entropy change

It is to be noted in Fig. 9.1 that when the reaction proceeds with a decrease in volume, i.e., decrease in entropy, the electromotive force decreases with temperature. In addition, at a constant temperature, this entropy decrease is an unavoidable heat loss during cell operation. Where there is an increase in entropy, e.g., $C+\frac{1}{2}O_2$, heat must be absorbed from the surroundings to maintain the cell temperature if heat losses within the cell are insufficient.

In high-temperature fused-salt cells, the electrode reaction for oxygen can be written as

$$\tfrac{1}{2}O_2 + CO_2 + 2e^- = CO_3^{-2} \tag{6}$$

and the electrode potential as

$$E = E^\circ - \frac{RT}{nF} \ln \frac{(a_{CO_3^-})}{(P_{O_2})^{1/2}(P_{CO_2})} \tag{7}$$

The carbon dioxide reduces cathode polarization by combining with the reduced oxygen, probably an oxide ion on the cathode surface, with the resulting carbonate ion migrating to the anode where it can react with an adsorbed proton. The net effect is the production of water vapor and carbon dioxide.

In high-temperature fused-electrolyte fuel cells operating above 500°C gaseous products are formed and the considerations of product effect are different from those in aqueous systems with constant electrolyte concentration. In almost all fused electrolyte cells to be discussed, the electrolyte is a mixture of fused carbonates, and the products are all gaseous. Under these conditions, using hydrogen as a representative fuel, Eq. (1) becomes

$$E = E^{\circ}_{H_2} - E^{\circ}_{O_2} - \frac{RT}{2F} \ln \frac{(P_{H_2O})}{(P_{H_2})(P_{O_2})^{1/2}} \tag{8}$$

and the potential depends on the pressure of water vapor in equilibrium with the reacting gases as well as on the pressure of the reactants.

The effect on the theoretical potential of the percentage of fuel consumed in the hydrogen-air cell is shown in Fig. 9.2 (10). If the cell is operated in a single stage with 97 percent fuel consumption, the output is proportional to *ABCD*. If the hydrogen is consumed in two stages, the first reacting 50 percent of the gas and the second an additional 47 percent, then the output is represented by *A′B′C′D′* plus *EBCC′* and approaches more closely the theoretical output. Thus for the high temperature fuel cells, greater efficiency of fuel utilization is obtained when the fuel is consumed in two or more stages; however, the increased complexity in operating with the fuel feed in series may outweigh this theoretical possibility for increased fuel efficiency.

Fuel cell efficiency

One of the major motivations in the development of the fuel cell is its great promise for increasing the efficiency of fossil fuel utilization. In the conversion of a fuel to electricity by means of a heat engine, a multistage process is required (Fig. 9.3). The fuel is first burned to give hot combustion gases. These are then used either directly or to generate steam as the working fluid. The heat engine or turbine drives an electric generator which produces electricity for distribution to the load. The fuel cell system consists simply of an electrochemical generator delivering direct current.

In the case of the heat engine, its thermal efficiency N_t is given by the work produced W divided by the calorific value of the fuel ΔH, or

$$N_t = \frac{W}{\Delta H_{25^\circ C}} \tag{9}$$

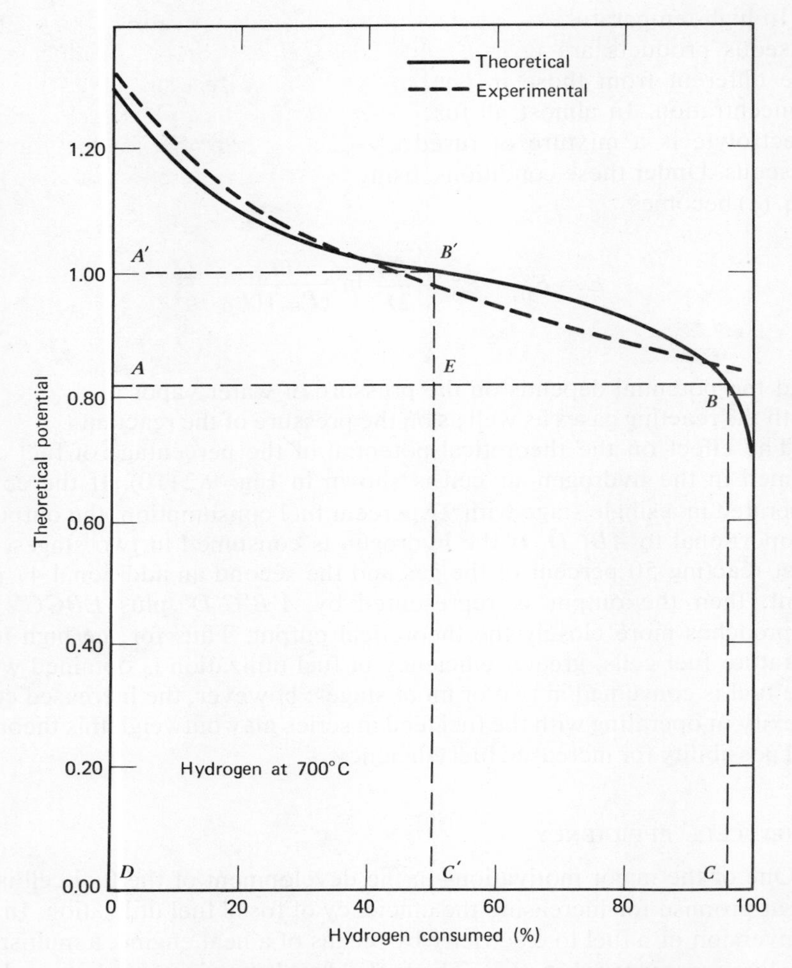

Figure 9.2 Theoretical potential versus percent gas consumed.

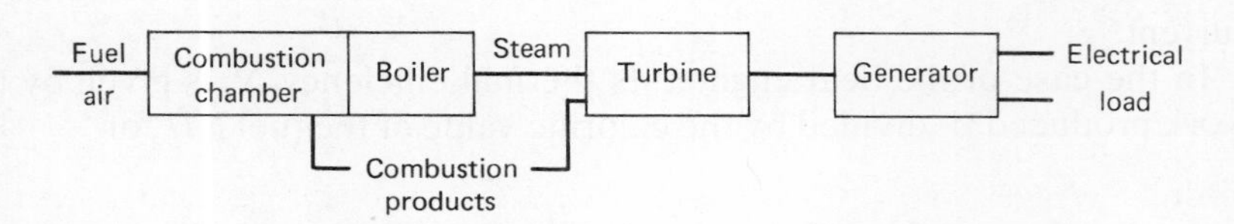

Figure 9.3 Conventional power generation.

The maximum amount of work done by a heat engine is determined by the Carnot cycle and given by

$$W = \Delta H_{25^\circ\mathrm{C}} \frac{T_1 - T_2}{T_1} \tag{10}$$

where T_1 is the inlet temperature and T_2 the outlet temperature in degrees Kelvin. The maximum Carnot efficiency N_c for a heat engine thus becomes

$$N_c = \frac{T_1 - T_2}{T_1} \tag{11}$$

Further, reduced by such factors as combustion inefficiency, heat, friction, and generator losses, the over-all efficiency N_c for most systems, is less than 0.35.

The efficiency of the electrochemical system N_E, which operates isothermally rather than through a temperature cycle, is determined by the fraction of the free energy ΔG, which is converted to useful work:

$$N_E = \frac{W}{\Delta G_{25^\circ\mathrm{C}}} \tag{12}$$

The useful work is proportional to the terminal voltage E_{terminal} and the free energy change is proportional to the reversible potential $E_{\mathrm{reversible}}$:

$$N_E = \frac{E_{\mathrm{terminal}}}{E_{\mathrm{reversible}}} \tag{13}$$

For comparison with Carnot limited methods of energy conversion, it is more appropriate to use the thermal efficiency (Eq. (9)). Again, the actual efficiency is less than this, due to incomplete utilization of fuel, energy for auxiliaries, heat losses in product removal, and, the fact that in some cases the electrode reaction is not the one having the maximum theoretical potential. Efficiencies of 60–80 percent have been obtained for fuel cells.

In general, the 35 percent upper performance level of Carnot-cycle limited systems is the minimum efficiency for fuel cell operation. This results from the power-voltage relationship for such circuits; since the maximum power from a fuel cell is obtained at 50 percent of the open circuit voltage, when the voltage-current curve is linear. A more detailed consideration of the thermodynamics of fuel cell efficiency has been given by Liebhafsky and Douglas (11). The special considerations of the efficiency of redox and regenerative fuel cells will be discussed when these systems are described.

ELECTRODES FOR FUEL CELL

Before describing specific fuel cells, a general discussion of electrode construction and material selection should be of value. The site of reaction in almost all primary batteries is the electrode-electrolyte interface. Fuel cells consuming gaseous fuels add a third phase to the reaction zone, that of the gaseous reactant. The reaction site, the intersection of the three phases (gas, liquid, solid), is illustrated in Fig. 9.4. To maximize the number of reaction sites, highly porous electrical conducting materials, such as carbon or metal, are used for the electrodes. The pores permit the gas to diffuse into the body of the electrode to the site of the electrochemical reaction at the solution-electrode interface.

The specific locus of the electrochemical reaction in these systems has been investigated by several groups. Weber et al. (12) have concluded that transport across the meniscus is more important than surface diffusion along the pores. The effective electrode-electrolyte interface extends well above the apparent meniscus into a thin film on the surface of the electrode. Measurements show a contribution to the current from points up to 17 mm above the liquid level (13). At high current density, transport through that portion of the meniscus closest to the liquid surface is of greater importance than the more extended part of the electrolyte capillary interface. In addition to similar experimental results, Bennion

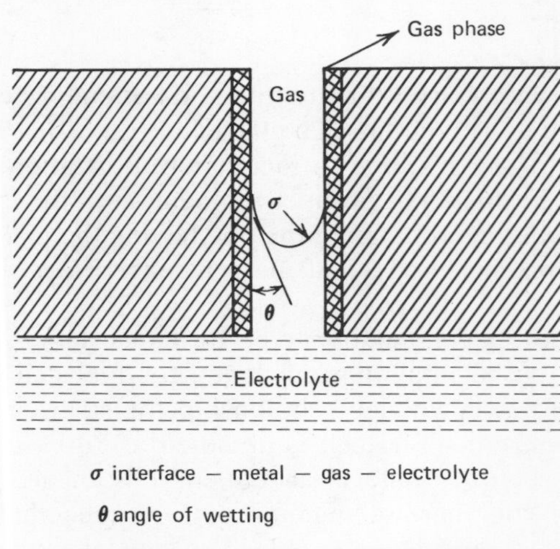

Figure 9.4 Idealized reaction interface.

(14) has developed a theoretical treatment of a one-dimensional model of the capillary-electrolyte interface with the reactant transported across the extended electrolyte film. He concluded that oxygen transport in the thin upper film is not the current-limiting process but that the current density distribution in this region arises from a balance between kinetic overpotential and resistance drop in the film.

References (12) and (13) and Fig. 9.4 describe the situation observed in experimental electrodes with hydrophilic surfaces. In the case of modern fuel cell electrodes, hydrophobic surface properties have been found necessary; a different electrode mechanism is needed to explain their operation. Several approaches have been used to maintain the three-phase boundary system and to prevent penetration of electrolyte into the porous electrodes. As in the zinc-oxygen cell (cf. Chap. 8), the gas-electrode material, e.g., carbon, should be relatively hydrophobic, but it still must be fortified by impregnation with a water-repellent substance. Davtyan (15) used paraffin for this purpose; Yeager (16) has described cathodes made of active carbon dispersed in an organic solvent such as benzene. To this was added 5–15 percent of a waterproofing agent, such as polystyrene or gum rubber. The dispersion of carbon and hydrophobic binder was then sprayed onto porous graphite or a metallic grid support maintained at a slightly elevated temperature. Haldeman and co-workers (17) prepared thin electrodes by pressing a mixture of teflon and powdered metal onto a metal screen support. Another approach is that of a graded-porosity electrode (18, 19). The coarser pore layer is on the gas side and the fine-pore layer on the electrolyte side. The latter, with the pressure of the reactant gas opposing the capillary wetting of the pores, prevents the flooding of the electrode with electrolyte.

A simplified construction of an electrode equivalent to the graded porosity electrode has been described by v. Sturm (20). The active part of the electrode is mechanically unstable, consisting of a powdered catalyst supported on the electrolyte side by a gas-tight fine-porous metallic layer. The catalyst layer, described as being equivalent to the inflexible working layer of a sintered double-porosity electrode, is retained within a skeletal body by an asbestos or other nonmetallic sheet wetted by the electrolyte. This technique has been used to prepare electrodes having an area of 300 cm^2.

Although the activity of most of the electrodes is decreased when flooded by electrolyte, porous metal electrodes that are not permanently damaged as gas electrodes by flooding have been developed (21). These recover their activity when the electrolyte is forced out of the pores. In addition, some fuel cells, such as those using liquid reactants or their solutions, operate in the flooded condition. Grens and Tobias (22) have examined polarization in such electrodes and have concluded that it varies within the pores. They have also shown that the relative resistivities of the electrode structural material and the electrolyte affect cell performance.

The porous electrodes are not only important for establishing the interface between the reactant, the electrode, and the electrolyte but make possible a large effective surface area, which, together with the effect of the metallic catalysts generally added makes large current densities possible. However, as pointed out by Adlhart and co-workers (23), catalyst accessibility as well as surface area must be considered.

The extensive studies on hydrogen overvoltage (see Chap. 2) on various metal surfaces gives considerable guidance regarding the performance and selection of the catalyst for the hydrogen electrode (24, 25). On metal surfaces of low hydrogen overvoltage, the reversible hydrogen potential is obtained and the potential-controlling step involves electron transfer.

The importance of hydrogen reactivity on the electrode surface carries over to organic and other fuels, such as hydrazine and ammonia. Studies on these fuels indicate that although the reactions are complex and the rate-controlling process has not been well defined, hydrogen adsorbed on the surface results from the chemisorption of the complex fuel molecule. Burshtein and others (26) have used the rapid reaction of adsorbed hydrogen as a measure of its surface concentration. Thus, in a comparison of the electrocatalytic oxidation of ethane and ethylene he and his co-workers show that a greater fraction of the surface is covered by hydrogen. His results agree with the mechanism proposed by Dehms and Bockris (27) for the oxidation of ethylene and helps explain its more rapid reaction. The greater reactivity of hydrogen than that of the carbon-containing species on the electrode surface leads to the existence of mixed potentials on these electrodes (see Chap. 2).

That the mechanism of electrochemical oxidation of hydrocarbons is at great variance with the direct chemical oxidation has been emphasized by Grubb and Michalske (28). At a selected voltage, 0.6 V against oxygen, the current density decreased with increasing hydrocarbon chain length, the reverse of gas phase reactivity. The importance of adsorbed water, or water-derived species, is suggested not only by the proposed mechanisms for the electrochemical oxidation of hydrocarbons but by the observation that even in high-temperature fuel cells, operating above 600°C, the presence of water leads to improved performance of hydrocarbon type fuels (9).

The investigation of the cathodic reaction of oxygen has been accelerated in an attempt to overcome or understand the kinetic factors that prevent achieving the full reversible oxygen potential in aqueous media. Even in the case of the catalytically active transition-metal surfaces, some peroxyl ion formation occurs in alkaline media and a metal oxide forms on the electrode surface (29). The existence of the oxide helps explain the failure to achieve the reversible oxygen potential at electrodes, except under very special laboratory conditions. According to Hoare (30), the interaction of oxygen with the electronic conductor leads to the existence of a mixed potential preventing the obtaining of the full oxygen potential. The formation of the peroxyl ion at carbon and metal electrodes (8) and (29) emphasizes the need for incorporating a catalyst to destroy HO_2^- at the cathode unless the active electrode material is itself a catalyst for this reaction.

LOW-TEMPERATURE CELLS

Hydrogen-oxygen cells operating below 250°C

Among cells operating at temperatures below 250°C greatest success and effort has been with those utilizing hydrogen and oxygen as reactants and potassium hydroxide as electrolyte.

Evans: Kordesch. The cell reported by Evans, Kordesch, and co-workers (31, 32) is among the most advanced of this group. The electrodes, of porous carbon, have been rendered water repellent. The hydrogen electrode contains metals from the platinum group to catalyze the anode reaction. These electrodes are highly resistant to poisoning. The oxygen electrode contains catalysts to decompose the peroxyl ion initially formed. Major improvements in electrode construction have been made (33). Before 1963, the electrodes were approximately 0.62 cm thick and fragile. The present electrodes have a thickness of approximately 0.09 cm and are much more resistant to shock and give better over-all characteristics. A schematic diagram for their preparation is shown in Fig. 9.5.

The electrolyte is a potassium hydroxide solution and a large range of concentrations, 6–14 *M*, has been used. The potential varies with electrolyte concentration. The open-circuit potential has a small negative temperature coefficient, and the electrode shows the theoretical dependence on pressure. The circulating gases also carry out the water produced, and this can be removed from the gas stream by a condenser system. Although the cell can be operated below room temperature and at atmospheric pressure, better performance is obtained at temperatures of 60–70°C and elevated pressure. The current-voltage curves for this cell (Fig. 9.6) show the performance of the cell and improvements achieved

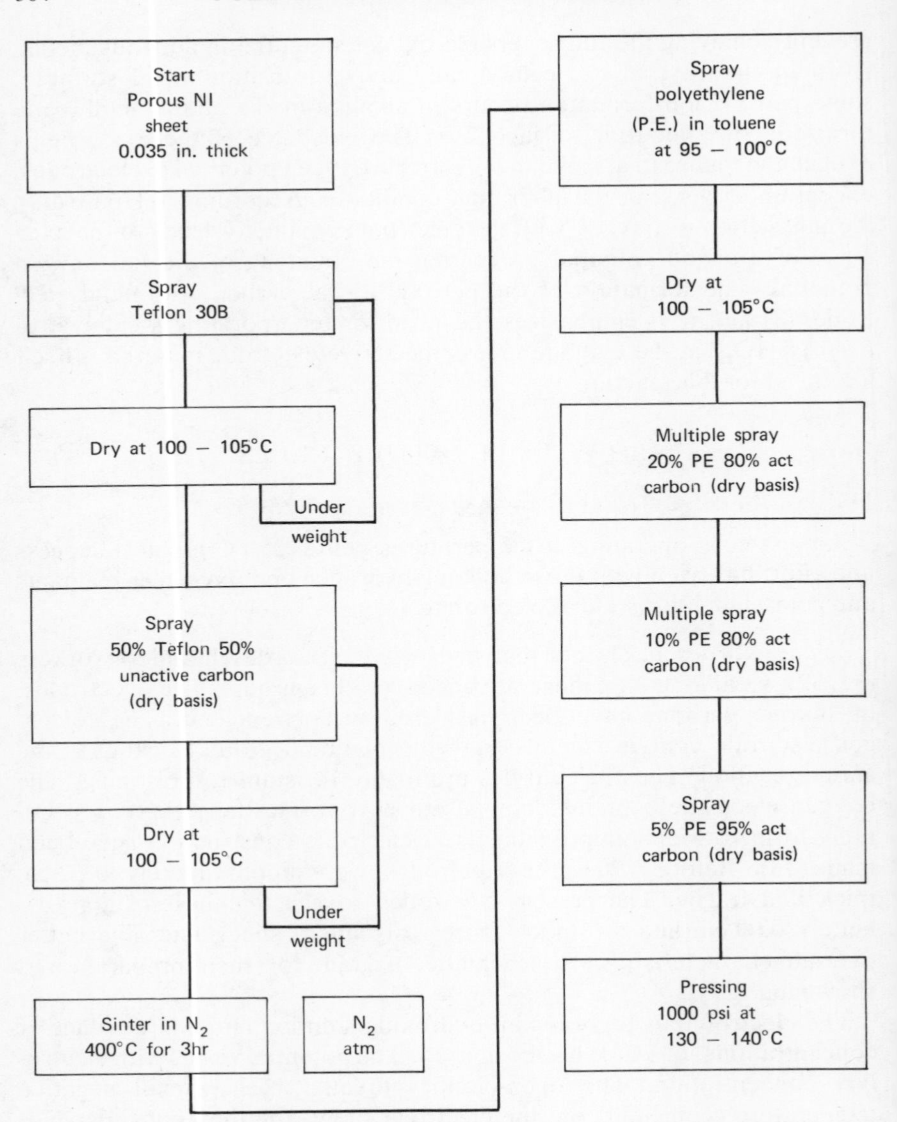

Figure 9.5 Process flow sheet for thin electrodes. From (32).

over a period of several years. As predicted by the Nernst equation, the output decreases when air is used in place of oxygen. The operation with air, especially at elevated pressures, requires constant purging of the cathode to prevent nitrogen build-up. Fuel cell batteries using the newer,

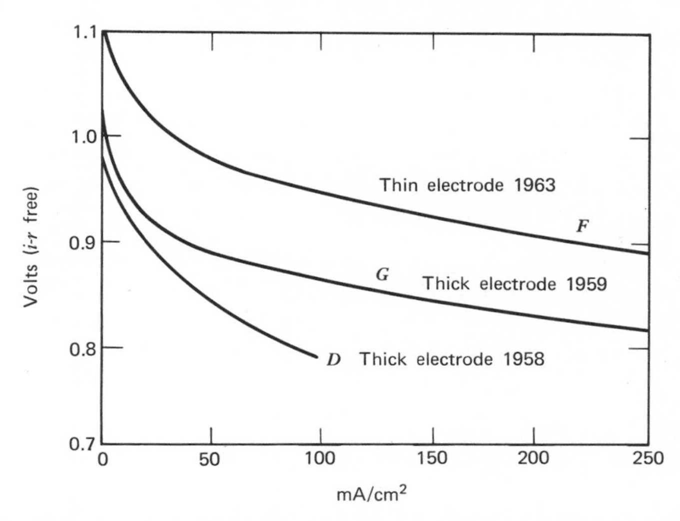

Figure 9.6 Performance parameters of Evans-Kordesch (Union Carbide) $H_2:O_2$ fuel cells at 60°C and atmospheric pressure.

thin electrodes have been built and operated for over 6000 hr. A very important characteristic of these batteries is their operation at low temperatures: at −5 to −10°C cells with these electrodes deliver 20 ma/cm² at 0.8 V.

Davtyan. A similar cell was reported in 1949 by Davtyan (15). This cell used an active-carbon, silver-catalyzed anode and a similar, nickel-catalyzed cathode. The cell was operated on air and hydrogen, and its performance is shown in Fig. 9.7.

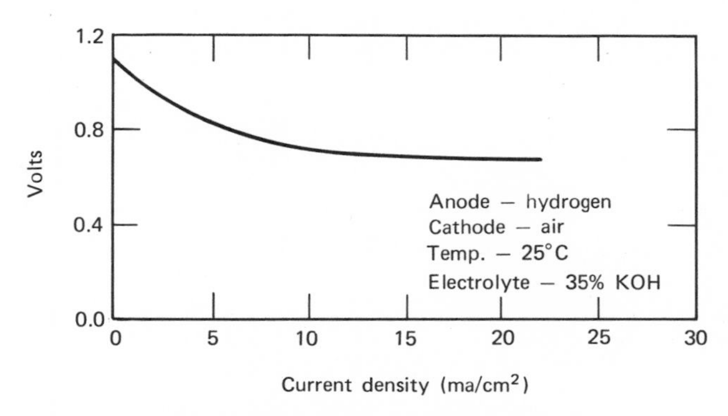

Figure 9.7 Performance of Davtyan hydrogen–air fuel cell.

Cells with Haldeman's Teflon: powdered metal electrodes (17) containing 9 mg Pt/cm^2 gave a potential of 0.77 V at 200 ma/cm^2 and 30°C, with a 0.04–0.05 increase of voltage at 70°C at the same current density. During the long-time testing of such electrodes with a nickel screen backing, it was shown that nickel oxide is formed at the cathode. This is considered to contribute to a decrease in performance of these cells, an effect that was not observed when a platinum screen backing was used.

Justi. In a hydrogen-oxygen fuel cell described by Justi and collaborators (19), the so-called DSK electrodes ("Doppelskellett-Katalysator-Electrode") were made of porous nickel 3 mm thick and were activated by a Raney-nickel catalyst. To decrease the loss of hydrogen by diffusion under pressure through the electrode and to minimize flooding, it was found necessary to use two layers of different pore size as Bacon (34) had demonstrated earlier. In most of the work, 6 *M* potassium hydroxide was used as the electrolyte.

In the first cells reported, the cathode was a pressed-carbon electrode of controlled porosity. Subsequent studies have led to DSK cathodes consisting of a nickel matrix supporting a silver catalyst (35). Excellent cell performance has been obtained with such electrodes at 80°C and a gas pressure of 3 atm (Fig. 9.8).

Platner. Another highly developed variation of the low-temperature alkaline hydrogen-oxygen cell is the capillary membrane fuel cell of Platner and co-workers at Allis-Chalmers (36). The major difference in this cell is that the alkaline electrolyte is retained in place by a membrane, an asbestos material saturated with a solution of potassium hydroxide. A range of concentrations has been studied and 35 percent KOH is con-

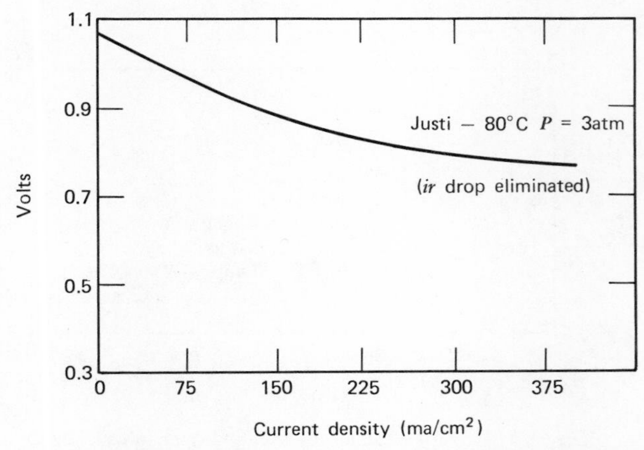

Figure 9.8 Performance of Justi metal–electrode hydrogen–oxygen cells.

sidered optimum. The electrodes are porous nickel with a heterogeneous pore volume of 85 percent, loaded with platinum and palladium as catalysts. The cells are approximately 0.15 cm thick. Operation at temperatures between 88 and 93°C gives approximately 0.88 V at a current density of 60 ma/cm².

A novel static method of water removal has been developed for use with the capillary membrane cell, based on the use of an additional asbestos sheet moistened with alkali behind the hydrogen feed space. The alkali concentration of asbestos sheet is greater than that of the cell, leading to moisture transport to the higher concentration. By maintaining a partial vacuum behind the external alkali-wetted asbestos, the water removal can be a continuous operation.

Attempts have been made to replace the expensive platinum catalysts and the difficult-to-prepare Raney nickel electrode of Justi. The borides of nickel, cobalt, and manganese have been investigated by several groups. Some characteristics of these materials have been summarized by Jasinski (37). A cell with nickel boride incorporated into the anode by mixing with Teflon and bonding to a metal screen, a platinum black cathode, and a 30 percent KOH electrolyte, operating at 100 ma/cm² and a temperature of approximately 90°C, had a voltage of 0.88.

Bacon. The work of F. T. Bacon and collaborators (18) on an elevated temperature (200–250°C) hydrogen-oxygen fuel cell did much to stimulate recent interest in the potentialities of direct electrochemical power sources. This work, initiated before World War II, has resulted in a highly developed, high-power density fuel cell system shown schematically in Fig. 9.9. As previously noted, the nickel electrodes are of a graded

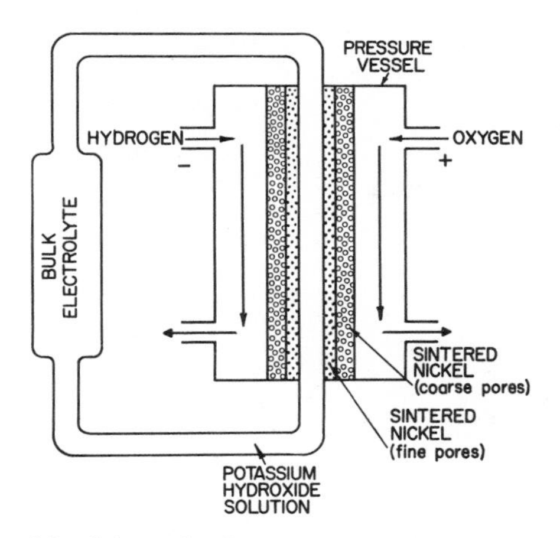

Figure 9.9 Schematic of Bacon hydrogen: oxygen fuel cell.

porosity. To increase its operating lifetime, the sintered nickel cathode is moistened with lithium hydroxide and air-oxidized. The electrolyte used is 37–50 percent potassium hydroxide. The usual operating conditions are 200°C and pressures of 300–600 psi. In operation, the hydroxide solution is circulated and the water formed is evaporated by passing the hydrogen feed over it. The excess water is condensed out of the hydrogen stream.

Although most of the original work was done on electrodes 5 cm in diameter, 40-cell batteries (Fig. 9.10) have been constructed with electrodes having a diameter of 25 cm. Figure 9.11 shows the current-voltage curve for a single cell.

To put the multicell battery into operation, it was first heated electrically using an external power source. Once at operating temperature and under load, it was cooled by cold air circulating around the battery.

From design estimates for a large battery delivering 44 kW, the cell weight was deduced as 50 lb/kW. The power per unit volume of the cell operating at 0.68 V was estimated as 8.2 kW/ft^3. This did not include control equipment.

Figure 9.10 Bacon fuel-cell battery (filter press type of 40 cells). (Courtesy of F. T. Bacon.)

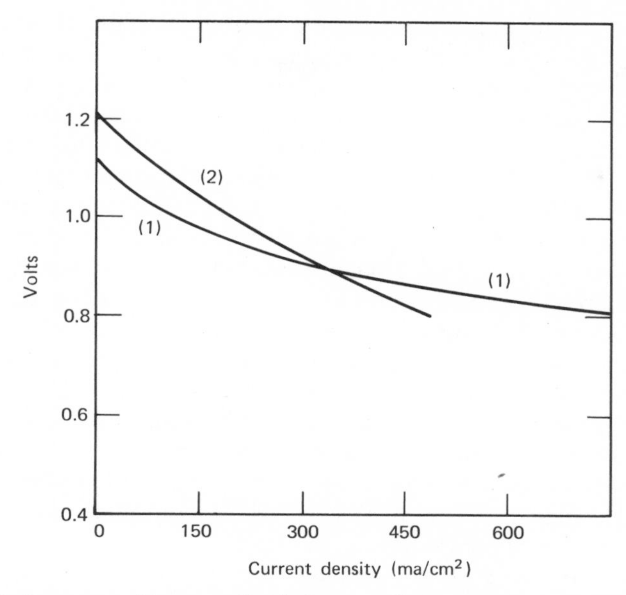

Fig. 9.11 Performance of bacon cell. (1) Bacon cell: $T = 200°C$; $P = 40$ atm. (2) Pratt-Whitney modification: $T = 250°C$; $P = 1$ atm.

In a low-pressure version of the same cell, the Pratt-Whitney Corporation (38) substituted an 85 percent solution of potassium hydroxide for the lower concentration of the Bacon prototype. This permits atmospheric pressure operation at a temperature of 250–260°C, with performance equivalent to that of the Bacon cell at 200°C, (cf. Fig. 9.11). An operating lifetime of over 1000 hr was reported in May 1965.

A low-temperature fuel cell operating in acidic media, reported by Grubb(39), uses cationic exchange membranes as the electrolyte. These membranes are based on polymers containing highly ionizable acidic groups, such as sulfonic acids.

The construction of the ion-membrane cell is shown schematically in Fig. 9.12. It was found desirable to

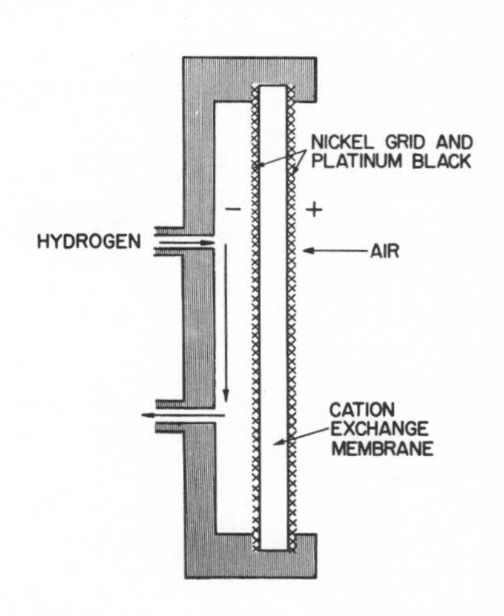

Figure 9.12 Schematic of membrane fuel cell.

keep the membrane thin to minimize the internal resistance. The electrode, in addition to being the catalyst, also served as the conductor. The specific catalysts used in the cells included copper, nickel and silver, and the group VIII metals, such as rhodium, palladium, iridium, and platinum(40). The cell has been operated on air or oxygen at the cathode and hydrogen at the anode. Figure 9.13 shows the performance of the membrane cell.

It is essential that the relative humidity of the air be maintained high enough to prevent the drying out of the membrane, a much less acute problem when operating on oxygen since no gas flows through the cell. The water produced drains out of the cell. Most of the studies were conducted using 11.4 cm² electrodes; however, 580 cm² electrodes also have been constructed and operated. In addition to the requirement for precious metal electrodes resulting from the use of an acidic membrane, this cell presents some difficulties. The temperature in the cell must be kept within restricted limits since the membrane is a thermoplastic; the humidity of the reactant gases must be carefully regulated, as previously noted; and a defect in the membrane which could lead to combustible hydrogen-oxygen mixtures must be avoided.

A variation of this battery is the dual-membrane fuel cell developed by Lurie, Berger, and Viklund(41). By using membranes at each electrode, it is possible to circulate the electrolyte, 3 *M* sulfuric acid, through the cell, reducing the problem of temperature control. At 30°C, this cell has the following voltage-current characteristics: at 0.64 V a current density of 78 ma/cm² and at 0.8 V, 28 ma/cm².

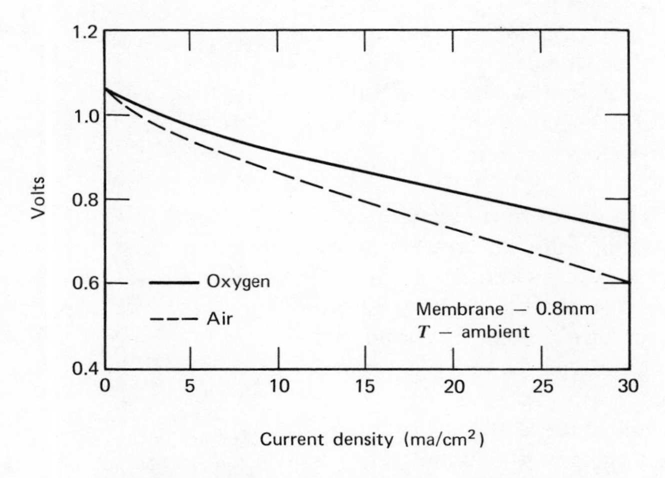

Figure 9.13 Performance of membrane fuel cell.

Miscellaneous Low-Temperature Cells. Elmore and Tanner (42) have reported the results of an interesting hydrogen-oxygen fuel cell which operated at temperatures between 125 and 200°C. Palladium foil, 0.0013 cm thick, was used as the anode. The cathode was prepared by mixing a catalytic metal, preferably silver, with a Teflon dispersion and placing the treated *powder* between silver screens. The electrolyte was a eutectic mixture of sodium and potassium hydroxide kept rigid with calcium hydroxide. At 150°C, an open circuit voltage of 1.15 was observed, with current densities of approximately 28 and 70 ma/cm^2 at voltages of 0.80 and 0.55, respectively.

A rigid acid-electrolyte cell also has been reported by these workers. The electrolyte was phosphoric acid gelled with silica powder. The electrodes were of carbon using either platinum or palladium as the catalyst for both anode and cathode. With hydrogen as the fuel, open circuit voltage of 0.82 was observed and current densities of 30 and 75 ma/cm^2 at 0.60 and 0.40 V, respectively, were obtained. In addition to hydrogen, the acid cell was operated with formic acid and with formaldehyde with voltages reduced by approximately one-half for a given current density.

Another acid system, based on a hydrated zirconyl phosphate ion-exchange electrolyte, has been reported by Dravnieks et al. (43). This material was placed between platinum screens which were backed with platinum foil. Using hydrogen and oxygen as the reactants, the open-circuit voltages were 0.93–1.04 V. The internal resistance was 10–25 $ohms/cm^2$ of cell. Though the over-all cell performance has not been studied in detail, it is reported that the electrodes showed a total polarization of approximately 0.15 V at 70°C and 0.20 V at 25°C at a current density of 10 ma/cm^2. The cell also has been operated on hydrazine hydrate as the fuel with various oxidizers, including oxygen, chlorine and nitrogen dioxide.

Fuel cells with hydrocarbons and other fuels (below 250°C)

As discussed in the section on electrodes, considerable study is being conducted on the direct oxidation of hydrocarbons in a fuel cell. Since alkali electrolytes would react with the carbon dioxide produced, decreasing the cell output, most efforts have followed Elmore and Tanner's (42) use of phosphoric acid. To date, most workers agree that platinum is the best catalyst for the hydrocarbon anode.

In view of the more corrosive nature of acid electrolytes, most electrodes used in these media have been constructed from noble metals. To decrease the quantity of these expensive metals, various substrates for the catalysts have been investigated. Both Haldeman (17) and Ruetschi and Sklarchuk (44) have shown that titanium is a suitable anodic substrate material in sulfuric acid. Ruetschi and Sklarchuk have also reported that tantalum and niobium electrode substrates, in acid electrolytes, have shown lifetimes of over six months.

Grubb and Michalski (45) have published a series of papers on hydrocarbon fuel cells with platinum-catalyzed electrodes and concentrated phosphoric acid electrolyte. With 14.6 *M* phosphoric acid electrolyte and propane as the fuel, a current density of 20 ma/cm^2 at a voltage of 0.48 (*i-r* loss included) was obtained. Within the experimental error of the analysis, the propane was quantitatively oxidized to carbon dioxide and water. Figure 9.14 shows the performance for methane and propane fuel cells. The requirement of temperatures greater than 100°C for obtaining good electrode performance led to the abandonment of sulfuric acid as an electrolyte since above this temperature there is direct reaction between the sulfuric acid and the hydrocarbon, leading to catalyst poisoning.

An attempt to avoid the current-density limitations of direct acid-electrolyte hydrocarbon fuel cells was made by Heilbronner and coworkers (46). The anode compartment contains a hydrocarbon-reforming catalyst, which catalyzes the reaction of a hydrocarbon with water to give hydrogen and carbon dioxide. The anode is a palladium-silver alloy that separates the hydrogen from the other gases by selective diffusion. The remainder of the cell is the same as the Pratt-Whitney modification of the Bacon cell. The utilization of the hydrocarbon depends on the catalyst bed thickness; the thicker the bed, the higher the utilization. However,

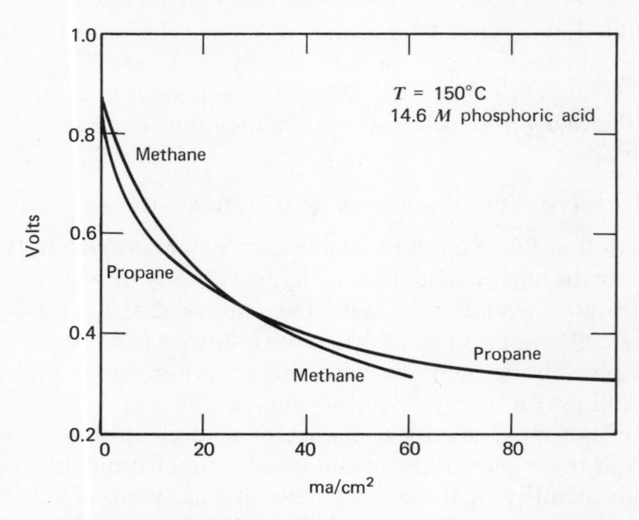

Figure 9.14 Performance of hydrocarbon fuel cell. From (45).

the current density is much lower for the thicker catalyst beds. For example, a bed depth of 0.15 cm gave a current density of 180 ma/cm² with 0.15 V of polarization. The fuel utilization was 20 percent. For a bed of 0.38 cm, the current density was 65 ma/cm² with a fuel utilization of 67 percent. The fuel used was octane.

More success has been achieved with methanol as a fuel than with hydrocarbons. Williams and co-workers (47) reported a voltage of 0.55 at a current density of 100 ma/cm² for a cell with a flooded anode, i.e., with methanol dissolved in circulating sulfuric acid electrolyte.

Ammonia as Fuel. Ammonia would be a relatively inexpensive fuel, but its direct utilization has long proved difficult. Eisenberg (48) employed a nonaqueous, fused-matrix electrolyte: the eutectic mixture of sodium and potassium hydroxide, kept rigid with magnesia. With nickel anodes and cathodes, the latter lithiated, the ammonia: oxygen couple showed a voltage of 0.5 at a current density of 60–70 ma/cm².

The improvement made with cells using aqueous electrolytes has been particularly striking. With electrodes catalyzed with platinum black, and KOH (54 percent) electrolyte, "performance levels . . . exceeded those reported for any fuels other than hydrogen or hydrazine" (48a). Even more significant was the finding that "iridium has a unique ability to dissociate ammonia in the gas phase and that fuel cell anodes containing iridium are capable of oxidizing ammonia at distinctly lower over potentials than those observed with platinum electrodes" (48b). Performance data [anode (Pt-Ir)] potentials versus hydrogen reference electrode at varying current density are shown in Fig 9.15. It is noted that addition of iridium makes possible a reduction in the total quantity of noble-metal catalyst required.

A further possibility is the use of a separate reactor containing iridium catalyst coupled with a conventional cell with platinized anode. For ammonia cracked at the comparatively low temperature of 410°C, the performance characteristics were essentially those of pure hydrogen.

Hydrazine as Fuel. The use of hydrazine as anodic reactant coupled with conventional oxygen cathodes has also led to fuel cells of good performance characteristics.

For the over-all cell reaction

$$N_2H_4 + O_2 \rightarrow N_2 + 2H_2O$$

the theoretical voltage is 1.57. However, since the observed open-circuit voltages are only of the order of 1 V, it may be assumed that hydrazine is first decomposed to nitrogen and hydrogen and that the hydrogen, ad-

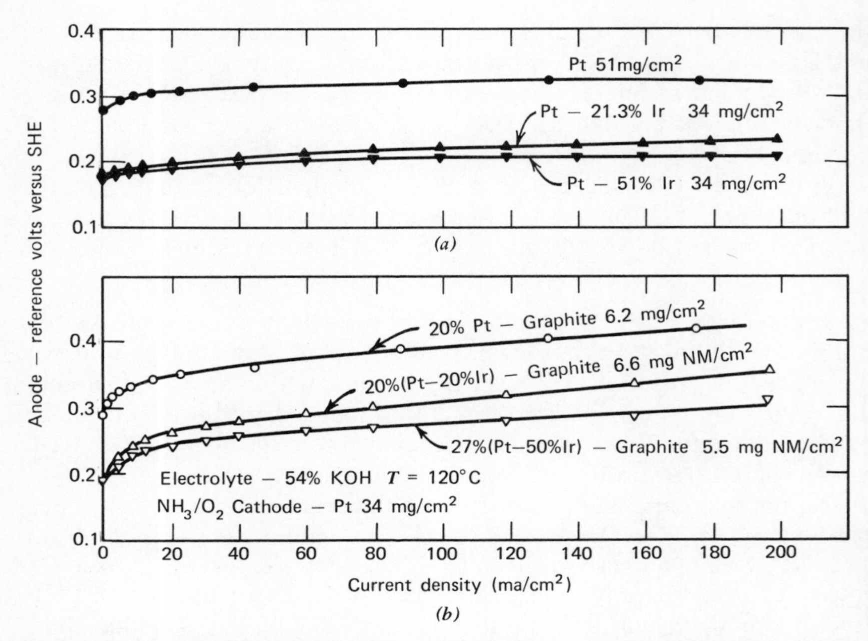

Figure 9.15 Performance of Pt/Ir anodes in NH_3 fuel cells, 120°C. (*a*) Unsupported blacks. (*b*) Pt/Ir supported on graphite. From (48b).

sorbed at the porous electrode, is the actual anodic species. Hydrazine therefore might be considered as a convenient means of packaging hydrogen rather than as a primary fuel.

Tomter and Antony (49) described cells of the capillary-membrane type, in which a 2–4 percent solution of hydrazine in 25 percent KOH was circulated behind the anode while oxygen was fed to the cathode. The water content of the circulating anode feed could be controlled by the addition of potassium hydroxide and the discard of the excess volume of feed solution. An operating voltage of 0.7 could be maintained for a current density of 140 ma/cm^2 at 70°C, or for 40 ma/cm^2 at 27°C. An 8-cell module, weight 10 kg, with 465 cm^2 electrodes, was constructed.

The continuous-feed hydrazine: air cell developed by Evans and Kordesch (50) is illustrated in Fig. 9.16. If the oxygen electrode contains catalysts such as noble metals which would lead to wasteful hydrazine decomposition, barrier membranes to prevent diffusion may be used at the electrode surface, or the cell electrolyte may be immobilized in a porous matrix. Hydrazine dissolved in 9.5 N KOH is circulated behind the anode, while an air flow at four times the theoretical requirement is maintained behind the cathode. Operating voltage characteristics are

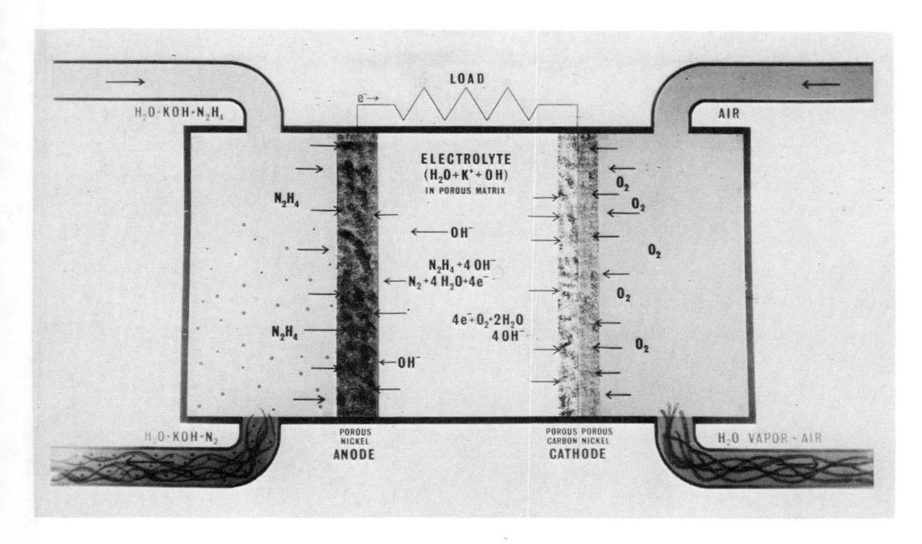

Figure 9.16 Union Carbide hydrazine: air fuel cell. (Courtesy of Union Carbide Corp.)

shown in Fig. 9.17. Coulombic anodic efficiencies in excess of 95 percent are observed in laboratory cells, and 75–95 percent can be attained in fully assembled operating units. A 300-W battery system, with accessories, is shown in Fig. 9.18.

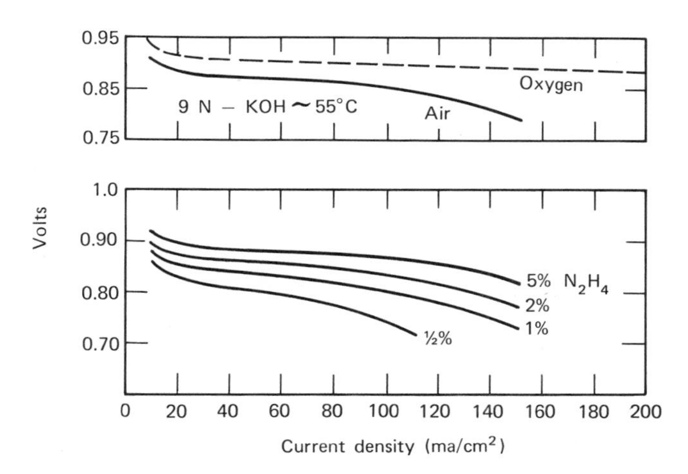

Figure 9.17 Performance characteristics of a hydrazine fuel cell. Upper: polarization curves of oxygen and air cathodes *vs* hydrogen electrode. Lower: Voltage of hydrazine: air cell as function of concentration of N_2H_4 in electrolyte.

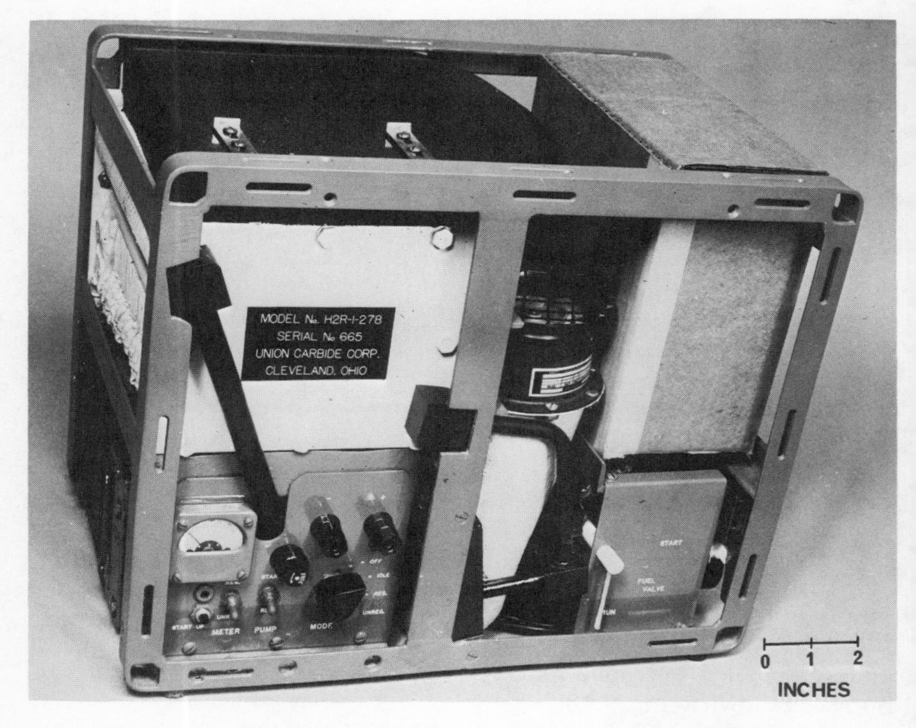

Figure 9.18 A 300-watt hydrazine fuel cell. (Courtesy of Union Carbide Corp.)

The Amalgam-Oxygen System. Although, in the sense of the definition presented, the amalgam-oxygen cell is a continuous-feed primary cell rather than a fuel cell, its characteristics are of sufficient interest to make a brief description appropriate for inclusion in this chapter. This cell as developed by Yeager (51) is schematically illustrated in Fig. 9.19. It consists of a liquid, sodium amalgam (0.3–0.5 percent Na) which flows down an iron anode electrode and a suitable high-current-density oxygen cathode. Since the product of the reaction is sodium hydroxide, the electrolyte must be diluted with water to prevent too large a build-up of the hydroxide concentration. The data of Fig. 9.20 were obtained in a small cell using a carbon cathode with an oxygen pressure of 1 atm, a 0.44 percent sodium amalgam, an electrolyte concentration of 5 M NaOH, and an electrode spacing of 0.3 cm.

Preliminary studies were made on a five-cell unit battery. The oxygen electrodes had an apparent area of 400 cm² per cell and the 0.16 cm thick steel electrodes had a somewhat larger area. The amalgam regeneration was regulated by using a reference amalgam electrode to determine the

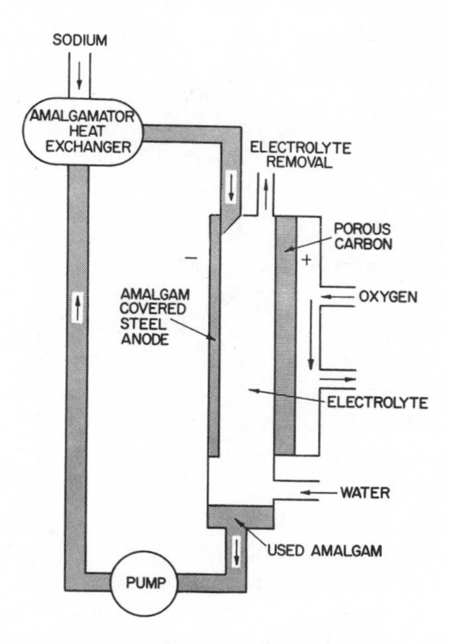

Figure 9.19 The amalgam–oxygen cell.

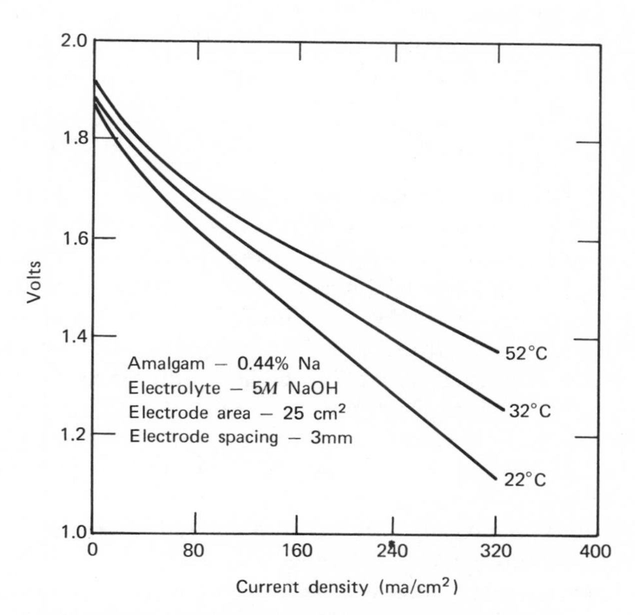

Figure 9.20 Performance of the amalgam oxygen cell.

extent of amalgam depletion. The lifetime and performance of this battery were determined by that of the oxygen electrode.

Yeager has estimated the following characteristics for a 10 kW system:

Open circuit voltage	2.05
Cell voltage	1.552
Current density	10.8 amp/dm^2
Electrolyte concentration	5.0 *M* KOH
Total weight of system	56.7 kg/kw
Sodium requirement	0.59 kg/kw-hr
Oxygen requirement	0.20 kg/kw-hr

A pilot plant model of the amalgam battery has been described by Miller (52). It consists of five large cells producing approximately 16.5 kw at 6 V. Two versions were constructed, one with flat plate cathodes and the other with tubular ones. The latter were nickel-silver prepared by the method of Duddy et al. (21). The operating conditions were similar to those given above, with optimum performance at 60°C. Figure 9.21 shows the construction of a tubular assembly. A single cell utilizing 37 of these, (each tube with an active length of 33 cm) had a cathode area slightly over 98 dm^2 (10 ft^2). The maximum output per cell was 3.3 kw (1.1 V and 3000 amp). Performance curves for both the flat and tubular plates are shown in Fig. 9.22. A photograph of the pilot-plant battery (5.9 kw) is shown in Fig. 9.23.

HIGH-TEMPERATURE FUEL CELLS (ABOVE 300°C)

The utilization of hydrocarbons and fuel gas in electrochemical converters has been most successful in cells operating above 550°C. The construction of the cell developed by Broers and Ketelaar (53) is shown in Fig. 9.24. The electrolyte disk, magnesium oxide impregnated with alkali carbonates, was covered on both sides with thin layers of metal powder. For the fuel electrode, platinum, platinized-iron, and nickel gave the best results. However, above 750°C, little difference was noted in the performance of different electrode materials. Silver was the preferred material for the oxygen electrode. Since, as in the low-temperature cells, the reaction site was at the gas : electrode : electrolyte interface, only a thin layer of powder, less than 1 mm thick, was needed.

Cells of this type have operated continuously for several months. The working temperatures were between 550 and 700°C and town gas, hydrogen, carbon monoxide, and natural gas were tested as fuels. The following open-circuit potentials were obtained for various fuel-air cells:

Fuel	H_2	CO	CH_4	C_2H_6
V at 800°C	1.00	1.10	1.04	1.07
V at 1000°C	1.00	1.01	1.04	1.08

In the operating cells the addition of product gases to the feed favorably affected operating voltage. The direct galvanic conversion of methane was inefficient at temperatures up to 770°C. However, performance improved when a mixture of water and methane was fed to the nickel anode. The following results have been obtained at 770°C for this fuel with 30 mole percent water in the feed (9):

Current density (ma/cm^2)	0	20	40	60
Terminal voltage (V)	0.98	0.80	0.63	0.50

Modified nickel electrodes have given good results at lower temperatures. The performance on town gas and carbon monoxide are shown in Fig. 9.25.

The studies of Chambers and Tantrum (10) utilized two types of cell construction. The first, similar to that of Broers and Ketelaar, retained the electrolyte—a eutectic of lithium and sodium carbonate—in a porous magnesium oxide diaphragm. Each electrode was 0.05 cm thick and made of zinc oxide containing silver. Batteries were built from the cells by sandwiching the disks between recessed metal plates arranged with tubes to feed in the gas and air and to remove reaction products. The other cell construction studied used a free molten electrolyte with double-porosity electrodes of the Bacon type. These electrodes were rigidly attached to metal backing plates. The electrolyte and electrode materials were the same as those used in the porous diaphragm cell. The working gas pressure in these cells was slightly above atmospheric. Cell performance was essentially independent of the method of construction.

The usual operating temperatures of these cells were between 550 and 700°C. With fuels such as hydrogen, carbon monoxide, and hydrocarbons—other than methane—there was little or no activation polarization. With methane at temperatures below 600°C, activation polarization was evident, and the power output was between one-half and two-thirds of that obtained with hydrogen. The performance curves for hydrogen and kerosene are included in Fig. 9.25.

A high-temperature solid-electrolyte cell has been investigated by Archer and co-workers (54). The electrolyte is a semiconductor, $(ZrO_2)_{0.9}$-$(Y_2O_3)_{0.1}$, which has a resistance of 9.0 ohm-cm at 1000°C. To construct the cell, porous coatings of platinum are applied to both sides of the 0.04 cm thick electrolyte. Using hydrogen as the fuel and air as the oxidant, the

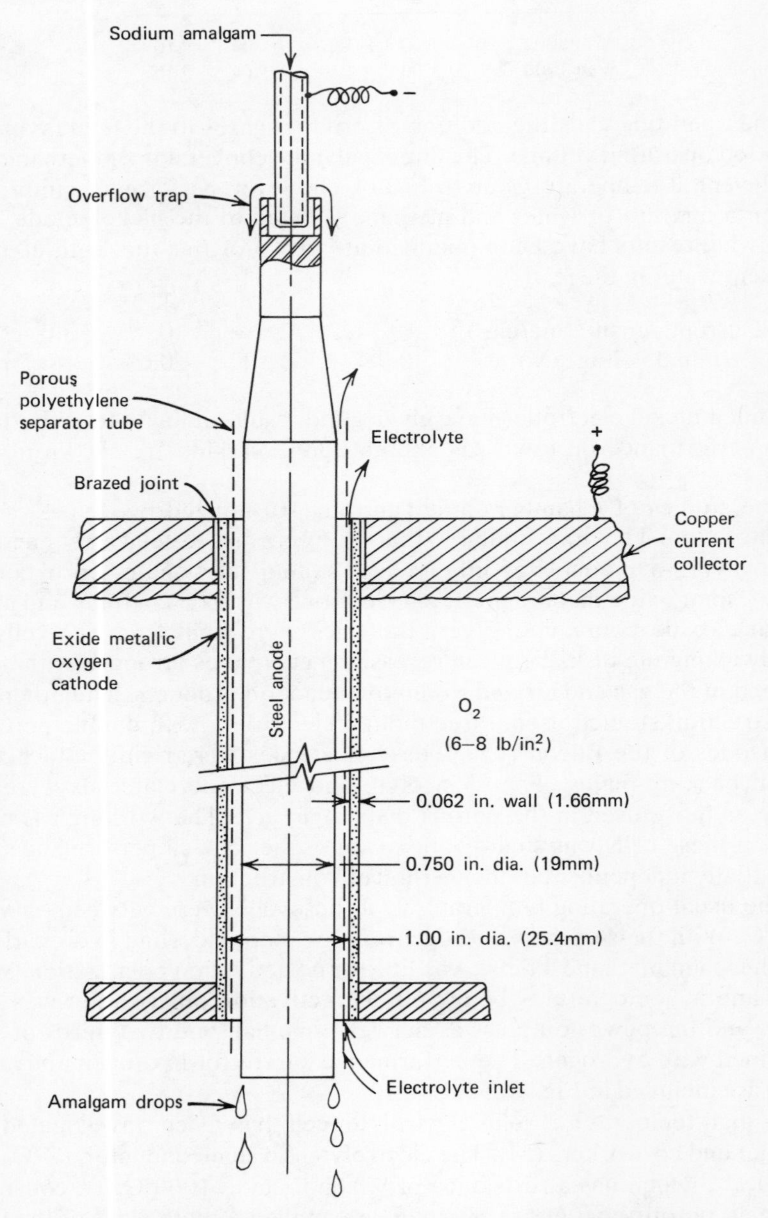

Figure 9.21 Tubular assembly for amalgam–oxygen cell. From (52).

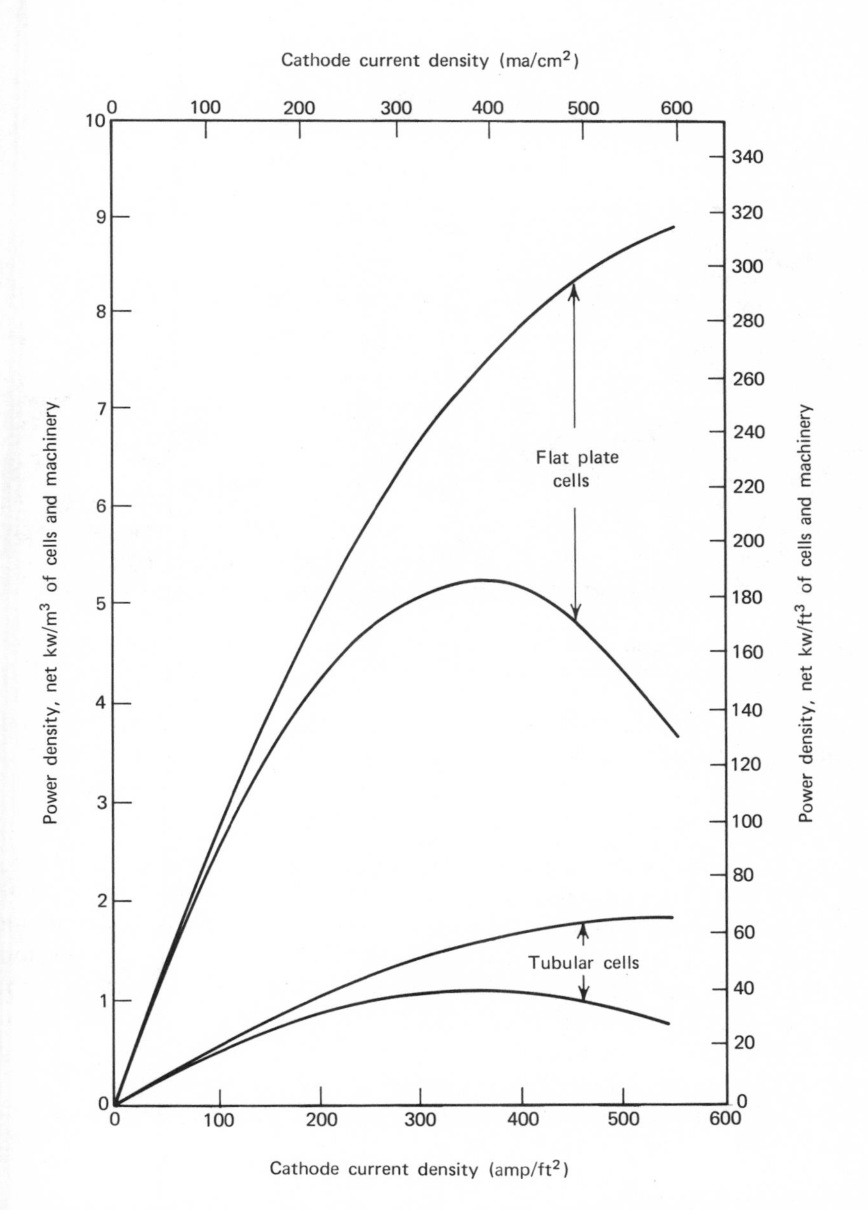

Figure 9.22 Performance range of flat and tubular anode amalgam-battery. (Sodium amalgam–oxygen battery power plant.)

Figure 9.23 Photograph of pilot plant for 5.9 kw amalgam–oxygen battery. Based on M. W. Kellogg design. (Photograph courtesy of U.S. Navy Marine Engineering Laboratory.)

polarization loss was 0.07 V at current densities up to 700 ma/cm^2; the cell voltage was 0.65 at this current density.

REDOX SYSTEMS

To avoid the decreased voltage of oxygen in the alkaline-electrolyte fuel cells and, initially, to utilize coal in a low temperature cell, Posner (55), in collaboration with Rideal, investigated the redox concept: an indirect system in which the anode and cathode reactants are regenerated externally to the cell by reaction with a conventional fuel and oxygen, respectively. (The construction of an idealized redox system is shown in Fig. 9.26.) The anode and cathode were separated from the anolyte and catholyte by a membrane. Porous carbon electrodes were used.

The electrode reactions and the EMF relationships for this cell system can be generalized as follows:

Negative electrode:

Internal to cell:

$$M^{+2} = M^{+3} + e^- \quad E_1 \tag{14}$$

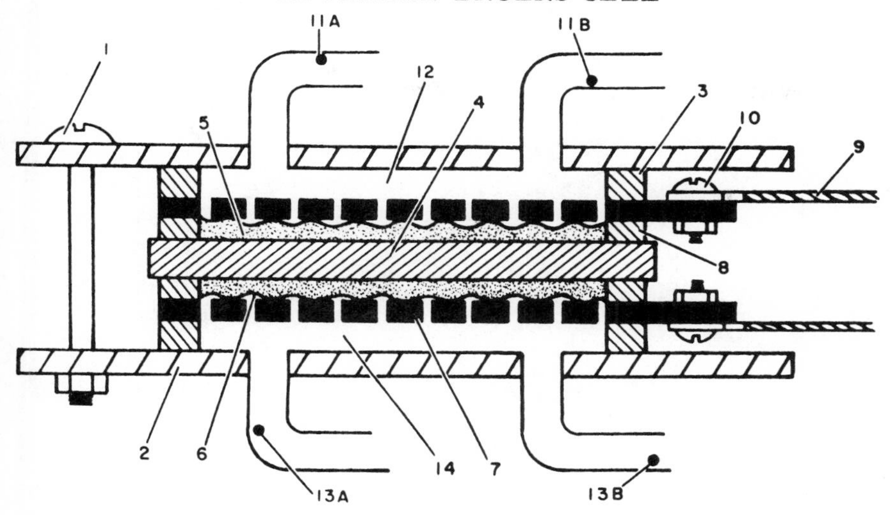

Fig. 9.24 Ketelaar–broers cell.

1. Clamping screw.
2. Nichrome steel covers.
3. Asbestos ring.
4. Electrolyte disk.
5. Electrode metal powder.
6. Wire gauze.
7. Perforated metal disk.
8. Mica ring.
9. Silver wire lead.
10. Screw.
11. A and B anodic gas inlet and outlet.
12. Anodic gas chamber.
13. A and B cathodic gas inlet and outlet.
14. Cathodic gas chamber.

External to cell:

$$4M^{+3} + C + 2H_2O = 4M^{+2} + CO_2 + 4H^+ \quad E_1{}^r \tag{15}$$

Positive electrode:

Internal to cell:

$$R^{+3} + e^- = R^{+2} \quad E_2 \tag{16}$$

External to cell:

$$4R^{+2} + O_2 + 4H^+ = 4R^{+3} + 2H_2O \quad E_2{}^r \tag{17}$$

Net reaction:

$$C + O_2 = CO_2 \quad E_3 \tag{18}$$

Thermodynamically, the following potential relations must prevail:

$$E_1{}^r > E_1, \quad E_2{}^r > E_2 \tag{19}$$

$$E_1 + E_2 < E_1{}^r + E_2{}^r \tag{20}$$

$$E_1 + E_2 < E_3 \tag{21}$$

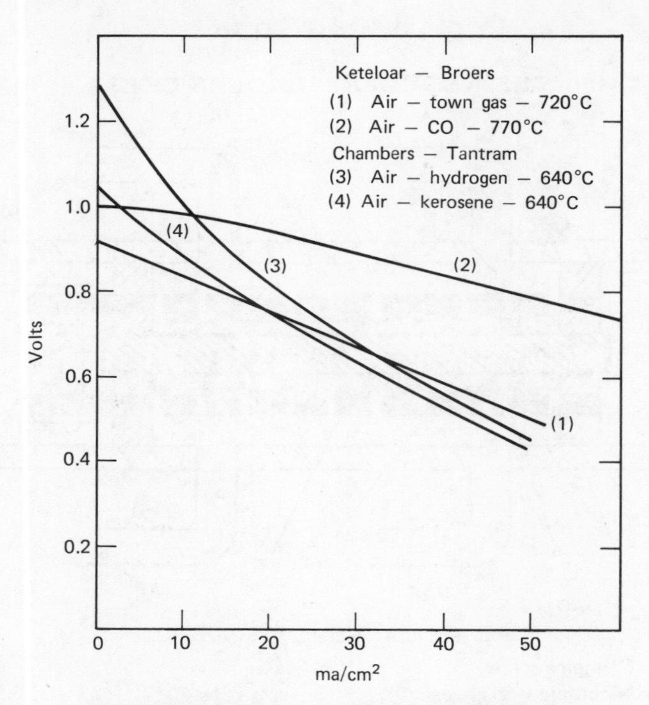

Figure 9.25 Performance of fused electrolyte cells.

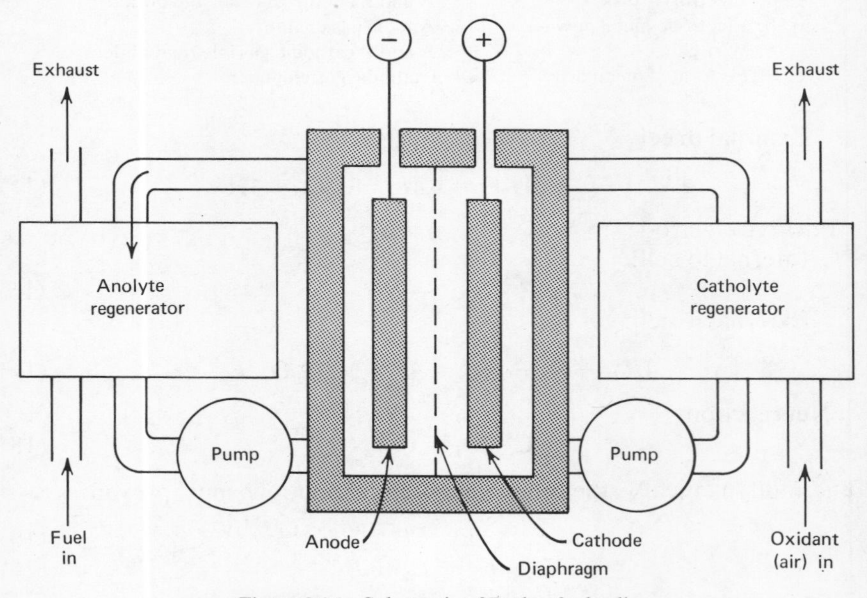

Figure 9.26 Schematic of Redox fuel cell.

The maximum thermodynamic efficiency for this system is given by Eq. (18):

$$\mu = \frac{E_1 + E_2}{E_3} \tag{22}$$

Two general types of redox fuel cells have been studied. One of them was a concentration cell, consisting of two ferrous-ferric couples, each having a different ratio of $Fe^{+2}:Fe^{+3}$. As expected from the Nernst equation, this system has a low voltage (0.25 V). In the second type, different couples are used in the anode and cathode compartments.

The best of the anodic materials reported by Posner (55) was stannous chloride in a hydrochloric acid solution. However, no efficient method for regeneration by reacting the stannic chloride product with coal was found. Greater success has been obtained with the cathodic reaction in which bromine in a solution of hydrobromic acid was electrochemically reduced to bromide, and subsequently regenerated at room temperature by oxygen in the presence of nitrogen dioxide as a catalyst (56). The performance of the Rideal-Posner stannous/stannic chloride-bromine/hydrobromic acid cell is shown in Fig. 9.27.

An alternative to the stannous-stannic couple, studied by Carson (57), is the titanous/titanyl ion couple:

$$2Ti^{+3} + 2H_2O \rightarrow 2TiO^{+2} + 4H^+ + 2e^- \tag{23}$$

The titanyl ion can be reduced with hydrogen in the presence of a catalyst such as palladium. A cell based on the regeneration of titanous ion and bromine has been operated for a period of 16 days (58). It had an open circuit voltage of 0.95 V.

CLOSED CYCLE AND OTHER REGENERATIVE SYSTEMS

Several approaches to closed-cycle electrochemical converters have been considered. All of these systems proposed to regenerate the cell reactants externally to the cell using an energy source: heat, light, or high-energy radiation. The basic operating cycle for thermal regeneration is illustrated in Fig. 9.28, using the lithium hydrogen reaction as an example. The electrochemical reaction is carried out at temperature T_2. The reaction product, or—if soluble in the electrolyte— its mixture in the electrolyte, is then thermally decomposed at T_1. The anode and cathode reactants are separated and returned to the cell.

It has been demonstrated by Liebhafsky (59) and Friauf (60) that, in a regenerative system, the energy efficiency is limited by the Carnot cycle.

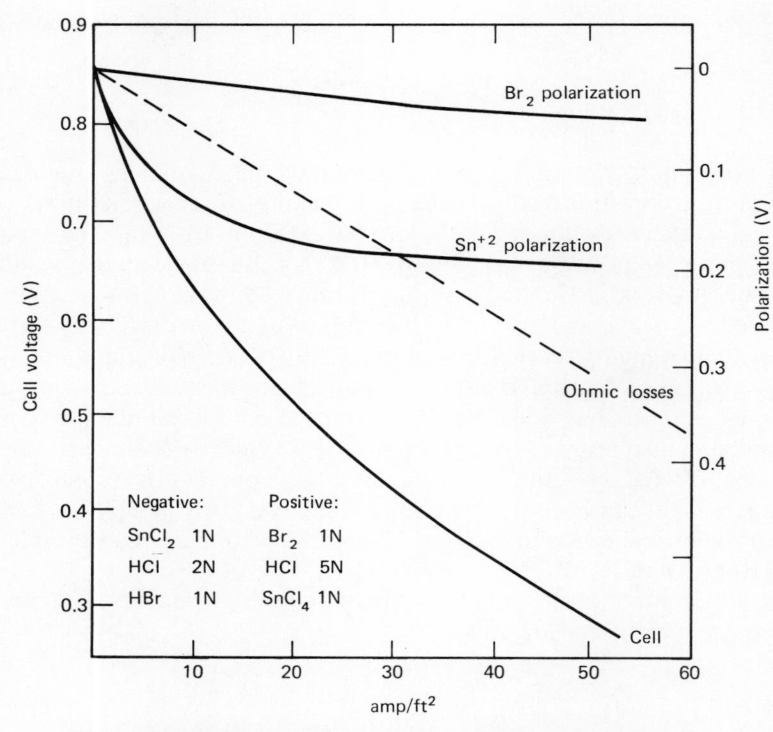

Figure 9.27 Performance curve for Rideal-Posner Redox fuel cell.

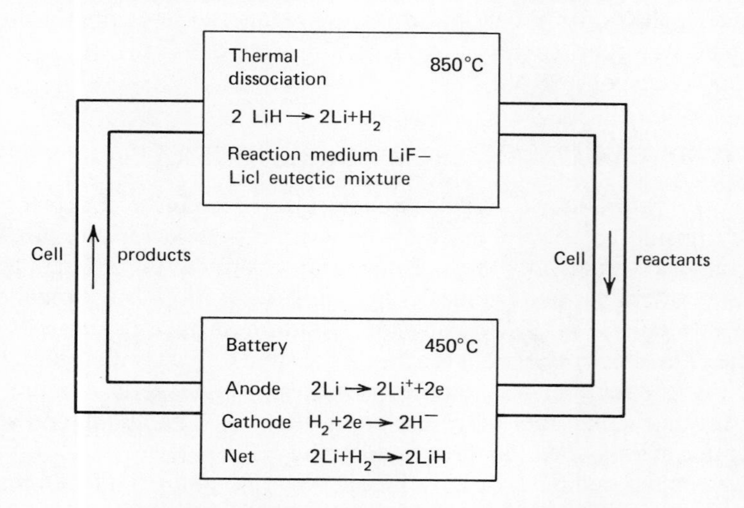

Figure 9.28 Schematic of thermal regenerative cell.

The over-all efficiency, assuming complete utilization of reactants, is given by

$$\mu = \frac{V}{V_{oc}} \frac{T_1 - T_2}{T_1} \tag{24}$$

A closed-cycle cell based on the alkali and alkaline earth hydrides has been investigated by Werner, Shearer, and co-workers (61, 62). Their studies included lithium and sodium as the anode and a hydrogen cathode with a fused mixed alkali halide electrolyte. Although this work proved the closed-cycle system in principle, it has not been developed into a complete power source.

McCully and co-workers (63) investigated a thermally regenerative system based on the cell reaction

$$TeCl_2 + 2CuCl_2 \xrightarrow{200°C} TeCl_4 + 2CuCl \tag{25}$$

with regeneration as follows:

$$\text{Anode:} \qquad TeCl_4 \xrightarrow{500°C} TeCl_2 + Cl_2 \tag{26}$$

$$\text{Cathode:} \qquad 2CuCl + Cl_2 \xrightarrow{200°C} 2CuCl_2 \tag{27}$$

The anode and cathode compartments of the cell were separated by a porous diaphragm and aluminum chloride was the primary component of the electrolyte. Current densities of 10 ma/cm^2 at 0.22 V were obtained. The regeneration of the tellurium dichloride caused considerable difficulty since its mixture with chlorine, resulting from the thermal decomposition of tellurium tetrachloride, could not be separated or cooled rapidly enough to prevent reversal of the reaction.

Another example of the thermally regenerative system is that reported by Agruss and Karas (64). This utilizes the difference in potential between two concentrations of potassium amalgam, the anode being the more concentrated. With an electrolyte of a mixture of potassium hydroxide, iodide, and bromide in a magnesium oxide matrix, and Kovar electrodes a current density of 100 ma/cm^2 at 0.5 V has been obtained. By distillation of the mercury, the original concentration of the amalgams can be restored.

A similar system, based on sodium amalgam, has been reported by Heredy and co-workers (65). The cell, contained in stainless steel and pressurized with argon, was operated in the temperature range of 470 to 510°C. The sodium concentration in the anode varied from 20 to 60 atom percent, that of the cathode from 0 to 15 atom percent. Cell potentials varied between 0.3 and 0.8 V. A fused sodium salt mixture retained in a porous ceramic was used as the electrolyte. With the best matrix-electrolyte combination current densities of 200 ma/cm^2 were achieved with no measureable electrode polarization.

The applicability of solar regeneration to fuel cells has been studied by McKee and collaborators (66). These authors have indicated that only a few chemical systems can meet the requirements for such a system. For laboratory study, they selected the reaction

$$2NO + Cl_2 \longrightarrow 2NOCl \tag{28}$$

Using an electrolyte of aluminum chloride and nitrosyl chloride, carbon electrodes, a pressure of 7–8 atm, and room temperature, a potential of 0.21 V was observed. At 0.1 V the current density was 1.5 ma/cm^2. Photochemical regeneration of the reactants from the product had a quantum yield of two. The over-all efficiency of such a system, based on the total solar spectral energy, was estimated as 7 percent.

Yeager et al. (67) described their results on a nuclear regenerative system. Both internal and external regeneration were considered. Feates (68) showed that internal regeneration is very inefficient. The cell reaction used was

$$Fe^{+3} + \tfrac{1}{2}H_2 = Fe^{+2} + H^+ \tag{29}$$

and the regeneration step in the presence of beta and gamma irradiation was

$$H^+ + Fe^{+2} = Fe^{+3} + \tfrac{1}{2}H_2 \tag{30}$$

The cell voltage observed was greater than 0.60 V, and total polarization of less than 0.15 V was obtained at a current density of 20 ma/cm^2. The over-all efficiency, based on power output and gamma radiation energy absorbed, was 3 percent.

BIOLOGICAL FUEL CELLS

Interest in biochemical fuel cells began in the late 1950s. Conceptually, in these cells, electrical energy is obtained directly or indirectly from biological systems. In the direct cells, an attempt is made to utilize biooxidation and bioreduction steps of the metabolic process at the anode and cathode, respectively. However, Blasco and Gileadi (69) doubted that a direct bioanode or biocathode has been experimentally demonstrated.

The current status of the indirect biological fuel cells has recently been reviewed by Acker (70). Most of these are based on the biological generation of an anodic reactant—hydrogen, ethanol, formic acid, and others. This approach has been stimulated by the desire for a closed ecological system for long-duration space travel. Blasco and Gileadi (69) reported than an anode based on the system formate: formic dehydrogenase, which produces hydrogen as the metabolic product, gave current-voltage characteristics almost identical to hydrogen gas in the same cell. Early results have not been encouraging for a practical system.

The use of biological systems for external generation of reactants for fuel cells appears to be more promising than internal generation. For example, May and co-workers (71) studied hydrogen production from carbohydrates. They investigated the action of escherichia coli and clostridium Welchii on glucose and maltose. Using a 10-liter reactor, the clostridium Welchii organism generated 45 liters of hydrogen in 8 hr when glucose was used as the substrate.

STATUS AND PROSPECTUS

In spite of the extensive effort, the only applications of fuel cells, as of the late 1960's, have been highly specialized. The major development effort has been for space-power sources. Table 1 gives the characteristics

TABLE 9.1. CHARACTERISTICS OF VARIOUS HYDROGEN-OXYGEN FUEL CELL BATTERIES

	Union Carbide[b]	Pratt Whitney (Apollo)	Allis-Chalmers	General Electric (Gemini)[e]
Power,[a] gross units (kw)	3.0	2.0[c]	2.0[d]	2.0[f]
Number of cells	34	31/module	33/module	32/module
Voltage	28	27–31	29	26.5–23.3
Weight (kg)	57	105	58	64
Volume (liters)	190	260	71	89
P_{H_2} (atm)	1	4.0	2.4	1.50
P_{O_2} (atm)	1	4.0	2.4	1.56
°C	70–80	250–260	88–93	25
Figure	29	30	31	32

[a]Includes parasitic power.
[b]Not a space system.
[c]Three modules.
[d]Two modules in parallel.
[e]Battery used in Gemini V.
[f]Two three-module units in parallel; three modules in each unit in parallel.
NOTE: Reactants and reactant storage not included.

Fig. 9.29 Union carbide fuel cell battery (thin electrode construction). (Courtesy Union Carbide Corp.)

of various hydrogen fuel cells, all but one of which have been designated for some space application, such as the Apollo and Gemini missions. The data given are based on company brochures and the components included vary from system to system. Therefore, these should not be used for a quantitative comparison of the different systems.

Another interesting application has been the use of the hydrazine fuel cell for the power source of a small submarine (72). The battery consisted of 50 cells connected in series with a total operating voltage of 36 under a load of 750 w.

The silent sentry developed for military use (73) is based on the hydrogen-air cell using carbon electrodes and lithium hydride as a hydrogen source. Among projected applications for fuel cells that have been discussed is the operation of remote relay systems for telephone service (74). Bacon (18) suggested the use of high current density batteries for operating electrical railways now using lead-acid storage batteries. Fuel cell standby power for hospitals and for other emergency power needs and for

Figure 9.30 Pratt and Whitney fuel cell battery (Apollo). (Modified Bacon cell.) (Courtesy of Pratt and Whitney Aircraft.)

industrial trucks and for railroad switch engines are possibilities.

Jenkins and Winn (75) have discussed the characteristics of a hypothetical fuel cell automobile based not only on advances in the power source itself but also on advances in direct current electrical motors, controls, and gears. They conclude, "in our view—fuel-cell traction for private vehicles will be a reality very soon." A more comprehensive study of the fuel-celled automobile and other petroleum-fueled vehicles

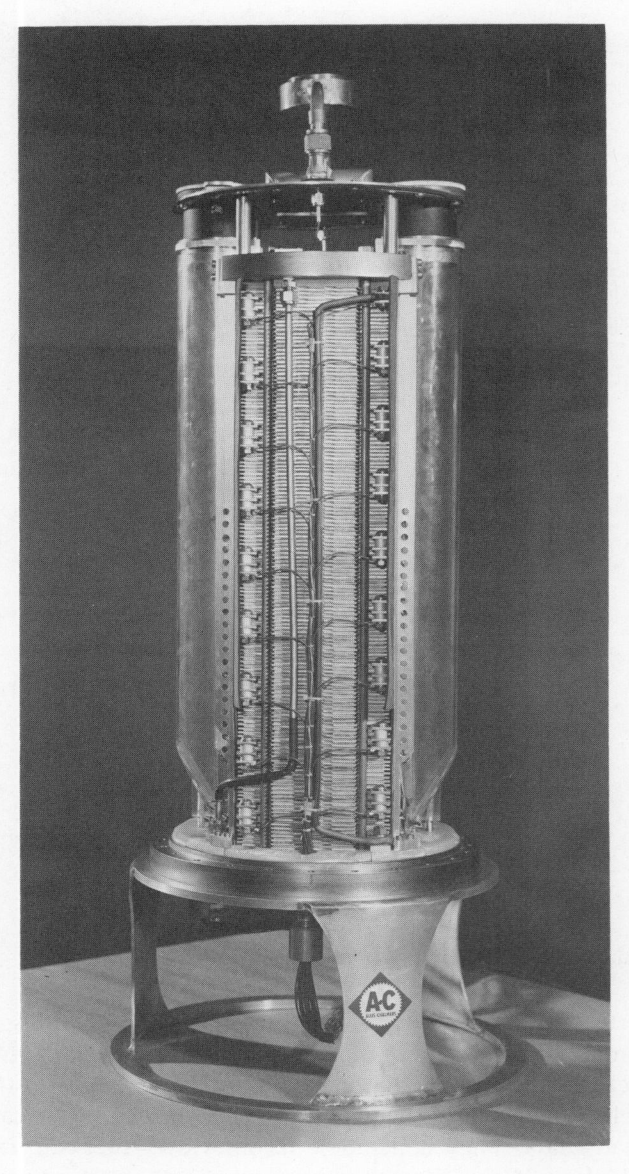

Figure 9.31 Allis-Chalmers fuel cell battery. (Courtesy of Allis-Chalmers Mfg. Co.)

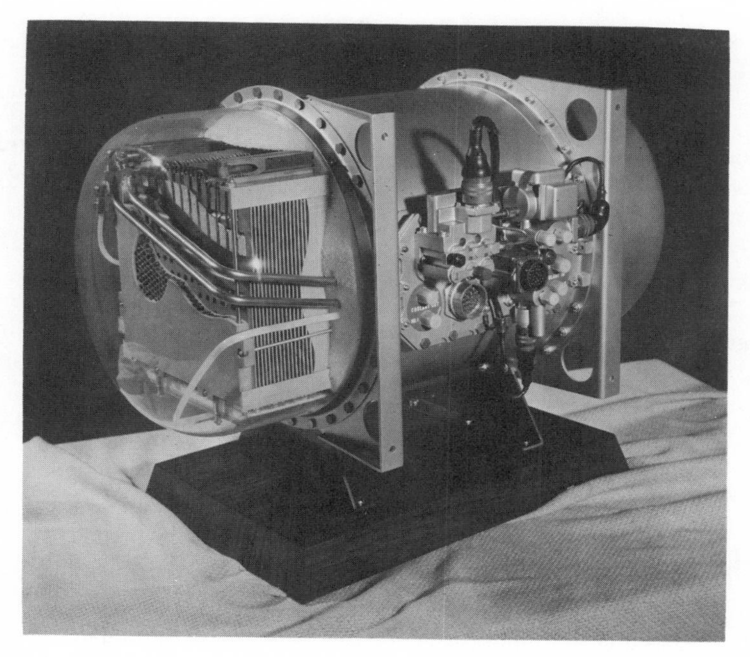

Figure 9.32 General Electric membrane fuel cell battery (Gemini). (Courtesy of General Electric Co.)

has been made by Szego (76). He is of the opinion that the high efficiency of fuel cells could make a major contribution to decreasing the cost of operating civilian traction vehicles and reduce the fuel logistic burden of the military. Among other advantages suggested for "economically feasible" fuel cells are: (*a*) a solution to the air pollution problem, (*b*) decreased ventilation requirements in tunnels and underground parking areas, (*c*) very low idling consumption, (*d*) elimination of auxilliary electric-power generators, and (*e*) high efficiency.

The hydrogen-oxygen system also has been investigated as a secondary cell (77, 78). Bacon suggested an application in which hydrogen and oxygen are generated in off-peak periods at central power plants and the fuel cell produces power during high demand periods. The work of Broers and Ketelaar (53) and the patents of Gorin (79) are based on the consumption of gasified coal in fuel cell batteries for central power production.

The commercial utilization of fuel cells has not yet been realized. Why and what are some of the prospects? There are a number of reasons for only limited application to date, largely related to the basic nature of fuel cells, their engineering, and their economics:

1. The operating characteristics of fuel cells are such that the rate of fuel consumption increased markedly with power generated, and the efficiency decreases. For a given system, power outputs per unit volume have a narrower range than that of the internal combustion engine. However, the fuel cell does have the unique characteristic of being the only power source whose efficiency is greatest at low power levels.

2. The low-temperature hydrogen-oxygen cells all require the removal of water. In the membrane cell, this is reported to be relatively simple, but in the caustic electrolyte cell it presents some difficulty. The proposed method of Kordesch (32) requires gas circulating equipment.

3. Most fuel cells show a decrease in current output at a given voltage with decrease in temperature, so that either external heating or a wasteful start-up period will be required for their use at low ambient temperature. Only two of the cells discussed, that of Clark et al. (33) and the hydrazine cell (49), operate effectively at 0°C.

4. The effect of carbon dioxide on the performance of the alkaline electrolyte cell is equally important: Even when operating on oxygen, carbon dioxide contamination of the electrolyte can decrease the operating lifetime and the effect is much more marked during operation on air. Carbon dioxide removal has been suggested, but this adds to the complexity of the system. The acid membrane cell has overcome this carbon dioxide problem but demands humidity control since both the conductivity and strength of the ion-exchange membrane are dependent on moisture content. This cell, on the other hand, has much poorer output per unit area than other hydrogen-oxygen systems and has a more limited temperature range of operation. A major factor is the higher resistance of the membrane; however, because of compactness, high power outputs per unit volume can be obtained. This unit and the capillary membrane cell are especially attractive for space application since they do not present the problem of liquid-gas separation under zero gravity conditions.

5. A major drawback of the hydrogen-oxygen system is fuel and oxidant storage. Both these gases have low boiling points, thus requiring either high pressure or cryogenic storage. Cryogenic storage offers the better possibility for weight minimization. For large installations, a container-to-fuel weight ratio of 3 : 1 for liquid hydrogen and 0.2 : 1 for liquid oxygen has been projected (80). Obviously, where air can be used, and refilling of the fuel storage system can be done at regular periods, only the container weight for the hydrogen need be considered.

6. The economics of the fuel cell appear to be its greatest hindrance to wider application at this time. The cost of fuel-cell batteries has been estimated by several groups. Batteries with platinum as the catalyst require a high investment in this scarce metal which was priced in 1966 at more than \$100.00/oz (\$3.50/g). The use of nickel boride (37), \$5.60/oz (\$0.20/g) instead of platinum black, could markedly reduce the catalyst cost for the anode; the cost of the cathode would be unaffected. Barak (81) has estimated the platinum cost for the Union Carbide (33) thin electrode, which requires 2 g platinum/ft^2 (0.22 g/dm^2), as \$5.50/ft^2 (\$0.59/dm^2). For the American Cyanamid (17) type of electrode, he estimates a catalyst cost of \$80.00/kw. The cost of the platinum catalyst and its scarcity make a more available, less expensive replacement catalyst a prerequisite to the general use of low temperature, and especially acid electrolyte fuel cells operating on hydrocarbon fuels.

Various alternatives to liquid or compressed hydrogen storage have been considered. Hydrogen, prepared by reforming of hydrocarbons—which has been highly developed by chemical industry—by the reforming of methanol, by the cracking of ammonia, and by the hydrolysis of water reactive chemicals are among these. Heffner and co-workers (82) have described a methanol reforming plant which operates at 260°C and can produce sufficient pure hydrogen for a 50 to 700 kw fuel cell power plant. Its weight is 19 kg/kw and its volume is 14.8 liters/kw. Its rate of hydrogen production can be varied from 2.28 kg/hr to 31.8 kg/hr in a period of 1.5 min. Purification of the hydrogen from the reformer is required for alkaline electrolyte cells. This is most readily accomplished by diffusion through a palladium or a palladium-silver alloy.

The production of hydrogen from another fuel poses the problems of operating a chemical unit in series with the demand of the battery or having some storage capacity in the system, and of overcoming the problems of start-up and shut-off. The storage as hydrogen producing chemicals, such as the hydrides or borohydrides, can give a more easily coupled hydrogen source, but at a very large increase in cost. All of these processes degrade the efficiency of the fuel-oxygen system, one of the major reasons for the development of fuel cells.

The high-temperature fuel cells, which can operate on hydrocarbons and air, have inherent difficulties. The internal resistance of the cells developed to date limit the range of power densities. The major problems, as in most high-temperature technology, are those of materials. The ceramic-type electrolyte system cannot undergo temperature cycling without the possibility of cracking. Seals to prevent gas leaks and the loss of electrolyte and the maintenance of long-lived electrical contacts also are problems. Start-up of these systems is a difficulty, and external heating is required. In installations of the kilowatt range, the temperature can be maintained by the heat losses during cell operation. Such systems offer the possibility of power generators whereby product combustible gases, cheap coal for conversion to coal-gas, or gasification of low-grade coal deposits is a convenient fuel source. They are not attractive for small, mobile, power plants for traction.

Since the late 1940s, there have been steady advances in electrochemistry, catalysis, and fuel-cell technology. These have led to varied approaches and better performance of fuel-cell energy converters. Sophisticated, well-developed hydrogen:oxygen systems have been constructed as power supply for vehicles (48a, 48b), and a hydrazine:air battery has been used to operate a motorcycle. However, the most promising applications have generally been limited to specialty military and space, e.g., the Gemini power sources. Cost of the fuel cell and poor storage characteristics of hydrogen were among the major drawbacks to broader utilization of the hydrogen-oxygen cell, the most advanced of the electrochemical converters. Promise of a broadened application of fuel cell technology centered in the discovery of acid-membrane cells and evidence of improvements in their power outputs. Along with this, limited success in the use of hydrocarbon in intermediate temperature cells, $150 \pm 25°C$, and methanol and hydrazine in low temperature cells, $50°C \pm 25°$, gave promise of ultimate success in the achievement of an electrochemical energy converter of broad applicability. It is the opinion of this writer that success with hydrocarbon fuels depends on the discovery of a moderate cost, low temperature catalyst that enables their direct electrochemical conversion with high efficiency. Once this has been achieved, the fuel cell will become one of the important power sources.

REFERENCES

1. Davy, H., *Nicholson's J.*, **144** (1802).
2. Grove, W. R., *Phil. Mag.*, III, **14**, 127 (1839).
3. *Id.*, *Phil. Mag.*, III, **21**, 417 (1842).
4. Becquerel, A. C., *Traite' d'electricité*, I, Paris (1855).
5. Baur, E., and Tobler, J., *Z. Elektrochem.*, **39**, 169 (1933).
6. Ostwald, W., *Z. Elektrochem.*, **1**, 122 (1894).
7. Baur, E., and Ehrenberg, H., *Z. Elektrochem.*, **18**, 1002 (1912).
8. Berl, W. G., *Trans. Electrochem. Soc.*, **83**, 253 (1943).
9. Broers, G. H. J., High Temperature Galvanic Fuel Cells (Dissertation), University of Amsterdam (8 July 1958).
10. Chambers, H. H., and Tantrum, A. D. S., *Fuel Cells*, Reinhold Publishing Corp., New York, p. 94 (1960).
11. Liebhafsky, H. A., and Douglas, D. L., *ibid*, p. 5.

12. Weber, H. C., Meisner, H. P., and Sama, D. S., *J. Electrochem. Soc.*, **109**, 884 (1962).
13. Fedotov, N. A., Veselousky, V. I., Rosenthal, K. I., and Mazitov, Y. A., *Batteries 2.* Fourth Int'l. Symp. on Batteries, Pergamon Press, London, p. 283 (1965).
14. Bennion, D. N., Phenomena at a Gas-Electrode-Electrolyte Interface (Ph.D. Dissertation), University of California (June 1964).
15. Davtyan, O. K., *The Direct Conversion of the Chemical Energy of Fuel Into Electrical Energy*, Academy of Sciences, Moscow (1947).
16. Yeager, E., *The Oxygen Electrode in Aqueous Cells*, Technical Report 12, ONR contract Nonr 2391(00), Dept. of Commerce P.B. No. 150 008 (1Aug. 1960). Cf. Berl (8).
17. Haldeman, R. G., Colman, W. P., Langer, S. H., and Barber, W. A., *Fuel Cell Systems*, Advances in Chemistry Series 47, p. 106, American Chemical Society, Washington, D.C. (1965).
18. Bacon, F. T., *Fuel Cells* (10), p. 51.
19. Justi, E., Pilkuhn, M., Scheibe, W., and Winsel, A., *Hochbelastbare Wasserstoff—Diffusion—Electroden für Betrieb bei Umgebungstemperatur und Niederdruck*, Akademie der Wissenschaft und der Literatur am Mainz, Nr 8 (1959).
20. Sturm, F. v., *Siemens Review*, p. 118 (March 1966).
21. Duddy, J. C., Ruetschi, P., and Ferrell, D. T., Proc. 16th Annual Power Sources Conference, p. 9 (May 1962).
22. Grens, II, E. A., and Tobias, C. W., *Ber. Phys. Chem.*, **68**, 236 (1964).
23. Adlhart, O. J., 19th Annual Power Sources Conference, p. 1 (May 1965).
24. Kortum, G., and Bockris, J. O'M., *Textbook of Electrochemistry*, Vol. II, Elsevier Publishing Co., New York, p. 400 (1951).
25. Ruetschi, P., and Delahay, P., *J. Chem. Phys.*, **23**, 195 (1955).
26. Burshtein, R. Ch., Tiurin, U. S., and Pslenichnikov, A. G., Fourth International Symposium on Batteries, Pergamon Press, London, p. 315 (1964).
27. Dahms, H., and Bockris, J. O'M., *J. Electrochem. Soc.*, **111**, 728 (1964).
28. Grubb, W. T., and Michalske, J., 18th Annual Power Sources Conference, p. 17 (May 1964).
29. Yeager, E., and Kozawa, A., *Kinetic Factors in Fuel Cell Systems: The Oxygen Electrode*, Sixth AGARD Combustion and Propulsion Colloquium, Cannes (March 1964), to be published by Pergamon Press.
30. Hoare, J. P., *J. Electrochem. Soc.*, **109**, 858 (1962).
31. Evans, G. E., Proceedings—*Thirteenth Annual Power Sources Conference*, p. 108, Power Sources Division, U.S. Army Signal Research & Development Laboratory, Fort Monmouth, N.J. (1959).
32. Kordesch, K., *Fuel Cells* (10), p. 11.
33. Clark, M. B., Darland, W. G., and Kordesch, K. V., *18th Annual Power Sources Conference*, p. 11 (May 1964).
34. Bacon, F. T., Br. Pat. 667,298 (1952).
35. Dihman, H. M., Justi, E. W., and Winsel, A. W., Preprints, *Symposium on Recent Advances in Fuel Cells*, American Chemical Society, p. B-139 (1961).
36. Platner, J. L., and Hess, P. D., *Static Moisture Removal Concept for Hydrogen-Oxygen Capillary Fuel Cells*, Allis Chalmers, Milwaukee, Wisc. (Sept. 1963).
37. Jasinski, R. S., *Fuel Cell Systems* (17), p. 95.
38. Cheney, E. O., Jr., Farris, P. J., and King, J. M., Jr., *Open Cycle Fuel Cell Systems for Space Applications*, Pratt and Whiteny Aircraft, New Haven, Conn. (no date).
39. Grubb, W. T., and Niedrach, L. W., *J. Electrochem. Soc.*, **107**, 131 (1960).
40. Grubb, W. T., U.S. Pat. 2,913,511 (1959).
41. Lurie, R. M., Berger, C., and Viklund, H., *J. Electrochem. Soc.*, **110**, 1173 (1963).

42. Elmore, G. V., and Tanner, H. A., *J. Electrochem. Soc.*, **108**, 669 (1961).
43. Dravnieks, A., Boies, D. B., and Bregman, J. L., 16th Annual Power Sources Conference, p. 4 (May 1962).
44. Ruetschi, P., and Sklarchuk, J., *Fuel Cell Electrodes for Acid Electrolytes*, Fourth Int'l. Symp. on Batteries, Brighton, England, Pergamon Press, London, p. 297 (1965).
45. Grubb, W. T., and Michalske, C. J., *Nature*, **201**, 287 (1964).
46. Heibronner, H. H., Levins, W. P., and Allison, J. W., Final Report, *Liquid Hydrocarbon Fuel Cell Development*, Contract No. DA-44-009-AMC-756(T), Department of Commerce, Clearing House, A.D. 613031 (27 Jan. 1965).
47. Williams, K. R., Pearson, J. W., and Gressler, W. J., *Low Temperature Fuel Batteries*, Paper presented at the Fourth Int'l. Symp. on Batteries, Pergamon Press, London, p. 337 (1965).
48. Eisenberg, M., *Proceedings of the 18th Annual Power Sources Conference*, p. 20 (May 1964).
48a. Simons, E. L., Cairns, E. J., and Surd, D. J., *Extended Abstracts*, Batt. Div., Electrochem. Soc. Vol. 13, Paper No. 349, p. 170 (1968); cf. Cairns, E. J., Simons, E. L., and Trevebaugh, A. D., *Nature*, **217**, 780 (1968).
48b. McKee, D. W., Scarpellino, A. J., Jr., Danzig, I. F., and Pak, M. S., *Extended Abstracts*, Batt. Div., Electrochem. Soc. Vol. 13, Paper No. 350, p. 174 (1968).
49. Tomter, S. S., and Anthony, A. Peter, *Hydrazine Fuel Cell System*, Allis-Chalmers Mfg. Co., Milwaukee, Wisc. (July 1963).
50. Evans, G. E., and Kordesch, K. V., *Science*, **158**, 1148–52 (1967).
51. Yeager, E., *The Sodium-Amalgam Oxygen Continuous Feed Cell*, Technical Report No. 14, Contract Nonr 2391(00), Dept. of Commerce, Clearing House, P.B. 158 555 (Dec. 1960).
52. Miller, K. D., *Symposium on Fuel Cells*, Swedish Academy of Engineering Sciences, p. 41 (1962).
53. Broers, G. H. J., and Ketelaar, J. A. A., *Fuel Cells* (10), p. 78.
54. Archer, D. W., Zahradnik, R. L., Sverdrup, E. F., English, W. A., Elikan, L., and Alles, J. J., 18th Annual Power Sources Conference, Atlantic City, N.J., p. 36 (May 1964); also *Fuel Cell Systems* (17), p. 24.
55. Posner, A. M., *Fuel*, **34**, 330 (1955).
56. Bingham, B. E. M., and Posner, A. M., *J. Am. Chem. Soc.*, **77**, 2634 (1955).
57. Carson, Wm. N., Jr., *Studies on a New Type of Fuel Cell*, Ohio State Univ. (1948).
58. Carson, W. N., Jr., and Feldman, M. L., Proceedings—*Thirteenth Annual Power Sources Conference*, U.S. Army Signal R&D Laboratory, p. 111 (1959).
59. Liebhafsky, H. A., *J. Electrochem. Soc.*, **106**, 1068 (1959).
60. Friauf, J. B., *J. Appl. Phys.*, **32**, 616 (1961).
61. Werner, R. C., Shearer, R. E., and Ciarlariello, T. A., Proceedings—*Thirteenth Annual Power Sources Conference*, U.S. Army Signal R&D Laboratory, p. 122 (1959).
62. Shearer, R. E., Mausteller, J. W., Ciarlariello, T. A., and Werner, R. C., Proceedings—*Fourteenth Annual Power Sources Conference,* U.S. Army Signal (R & D) Laboratory, p. 76 (1960).
63. McCully, C. R., Rymarz, T. M., and Nicholson, S. B., *Abstracts of 149th Meeting of the American Chemical Society*, p. 10L (April 1965).
64. Agruss, B., and Karas, H, *ibid*, p. 622.
65. Heredy, L. A., Iverson, M. L., Ulrich, G. D., and Recht, H. L., *ibid*, p. 11L.
66. McKee, W. E., Findl, E., Margerum, J. D., and Lee, W. B., (62), p. 68.
67. Yeager, J. F., Sixteenth Annual Power Sources Conference, p. 39 (May 1962).
68. Feates, F. S., *Trans. Far. Soc.*, **56**, 1671 (1960).

69. Blasco, R. J., and Gileadi, E., *Advanced Energy Conversion*, **4**, 179 (1964).
70. Acker, R. F., *Developments in Industrial Biology*, **6**, 260 (1964).
71. May, P. S., Blanchard, G. C., and Foley, R. T., *18th Annual Power Sources Conference*, p. 1 (1964).
72. Hall, J. B., and Antony, A. P., *Fuel Cell Power Plants for Research Submarines*, Paper presented at the A.S.M.E. Underwater Technology Conference, New London, Conn. (May 1965).
73. *Chem. and Eng. News*, **35**, 25 (1957).
74. Roberts, R., *Electrochem. Soc.*, **105**, 428 (1958).
75. Jenkins, J., and Winn, R. E., *New Scientist*, **24**, 222 (1964).
76. Szego, G. C., *Economics, Logistics and Optimization of Fuel Cells*, Extract de la Revue, E.P.E., Vol. 1, No. 2 (1965).
77. Bone, J. S., Proceedings – *Fourteenth Annual Power Sources Conference*, U.S. Army Signal R&D Laboratory, p. 62 (1960).
78. Podolny, W. H., *ibid*, p. 64 (1960).
79. Gorin, E., U.S. Pats. 2,570,543 (1951); 2,581,650 (1952); 2,581,651 (1952).
80. Alexander, L. R., Moos, A. M., Rapp, K. I., and Sommer, R. C., *Continuous Feed Fuel Cell Systems*, Wright Air Development Center Report 57-605, Dept. of Commerce, Clearing House, A.D. 276,524 (Sept. 1957).
81. Barak, M., *Advanced Energy Conversion*, **6**, 29 (1966).
82. Heffner, W. H., Veverka, A. C., and Skaperdas, G. T., *Fuel Cell Systems*, (17), p. 318.

10

Low-Temperature Nonaqueous Cells

John M. Freund and William C. Spindler

Laboratory studies of primary cells and batteries utilizing nonaqueous electrolytes are reviewed in this chapter. The survey is limited to highly conductive solutions that are stable at normal or subzero temperatures, leaving high-temperature electrolytes, such as fused salts, for discussion in other chapters. Electrolytes of very low conductivity, such as gels, waxes, polymers, and ionic-conduction crystals, are also discussed elsewhere in this book.

The nonaqueous battery field is rapidly expanding and appears promising as a new field of experimental investigation. Most of the results of early studies, however, were either negative or inconclusive. Much work is being done under government contract for the development of new, special-purpose batteries capable of discharging equally well at low temperatures (220°K or −63°F) and room temperature. Most of the systems studied were intended for use in reserve battery[1] designs, sometimes known as deferred-action, remotely activated, or dry-charged batteries.

Since water is cheap, readily available, and an excellent solvent for conducting solutions, one may ask why there is any interest in using nonaqueous electrolytes for practical purposes. The answer is that aqueous electrolytes impose several limitations on the low-temperature operation of batteries. First, the conductivity of aqueous electrolytes decreases rapidly as the temperature is lowered, with the result that at very low temperatures the internal resistance of cells increases to the point at which equipment performance is affected. Second, polarization, particularly at the positive electrode plates, becomes significant at

[1]A reserve battery design simply requires an incomplete, inactive battery assembly, with provision for activation when the user is ready. Details will be given later of two different activation schemes. One depends on the vapor transfer of the solvent into the cell to form a conducting solution with the dry electrolyte salt already in the cell. The other transfers a liquid into the cell, either the solvent or the electrolyte solution.

increasingly lower current density as the temperature is lowered, due to the decrease in both chemical reaction and diffusion rates. Third, few aqueous electrolyte solutions are conductive at temperatures below 233°K (−40°F), and many of these are unsuitable for battery operation at higher temperatures. These limitations have made it necessary to resort to oversize batteries or to temperature-conditioned environments, or else simply to accept a degradation in performance.

A possible method of avoiding such compromises is to use a solvent other than water for the electrolyte, with the objective of producing cells than can be discharged efficiently at temperatures as low as 220°K (−63°F). This entails judicious selection of solvent, solute, and electrodes on the basis of a limited amount of usable information in the literature. It also makes necessary an experimental program in which the apparatus must be more elaborate than is needed in aqueous electrochemistry, in order to avoid contamination by atmospheric moisture (or even oxygen) and to protect personnel from toxic vapors.

In selecting nonaqueous solvents for battery use, a number of characteristics must be considered. The chemical literature is replete with studies of inorganic and organic reactions and the physical chemistry of substances in nonaqueous solvents. Much of this has been compiled, so that the reader who wants a general background or is looking for reference material in the field can readily obtain such information [see (14, 24, 4, 11, 47, 53)]. However, of the twenty or more solvents discussed, only a few seem potentially suitable for use at low temperatures.

The formation of conducting solutions is a fundamental requirement in determining the usefulness of a solvent for battery purposes. Two other requirements are that electrode processes occur at a useful rate and that the electrode potential developed be reasonably large. The Nernst-Thomson principle relates the power of a solvent to dissociate ions in a crystal to its dielectric constant. In addition to the dielectric constant, the tendency of polar molecules to solvate ions plays an important role in the solution process. There has been a tendency to evaluate the potentiality of a solvent in terms of its dielectric constant alone. This has led to neglect of nonaqueous solvents that later proved promising. Liquid ammonia is an example. Its dielectric constant is relatively low compared to that of water. Ammonia, however, forms solvates far more readily than does water, with the result that many substances are soluble, although perhaps the species in solution are principally charged molecular aggregates. The viscosity of ammonia is so low, even at low temperatures, that these ions are fairly mobile, and high conductance results.

Three representative solvents, NH_3, SO_2, and acetonitrile, and cells utilizing these, will be discussed in the following pages. Table 10.1 lists

TABLE 10.1. PROPERTIES OF WATER, LIQUID AMMONIA, LIQUID SULFUR DIOXIDE, AND ACETONITRILE

Property	Solvent			
	H_2O	NH_3	SO_2	CH_3CN
Freezing point (°K)	273	195	200	227
Boiling point (°K)	373	240	263	355
Critical temperature (°K)	647	140	430	548
Critical pressure (atm)	218	112	78	48
At 298°K				
Vapor pressure (atm)	0.031	0.99	0.38	0.12 (est)
Density (g/cm³)	0.997	0.603	1.37	0.777
Latent heat of vaporization (cal/mole)	10,500	4,750	5,720	–
Dielectric constant	79	17	12	36
Viscosity (centipoises)	0.894	0.135	0.26	0.345
Heat capacity (cal/mole/°K)	18.0	19.3	21.1	22
At normal boiling point				
Density (g/cm³)	0.958	0.682	1.46	–
Latent heat of vaporization (cal/mol)	9,710	5,580	5,960	7,120
Dielectric constant	56	22	17 (est)	–
Viscosity (centipoises)	0.284	0.266	0.43	–
Heat capacity (cal/mole/°K)	18.2	18.2	20.2	–

some physical properties of general interest. To facilitate comparison, data are given at 298°K (77°F) as well as at the normal boiling point of each solvent.

LIQUID AMMONIA

GENERAL PROPERTIES

Pure ammonia is a substance that has abnormally high boiling and freezing points, heats of vaporization and fusion, heat capacity, and dielectric constant. In these respects it is similar to water and hydrogen fluoride, all three being abnormal when compared with the hydrides of other members of groups V-B, VI-B, and VII-B, respectively, of the Periodic System. Properties of the liquid that are particularly important for battery applications are the low viscosity and density, together with the small temperature coefficient of the former. Autoionization is less than that of water:

$$2NH_3 \leftrightharpoons NH_4^+ + NH_2^- \qquad K_i = 10^{-33} \text{ at } 220°K$$
$$K_i = 10^{-27} \text{ at } 300°K$$

Ammonia is a stronger proton acceptor than water, which greatly influences its solution chemistry. Strong reducing agents are stable in ammonia, but only weak oxidizing agents dissolve without decomposition of the solvent.

Solubilities

An extensive bibliography of the literature on physical and chemical properties of ammonia solutions was compiled by Heuberger and Botter (21), and a monograph on the same subject has been published (25).

The solubility of crystalline substances can be described qualitatively as follows. Salts that form stable hydrates tend to form stable ammoniates, the solubility depending in large part on the anion. Most thiocyanates, perchlorates, nitrates, and nitrites are highly soluble, whereas most fluorides, sulfides, hydroxides, and polyvalent oxy-acid salts are insoluble. The iodides, bromides, and chlorides are only slightly soluble, with solubility decreasing markedly from iodide to chloride.

Ammonium salts are acids in liquid ammonia; they tend to be more soluble than the corresponding metal salts. The ammonium ion has a strong affinity for additional ammonia, so much so that the volume of moderately dilute solutions of an ammonium salt is appreciably less than the volume of ammonia used to prepare them. The ammonium salts are strongly deliquescent toward ammonia vapor, a property that influenced the development work on ammonia vapor-activated batteries. Silver salts are unusually soluble, whereas cadmium salts are insoluble regardless of the anion. The bases (amides, imides, and nitrides) are virtually insoluble with the exception of potassium, rubidium, and cesium amides. Many substances form insoluble ammoniates or dissolve by reaction with the solvent. These include sulfur, iodine, and some heavy metal salts.

The most remarkable solutions in liquid ammonia are those of the alkali and earth alkali metals, which almost instantly dissolve on contact to form metastable blue or bronze solutions of high conductivity. They appear to contain ammoniated metal ions and electrons. Jolly (27) has reviewed the structure and kinetics involved, Pray (39) has discussed some of their chemical reactions, and Wong (54) measured physical properties of $Ca\text{-}NH_3$ solutions.

Conductivity

Much of the literature on the conductivity of ammoniacal solutions is concerned with low concentrations, and the applicability of the Debye-Hückel and Onsager relationships, of Walden's theory, or of other limiting laws. Franklin (13) gives conductance data for NH_4NO_3 and KNH_2 solutions at concentrations as high as 8 *M* at 240°K (−27°F). Foote and Hunter (12a) made measurements on highly concentrated NH_4SCN solutions (25 to 40 mole percent) at 273°K (32°F) in the course of their vapor-pressure investigations. Conway (9) tabulated data by Pleskov covering solutions at 233°K (−40°F) of 18 compounds at moderate concentrations (less than 1 *N*), which are probably more dilute than battery electrolytes. Presgrave, Corson, Hendler, Bryant, and others, have measured the specific conductivities of several ammoniacal electrolyte solutions [Freund, Bryant, and Schliff (15)]. Figure 10.1 presents

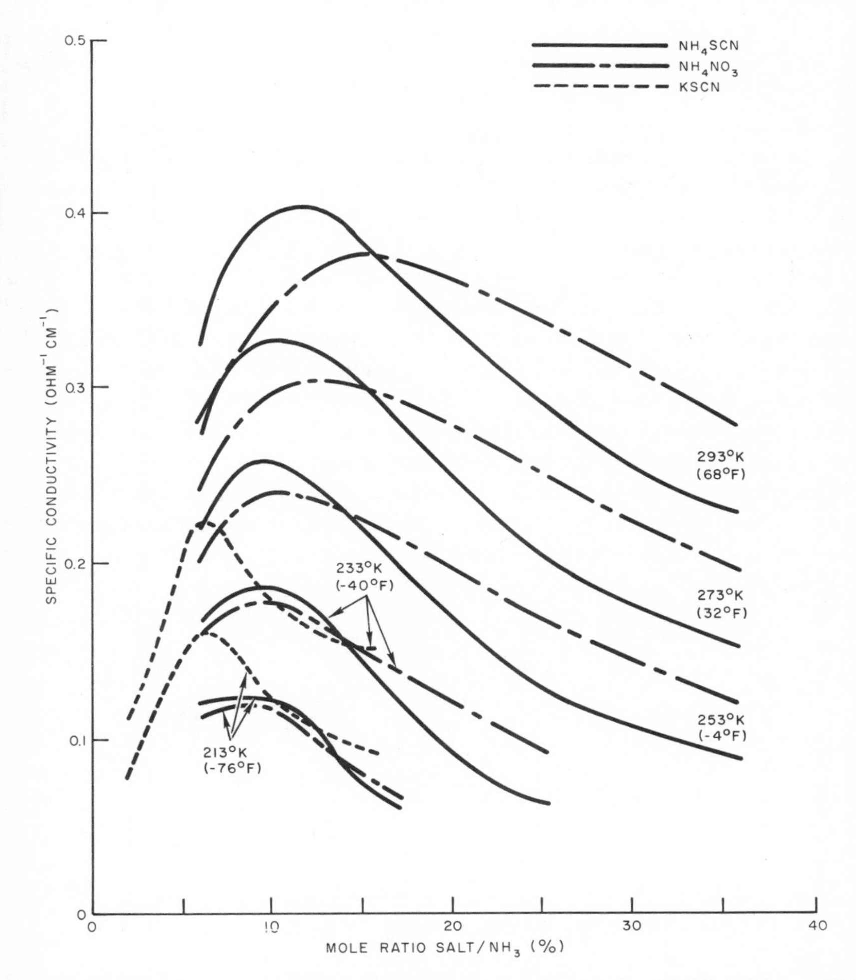

Figure 10.1 Specific conductivity versus concentration from 213 to 293°K (−76 to +68°F) of NH_4SCN, NH_4NO_3, and KSCN in NH_3.

specific conductivity values at several temperatures for NH_4SCN, NH_4NO_3, and KSCN solutions as a function of concentration.

Values of specific conductivities of ammoniacal solutions are compared with those for aqueous solutions of H_2SO_4 and KOH in Table 10.2. The variation with temperature of the concentration at which maximum conductivity occurs is greater in ammonia than in water. This is presumably a concomitant of the much larger thermal expansion of ammonia. As expected from the low temperature coefficient of viscosity, the specific conductivity of ammoniacal solutions is less affected by temperature than that of aqueous solutions.

Electrode potentials

Many investigators have studied electrode potentials in liquid ammonia. Jolly (26) evaluated the literature and computed a number of standard electrode potentials from free energy data. Table 10.3 presents a few of his values, with the sign changed to conform to the 1953 Stockholm recommendation.

Rosztoczy (41) (1966) discussed both calculated and measured values reported by a number of investigators. For Pb, Cd, and Zn he found good agreement. For Ca, however, values ranged from 1.73 to 2.44 V. He reasoned that 2.185 V is the best present value for Ca/Ca^{+2}, in thiocyanate-ammonia solution at 239°K, relative to the $Pb/Pb(NO_3)_2$ reference electrode. Thiocyanate ion was apparently necessary to minimize corrosion and stabilize the electrode potential.

Efforts to use Ca in practical galvanic cells have not yet been successful, owing to excessive reactivity in the less concentrated solutions and over the wider temperature range normally required in battery studies.

Table 10.2. Specific Conductivity of Ammonia Solutions[a] ($ohm^{-1}\ cm^{-1}$)

Solution	Conductivity		
	213°K −76°F	233°K −40°F	293°K 68°F
32% KSCN in NH_3	0.160	0.223	—
32% NH_4NO_3 in NH_3	0.119	0.177	0.332
31% NH_4SCN in NH_3	0.122	0.185	0.391
31% H_2SO_4 in H_2O	—	0.126	0.825
31% KOH in H_2O	—	0.055	0.55

[a]Concentrations, in weight percent solute, chosen for maximum conductivity at the lowest temperature tabulated.

TABLE 10.3. COMPUTED ELECTRODE POTENTIALS IN LIQUID AMMONIA AT 298°K (77°F)

Acid Solutions		Basic Solutions	
Electrode	$E°$ (V)	Electrode	$E_B°$(V)
$Li = Li^+ + e^-$	−2.34	$Ca + 2NH_2^- = Ca(NH_2)_2 + 2e^-$	−2.83
$Ca = Ca^{+2} + 2e^-$	−2.17	$Li + NH_2^- = LiNH_2 + e^-$	−2.70
e^- (am) $= e^-$	−1.95	$Na + NH_2^- = NaNH_2 + e^-$	−2.02
$Mg = Mg^{+2} + 2e^-$	−1.74	e^-(am) $= e^-$	−1.95
$Zn = Zn^{+2} + 2e^-$	−0.54	$Zn + 4NH_2^- = Zn(NH_2)_4^{-2} + 2e^-$	(−1.8)
$Cd = Cd^{+2} + 2e^-$	[−0.2]	$\frac{1}{2}H_2 + NH_2^- = NH_3 + e^-$	−1.59
$Ni = Ni^{+2} + 2e^-$	[−0.1]	$3NH_2^- = \frac{1}{2}N_2 + 2NH_3 + 3e^-$	−1.55
$Fe = Fe^{+2} + 2e^-$	[0.0]	$Pb + 3NH_2^- = Pb(NH_2)_3^- + 2e^-$	(−1.4)
$\frac{1}{2}H_2 + NH_3 = NH_4^+ + e^-$	0	$3Hg + 6NH_2^- = Hg_3N_2 + 4NH_3 + 6e^-$	(−1.1)
$4NH_3 = \frac{1}{2}N_2 + 3NH_4^+ + 3e^-$	0.04	$Ag + 2NH_2^- = Ag(NH_2)_2^- + e^-$	(−1.0)
$Pb = Pb^{+2} + 2e^-$	0.28	$OH^- + NH_2^- = \frac{1}{2}O_2 + NH_3 + 2e^-$	−0.06
$Cu = Cu^{+2} + 2e^-$	0.40	$I^- = \frac{1}{2}I_2(s) + e^-$	1.26
$Hg = Hg^{+2} + 2e^-$	0.67	$Br^- = \frac{1}{2}Br_2(1) + e^-$	1.73
$Ag = Ag^+ + e^-$	0.76	$Cl^- = \frac{1}{2}Cl_2(g) + e^-$	1.91
$I^- = \frac{1}{2}I_2(s) + e^-$	1.26	$F^- = \frac{1}{2}F_2(g) + e^-$	3.50
$Br^- = \frac{1}{2}Br_2(1) + e^-$	1.73		
$Cl^- = \frac{1}{2}Cl_2(g) + e^-$	1.91		
$F^- = \frac{1}{2}F_2(g) + e^-$	3.50		

NOTE: [ ] indicates probably ±0.1 V; ( ) indicates possibly ±0.5 V.

Data on activity coefficients in liquid ammonia are sparse. Ritchey and Hunt (40) determined the activity coefficients of NH_4Cl solutions at 298°K. They found that at concentrations in the range of battery operation (approximately 4–8 molal) the activity coefficient of NH_4Cl is about 0.01. The large vapor-pressure depression of such concentrated solutions makes the solvent activity significantly less than unity. Use of the standard potentials in ammonia to estimate the potential of real cells is therefore likely to result in appreciable error. Table 10.4 lists observed open-circuit potentials of various cells as reported by Adlhart (1) or measured by Freund and co-workers (15).

ANODES

In principle the dissolution process at the anode in ammoniacal electrolytes is identical with that in aqueous solutions, the dissolved species being an ammoniated cation instead of a hydrated cation. Therefore, one would expect the electrochemical behavior of the anode reaction in the two solvents to be similar, allowing for the effect of differences in solvation energy and solubility of salts formed with the anions present.

Magnesium, zinc, and lead are the metals commonly used as anodes in cells with ammoniacal electrolytes. They are readily available, easy to

TABLE 10.4. OBSERVED OPEN-CIRCUIT POTENTIALS OF VARIOUS CELLS AT 233°K (−40°F)

Cell	Volts ±0.05
$Mg/NH_4SCN/PbO_2$	2.95
$Zn/NH_4SCN/PbO_2$	1.90
$Pb/NH_4SCN/PbO_2$	1.10
$Pb/NH_4NO_3/PbO_2$	1.10
$Mg/NH_4SCN/MnO_2$	2.55
$Mg/NH_4SCN/PbSO_4$	1.85
$Mg/KSCN/HgSO_4$	2.40
Mg/KSCN/S	2.25
Mg/KSCN/C	2.35
$Mg/KClO_4/C$	2.35

handle, and operate with little polarization at high current densities. Lithium and its alloys with aluminum are good anodes, but they are more difficult to obtain and manipulate. Other metals have been used in cells, but simply as substitutes for one of the foregoing metals rather than in combination with possibly more suitable electrolytes.

Magnesium. Acid electrolytes[2] attack magnesium rapidly, even at 220°K (−63°F), with copious evolution of hydrogen. The reaction rate decreases markedly, however, as the concentration of ammonium salt is increased above 25 mole percent or as impurities such as water are decreased. Freezing-point data show that NH_4X type salts, where X is a univalent anion, form triammoniates. The protons in solutions of 25 mole percent NH_4X, or more concentrated solutions, are apparently coordinated with all the available NH_3 molecules, with the result that the activation energy barrier for solvation of the metal ions is increased and the solution rate of the metal is decreased. Analysis of residues from $Mg/NH_4SCN/MnO_2$—C vapor-activated cells showed that their magnesium content correlated well with the sum of the amounts of magnesium produced by corrosion and electrochemical solution. In electrolytes containing thiocyanate ion, part of the attack on Mg may be due to the reduction of thiocyanate to sulfide and cyanide. Evidence is given by the odor of H_2S emanating from residues of the $Mg/NH_4SCN/MnO_2$—C cells as they evaporate.

[2]Ammonium salt solutions, in which NH_4^+ corresponds to the OH_3^+ of aqueous electrolytes. In both cases, since neither solvent is appreciably ionized, acidity depends on the nature and dissociation of the solute.

In neutral electrolytes, such as KSCN, magnesium corrodes slowly, forming the insoluble $Mg(NH_2)_2$ and evolving hydrogen and H_2S. The rate of sulfide formation appears qualitatively to be less than in NH_4SCN solutions. This may be due simply to the fact that in the acid solutions enough heat is released by the hydrogen-evolution process to increase the temperature at the electrode, thereby accelerating the thiocyanate reduction.

In liquid cells containing KSCN or $KClO_4$ electrolyte, the magnesium potential versus a zinc reference electrode is 1.00 ± 0.05 V at 233°K (−40°F). At current densities up to 30 ma/cm², there is no significant change in this potential. Adlhart (1) states that in Mg/KSCN/$HgSO_4$—C cells magnesium dissolves electrochemically in dilute solutions with high anode efficiency at current densities up to 250 ma/cm². Polarization at 223°K (−58°F) is shown on Fig. 10.2.

Zinc. Zinc is attacked by NH_4SCN solutions more slowly than is magnesium, as one would expect from the relative positions of the two metals in the electrode potential series. The evolution of hydrogen seems to be the only significant corrosion process; there is no evidence of thiocyanate reduction by zinc. The limited polarization data for zinc anodes are inconclusive. Vapor-activated cells at normal room temperature with zinc anodes have been discharged at current densities as high as 100 ma/

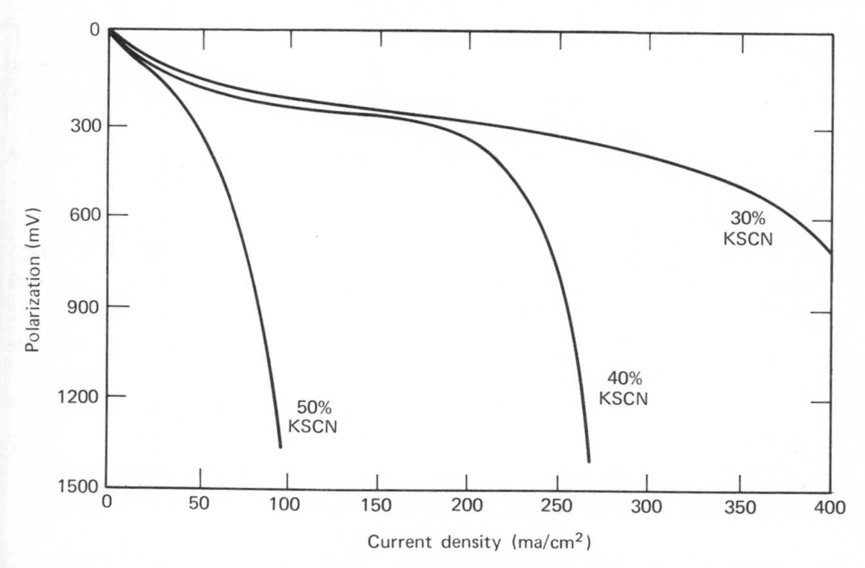

Figure 10.2 Effect of current density on polarization of Mg in KSCN-NH_3 electrolytes at 223°K.

cm^2 without indications of anode limitation. However, at 233°K (−40°F) performance is poorer than expected, even after considering the effect of increased cathode polarization. There is strong evidence that the presence of water, even in small amounts, may be deleterious to the zinc anode. This has not been investigated using cells with nonoxide cathodes. There is no visible reaction of zinc with KSCN. The use of zinc cells with neutral electrolytes has not been studied.

Lead. Lead is relatively inert in either acid or neutral electrolytes. The low potential of the lead anode requires coupling with a strongly oxidizing cathode or depolarizer to realize a useful cell. It is possible to use NH_4NO_3 as an electrolyte with lead anodes, as the nitrate ion is not reduced to nitrite by lead. Zinc and magnesium, on the other hand, are sufficiently active to effect this reduction.

Cathodes

Cathodes have been classified in a number of ways. They are sometimes described in terms of mechanical structure (such as film type, bobbin or volume type, or inert-with-soluble-oxidant type) or in terms of the type of electrode process (oxide type, electrode of the first or second kind, etc.). All these categories have been used in ammonia-battery studies.

To some extent, research programs have been limited to cathode materials selected for their availability rather than for their suitability for use in ammoniacal systems. This was due largely to the end-use objectives of the sponsoring agencies, involving a very high-rate, low-capacity service at an energy density in the order of 8 J/g (1 w-hr/lb), with no activated-stand requirement. These conditions permitted the use of high-valence oxides, such as MnO_2 or PbO_2, as cathode materials.

The cathode process involves transfer of protons from ammonium ions to oxygen atoms and therefore an acid electrolyte. Here a high rate of anode corrosion is tolerable because of the short service life and absence of an activated-stand requirement. The cathode behavior of oxides in ammoniacal acids is similar to that in aqueous acids. Concentration polarization occurs at lower current density in ammoniacal electrolytes because transport of protons to the cathode surface by a Grotthuss-type conduction mechanism[3] does not take place.

[3]A mechanism proposed to account for unusually high conductivity in certain aqueous electrolytes, in which proton transfer occurs by the shifting of hydrogen bonds along a chain of water molecules. Thus by only shifting hydrogen bonds, a proton can effectively move the length of a chain of water molecules with only a slight shift of atoms. The normal conduction processes also occur. See Meredith (33).

The exact mechanism of the cathode process with MnO_2 or PbO_2 electrodes has not been established with any certainty. Qualitatively, after discharge of PbO_2 cathodes electrodeposited on stainless steel, there appears to be a deposit of PbO. The problem of analyzing the reaction products is complicated by the absorption of atmospheric moisture on the electrodes, hydroxy salts of lead being precipitated with consequent confusion of the results.

Scribner (45) mentioned a Pb/PbO_2 cell for a low-rate application that delivered 1.5 ma/cm^2 for 96 hr at 0.6 V at 220°K (−63°F).

Minnick and Presgrave (36) described cells of the type

$$M_1/M_1X/M_2X/M_2$$

$$M_1/M_1X/NH_4X/M_2$$

where M_1 is an alkali or alkaline earth metal, magnesium, zinc, aluminum, beryllium, or manganese and M_2 has a potential at least 0.75 V higher, such as Pb, Hg, or Fe.

The large temperature coefficient of performance of oxide cathodes has led to a study of other reducible substances. Two classes of these have been discussed by Adlhart (1) and Minnick (34): insoluble heavy-metal salts, and sulfur or sulfur compounds. The insoluble heavy-metal salt cathode is an analog of the AgCl cathode in aqueous systems in that the metal ion is reduced to free metal. However, solubility relationships in ammonia are such that the anion does not usually go into solution, but is reprecipitated as a salt of a cation in the electrolyte. Mercury, lead, and silver sulfates have been tried, according to Minnick. Adlhart's curves for magnesium-mercuric sulfate cells are presented in Fig. 10.3.

The small effect of temperature on capacity and potential is an outstanding characteristic of these cells. Adlhart estimates that a practical battery, exclusive of battery case and activator, can be built to deliver 208 J/g (26 w-hr/lb) with $HgSO_4$, 150 J/g (19 w-hr/lb) with $PbSO_4$ at a 10- to 20-min rate, for temperatures between 223 and 323°K (−58 and 122°F) (the limits of his investigation).

Murphy (38a) reported that by substituting AgCl for $HgSO_4$, cell capacity was increased and heat of ammoniation during activation was decreased. Silver chloride-carbon cathodes can be rapidly produced by a hot dry-pressing method (similar to the sulfur-carbon cathode process previously described), which was evaluated in a program of statistically designed cell tests described by Almerini (2). Magnesium-mercuric sulfate cells with KSCN electrolyte can be discharged for over 72 hr below 200°K (−100°F), as noted in an ammonia battery progress report

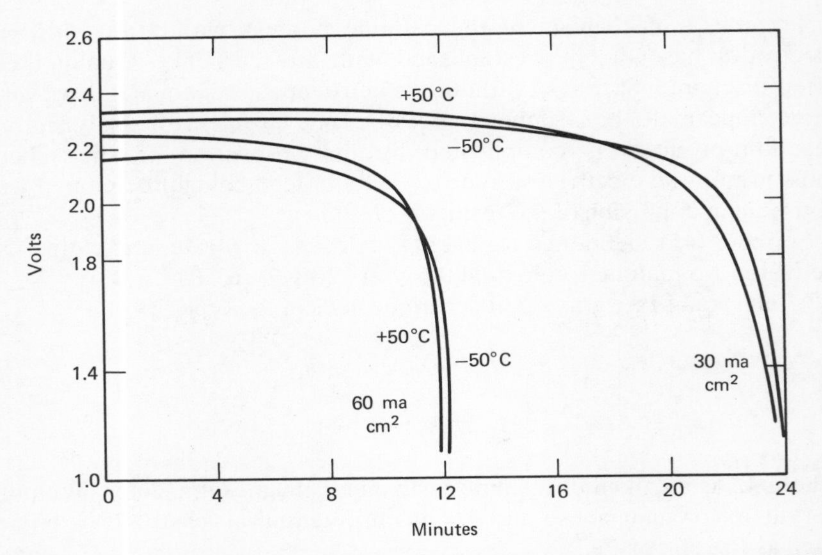

Figure 10.3 Operating characteristics of Mg/KSCN-NH_3/$HgSO_4$ cells at 233 and 323°K (−60 and +112°F).

by Spindler (49). At a current density of about 1 ma/cm², based on plane anode area, cell potential remains above 1.5 V [Schwartz (44)].

The use of sulfur as the cathodic reactant was proposed as a means of achieving a cathode reaction as simple as the metal-dissolution process at the anode: $S^o + 2e \rightarrow S^{-2}$. Sulfur dissolves slowly in liquid ammonia. The reaction

$$10S + 16NH_3 \leftrightharpoons N_4S_4 + 6(NH_4)_2S$$

proceeds only to a small extent, if at all [Zipp (55)]. Raman spectra show the presence of the S_8 ring in solution. It is believed the sulfur tetranitride, N_4S_4, undergoes solvolysis, forming NH_4^+ and $S_2N_2 \cdot NH_2^-$ ions. On the hypothesis that the $(NH_4)_2S$ takes no useful part in the reaction, solutions of N_4S_4 have been tried. This material is somewhat dangerous to handle and difficult to procure in the solid state. These disadvantages outweigh any slight advantages in performance. The mechanism by which the sulfur molecule, S_8, is reduced is still obscure, although there is some indication that the thiocyanate ion or sulfide films on the cathode surface may be involved. Sulfur solution does not react with magnesium, so that rapid self-discharge does not occur even in the absence of a separator.

Sulfur cathodes made by a dry-pressing technique are described by Mount (38). A mixture of 60 percent or more of sulfur with carbon black or acetylene black is compacted on an expanded metal grid enclosed within an envelope of separator material. Pressing time may vary from $\frac{1}{2}$ to 5 min, and pressure from 5000 to 15,000 lb/in². Heating the mix to as high as 340°K (150°F) is permissible, but room temperature operation is preferred.

Organic substances as cathodic oxidants, which have shown promise in aqueous solutions, were investigated by Harris and Matson (19). In concentrated, strongly acid ammonia solutions, meta-dinitrobenzene was found to be reducible in two steps of four electrons each. The para and ortho isomers, however, could be reduced by a total of twelve electrons, in two steps, one of four electrons followed by one of (Fig. 10.4) eight electrons. These results may be compared with a recent (1965) study by Jackson and Dereska (23), who reported that in aqueous solution

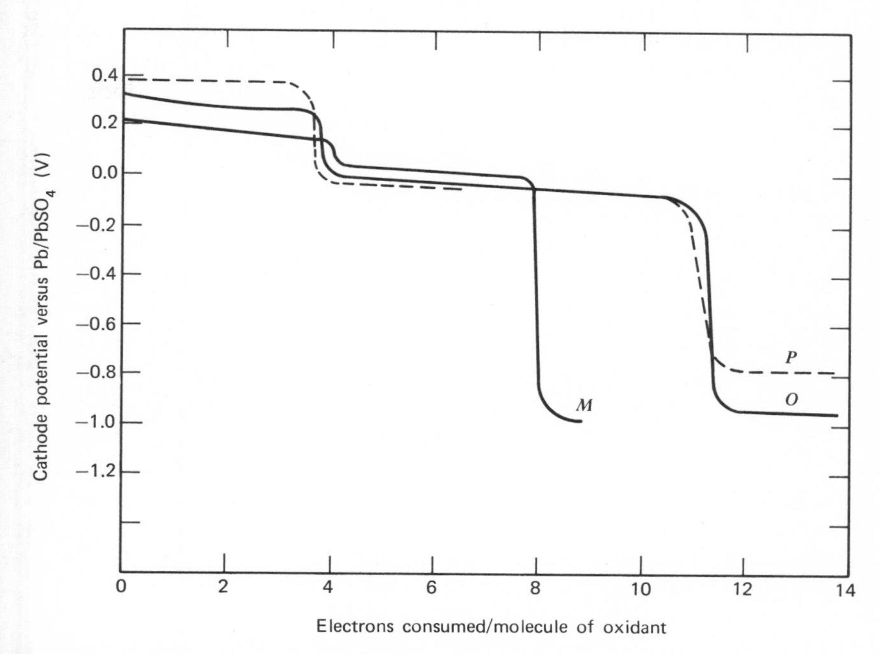

Figure 10.4 Reduction steps for dinitrobenzene isomers. M = meta, P = para, O = ortho, Test conditions:

Current = 2.0 ma (0.5 ma/cm² or less)
Temperature = 300°K
Electrolyte = NH_4SCN saturated with NH_3 vapor
Cathode = stirred Hg pool

of meta-dinitrobenzene below *p*H 2, two steps of six electrons each occurred, while above *p*H 3, only two steps of four electrons each occurred.

In neutral ammonia solutions, Harris (18) reported that reduction of meta-dinitrobenzene proceeded in one fairly well-defined step of two electrons followed by rapid potential drop and two or more poorly defined steps to a total reduction of not more than six electrons. Kuwana (31) reviewed the general difficulties encountered in the study of organic electrode reactions, and Darlington and Kuwana (10) reported studies in progress of the formation of free radicals during electroreduction of nitrobenzenes in lithium nitrate ammoniate solutions.

SULFUR DIOXIDE

General properties

One of the most thoroughly studied nonaqueous solvents is sulfur dioxide. General properties of SO_2 were given in Table 10.1, where they can be compared with those of water and ammonia. Sulfur dioxide is an example of a nonprotonic solvent, i.e., one that contains no hydrogen. Its self-ionization is slight and the specific conductance of 4×10^{-8} ohm^{-1} cm^{-1} at 263°K (14°F) is close to that of water. Cady and Elsey (8) were the first to propose that dissociation occurs as

$$SO_2 \leftrightharpoons SO^{+2} + O^{-2}$$

or

$$2SO_2 \leftrightharpoons SO^{+2} + SO_3^{-2}$$

This mechanism has been confirmed in the researches of Jander and his associates (24), who showed that reactions in SO_2 could be systematized on this foundation. Solvate formation, metathesis, neutralization, solvolysis, amphoterism, oxidation-reduction, and complex-compound formation have all been observed. Compounds containing the thionyl group, SO^{+2}, e.g., $SOCl_2$, are the acids of sulfur dioxide chemistry. The sulfites are the bases (as are the pyrosulfites, $S_2O_5^{-2}$, which can be regarded as a doubly solvated oxide ion).

Solubility

Konecny (29) reviewed the literature on liquid SO_2 with a view toward its possible use as a solvent for battery electrolytes. Inorganic compounds that dissolve in amounts exceeding 0.1 mole/kg SO_2 at 273°K (32°F) include $SbCl_3$ (0.575), LiI (1.49), NaI (1.0), KI (2.49), NH_4I (0.58),

KSCN (0.502), NH_4SCN (6.16), $NH_4C_2H_3O_2$ (0.141), and $TlC_2H_3O_2$ (0.285). $AlCl_3$ is soluble to the extent of 0.6 mole/kg SO_2. (The alkali metal halides have a negative temperature coefficient of solubility.) Substituted ammonium and sulfonium compounds are also fairly soluble. There are inconsistencies among the data of various authors which may be due in part to the marked influence of traces of water on the solubility of many salts. As an example, Jander states that $Co(SCN)_2$ is wholly insoluble, whereas $Co(H_2O)_2(SCN)_2$ forms a blue solution, and suggests that this effect be used to determine the moisture content of liquid sulfur dioxide.

Covalent compounds and elements tend to be more soluble than ionic compounds. Bromine, boron trichloride, carbon disulfide, phosphorus (III) chloride, phosphorus(V) oxychloride, arsenic(III) chloride, iodine monochloride, and thionyl compounds are miscible with SO_2 in all proportions. Iodine monobromide, phosphorus(III) bromide, and Group IV tetrahalides are likewise quite soluble. Many of the latter separate into two liquid phases at lower temperatures. Schaschl and McDonald (43) found that a mixture of one mole of I_2 with three moles of Br_2 forms a dark brown liquid that has a markedly higher conductivity than either of the elements alone. They postulated the substance as IBr_3.

Conductivity

Nearly all the conductivity data in the literature on SO_2 solutions are in terms of equivalent conductance, and for dilute or moderately concentrated solutions. Examination of Jander's (24) plots of conductance as a function of dilution shows that the specific conductivity of solutions of salts, soluble to the extent of 1 mole/liter, does not exceed 0.09 ohm^{-1} cm^{-1} at 273°K. A 1 *M* potassium iodide solution has specific conductivities of approximately 0.036 and 0.047 $ohm^{-1}cm^{-1}$, at 240 and 273°K, respectively. By way of comparison, at 240°K a 1 *M* solution of KI in ammonia has a specific conductivity of approximately 0.130 $ohm^{-1}cm^{-1}$, while 1 *M* aqueous solutions have specific conductivities at 273°K of 0.082 $ohm^{-1}cm^{-1}$ and at 291°K of 0.127 $ohm^{-1}cm^{-1}$.

Cells

Few published reports have appeared on studies of sulfur dioxide cells. Schaschl and McDonald (43) used sodium or magnesium as anodes because of the high potential and relatively low cost of the metals. Today, the use of calcium would certainly be considered. The choice of cathode materials was limited because of solubility problems. Iron was selected as an inert cathode, and soluble oxidants were used for the cathode reaction. Among the more promising of those tried were $I_2 + KI +$ KCNS: $KI + Br_2$; $NH_4SCN + I_2 + FeCl_3$; $NH_4SCN + KI + FeCl_3$; $KI + IBr_3$; $FeCl_3 + IBr_3$; $FeCl_3 + IBr_3 + NH_4SCN$. These were dissolved in SO_2 to form the electrolyte as well as the cathodic reactant.

Schaer (42) summarized his doctoral thesis studies as follows

A liquid sulfur dioxide cell, such as $K/KCl \cdot SbCl_5$ (0.25 molar) $+ IBr_3$ (15% wt) $+ SO_2$ (1) carbon matrix is shown to have an open circuit potential greater than 3.70 V and an initial potential of 3.15 V, decreasing to 2.40 V in 10 minutes at a current density of 20 ma/cm^2 at a temperature of 293°K (68°F). The anode and cathode are in a parallel arrangement and are separated by a distance of 0.15 cm, thus simulating the actual conditions in a wafer type battery. The battery cell is tested at four temperatures in the range of 230°K to 293°K (−47°F to 68°F). The performance decreases at the lower temperatures.

The individual types of polarization, such as activation, concentration, and *i-r* drop are isolated. Activation polarization is evaluated for the Br^0-Br^- cathode and the Zn^0-Zn^{+2} anode by the double pulse galvanostatic method. The rate constant, k_s, for the Br^0-Br^- reaction is evaluated to be 0.23 cm/sec at 253°K. Concentration polarization is mathematically shown to be very small at the current densities studied (less than 100 mV at 20 ma/cm^2). The *i-r* drop in the electrolyte contributes appreciably to the total polarization of the cell. However, the largest single factor in the total polarization of the cell is the corrosion film formed on the anode surface by the reaction of the anode with the liquid sulfur dioxide electrolyte. Oxidizing agents, such as water, anhydrous ferric chloride, and anhydrous antimony pentachloride appear to alleviate this polarization somewhat.

Polarization data are tabulated for the dissolution of zinc, magnesium, calcium, lithium, sodium, and potassium. Data are also tabulated for the reduction of IBr_3 on porous stainless steel, flat stainless steel and a porous carbon matrix.

ACETONITRILE

The high boiling point of acetonitrile, or methyl cyanide (see Table 10.1), is an obvious advantage over NH_3 or SO_2. Many inorganic compounds dissolve in acetonitrile: Audrieth and Kleinberg (4) list 25 as readily soluble, 15 as soluble, 12 as slightly soluble, and 9 as relatively insoluble. A number of solvates have been isolated.

Arcand and Tudor (3) give the specific conductivity of 0.541 *M* KSCN solution in CH_3CN as 0.0218 ohm^{-1}cm^{-1} at 298°K and 0.0096 ohm^{-1} cm^{-1} at 233°K. Solomon (48) gives values at room temperature for several saturated and concentrated solutions, ranging from 4.8×10^{-5} ohm^{-1}cm^{-1} for saturated KCl to 0.042 ohm^{-1}cm^{-1} for saturated KSCN. (Numbers in his Fig. 3 should be multiplied by 10^{-2}.) Horowitz and Breitner (22) give compositions and freezing points of many electrolytes based on acetonitrile or propionitrile with Lewis acids, (e.g., aluminum chloride) as salts. Equal parts of acetonitrile and propionitrile with 20 weight-percent aluminum chloride freezes at 197°K and has a conductivity of 0.03 ohm^{-1}-cm^{-1}. At room temperature, a Ca/PbO_2 couple in a 20 percent solution of aluminum chloride in acetonitrile had an open-circuit potential of 3.6 V, but no data were given at low temperature or on performance of cells under discharge.

Audrieth and Kleinberg (4) cite the data in Table 10.5 (9) on potentials in acetonitrile at 298°K. Arcand and Tudor (3) discharged some modified Leclanché cells at room temperature, using a KSCN in CH_3CN electrolyte. It was necessary to add $ZnCl_2$ or NH_4SCN to the mix in order to sustain the discharge, but even in these cells the potential was 0.1–0.15 V lower than that of a standard Leclanché cell. Low-temperature performance was not reported.

Solomon (48) describes a cell using a calcium anode, silver chloride cathode, and a fibrous separator, with an undisclosed "Arctic" electrolyte which probably was 20 percent $AlCl_3$ in acetonitrile. The cell case was made of polyethylene. Discharge data are shown on Fig. 10.5. Pulses or brief high rate drain are permissible provided the cell is given an adequate rest period thereafter. Shelf life data are reported for seven weeks only, after which no loss of capacity was noted. It should be pointed out that the nonprotonic nature of the electrolyte permits it to remain in contact with the calcium anode without galvanic displacement of hydrogen.

TABLE 10.5. POTENTIALS IN ACETONITRILE AT 298°K (77°F) (System: M/0.01 *N* MX in CH_3CN/0.01 *N* $AgNO_3$ in CH_3CN/Ag)

Half Cell	Cell Potential (V)
Li/LiCl	2.476
K/KI	2.3443
Rb/RbI	2.3275
Cs/CsI	2.2723
Na/NaI	2.2592
$Ca/Ca(NO_3)_2$	2.140
$Zn/ZnCl_2$	0.975
Cd/CdI_2	0.7031
Cu/CuCl	0.6065
$Pb/PB(ClO_4)_2$	0.3544
Pt, H_2/HCl	0.2324
$Hg/HgBr_2$	0.0231

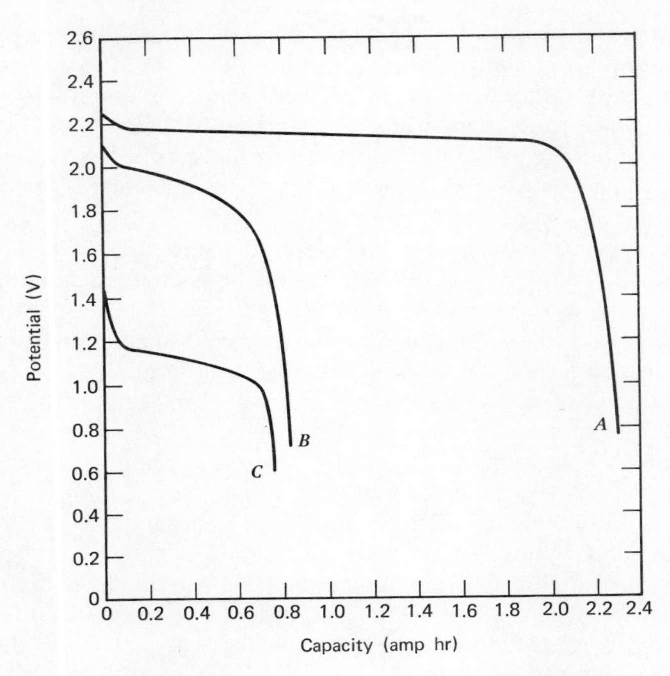

Figure 10.5 Discharge curves of Arctic Cells at 297°K. A = 100 ma (0.8 ma/cm²), B = 1.0 amp (8 ma/cm²), C = 6.0 amp (46 ma/cm²). Cell weight = 126 g.

OTHER SOLVENTS

Among other inorganic solvents, HF is often suggested, but no one has yet developed a practical cell. Properties of HF as a solvent were reviewed by Simon (45a).

Pyridine has been used for nearly a century as a solvent in electrodepositing metals, but there is no evidence of use in galvanic cells.

Interest in organic solvents of the cyclic ester type, such as propylene carbonate, originated in the graduate studies of Harris (18) under the direction of Tobias at Berkeley. Organic solvent electrolytes usually have high viscosity and low conductivity, differing from aqueous or ammonia electrolytes by 1 to 3 orders of magnitude. A number of organic solvents are currently under study, principally by industrial and government laboratories for possible application in high energy density cells. Organic solvent cells cannot be discussed adequately in a chapter on low temperature primary cells. As much work has been done on secondary applications as on primary, again putting the subject into a broader field of interest than can be encompassed here. Progress in studies for battery applications has been reviewed frequently [see Braeuer (6)].

Bauman (5) described the development of a Li/CuF_2 cell, which utilized an electrolyte of lithium perchlorate in propylene carbonate solvent. Performance of cells of this type (organic electrolyte, light metal anode, inorganic halide, or oxide cathode) is frequently reported in excess of 800 J/g (100 w-hr/lb) for low-rate discharge times of about 200 hr within the potential range of 1.5–3.5 V.

For low-temperature operation, a $Li/CuCl_2$ cell was reported by Kuppinger and Eisenberg (30). Performance was fairly uniform over the temperature range from 345 to 233°K (160 to −40°F), and even at 220°K (−65°F) the cell could deliver appreciable current. Construction was a conventional flat plate type similar to the Zn/KOH/AgO (discussed elsewhere in book). Experimental model batteries with 15 cells in series were discharged for 30 min at 6a [Knight (28)]. Load potential varied from 33 to 27 V, and energy density was around 160 J/g (20 w-hr/lb). Factors believed to contribute significantly to the extended performance of this type of cell are (*a*) porous $CuCl_2$ cathode containing silver and carbon powders, (*b*) porous anode, (*c*) mixed solvent of propylene carbonate and nitromethane, and (*d*) thin structures (entire cell, electrodes and electrolyte, was less than 1.0 mm thick).

RESERVE BATTERIES

In one type of vapor-activated reserve battery, a dry deliquescent electrolyte salt is placed between the electrodes. A solvent is stored separately until battery activation is required, when it is released as a vapor into the assembly. Condensation proceeds as a result of deliquescence of the dry component on contact with the vapor or as a result of an actual chemical reaction. In either event, an electrolyte solution forms and the battery is activated.

Mott-Smith (37) invented a different vapor-activated battery of the reaction type in 1943, although his patent was not issued until 1960. Hydrogen chloride gas, released from storage under pressure in a frangible vessel, reacted with a water-glycerol solution absorbed in a blotting-paper vehicle predisposed between zinc and lead dioxide electrodes, to form a hydrochloric acid electrolyte.

In a boron trifluoride cell [Evans (12)], still another type of reserve, activation was obtained from coordination of BF_3 with water of crystallization in the hydrated electrolyte, $Ba(OH)_2 \cdot 8H_2O$), to yield fluoboric acid.

As a historical note, the Upward cell of 1876 was a $Zn/ZnCl_2/C$ assembly activated with chlorine [Tomassi (51)]. Heavy-duty chlorine-activated cells with Zn, Mg, Al, and Fe anodes have been developed [Heise, Schumacher, and Cahoon (20)] and are also described by Vinal (52).

A chlorine-activated cell, if stored dry, requires several days of pretreatment with water vapor before it can be activated with chlorine for discharge [Langworth, Randolph and Lawson (32)]. Since both the BF_3 and Cl_2 cells are essentially aqueous systems, they will not be discussed further.

The use of sulfur dioxide for a vapor-activated cell was surveyed by Konecny (29), who concluded that there was little advantage compared to ammonia.

Ammonia vapor-activated (AVA) batteries

Practical vapor-activated reserve batteries have successfully been produced. Condensation of ammonia vapor on deliquescent salts to form an electrolyte solution is a purely physical process. Though in principle the thermodynamic and kinetic equations for mass transport can be used to describe what happens, enough variables enter into the activation process to render a theoretical analysis impractical. As a result, the development of ammonia vapor-activated (AVA) batteries proceeded in a fairly empirical fashion.

A description of the design and characteristics of AVA batteries was published by Gleason, Freund, Minnick, and Meyers (16). Cells were principally of the wafer type (Fig. 10.6). Plates were blanked from strip stock, either of Pb or Zn anode metal, or of Pb and PbO_2 electrodeposited on each side of stainless steel base. One or more layers of cellulose or glass fiber papers, impregnated by immersion in an aqueous solution of the desired solute and then air dried, were placed between the electrodes. Some Leclanché type structures were also built. [Typical cell potentials were given in Table 10.4.]. Hale (17) described several design changes that markedly improved the performance of $Zn/NH_4SCN/PbO_2$ cells over the data given by Gleason *et al.* An increase of salt content per unit area was achieved by using a more porous vehicle. Introduction of a thin, dense, sized, cellulose separator at the anode interface limited diffusion of ammonium ions toward the anode, keeping them available for the cathode reaction and at the same time decreasing anode corrosion. The combined effect of these changes was an improvement in cell potential from 1.4 to 1.75 V at 24 ma/cm^2 and a doubling of capacity.

Performance of AVA batteries in laboratory tests often exceeded specifications, but at temperatures below the solvent boiling point (240°K, −27°F), performance was frequently unpredictable. Several factors probably were responsible for this:

1. The solutions formed were of variable concentration within any cell and also from one cell to another. This effect was accentuated at

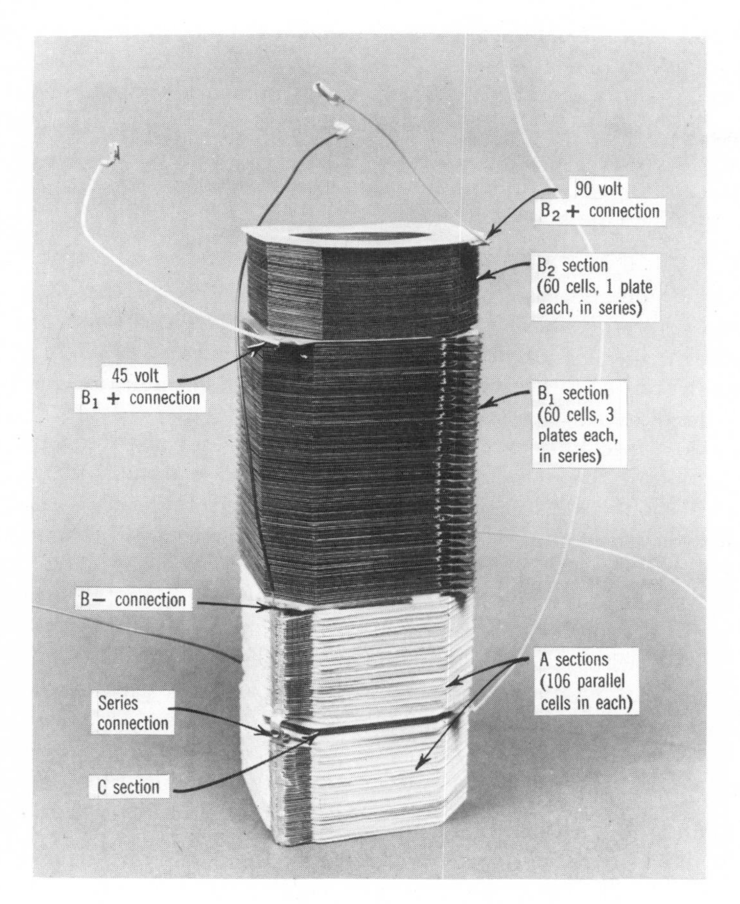

Figure 10.6 Complete cell stack for prototype AVA Battery.

low temperature, where the pressure difference driving the condensation process was diminished. Ammonia vapor would also condense outside the cells as the resistance to vapor flow into the cells increased. The net effect was the formation of less conductive, high-concentration solutions (Fig. 10.1).

2. The transfer of ammonia from reservoir to cell is in essence a batch-distillation process. The liquid ammonia supply, in the absence of added heat, will evaporate as long as there is a pressure difference between liquid surface and outside environment. Latent heat is supplied by cooling of residual liquid, particularly near the liquid surface. On the other hand, as ammonia condenses in the cells, latent heat is released and,

in addition, heat of solution is evolved. Heat from both of these sources raises solution temperature and vapor pressure. Dilution during the condensation process also increases vapor pressure. Consequently, the pressure difference during the mass transfer process rapidly diminishes. If a pyrotechnic heat source is placed in the reservoir, transfer of heat to liquid ammonia is difficult to accomplish. This is primarily due to formation of a vapor film on the heater surface, which markedly reduces the heat transfer coefficient.

It was concluded that for high-rate, short-life service, the advantages of vapor activation were not sufficient in the face of all these problems to justify further effort. Liquid-activated ammonia battery designs thus began to receive the attention of those engaged in high-rate ammonia-battery development.

Some low-rate battery requirements may be met advantageously by use of a modified ammonia vapor activation. Service of this type generally allows relatively long activation time, often several minutes. In such batteries provision may be made for an external ammonia supply, which is attached to the battery only during the activation process. By cooling the battery or slowly heating the ammonia supply, a temperature differential can be established that permits distillation of sufficient ammonia to meet the needs of the battery. Under these conditions use of deliquescent ammonium salts is not necessary, and nonvolatile salts such as KSCN may be used as the solute.

Ammonia liquid-activated (ALA) batteries

By maintaining pressure on ammonia throughout the discharge interval, both to improve performance of cells immediately upon activation and to minimize decline in performance due to expulsion of electrolyte from cells back into the ammonia reservoir, new battery designs were developed on the principle of liquid rather than vapor activation. The liquid ammonia solvent or electrolyte stored in the reservoir was transferred into the cells by applying a pressure on the reservoir greater than the vapor pressure of the liquid.

Typical appearance of experimental model ammonia batteries built by G. & W. H. Corson, Inc., for various military requirements shown in Fig. 10.7.

Minnick (34) presented discharge data for a one-cell battery having a Mg anode, KSCN electrolyte, and sulfur-carbon cathode. Performance at maximum and minimum test temperature is shown in Fig. 10.8. Open-circuit potential of the cell was 2.2 V. At 250 ma, it discharged for more than 15 min above 1.8 V, and at 50 ma, over 2 hr. Activation was

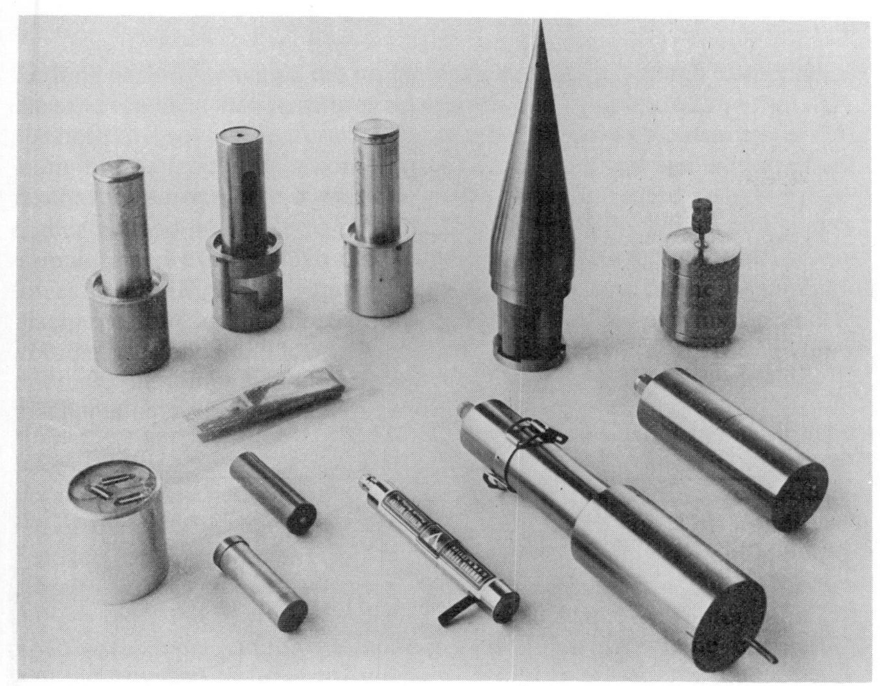

Figure 10.7 Pictorial view of early model ammonia batteries. (Courtesy of Livingston Electronic Corp., subsidiary of G. & W. H. Corson, Inc., Plymouth Meeting, Penna.)

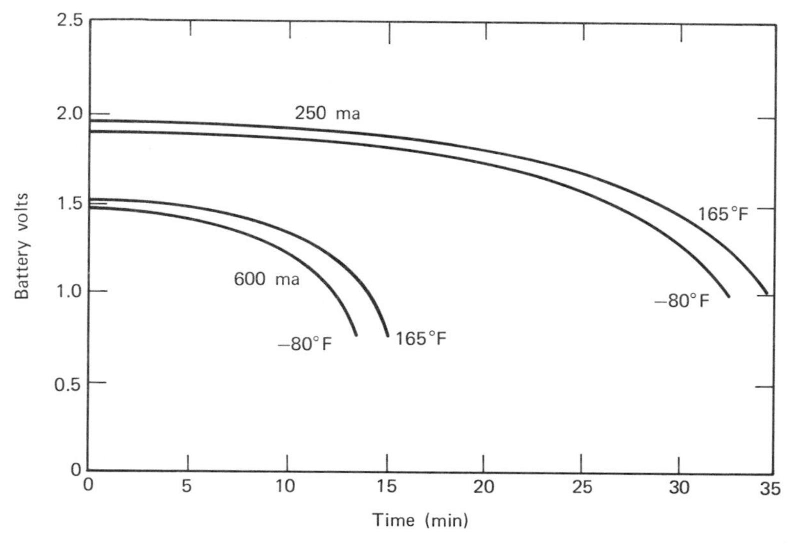

Figure 10.8 Discharge curves of type N-2 Battery with $Mg/KSCN\text{-}NH_3/S$ cells. Anode area = 14 cm^2, bobbin volume = 4.5 cm^3.

accomplished by either an electric or a pneumatic attachment that provided means for transferring ammonia into the cell. Mercuric sulfate was also used in this battery to obtain higher voltage and current, but performance tended to be erratic because of ammoniation heat effects.

Activation time for this type of ALA battery usually was less than 1 sec after opening the ammonia reservoir. The small effect of temperature on voltage and capacity is noteworthy (Fig. 10.8).

Similar cells are also being used in a battery designed by Corson for rocket telemetry. A typical assembly of cells for this purpose is shown in Fig. 10.9. Performance rating is 6 V at 300 ma for 30 min over the temperature range 222–300°K (−60–+80°F).

Typical designs for fuze applications were described by Minnick (35) and by Spindler and Pritchard (50). A production model before (*a*) and after (*b*) activation is shown in Fig. 10.10. Volume is 180 cm^3 and mass 590 g. Operating characteristics are 9 V at 15 ma over the temperature range of 220–345°K (−63–160°F). Activation is accomplished in 315 ± 110 ms,[4] and discharge capability may exceed 100 hr. Cells have Mg anodes, KSCN electrolyte, and meta-dinitrobenzene cathodes. Livingston Electronic Corporation was producing this battery in 1966 at a rate of 50/day for an Air Force fuse application. Details were reported by Smith and Tierney (46). This is believed to be the first model liquid ammonia battery to be continuously produced in quantity for field use, and to meet performance specifications.

A smaller and different kind of ammonia battery (Fig. 10.11) was designed and developed for Picatinny Arsenal by Livingston,[5] for activation by an impact force. Performance rating is 16V at 35 ma for 1 min, and the battery can be activated in less than 35 ms. Cell materials and operating temperature range are the same as for the preceding battery. Volume is 15 cm^3 and mass is 50 g. A brief description is given in the review paper by Spindler (49).

FUTURE PROSPECTS

Of the nonaqueous solvent cells studied to date, battery development is furthest advanced in the reserve-activated, liquid ammonia designs. Batteries now in production are meeting specifications not only at low temperature but across the entire environmental range from 220 to 350°K (−65 to 165°F). Discharge times, which now reach 100 hr, may be

[4]ms = milliseconds

[5]A subsidiary of Honeywell, Inc. as of 1966. G. & W. H. Corson, Inc., Eastman Kodak, and Eagle-Picher, the principal contractors during the early years of NH_3 battery R & D, have completely discontinued work in this field.

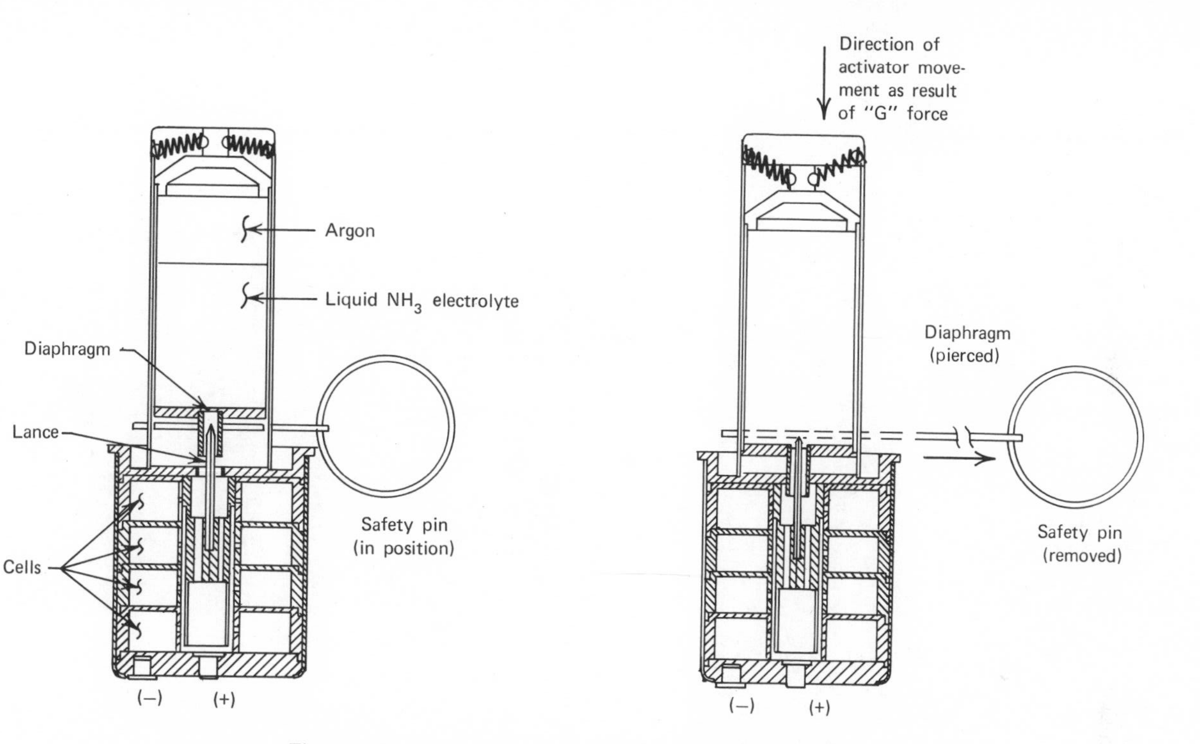

Figure 10.9 (*a*) Assembly for rocketsonde telemetry battery.

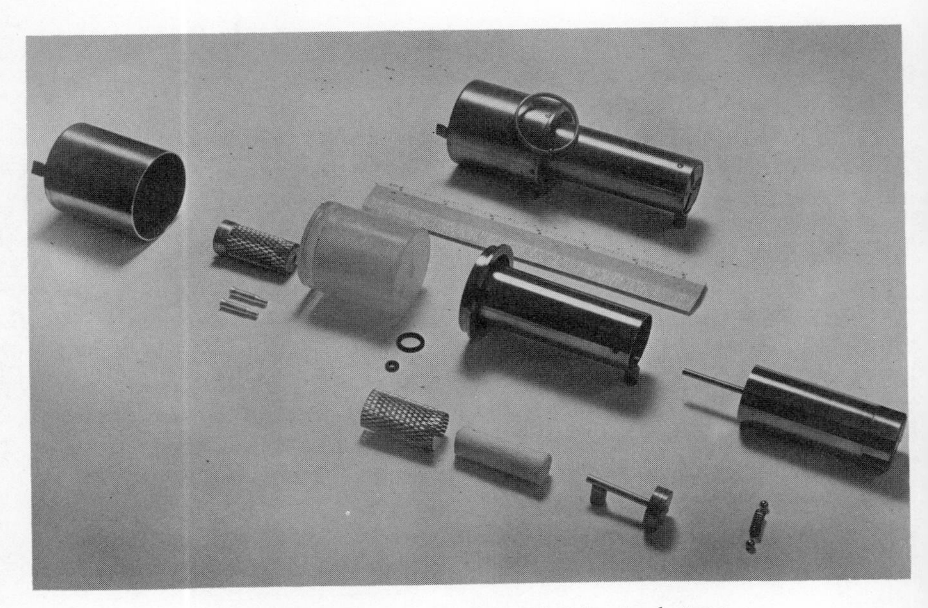

Figure 10.9 (*b*) Parts for rocketsonde telemetry battery.

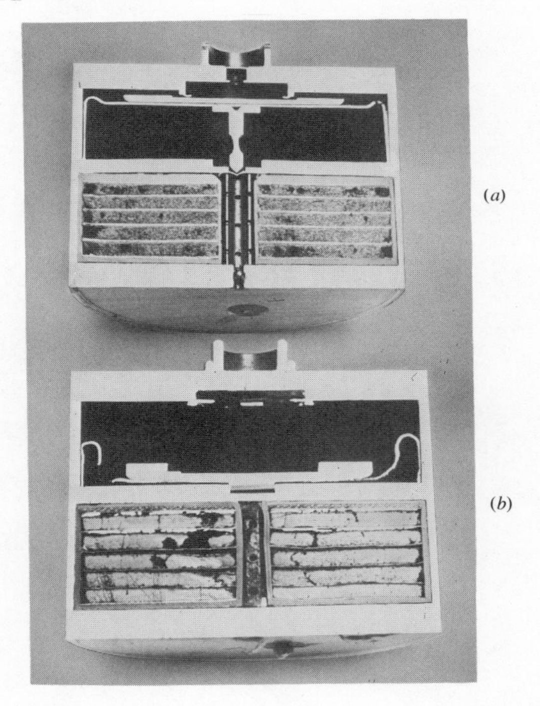

(*a*)

(*b*)

Figure 10.10 Production-model ammonia battery, before (*a*) and after (*b*) activation.

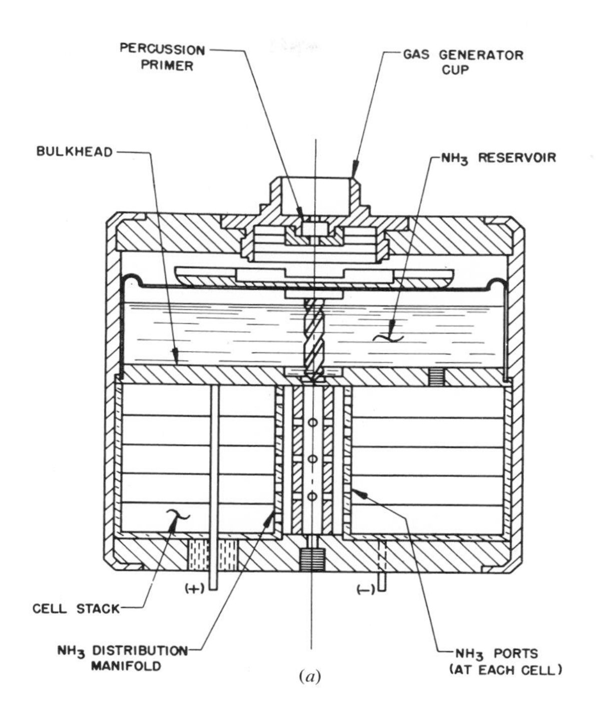

(*a*)

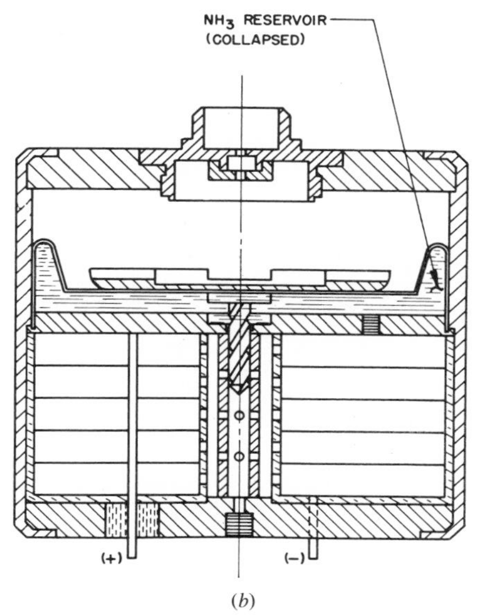

(*b*)

Figure 10.10 (*continued*)

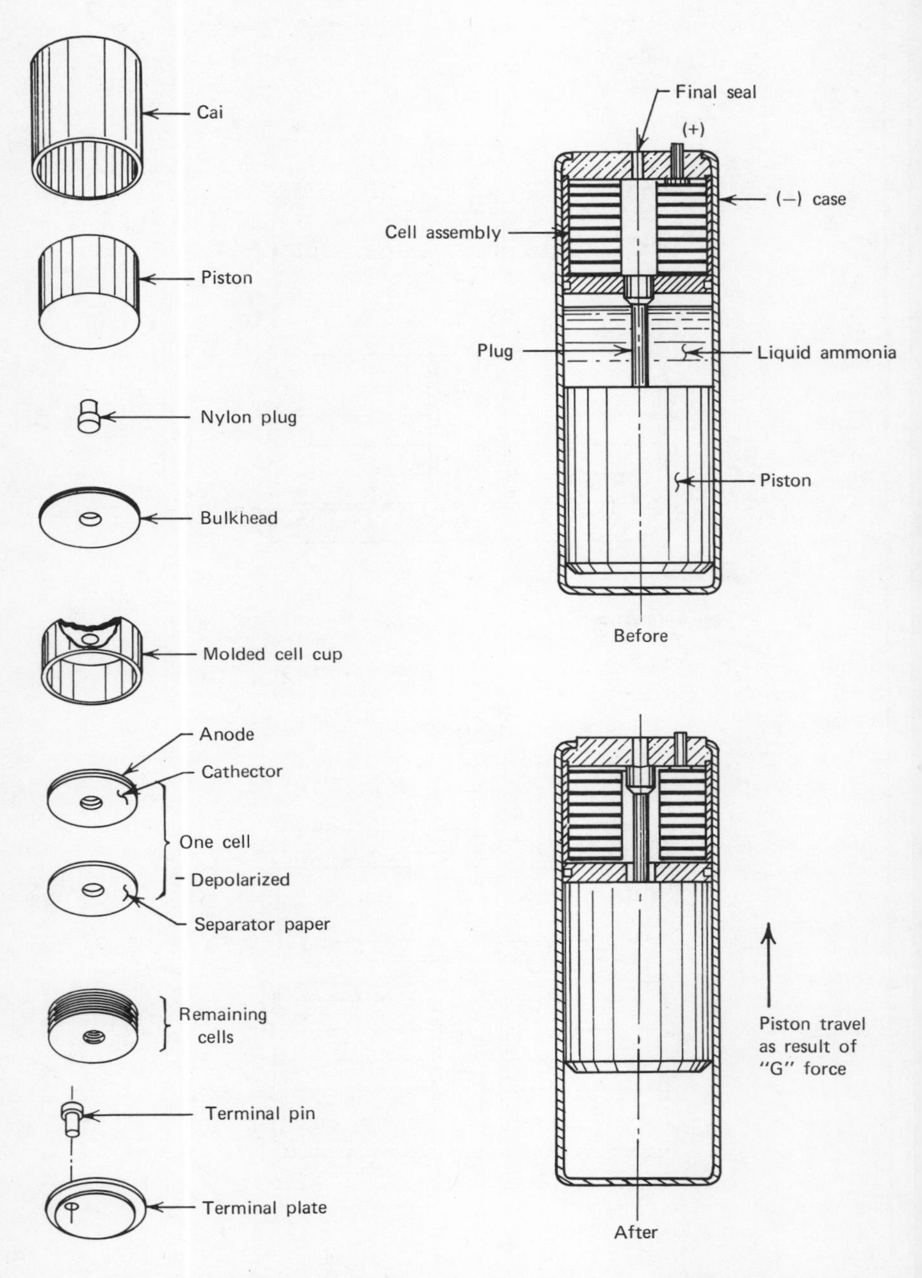

Figure 10.11 Small ammonia battery activated by shock; cross section and exploded views.

extended to 1000 hr. Complete reserve-activated designs as small in volume as 1 cm^3 may soon be produced in large quantities. Primary, nonreserve models may become feasible, as may fuel cells that will operate at low temperature, in the range 200–300°K.

Organic solvent cells are being developed rapidly, particularly with propylene carbonate. Many companies are on the threshold of having cells which will deliver more than 1600 J/g (200 ω-hr/lb), at low rate. But high rate and low temperature [to 233°K (−40°F)] performance will also be developed, perhaps exceeding that of silver-zinc batteries in the 5–30 min military application range.

A principal difference in future prospects for nonaqueous cells compared to aqueous is that there are fewer limitations on the new combinations of materials that can be tried. Nonaqueous cells can be anticipated which will overcome major barriers that have stalled progress on aqueous systems—energy density surpassing 800 J/g (100 w-hr/lb), shelf life exceeding 5 years without serious degradation, and good performance in the low temperature range [220–280°K (−65 — +40°F)]. In the aqueous field, it is difficult to foresee more than minor improvements in existing cells. It seems almost certain that within a decade, nonaqueous batteries will outperform in energy density silver-zinc batteries by a factor of 2 or 3 and lead-acid batteries by a factor of 5 or 10, and they have additional advantages in meeting a greater variety of environmental operating conditions as well.

The words of Brenner (7) are still timely. In a memorial address on the subject of organic solvent electrolysis, he suggested that battery engineers might profit by a broader knowledge of research in related fields of electrochemistry, particularly with respect to nonaqueous solvents:

> The electrolysis of nonaqueous media is probably the most neglected field of inquiry in both electrochemistry and the chemistry of nonaqueous solutions. At the present time most of the knowledge of electrolysis has been obtained from aqueous solutions, and many of the phenomena observed for water may be specific for that solvent. The study of the electrolysis of a large number of other solvents will lead to generalizations that cannot be obtained from a study of a single solvent....

REFERENCES

1. Adlhart, O., High Rate Ammonia Systems with Magnesium Anodes, Proc. 14th Annual Power Sources Conference, Army Signal Laboratory, Fort Monmouth, N. J. (1960).
2. Almerini, A. L., Progress Report on the Hot Dry Pressed Silver Chloride Cathode for Liquid Ammonia Systems. In Supplement to 6th Symposium on Ammonia Batteries, NOLC Report 597, Naval Ordnance Laboratory, Corona, Calif. AD-434,022 (30–31 Jan 1964).

3. Arcand, C. M., and Tudor, J. R., Nonaqueous Solvent Electrochemical Systems, Proc. 11th Annual Battery Research and Development Conference, Army-Signal Laboratory, Fort Monmouth, N. J. (1957).
4. Audrieth, L. F., and Kleinberg, J., *Nonaqueous Solvents*, John Wiley and Sons, New York (1953).
5. Bauman, H. F., Development of Lithium Cupric Fluoride Batteries, Proc. 20th Annual Power Sources Conference, Army Signal Laboratory, Fort Monmouth, N.J. (1966).
6. Braeuer, K. H, Organic Electrolyte High Energy Density Batteries Status Report, Proc. 20th Annual Power Sources Conference, Army Electronics Command, Fort Monmouth, N. J. (1966).
7. Brenner, A., "Electrolysis of Organic Solvents with Reference to the Electrodeposition of Metals," *J. Electrochem. Soc.*, **106**, 148 (1959).
8. Cady, H. P., and Elsey, H. M., *J. Chem. Educ.*, **5**, 1425, (1928).
9. Conway, B. E., *Electrochemical Data.*, Elsevier Publ. Co., New York (1952), quoting Pleskov, V. A., *Acta Physicochim USSR*, **5**, 509 (1936).
10. Darlington, K., and Kuwana, T., Electrical and ESR Studies in Lithium Nitrate Ammoniate Solutions, Ext. Abs. EL-ORG. Div. Vol. 2, Electrochem. Soc. Spring Meeting, San Francisco (1965).
11. Emeleus, H. J., and Anderson, J. S., *Modern Aspects of Inorganic Chemistry*, Van Nostrand, New York (1954).
12. Evans, G. E., Boron Trifluoride Batteries, Proc. 11th Annual Battery Research and Development Conference, Army Signal Laboratory, Fort Monmouth, N. J. (1957).
12a. Foote, H. W., and Hunter, M. A., "Equilibrium in System NH_3-NH_4 CNS," *J. Am. Chem. Soc.*, **42**, 69–78 (1920).
13. Franklin, E. C., *Z. Physik Chem.*, **69**, 272 (1909).
14. Id., The Nitrogen System of Compounds, *Am. Chem. Soc. Monograph 68*, Reinhold, New York (1935).
15. Freund, J. M., Bryant, V. M., Jr., and Schliff, J. B., Final Report on Basic Research Investigations for Ammonia Vapor Activated Batteries, Eastman Kodak, Army-Signal Laboratory, Fort Monmouth, N. J., Contract DA-36-039 SC-74902, AD-235,814 (1959).
16. Gleason, H. S., Freund, J. M., Minnick, L. J., and Meyers, W. F., "Ammonia Vapor Activated Batteries," *J. Electrochem. Soc.*, **106**, 157 (1959).
17. Hale, T., Ammonia Vapor Activated Batteries, Proc. 12th Annual Battery Research and Development Conference, Army Signal Laboratory, Fort Monmouth, N. J. (1958).
18. Harris, W. S., The Electroreduction of Nitrobenzenes in Liquid Ammonia Solutions, 6th Symposium on Ammonia Batteries; Naval Ordnance Laboratory, Corona, Calif., NOLC Report 597, AD-433,973 (30–31 Jan. 1964).
19. Harris, W. S., and Matson, G. B., The Electroreduction of Nitrobenzenes in Acid Liquid Ammonia Solutions, NAVWEPS Report 7241, Naval Ordnance Laboratory, Corona, Calif., AD-293,446 (1 Dec. 1962).
20. Heise, G. W., Schumacher, E. A., and Cahoon, N. C., "A Heavy-Duty Chlorine-Depolarized Cell," *J. Electrochem. Soc.*, **94**, 99 (1948).
21. Heuberger, and Botter, *Bibliography of Ammonia Solution Chemistry*, No. 4, CEA, Centre D'Etudes Nucleaires, Saclay, France (1959).
22. Horowitz, C., et al., Anhydrous Electric Battery, U.S. Pat. 3,098,770 (1963).
23. Jackson, G. W., and Dereska, J. S., "An Electrode Mechanism for the Electrochemical Reduction of M-Dinitrobenzene," *J. Electrochem. Soc.*, **112**, 1218, (1965).
24. Jander, G., *Die Chemie in wasserähnlichen Lösungsmitteln*, Springer, Berlin (1949).
25. Id., *Chemistry in Anhydrous Liquid Ammonia*, Vol. I, Part 1, *Inorganic and General Chemistry in Liquid Ammonia*, Interscience, New York, London; Friedr. Vieweg & Sohn, Braunschweig (1966).

26. Jolly, W. L., "Interpreting Liquid Ammonia Chemistry with Thermodynamics," *J. Chem. Educ.*, **33**, 512 (1956).
27. Id., The Structure and Kinetics of Metal-Ammonia Solutions, in *Solvated Electron*, Advances in Chemistry Series 50, R. F. Gould, Ed., Am. Chem. Soc., Washington D.C. (1965).
28. Knight, M. A., Private communication (1966).
29. Konecny, J., A Survey of the Potential of Sulfur Dioxide Reserve Battery Activation, Navord Report 4639, Naval Ordnance Laboratory, Corona, Calif. (1958).
30. Kuppinger, R. E., and Eisenberg, M., Nonaqueous Organic Electrolyte Batteries for Wide Temperature Range Operation, Ext. Abs. Batt. Div. Vol. 11, Electrochem. Soc. Fall Meeting, Philadelphia (Oct. 1966).
31. Kuwana, T., Approaches to the Elucidation of Organic Electrode Reactions, 6th Symposium on Ammonia Batteries, NOLC Report 597, Naval Ordnance Laboratory, Corona, Calif., AD-433,973 (30–31 Jan. 1964).
32. Langworthy, E. M., Randolph, C. L., and Lawson, H. E., Chlorine-Depolarized Battery, Proc. 11th Annual Battery Research and Development Conference, Army-Signal Laboratory, Fort Monmouth, N. J. (1957).
33. Meredith, R. E., Electrolytic Conductivity, *Encyclopedia of Electrochemistry*, C. A. Hampel, Ed., Reinhold, New York (1964).
34. Minnick, L. J., Sulphur and Sulphur Compounds in Ammonia Batteries, Proc. 14th Annual Power Sources Conference, Army Signal Laboratory, Fort Monmouth, N. J. (1960).
35. Id., Ammonia Batteries, Part 1, Reserve Liquid Ammonia Fuze Battery, Proc. 17th Annual Power Sources Conferences, Army-Signal Laboratory, Fort Monmouth, N. J. (1963).
36. Minnick, L. J., and Presgrave, C., Electric Current Producing Cell and Method of Producing Current Using the Same, U.S. Pat. 2,863,933 (1958).
37. Mott-Smith, L. M., U.S. Pat. 2,921,974 (1960).
38. Mount, T. L., Cathodes for Liquid Ammonia Batteries, U.S. Pat. 3,082,284 (1963).
38a. Murphy, J. J., New Primary Battery Systems, 3rd Biennial Aerospace Power Systems Conference, Army Electronic Laboratories, Fort Monmouth, N. J. (1964).
39. Pray, A. R., Nonaqueous Chemistry, *Comprehensive Inorganic Chemistry*, Vol. 5, Sneed, M. C., and Brasted, R. C., Eds., Van Nostrand, Princeton, N. J. (1956).
40. Ritchey, H. W., and Hunt, H., *J. Phys. Chem.*, **43**, 407, (1939).
41. Rosztoczy, F. E., and Tobias, C. W., "Half-Cell Potentials of the Ca/Ca2+ Couple in Thiocyanate Liquid Ammonia Solutions," *Electrochim. Acta*, **11**, 857 (1966).
42. Schaer, M. J., A Liquid Sulfur Dioxide Battery, Dissertation, Oregon State University, Corvallis (1965).
43. Schaschl, E., and McDonald, H. J., "A Low Temperature Liquid Sulfur Dioxide Cell," *Trans. Electrochem. Soc.*, **94**, 299 (1948).
44. Schwartz, H. J., et al., Batteries and Fuel Cells, Space Power Systems Advanced Technology Conference, NASA-Lewis Research Center, Cleveland, Ohio, NASA Report SP-131 (1966).
45. Scribner, K. R., Ammonia Systems and the AVA Battery, Proc. 13th Annual Power Sources Conference, Army-Signal Laboratory, Fort Monmouth, N. J. (1959).
45a. Simons, J. H., Hydrogen Fluoride, *Fluorine Chemistry*, Vol. I, Academic Press, New York, pp. 225–259 (1950).
46. Smith, H. R., and Tierney, B. C., Ammonia Batteries, Proc. 19th Annual Power Sources Conference, Army-Signal Laboratory, Fort Monmouth, N. J. (1965).
47. Sneed, M. C., and Brasted, R. C., Eds., *Comprehensive Inorganic Chemistry*, Vol. 5 (including Sisler, H. H., Nitrogen, Phosphorus, Arsenic, Antimony, and Bismuth, and Pray, A. R., Nonaqueous Chemistry), Van Nostrand, Princeton, N. J. (1956).

48. Solomon, F., A − 100°F Nonaqueous Battery, Proc. 12th Battery Research and Development Conference, Army-Signal Laboratory, Fort Monmouth, N. J. (1958).
49. Spindler, W. C., Ammonia Batteries, Proc. of Advances in Battery Technology Symposium, Electrochem. Soc., AD-624, 768 (4 Dec. 1965).
50. Spindler, W. C., and Pritchard, R. J., Liquid Ammonia Reserve Battery for Navy Missile Fuzes, *Nava. Research Reviews,* 17, 8 (March 1964).
51. Tomassi, D., *Traité des Piles Électriques,* Paris (1889).
52. Vinal, G. W., *Primary Batteries,* John Wiley and Sons, New York (1950).
53. Waddington, T. C., Ed., *Nonaqueous Solvent Systems,* Academic Press, New York (1965).
54. Wong, W., Selected Physical Properties of Solutions of Calcium in Liquid Ammonia, Dissertation, Univ. of California, Berkeley (1966).
55. Zipp, A. P., Spectral Properties and Conductance of Solutions of Sulfur in Liquid Ammonia, Dissertation, University of Pennsylvania (1964).

11

Semidry and Solid-Electrolyte Batteries

John N. Mrgudich and Demetrios V. Louzos

Spectacular progress has been made by the electronics industry in decreasing the size of electronic components and circuits. Modern advanced integrated microcircuit technology can pack hundreds of switching circuits into less than 0.05 cm^2 of silicon-chip area, and circuit boards which can be threaded through the eye of a needle are the subject of industrial exhibits and advertisements.

In many cases this circuit miniaturization has resulted in tremendous reductions in circuit power requirements. A 1968 RCA announcement (1), for example, describes metal-oxide semiconductors operating on nanowatts in the quiescent state, using greater power only during the instant of information switching. Circuits and applications requiring only a few microwatts of power at microampere drains are becoming rather commonplace.

Large batteries can, of course, be used to power these smaller devices and, indeed, have the advantages of lower over-all cost and increased reliability. However, the use of unnecessarily large batteries can be as incongruous as installation of 20-gallon gasoline tanks on motorcycles. Special occasions may arise where a large reservoir of stored energy is needed, but generally the power source should be appropriately matched to the energy requirements of the operating device. To this extent it seems inevitable that these emerging smaller, more sensitive electronic circuits will call for equivalent miniaturization of the batteries required to power them.

MINIATURIZATION TRENDS: COMMERCIALLY AVAILABLE BATTERIES

The primary battery industry has been responsive to the trend toward smaller and more efficient electronic components. In the Leclanché cell area the smallest cell in 1940 was the Burgess NS (2.3 cm^3). This was followed by the Winchester flat Leclanché F15 cell (0.63 cm^3) during the early 1950s. The late 1950s saw introduction of modified Leclanché cells such as the Eveready 201 "wrist-watch" battery (0.31 cm^3, including packaging). Consisting of a unitized cell encapsulated in a leakproof metal container, this 201 exhibits 95 percent current maintenance after 12 months' storage at 21°C. At low current drains (0.03–0.30 ma/cm^2) its output is at a relatively high level of about 0.3 w-hr/cm^3. These values are based upon completely packaged cells.

In the conventional Ruben-Mallory (RM) alkaline zinc : mercuric oxide system, we find an even sharper miniaturization trend. The smallest RM cell commercially available in 1952 had a volume of 1.15 cm^3. The RM-312 cell of the late 1950s had a volume of 0.14 cm^3. Its shelf life was about 2 years at 21°C and its output was a respectable 0.28 w-hr/cm^3 at a relatively high drain of 13 ma/cm^2. The RM-212 cell of 1966–67, released as the "world's smallest primary battery," has a volume of only 0.085 cm^3. Its output is about 0.23 w-hr/cm^3 at an average drain of 9.5 ma/cm^3.

In the rechargeable sealed nickel-cadmium system, the smallest cell now available is the Eveready B20. It has a displacement volume of 0.49 cm^3. A rough estimation of output (continuous drain to 0.9 V) yielded a value of 0.049 w-hr/cm^3 per cycle at an initial drain of 5.2 ma/cm^2. This can be repeated for several hundred cycles. The rechargeable feature compensates for the relatively low, single-discharge, specific output value; it also compensates for a lower shelf life in the charged state than that of conventional primary cells. The progress in Leclanché, mercuric oxide, and sealed Ni-Cd batteries indicates that the battery industry can probably meet any normal future requirement for miniature cells having volumes in the range of 0.05–0.1 cm^3.

We began this chapter with electronic circuits that would pass through the eye of a needle; the following sections will describe possible development avenues that might lead to cells in the volume range of 0.001–0.05 cm^3, or some 200–1000 cells/cm^3.

BASIC PROBLEMS ASSOCIATED WITH BATTERY MINIATURIZATION

CHARGE RETENTION: SHELF LIFE

All batteries, being intrinsically unstable, are susceptible to deterioration from the moment they are assembled. Any of a large number of design, manufacturing, or storage errors can result in poor resistance to such shelf deterioration. Among these we find factors such as wasteful anode corrosion, side reactions within the cathodic mass or between cathode and electrolyte, side reactions between the separator and the cathode and/or anode, internal electronic shorting from anode to cathode, inadequate sealing, electrolyte leakage between adjacent cells of multiple-cell batteries, and external electronic leakage across nominally insulative sealing rings or encapsulating materials. All of these problems become increasingly serious as manufacturing tolerances are tightened to meet the demands of miniaturization. In addition, since these tiny cells, by definition, contain only small amounts of active materials, a small absolute shelf loss, unnoticeable in large cells, can cut sharply into an output capacity which even ideally is already small. Thus good shelf life is an important criterion for acceptable miniaturization-potentiality of any primary battery system under consideration.

Charge delivery: internal losses on discharge

Although good charge retention is a necessary condition for miniaturization, it alone is not sufficient; the battery system under consideration must also be capable of efficient delivery of this retained charge when such delivery is called for by the external circuit. In a fundamental sense, high efficiency of charge delivery can be realized only if internal resistance and polarization losses within the discharging battery are small. Both of these losses cause decreased closed-circuit voltage of the discharging batteries and, to this extent, produce a decrease in efficiency of utilization of originally incorporated active material. Small cells with their small amount of starting active material can ill afford this decrease in efficiency.

An important factor determining the magnitude of these losses is the freedom with which ions can move within the cell, e.g., the ease in which a positive, anodically generated ion can move from the anode through the electrolyte toward and into the cathodic material. In conventional cells, ions move through liquid electrolytes, therefore the design engineer must ensure the presence of ample liquid conduction paths, including paths within the cathode. Another factor is the ease with which electrons from the external circuit can enter into and become available for reduction of the cathodic mass. Since solutions are insulative to electronic flow, the cathode mass cannot be completely inundated with liquids; it must possess "dry" access paths. This is often accomplished by incorporation of an inert material such as electronically conductive carbon into the cathodic mass.

Cost of materials

Material costs per cell obviously decrease as cell sizes become smaller and smaller. To this extent the design engineer may turn, if necessary, to materials such as silver, gold, palladium, and even platinum. One gram of platinum in the form of an evaporated film 2000 Å thick will cover some 2400 cm^2 of surface or (if platinum costs \$ 4.00/g) \$ 0.17/100 cells each of 1 cm^2 cross-sectional area.

Feasibility of manufacture

Since hand-assembly of the tiny cells is impractical, ease of mechanization of the assembly process becomes an important factor determining the miniaturization potential of any system. Mechanization presents no serious problems in the handling of *solid* components of the cell system. Many techniques (evaporated or sputtered films, chemical deposits, rolled sheets, painted or sprayed layers, extruded sections, compressed pellets, etc.) are available to the process engineer to position small amounts of solid materials within the tiny cells.

More serious problems are associated with control of *liquid* electrolyte and mix-wetting solutions and, especially, of proper containment of these liquid phases during subsequent storage and ultimate discharge of the assembled battery. An excessive amount (or inadequate distribution of a proper amount) of solution leads to difficulties involving intercell leakage and poor shelf life. A shortage of added solution leads to increased internal resistance and higher polarization losses. Loss of moisture during storage (i.e., poor sealing) is recognized as one of the major factors leading to poor shelf life.

These difficulties of electrolyte control and containment become progressively more pronounced as cell size decreases and, to a major degree, determine the extent to which a given battery system can be miniaturized without prohibitive sacrifice of either charge retention or efficiency of delivery of retained charge. The problem is so important that we can classify known methods for manufacturing miniature batteries on the basis of how each method controls and contains the ionically conductive electrolytes and mix wetters. Thus conventional Leclanché and mercuric oxide batteries would be classified as systems using "partially immobilized electrolyte." In the following sections we shall discuss battery systems using only a trace amount of electrolyte (DeLuc, Zamboni and other piles) and a strongly immobilized electrolyte (the "wax" battery), and batteries using electrolytes in which either the cation or anion are immobilized by virtue of lattice forces (ion-exchange batteries, solid-electrolyte batteries).

BATTERIES WITH TRACE ELECTROLYTE

One approach to alleviation of the manufacturing problems of liquid control and containment in very small cells involves a drastic reduction in the amount of liquid used. Such reduction yields adequate shelf life but usually increases polarization losses to a prohibitive degree.

THE DELUC COLUMN

This example (2) of a possible method of making batteries in the future is concerned with a system described over 150 years ago consisting of a large number of disks of zinc and gilt Dutch paper (paper coated with copper) alternately placed upon each other. Operation apparently depended upon the presence of trace moisture in the paper since, "a column composed of paper disks thoroughly dried had little power." The presence of too much moisture, however, resulted in "fatal deterioration." With proper moisture content, shelf life was spectacularly good: Mottelay (3) reported that "an apparatus comprising 20,000 groups of silver, zinc and double disks of writing paper . . . in the Clarendon Laboratory at Oxford rang ten small bells continuously for over forty years."

THE ZAMBONI PILE

Zamboni's battery (4), described in 1812, is considered an improvement over deLuc's in that Zamboni discarded the zinc plates and used "only discs of paper having one side tinned and the other coated with a thin layer of black oxide of manganese pulverized in a mixture of flour and milk." There was no deliberate addition of electrolyte: operation was apparently dependent upon the presence of residual starting moisture or that absorbed from air.

The outstanding feature of the deLuc and Zamboni concepts is their simplicity of assembly, which may involve merely the preparation of large, thin sheets of anode, separator, cathode, and intercell connector (possibly a single 4-ply sheet) followed by a punching or cutting operation and final mechanical stacking and endcapping of substantially dry disks into appropriate insulating tubes.

Unfortunately, internal resistance and polarization losses in such electrolyte-starved batteries are extremely high and result in infinitesimally small current delivery even on dead short. Because of this shortcoming, interest has remained relatively dormant for over a century.

THE ELLIOTT PILE

Interest was revived during the early part of World War II when development of a new infrared image converter tube precipitated a need by the British Services for a high-voltage (3000 V) portable battery capable of delivering small currents (10^{-9} amp). Elliott (5) in 1948 described the manufacturing details—and difficulties—of an MnO_2-gelatin/$ZnCl_2$ solution/paper/Zn-foil pile developed for this application. The average fresh voltage was 0.9 V/cell, stabilizing to 0.7 V after a few months. The average cell volume was about 0.2 cm^3 (1.9 cm diameter, 0.007 cm thick).

Internal resistance and polarization losses on discharge were quite high. For example, Howard (6) showed that a 575-V Elliott pile (comprised of 800 cells) exhibited an initial voltage drop of 100 V when placed across a 10^{11}-ohm resistance. The internal resistance r is given by

$$r = (E - V) \times \frac{R}{V} \tag{1}$$

where E is the open-circuit voltage and V is the closed-circuit voltage immediately after connection across a resistance R. By applying this expression, the 100-V drop of the Elliott pile is found to correspond to an internal resistance of 2.1×10^{10} ohms/800-cell battery, or 2.6×10^7 ohms/cm^2 cross section of the individual cell. This high internal resistance can be reduced by increasing the amount of electrolyte incorporated but, as Elliott points out, this "always reduced the shelf-life greatly."

THE HOWARD PILE

A specific effort was made by Howard (6) to improve over-all internal ionic mobility without sacrifice of shelf life. His approach involved use of a moist cathodic mix, gelatin or carboxymethyl cellulose immobilization of electrolyte, incorporated ethylene glycol for low temperature operation, vacuum-evaporation of anode material, and the use of electronically conductive resin-carbon paints for intercell connection. Of various combinations that were tried, the one that best met his requirements was a carbon-MnO_2/$AlCl_3$ in carboxymethyl cellulose gel/paper/evaporated Al/carbon paint system. The voltage was about 1.4 V/cell. The cell volume was 0.56 cm^3 (1.9 cm diameter, 0.024 cm thick).

The initial voltage drop of a 138-V battery (100 cells) across a 9×10^7 ohm resistance was 17 V, corresponding to an internal resistance of 1.26×10^7 ohms/battery, 1.26×10^5 ohms/cell, or roughly 4.5×10^5 ohms/cm^2 cross section of the individual cell. This specific resistance, which is about 1/20 that of the Elliott cell, illustrates the improvement that can be made.

Despite this improvement in internal resistance, much remained to be done to improve output capacity. For example, on the basis of discharge curves given by Howard, it is possible to show that the watt-hour output (continuous drain to 0.9 V/cell) is only 0.0036 w-hr/cm^3 at a low current drain of 0.0024 ma/cm^2. By way of comparison, the 0.085 cm^3 RM-212 cell delivers 0.23 w-hr/cm^3 at 9.5 ma/cm^2.

Other recent types

Interest in the starved-electrolyte, Zamboni-type approach continues, as shown by a series of relatively recent patents. Victoreen (7) in 1954 proposed a battery of the type PbO_2/electrolyte/Pb in which the small electrolyte disks are of glycerine-plasticized gelatin containing trace amounts of an ionizable material (such as NH_4Cl, KOH, or H_2SO_4) or of paper impregnated with a thermosetting plastic and an electrolyte. The assembled stack is subjected to heat and endwise pressure and the peripheral edges are then ground clean, e.g., in a centerless grinder. Battery dimensions of 0.3–0.6 cm diameter and 2.5–5.0 cm length are reported. In a proposed modification, the electrolyte disks are completely dry, being conditioned for use by exposure to a moist atmosphere. No discharge data are reported, but it seems obvious that polarization losses will limit the application to extremely low drains.

Reiner (8) in 1955 proposed the use of slightly hydroscopic tissue paper in cells of the type Zn-dust/paper/MnO_2-powder, where an adhesive previously applied to the paper holds the elements together. The individual cells, some 0.05 cm thick, are stacked to yield 2000–4000 V batteries to operate a pendulum clock.

Bjorksten (9) in 1954 described a rather novel trace-electrolyte "printed" battery. In this an "ink" (composed of a magnetic powder such as Fe, which is dispersed in a vehicle of linseed oil, a volatile solvent, a humectant, and some appropriate electrolyte solution) is printed on paper or a thin plastic film and passed through a magnetic field so that the metallic particles make electronic contact with each other while oriented parallel to the sheet. After drying, a second "ink" containing a different magnetic powder such as Ni is similarly dispersed, printed, oriented, and dried onto the first layer. The pair of layers constitute a cell of the type Ni/immobilized electrolyte/Fe. The printing process may be repeated any number of times to build up a thin, lightweight, high-voltage battery. No operational data are given in the patent.

Summary

The obvious advantage of the trace-electrolyte approach to battery miniaturization is manufacturing simplicity. The major problem is a familiar one: although the use of trace amounts of electrolyte leads to ease of handling and good shelf life, electrolyte starvation results in abnormally high internal resistance and large polarization losses during discharge. These losses can be decreased by increasing the amount of electrolyte used but at a noticeable sacrifice of shelf life.

BATTERIES WITH STRONGLY IMMOBILIZED ELECTROLYTE

Another approach to the fundamental problem of improved charge retention in miniature batteries through better control and containment of electrolyte involves stringent physical immobilization of the electrolyte. As we shall see, such immobilization does improve shelf life, but at the cost of increased polarization losses during discharge.

Immobilization in a cooling wax

A good example of extreme electrolyte immobilization is found in the "wax" battery described by Wood (10). Assembly begins by melting a quantity of partially polymerized polyethylene-glycol wax and adding 5 percent by weight of solid $ZnCl_2$. While still molten, a portion of this $ZnCl_2$-wax mixture is painted onto one side of a sheet of filter paper. Manganese dioxide is added (50 percent by weight) to the other portion of the $ZnCl_2$-wax mixture and, after mixing, a thin layer of this is painted onto the other side of the paper. Cell assembly is completed by placing a sheet of Zn foil on the $ZnCl_2$-wax side and a conductive vinyl film on the MnO_2-wax side. The resultant multiplex sheet is then hot-pressed to remove excess wax. After trimming to remove roughened edges, the sheet is run through a tungsten-carbide punch press and the resulting circular disks are stack-assembled into a battery. The final block is then sealed against moisture absorption by several coats of lacquer or neoprene-base adhesive applied while the stack is in a pressure jig.

Wood prepared 25-cell batteries using 1.27 and 0.635 cm diameter disks punched from a multiplex sheet approximately 0.034 cm thick (the smallest cell volume reported was 0.011 cm^3). Shelf life was reasonably good in that most batteries showed only a 5 percent drop in open circuit voltage after 3–5 years storage at room-temperature in a silica-gel desiccator. The internal resistance, however, was quite high, about 4×10^7 ohms/cm^2 cell cross section. Polarization losses also seemed high even on very light discharge drains. Attempts to decrease the high internal resistance by use of more "tacky" wax resulted in erratic shelf life due primarily to attack on the final lacquer seal and subsequent intercell electrolyte leakage.

Immobilization in a polymerizing matrix

Certain polymers containing polar groups (e.g., hydroxyl, carboxyl, acrylyl, carbonyl, sulfonyl) are capable of polymerizing in the presence of aqueous solutions of dissolved salts, thereby physically immobilizing the aqueous solution. Furthermore, this polymerization can take place in the presence of suspended solids and various plasticizing agents, hence permitting wide flexibility of polymer composition, which results in an opportunity to tailor physical properties.

Kirkwood and West (11) described miniature Daniell-type Cu/$CuCl_2$/ $ZnCl_2$/Zn cells in which Cu and Zn sheets were coated, respectively, with $CuCl_2$ and $ZnCl_2$ solutions containing polyvinyl alcohol and glycerine. The dried films were only 0.0025 cm thick. The coated metal sheets were laminated face to face at 70°C under a pressure of 35 kg/cm^2 (500 psi) to yield a 1.15-V cell. In another experiment involving large (48 cm^2) electrodes of $CuSO_4$ on Cu and $ZnCl_2$ on Zn it was found that the initial open circuit voltage of 0.81 V dropped to 0.80 V when the cell was placed across an 8 megohm resistor. This corresponds to a high internal resistance of some 5×10^6 ohms/cm^2 of cell cross section. Furthermore, data were presented indicating significant polarization at current drains as low as 6×10^{-5} ma/cm^2.

Another example of electrolyte immobilization through use of a polymerizing matrix is given in a series of three interrelated patents. The first, the Corren patent (12), describes the *electrolyte-separator* sheet as a 0.007 cm fiberglass cloth built up to a final thickness of 0.02 cm by successive coating and baking (65°C) with a solution of polyvinyl alcohol, dimethanol urea, NH_4Cl, and water followed by a final 1-hr bake at 120°C. The second patent, issued to Zablocki (13), describes a *battery* using a 0.0025 cm depolarizer coating of MnO_2, carbon black, polyvinyl alcohol, and water; this was baked at 120°C on the Corren electrolyte sheet. The dry duplex sheet was soaked for 24 hr in a solution of NH_4Cl, glycerine, and water and then towel-dried before die-cutting into disks. The Louis patent (14) describes the *molded plastic parts* designed particularly to minimize electrolyte leakage when these disks are used in the assembly of a battery of the type: Zn/electrolyte sheet/MnO_2, C, polymer/conductive plastic.

Combining the patent information with data contained in a final report under Signal Corps Contract DA-36-039-sc-64463 dated September 15, 1957, it is concluded that batteries can be made from rather small cells (0.1 cm^3) with fairly high voltage (1.2 with Zn and 1.7 with Mg) and fairly low internal resistance (estimated at about 1000 ohms/cm^2 of cell electrode area). Shelf life of the order of 2–3 years is anticipated. Further work is required to confirm and crystallize these conclusions.

SUMMARY

The idea of incorporating anode, cathode, and electrolyte into thin, easily handled wax or plastic films whose physical and electrical properties can be optimally balanced to meet the peculiar and complex requirements of miniature batteries is much too promising to be dismissed lightly. The approach remains as a potentially fruitful development area.

BATTERIES WITH LATTICE-IMMOBILIZED IONS

Ion-exchange membrane batteries

The basic principle of ion exchange was announced over 100 years ago by Thompson (15). He reported that percolating a solution of $(NH_4)_2SO_4$ through soil resulted in an effluent rich in Ca^{+2} ions but containing no NH_4^+ ions. It was immediately apparent that NH_4^+ ions had been absorbed by the soil through "exchange" for Ca^{+2} ions. This exchange phenomenon implies that cations in soil are more mobile than anions and leads to the interesting thought that ion-exchange materials might be used as battery electrolytes in which one type of ion is relatively mobile while the other is immobilized in the ion-exchange lattice. Our interest is sharpened in the hope that such selective ionic immobilization might simplify the electrolyte control and containment problem encountered in miniature batteries, without serious interference with internal ionic movement of the ions taking part in the cell reaction.

We cannot here enter deeply into modern ion-exchange technology. It is an active field, with excellent reference texts available: F. C. Nachod and J. Shubert (16); J. A. Kitchener (17); R. Kunin (18); J. E. Salman and D. K. Hale (19); and R. Kunin (20). Much of the momentum for this activity is due to an increasing ability to synthesize a large variety of membranes with highly specific exchange properties. Generally, all such materials are high molecular weight polymers containing certain ionic groups as integral and attached parts of the polymer structure. A cation exchange resin is one in which the attached and immobilized groups are acidic (phenolic, sulfonic, carboxylic, etc.) with an equivalent number of loosely held, relatively mobile cations (H^+, Zn^{+2}, Ag^+, etc.). Conversely, an anion exchange resin contains attached nitrogen groups with an equivalent number of relatively mobile anions such as OH^-, Cl^-, etc. Membranes can be made by casting, by incorporating a finely-ground resin into an inert plastic matrix, or by impregnation of paper or other supporting films with polymeric solution followed by resin polymerization. A versatile feature of such membranes is that the mobile ions can be easily replaced; e.g., the H^+ ion in a cationic membrane can be replaced with metallic ions such as Cu^{+2}, Ag^+, or Zn^{+2} merely by soaking the membrane in a strong solution of the appropriate metallic salt. The procedure can be reversed by soaking the membrane in an acid solution. The membranes can be made quite thin (about 0.006 cm) with good stability and insolubility. Moreover, they exhibit remarkably high ionic conductivities with the added feature that, under appropriate conditions, this conductivity can be due solely to the transport of only one ionic species.

Such membranes have been used as the separating electrolyte in concentration cells by Meyer (21) and in Daniell cells by Juda and McRae (22). Ion-exchange materials have also been used as cell depolarizers; Robinson (23), for example, has used a permanganate anion-exchange resin as the depolarizer in an analogue of the Leclanché cell.

Pauli (24) has described a multiplicity of cell arrays using various types of ion-exchange membranes. The versatility of such constructions is shown in Fig. 11.1. In Fig. 11.1*a* the current-generating mechanism involves solution of the Zn, releasing electrons to the external load and Zn^{+2} into the Zn^{+2} ion-exchange membrane. The Zn^{+2} ions so released displace Zn^{+2} ions from the Zn^{+2} ion-exchange membrane into the Ag^{+} ion-exchange membrane; in turn displacing Ag^{+} ions which plate out onto the Ag cathode with simultaneous absorption of the electrons from the external circuit. The capacity of the cell is determined by the number of Ag^{+} ions in the Ag^{+} ion-exchange membrane. The shelf life is determined by the thickness of the Zn^{+2} ion-exchange membrane; the battery discharges internally when Ag^{+} ions from the Ag^{+} ion-exchange membrane diffuse through the Zn^{+2} ion-exchange membrane and react directly with the Zn plate. (Pauli suggests that this diffusion-dependent shelf life can be improved by use of bone-dry membranes followed by water-activation immediately prior to battery use.) Figure 11.1*b* illustrates a

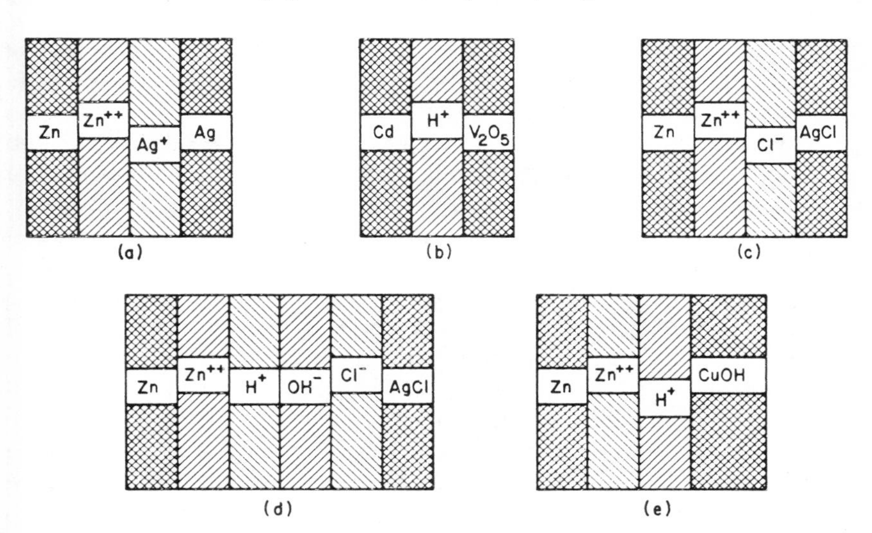

Figure 11.1 Schematic drawings of various types of ion-exchange battery systems (after Pauli).

V_2O_5-depolarized ion-exchange membrane cell. Solution of Cd releases Cd^{+2} ions, which displace H^+ from the H^+ ion-exchange membrane, these being then adsorbed through reduction of the V_2O_5. Figure 11.1*c* illustrates an array in which generated Zn^{+2} ions combine with Cl^- ions of the Cl^- ion-exchange membrane, generating $ZnCl_2$ at the interface, the consumed Cl^- being replaced by reduction of the AgCl cathode to metallic Ag. Another modification is shown in Fig. 11.1*d*, where generated Zn^{+2} ions replace H^+ which combine with OH^- ions replaced by cathodically generated Cl^- ions. Water is formed at the interface between the H^+ and OH^- membranes. Still another array, Fig. 11.1*e*, shows generated Zn^{+2} ions replacing H^+ ions which combine with OH^- generated by the reduction of cathodic CuOH, water being formed at the interface between the H^+ ion-exchange membrane and the CuOH plate.

In related patents, Schwartz and Frankline (25), working with a cell of the type Mg/H^+ ion-exchange membrane/Ag, used dimethyl formamide instead of water as the wetting agent, adding azobenzene to lower polarization at the silver electrode. An operating temperature range extending from -65 to $+188°C$ was reported. Batteries were activated just prior to use.

Harding (26) described an interesting cell using an electron-exchange material. Such materials, difficult to prepare and usually of low molecular weight, are capable of oxidation by Fe^{+3}, $Fe(CN)_6{}^{+3}$, I_2, etc., and once oxidized may be reduced. Harding pressed a small amount (0.1 g) of electrolyte-moistened electron-exchange material (vinyl hydroquinone polymer) between 2 in^2 Pt electrodes, passed a "charging" current, removed the applied voltage (2.0 V), and noted that the now-charged array would develop 0.1 amp at 1.8 V.

The most extensive work on ion-exchange batteries has been by Grubb (27), who described rechargeable cells of the type Zn/Zn^{+2} ion-exchange membrane/Ag which can be charged to yield a cell of the type Zn/Zn^{+2}/Ag^+/Ag (see Fig. 11.1*a*). The lifetime in the charged state is short (less than 24 hours). The capacity is limited by the maximum number of Ag^+ ions that can be held by the membrane (about 1 equivalent per liter volume of membrane). Grubb also described a Leclanché system comprising Zn/Zn^{+2} ion-exchange membrane/MnO_2. The internal resistance can be quite low, as evidenced by a dead-short flash current of some 3 ma/cm^2 of cell cross section using a 0.007 cm thick membrane.

All of the ion-exchange applications to date require that the membrane be moist since, in the "bone-dry" state, all exhibit high ionic resistance. This requirement subjects all ion-exchange membrane batteries to the problem of long-term containment of moisture during cell storage. Until this difficult problem has been solved, ion-exchange batteries offer little promise from the standpoint of truly miniature primary batteries. However, there is a possibility that solvating agents other than water may be used (notably aliphatic alcohols) and that these will present less formidable containment difficulties.

Solid-electrolyte batteries

It has long been known that many solids are electrically conductive, especially at high temperatures. In some cases, this conductance is due solely to the transport of a single cation (e.g., Ag^+ ions moving through solid AgI) or more than one cation (e.g., Ag^+ ions and Hg^{+2} ions) moving through solid Ag_2HgI_4. In other cases, the conductance may be anionic (e.g., Cl^- ions moving through solid $PbCl_2$ or O^{-2} ions moving through solid ZrO_2 containing 15 mole percent of CaO); it may be mixed cationic and anionic (e.g., K^+ ions and I^- ions through solid KI at high temperatures); it can be mixed electronic-cationic (e.g., Ag^+ ions and electrons through solid Ag_2S at room temperature); or it may be purely electronic at room temperature changing to purely cationic at high temperature (e.g., the cuprous halides). Thus a wide range of choices is available in solids for possible use in batteries.

With respect to miniature batteries, the significant point is that a solid, purely cationic conductor, such as AgI with its relatively mobile Ag^+ ions moving through a fixed I^- ion lattice, is analogous to an ion-exchange resin highly specific for Ag^+ ions with the important difference that no polar (liquid) solvent is required to supply Ag^+ ions. To this extent, solid ionic conductors appear promising from the standpoint of substantial simplification of troublesome problems of electrolyte control during manufacture of miniature batteries and of electrolyte containment during subsequent battery storage. In addition, manufacture of "all-solid" batteries appears amenable to easy mechanization, including the intriguing possibility of an "evaporated-film" battery; such batteries may be capable of pressure-packaging or hard encapsulation to yield extremely rugged assemblies; operating temperature ranges may be extended since we need no longer worry about the drastic cut-off in battery performance associated with freezing or boiling-off of liquid phases essential to conventional battery systems; and such "all-solid" batteries would be truly "leakproof."

Early Solid-Electrolyte Battery Systems. Attempts to use solid ionic conductors specifically as the electrolytes of practical and useful primary batteries are of relatively recent origin. The earliest reference is a patent issued to Ruben (28) in 1935. This describes a cell employing a flat electrode of Cd, Al, or Zn in direct contact with a pellet of compressed Cr_2O_3 powder backed by a sheet of Ni. The cell had a voltage of 1.6 V, but internal resistance and polarization losses on discharge were so high that "a current of only several microamperes was sufficient to reduce the potential to a negligible value." Little interest was engendered by the Ruben disclosure and it remained for Sator (29), Lehovec and Broder (30), and van der Grinten (31) to initiate several serious efforts to develop useful solid-electrolyte batteries.

It was apparent from the start of this early work that the room-temperature resistance of solid electrolyte systems would be quite high. An often-cited reason for this was that at that time (about 1956) even the most conductive solid electrolytes (i.e., AgI and Ag_2HgI_4) had room-temperature specific resistance (at 10^4–10^7 ohm-cm) which were at least four to seven orders of magnitude higher than those of most liquid solutions used as electrolytes in conventional primary batteries. It was realized, however, that this serious shortcoming would lose some of its significance if very thin films of solid electrolyte could be used to separate active anode and cathode materials. For example, a specific electrolyte resistance of 10^4 ohm-cm would reflect as an electrolyte resistance of only 100 ohms/cm^2 of cross section if an electrolyte thickness of 0.1 mm could be used. Furthermore, there was at that time a need for a small, rugged, stable, high-voltage, low-current power source to maintain a charge on a capacitor whose stored charge could be made available for momentary high-current bursts.

Thus several concerns became interested in production-development of solid-electrolyte miniature batteries with the result that by 1959, following a series of patents by Lehovec (32), Broder (33), Bradshaw and Shuttleworth (34), Shorr (35), Hack, Shapiro, and Shorr (36), Louzos (37), and Evans (38), five major battery systems emerged. These were discussed by Shapiro (39) and are summarized in Table 11.1. During the interval from 1957 to 1960, interest was concentrated on development of pilot-plant production techniques, usually under Industrial Preparedness Study contracts, to solve design and production problems. By 1960 most of the concerns involved were in a position to supply solid electrolyte

TABLE 11.1. BASIC CHARACTERISTICS OF EARLY SOLID ELECTROLYTE SYSTEMS

Type	Open-Circuit Voltage	Studied by
$Ag/AgBr/CuBr_2$	0.74	General Electric
$Ni/SnSO_4/PbO_2$	1.2–1.5	P. R. Mallory
$Ag/AgI/V_2O_5$	0.46	Union Carbide
$Ag/AgBr\text{-}Te/CuBr_2$	0.80	Patterson-Moos
$Ag/AgCl/KICl_4$	1.04	Sprague Electric

batteries to the military or civilian markets, or to furnish prototype batteries for test and evaluation by potential users.

This development work confirmed the predicted advantages associated with solid electrolyte batteries. Miniaturization approaching 100 V/cc was indeed possible. Despite this small cell size, shelf life was good at an estimated 10–20 years. Manufacturing techniques were amenable to a high degree of mechanization (see especially the discussion below of Union Carbide system). Some battery systems were operational and apparently unharmed at temperatures approaching 100°C. Packaging could be exceptionally sturdy and rugged.

This development work also brought into bold relief a major disadvantage: all systems exhibited very high internal resistances and high polarization losses during discharge under room-temperature conditions. None of the developed systems could be considered as useful power sources for drains in excess of about 1 $\mu a/cm^2$ of cell cross-sectional area. In the meantime, the original battery-capacitor application disappeared as a large-volume market for such batteries. Thus there seems to be no current need for such early state-of-the-art batteries and accordingly all production facilities have been dismantled.

Despite the dampening effect of this semidemise of early production effort, continuing research (sponsored initially by the U. S. Army Electronics Command as a modest in-house program in its Institute for Exploratory Research, and later supplemented by increased activity on the part of several industrial concerns) has demonstrated several approaches to increasing the room-temperature ionic conductivity of silver iodide, has demonstrated feasibility of rechargeability of solid electrolyte batteries, has developed new solid electrolytes of much greater conductivity than silver iodide, has shown progress toward rechargeable "all-thin-film" batteries whose thickness is measured in microns, and, of possibly greater potential significance, has created a new family of solid ionic devices, sensors and transducers. Before discussing these interesting new developments, it is appropriate that we here describe the salient features of the early (but now dormant) solid electrolyte systems since the lessons learned there will undoubtedly be applicable to efficient exploitation of the newer research findings.

THE GENERAL ELECTRIC $Ag/AgBr/CuBr_2$ SYSTEM. The anode reaction in this system involves electrochemical dissolution of Ag metal into Ag^+ ions and electrons. The cathodically active material is Br_2 vapor generated by dissociation of $CuBr_2$ into CuBr and Br_2. The reaction product is AgBr (the electrolyte), which is generated within the cathodic mass and undoubtedly serves to make this mass exhibit increasing ionic conductivity during discharge.

Construction details are given in patents issued to van der Grinten (40) in 1957 and van der Grinten and Mohler (41) in 1960. As an example of an apparently preferred assembly, the 1960 patent describes a 43-cell, 31-V battery (Fig. 11.2). A thin sheet of 0.0013 cm Ag foil is cleaned and tightly folded on itself. It is then heated in Br_2 vapor to build up a layer of AgBr some 0.0037 cm thick. The folded layer is then opened and the uncoated side painted over with a thin (0.0038 cm) layer of an impervious, electronically conductive material which serves as the intercell connector. Disks of 0.635 cm diameter are punched out of this coated sheet and a Teflon washer some 0.0089 cm thick is centered on the AgBr-coated face of each disk. A dry-powder mixture of $CuBr_2$ and carbon black (the carbon black imparts necessary electronic conductivity to the $CuBr_2$ cathodic mix) is then pressed into the opening defined by the washer opening. The resulting intercell connector/Ag/AgBr/$CuBr_2$ + C cells are then stack-assembled into an insulating cylinder fitted with a bottom-terminal assembly, the top-terminal assembly is added, and the stack is sealed under pressure.

Wagner (42) has given the dimensions and some operational characteristics of a 127-cell, 94-V General Electric battery. These are given in Table 11.2. The internal resistance values were computed by means of Eq. (1) using initial voltage-drop data given by Wagner.

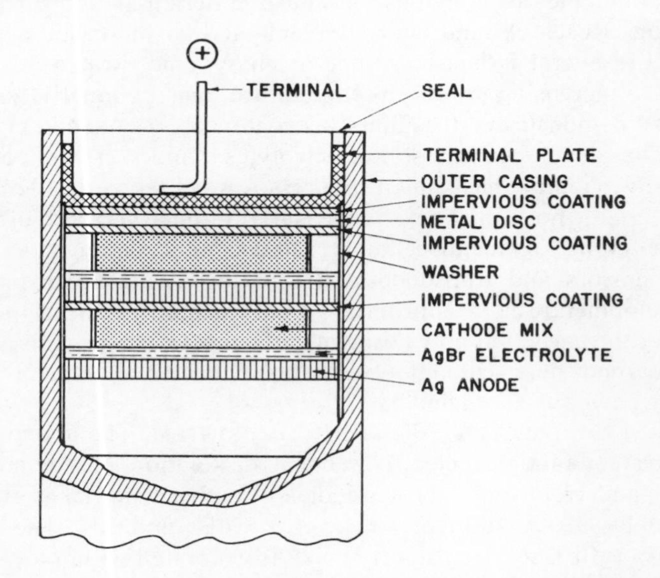

Figure 11.2 Sectional drawing of the General Electric Solid-Electrolyte Ag/AgBr/$CuBr_2$ cell.

TABLE 11.2. CHARACTERISTICS OF EARLY SOLID-ELECTROLYTE SYSTEMS

Characteristic	General Electric	Mallory	Union Carbide	Patterson-Moos	Sprague Electric
Typical battery voltage (V)	94	30	95	95	150
Number of cells per battery	127	20	237	120	144
Average cell voltage (V)	0.74	1.5	0.40[a]	0.79[a]	1.04
Battery height (cm)	2.46	3.25	2.54	3.18	4.45 H[b]
Battery diameter (cm)	0.85	1.27	0.725	0.95	2.86 W[b]
Battery thickness (cm)	–	–	–	–	1.27 T[b]
Battery cross section (cm^2)	0.57	1.27	0.41	0.71	3.63
Battery volume (cm^3)	1.40	4.12	1.04	2.26	16.2
Effective cell volume (packaged) (cm^3)	0.011	0.21	0.0044	0.019	0.112
Volts per cm^3	67	7.3	91	42	9.3
Volts per cm length	38	9.2	38	30	34
Battery flash current (μa)	4	40	–	22	10–20
Flash current per cm^2 cell anode area (μa)	7	32	–	30	15–30
Battery internal resistance (ohms)	5.4×10^7	4.8×10^7	10×10^7	–	3×10^7
Internal resistance per cell (ohms)	4.2×10^7	2.4×10^6	4.2×10^5	–	2×10^5
Internal resistance per cm^2 of cell anode	2.4×10^5	3.0×10^6	1.7×10^5	–	6×10^4
Available charge per battery (coulombs)	1	–	0.34	–	10[c]
Time to 90% charge a capacitor (secs)					
0.01 μF	3	–	–	–	–
0.05 μF	25	–	–	–	3
0.15 μF	–	–	–	6.5	8
0.25 μF	100	–	–	–	15
2.0 μF	–	–	–	90	–

[a]Deviation from single cell ocv may be ascribed to electrode formulation modifications.
[b]Rectangular block Sprague cell.
[c]Five coulombs to $KICl_4$ endpoint and five more to $KICl_2$ endpoint.

A possible shortcoming of the General Electric system is, as Luckey and West (43), Weininger (44), and others have shown, that AgBr in the presence of excess Br_2 may exhibit some electronic conductivity. This small, partial, but continuous internal drain would not reflect as a drop in open-circuit voltage as long as excess $CuBr_2$ is present in the cathodic mass to maintain voltage but would inevitably limit shelf life.

THE MALLORY INCONEL/$SnSO_4$/$Ba(MnO_4)_2$ SYSTEM. Three patents by Ruben (45) serve to describe evolution of the Mallory "Solidion" system. The earliest of these describes arrays in which an anode of Ni or Ni alloys is combined with a compressed pellet of $SnSO_4$, against a cathode of PbO_2 electrodeposited on iron strip or compressed PbO_2 powder containing small amounts of binder material. In the second patent, an anode of INCONEL (an 80 Ni-14 Cr-6 Fe corrosion-resistant alloy) combined with pellets of either $Bi_2(SO_4)_3$ or $SnSO_4$ is used with the same type of PbO_2 cathode. The third, and apparently applicable procedure, involves an INCONEL anode, a $SnSO_4$ electrolyte, and a cathode consisting of a compressed mixture of 90 percent $Ba(MnO_4)_2$ and 10 percent graphite. None of the patents gives information concerning the over-all cell reaction.

In a typical assembly the anodes were 0.635 cm disks blanked out of 0.013 cm INCONEL strip. The electrolyte pellets were made by pressing chemically pure $SnSO_4$ under 680 kg (1500 lb) pressure into 0.635 cm disks 0.058 cm thick. The cathode pellets were prepared by compressing a uniform mixture of 90 percent $Ba(MnO_4)_2$ and 10 percent graphite under 680 kg (1500 lb) pressure into 0.635 cm diameter pellets 0.041 cm thick. Actual battery assembly consisted of dropping the cell elements [in the order: INCONEL, $SnSO_4$, $Ba(MnO_4)_2$] into an insulating casing equipped with a bottom-terminal plate and final insertion of the top-terminal plate. Adequate assembly pressure was obtained by pressing against the top-terminal plate until the "internal resistance" of the battery to 1000-cycle a.c. was reduced to 200–300 ohms/cell. The assembled, pressurized unit was then vacuum-impregnated with microcrystalline wax or encapsulated in plastic. No intercell connector separated the $Ba(MnO_4)_2$ from the INCONEL anode of the adjacent cell.

The characteristics given in the third column of Table 11.2 were developed from data available in Battery Data Manual, Mallory Battery Company (Cleveland, Ohio), p. 315 (in effect in 1960). This manual indicated that 50-, 100-, 150-, and 200-V batteries were available at that time.

THE UNION CARBIDE $Ag/AgI/V_2O_5$ SYSTEM. This battery system is described in a series of four patents issued to Louzos (37), Evans (38), Buchinski et al. (46), and Richter et al. (47). The discharge mechanism consists of transport of Ag^+ ions through the AgI followed by absorption, together with electrons, by the cathodic mass through formation of a reduced phase of vanadium oxide.

The Louzos patent describes the basic components: a duplex electrode of Ag electrodeposited or laminated onto a sheet of Ni; an electrolyte of AgI powder whose conductivity is increased by subjecting it to a pressure of about 1120 kg/cm² (16,000 psi); a cathodic depolarizer of V_2O_5 containing acetylene black to impart electronic conductivity. Louzos used essentially a hand-assembly technique to bring these components together.

The Evans and the Richter et al. patents translate major features of the Louzos disclosure into a compact assembly procedure and add several interesting features. Figure 11.3, a flow sheet of the assembly procedure, is a good illustration of the mechanization possibilities associated with manufacture of solid-electrolyte batteries. Assembly begins with a bimetallic strip of silver and stainless steel which is sub-

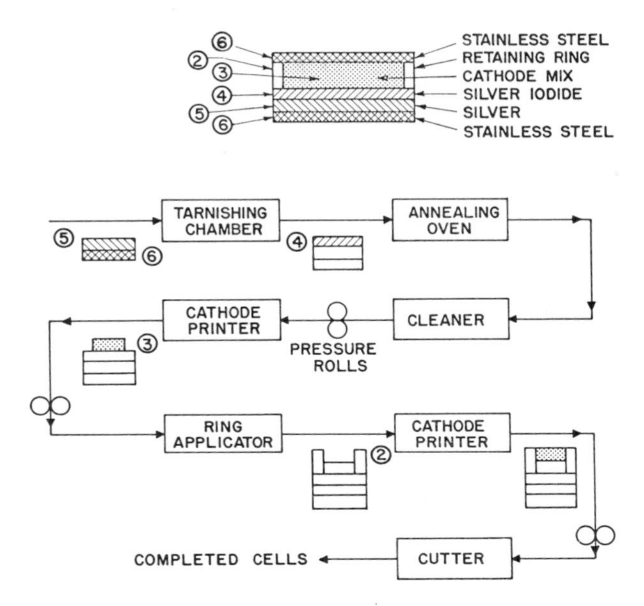

Figure 11.3 Flow sheet illustrating simplicity of mechanized fabrication of Union Carbide Ag/AgI/V_2O_5 Solid-Electrolyte Battery.

jected to the action of iodine vapor carried in a nitrogen stream. (I_2 at a partial pressure of 10–50 mm of Hg at a temperature of about 230°C.) This tarnishing reaction places a thin film of AgI rich in iodine upon the silver surface. Such nonstoichiometric AgI_x (where $x > 1$) exhibits some electronic conductivity and must be annealed for 15 min to 3 hours at 150–200°C (the annealing is complete when the electronic resistivity is greater than 10^{13} ohm-cm, at which point the AgI shows no visible darkening when exposed to ordinary room lighting). The strip is then cleaned to remove any deposit on the stainless steel side and passed through a roller to compact the AgI. A coating of colloidal graphite can be placed on the stainless steel side to safeguard against corrosion when it (later) contacts the anode of an adjacent cell. A portion of the cathode mix is then printed upon the AgI side of the strip. A unique and important feature of the Evans cathode mix is that it contains 5 parts by weight of AgI (to impart ionic conductivity to the cathodic mass) in addition to 2 parts of micronized graphite (to impart electronic conductivity) to each 10 parts of V_2O_5. After drying at about 85°C for 15 min, the mix is compacted to the AgI at 700–1400 kg/cm² (10,000–20,000 psi) and a mix-retaining member of a suitable resin is applied. A second deposit of cathodic mix (which may or may not have the same relative proportions of AgI-graphite-V_2O_5 as before) is applied and again compacted after drying. The cell is now complete and may be cut from the strip and individual units stacked to form the battery.

The Buchinski et al. patent teaches that a small amount of Cu (0.05–0.15 percent) and Si (0.01–0.15 percent) alloyed into the silver before tarnishing beneficially increases the adherence of the resulting AgI film. (This improved adherence may be important in the fabrication of all thin-film batteries to be discussed later).

The characteristics shown in the fourth column of Table 11.2 were supplied by Union Carbide. This organization also supplied voltage-time discharge curves from which it was possible to compute output parameters for continuous drain to a voltage endpoint corresponding to 60 percent of initial open-circuit voltage (i.e., equivalent to a 0.9 V endpoint for a 1.5 V Leclanché cell). These output parameters were 15.5×10^{-5} w-hr/cm^3 at an initial drain of 0.009 μa, 12.9×10^{-5} at 0.09 μa, and 4.8×10^{-5} at 0.9 μa.

Louzos (48) later reported that the open-circuit voltage of his research cells dropped by only 0.03 V (from 0.46 to 0.43) after nearly 8 years of storage and that 250-V production cells (initially 56 V/cm^3, 920 V/in^3 and 14.2 V/g, 400 V/oz), showed comparable stability after 3 years of storage. (The production cells referred to by Louzos were somewhat thicker than those whose characteristics are given in Table 11.2.)

THE PATTERSON-MOOS Ag/AgBr-Te/$CuBr_2$ SYSTEM. The distinguishing feature of this system, under the trademark of DYNOX, is the incorporation of about 5 percent of tellurium into the AgBr electrolyte to increase its conductivity. Defining cell resistance as the external load required to reduce the terminal voltage of to one-half the open-circuit voltage, Lieb (49) showed that the addition of Te can reduce electrolyte resistance by a factor of about 10. The patent describes a preferred assembly consisting of fusing together an appropriate weight mixture of AgBr and Te and flame-coating or pressure-laminating a thin film of this electrolyte onto a sheet of Ag, followed by printing of a cathodic layer from a thick, pastelike suspension of colloidal graphite dispersed in a concentrated aqueous solution of $CuBr_2$.

The characteristics shown in the fifth column of Table 11.2 were taken from an engineering brochure supplied by Patterson-Moos. Batteries of 190, 380, and 950 V were available in 1960.

In view of the reported mixed ionic-electronic conductivity of the silver tellurides (see Friauf (50), pp. 9–72) one would expect that the use of Te as a conductivity-imparting addition agent might be hazardous. Lieb recognized this danger of a possibly electronically conductive electrolyte and was careful to limit the Te concentration in AgBr to less than 10 percent. Another safeguard is that, even if trace electronic conductivity does exist, the reaction product of such partial, initial internal discharge would result in the formation of a "healing" AgBr film which, barring subsequent long-term diffusion of Te from the relatively Te-rich electrolyte, would block further internal electron transport. That is confirmed by a recent private communication from Patterson-Moos that some of their 7-year-old batteries are still operative.

THE SPRAGUE ELECTRIC $Ag/AgCl/KICl_4$ SYSTEM. We shall describe this system in somewhat greater detail since its development illustrates that future advances can profit by recognizing the interplay of electrochemistry, modern solid-state theory, and conventional battery technology. The following information was made available to us by Dr. D. M. Smyth of Sprague Electric. Additional details are available in Lehovec (51), Smyth (52), Smyth and Cutler (53), Smyth (54), Smyth and Shirn (55), and Sprague Electric Final Contract Reports (56).

The system is basically of the type $Ag/AgCl/Cl_2$ in which the control and confinement problems associated with the use of free cathodic halogen are minimized through adoption of a tightly sealed cylindrical cell design and use of solid $KICl_4$, which furnishes Cl_2 vapor at a partial pressure of only 0.6 mm at 25°C by virtue of the two-step dissociation, first into $KICl_2$ and Cl_2 followed by dissociation of the $KICl_2$ into KCl and ICl at a partial pressure of 0.2 mm at 25°C. The initial open-circuit voltage of 1.04 V at 25°C corresponds to that calculated for Ag reacting with $\frac{1}{2}Cl_2$ at 0.6 mm pressure and the open-circuit voltage of 0.89 V after one-half full discharge corresponds to that of an Ag/AgCl/ICl cell. The total cathode vapor pressure increases with temperature, reaching one atmosphere at 124°C.

The system has certain obvious advantages and among these we find: (*a*) use of a fairly active halogen results in a higher-cell voltage; (*b*) the relatively high halogen pressure permits faster reaction with incoming silver ions, thus tending to reduce one component of cathodic polarization; (*c*) this high halogen pressure also leads to almost instantaneous recovery of open-circuit voltage even after prolonged discharge; (*d*) the reaction product, AgCl, is almost the same as the starting electrolyte which, unlike many cell discharge products (e.g., the reduced vanadium oxide phase of the Union Carbide system), does not seriously block cathodic performance and the generation of AgCl may actually impart some beneficial ionic conductivity to the cathodic mix; (*e*) the generated AgCl also serves to self-heal any pinholes that may exist when the AgCl electrolyte is initially deposited by chlorination of the silver anode surface.

The system also has certain disadvantages: (*a*) the extrapolated room-temperature ionic resistivity of AgCl is considerably higher than that of AgBr or AgI; (*b*) AgCl in the presence of chlorine vapor seems to exhibit some electronic conductivity; (*c*) the relatively high halogen vapor pressure of $KICl_4$ at elevated temperatures necessitates exceptionally good sealing.

Sprague considered the advantages of its system important enough to warrant a serious effort to solve these problems of halogen gas containment, high electrolyte resistivity, and trace electronic transport in the electrolyte. They began by abondoning the "stacked wafer" design employed by the four other major systems. Sprague used a silver anode can whose configuration is illustrated in Fig. 11.4. These cans (0.843 cm high, 0.208 cm diameter, 0.0287 cm^3 volume) were then chlorinated at elevated temperature to form an electrolyte film of AgCl about 10 μ thick. The can was partially filled with the cathode mix consisting of $KICl_4$, carbon black (to impart electronic conductivity of the mix), and KEL-F 90 grease made by Minnesota Mining and Manufacturing Company. The KEL-F was added to suppress the electronic conductivity of the AgCl in the presence of chlorine vapor. The cell was then tightly sealed with a compression-fitted Teflon plug pierced by a tantalum wire, which served as the positive terminal. This completed fabrication of the individual cell.

Battery assembly involves series connection of individual cells by mechanically crimping the tantalum terminal wire of one cell into the anode crimp socket of the adjacent cell. Battery strings of the desired voltage were thus built up and these could be folded into compact struc-

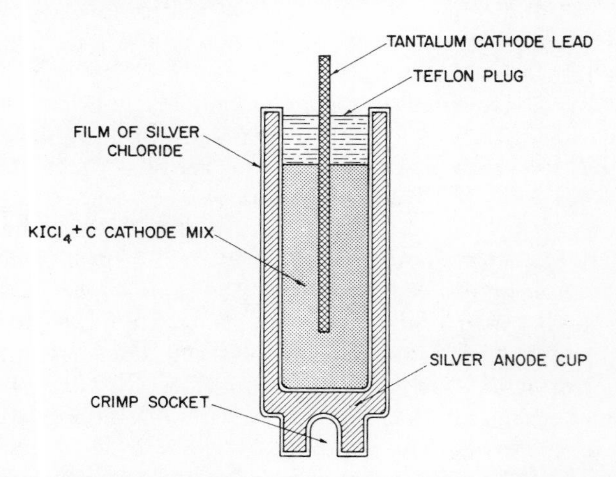

Figure 11.4 The Sprague Cylindrical Solid-Electrolyte Cell. Series assembly consists of mechanical crimping of the tantalum cathode lead of one cell into the crimp socket of the next cell.

tures in which adjacent cells were separated by interleaved Teflon foil. Battery assembly was completed by insertion of the cell strings into a container and vacuum-impregnation with an appropriate encapsulating agent. It is obvious that this approach allows for considerable flexibility in battery shape.

The internal resistance of this cell system is almost completely controlled by the ionic resistance of the AgCl electrolyte, which, in turn, is determined by the concentration of ionic charge carriers, their mobility, and their charge. In the temperature range of interest for battery applications, the charge carriers in AgCl are Ag^+ ions moving through cation vacancies within the AgCl lattice or along vacancies on the surfaces of the AgCl grains. Their mobility is fixed primarily by the temperature. The concentration of cation vacancies, however, can be increased through application of a concept generally attributed to Koch and Wagner (57) and specifically applied to AgCl by Ebert and Teltow (58). This involves deliberate addition of a multivalent cation such as Cd^{+2} to the host AgCl lattice. Each divalent Cd^{+2} ion electrostatically replaces two monovalent Ag^+ ions, but occupies only one lattice position, thus creating a Ag^+ vacancy. Sprague added 0.02 percent Cd to the "high fine" silver used to make the anode can; the subsequent chlorination step yielded AgCl electrolyte "doped" with conductivity-imparting $CdCl_2$. Such doped electrolyte had a measured resistance which was one-fourth that of the undoped electrolyte material. Note that the silver chloride generated by the discharge reaction will also contain this conductivity-imparting $CdCl_2$.

Defining cell internal resistance as equal to the external load required to drop cell voltage to one-half of the open-circuit voltage, Sprague reported that cell resistance was 0.025 megohm at 60°C, increasing to 0.054 at 40°C, 0.11 at 25°C, 0.36 at 0°C, 1.2 at −20°C, and 5.1 megohms at −40°C. Sprague also submitted the discharge curves shown in Fig. 11.5. Referring to the 10^6 ohm curve, it will be noted that the *initial* closed circuit voltage ν is 0.98. Using this value in Eq. (1), we find that the starting cell internal resistance at 25°C is $(1.04–0.98) \times 10^6/0.98$, or 6.1×10^4 ohms, compared to the higher value of 11×10^4 ohms based on the Sprague difinition of internal resistance.

The general characteristics of a rectangular 144-cell 150-V Sprague battery are given in Table 11.2. With respect to shelf life, a recent (August 1968) personal communication from Dr. Smyth indicated that the open-circuit voltage of a packaged 150-V battery showed no perceptible voltage drop after some 8 years of storage at room temperature.

Even though the Sprague approach resulted in probably the best solid electrolyte system available in 1960, its performance in terms of the

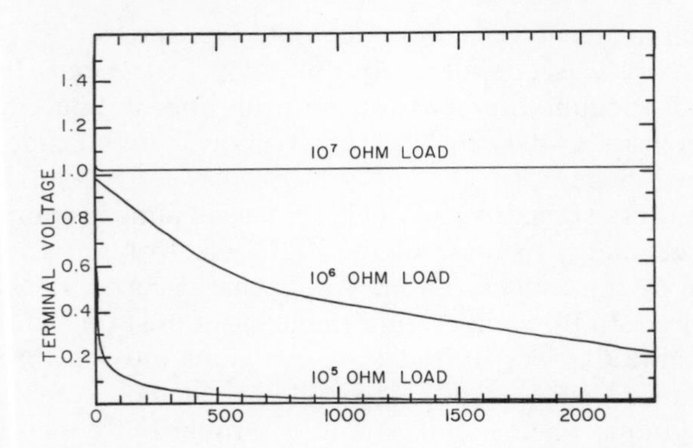

Figure 11.5 Typical room-temperature curves for the Sprague $Ag/AgCl/KICl_4$ Cell.

conventionally accepted yardstick of watt-hours per cubic centimeter was discouragingly low. For example, using the 10^6 ohm discharge curve of Fig. 11.5 and graphically integrating the output to a 0.6-V cutoff (comparable to a 0.9-V cutoff for a 1.5-V Leclanché cell), one obtains a value of 3.9×10^{-4} w-hr for a packaged cell volume of 0.112 cm^3. This reflects as a specific output of 3.5×10^{-3} w-hr/cm^3 at an average drain of about 0.8 μa/cell. This is far below that of the better liquid-electrolyte battery systems at 0.25–0.35 w-hr/cm^3 at milliampere drains.

The Common Limitations of Early Systems. The preceding discussion of early systems demonstrates that the solid electrolyte approach to battery miniaturization did indeed exhibit the predicted advantages of ease of fabrication, rugged construction, good shelf life, and extended temperature range of operation. These real advantages, however, were negated by very low specific outputs.

There is ample evidence that these low output currents are due in great part to the low room-temperature ionic conductivity of even the most conductive solid electrolytes known at that time. Dramatic proof that increased solid electrolyte conductivity would substantially improve output current was given by Weininger (59), who took advantage of the fact that at 145.8°C AgI exhibits a phase change to yield a body-centered cubic structure whose ionic specific conductance is 1.3 ohm^{-1} cm^{-1}. A schematic drawing of an early Weininger "tube" cell designed to operate above this transition temperature is shown in Fig. 11.6. One such cell having a volume of about 0.1 cm^3 could deliver 3000 μa while showing a voltage drop of only 0.12 V (from 0.67 to 0.55). Such cells, however, could not be discharged at temperatures much higher than 170°C because the build-up of iodine pressure caused the tantalum tube to burst. Using

CsI_4 as a source of low-pressure iodine and discharging through a 10,000-ohm resistor at 300°C, Weininger obtained the nearly ideal discharge curve shown in Fig. 11.7. The initial short-circuit current of this type of cell at 300°C was about 290 ma/cm² of imbedded silver anode surface. Useful output was a respectable 79 percent of the theoretical limit imposed by the weight of incorporated active material. Earlier work by Weininger (60) using "bead" cells of the general type Ag/Ag halide/halogen vapor designed to operate above 145°C also reflected the sharp improvement that results from increased electrolyte conductivity.

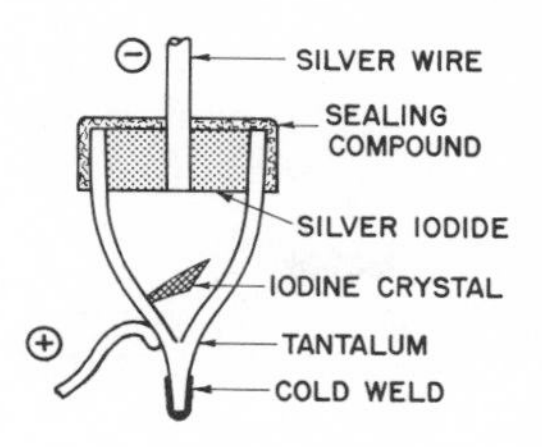

Figure 11.6 Cross section of the Weininger $Ag/AgI/I_2$ Tube Cell (inert at room temperature).

It should be pointed out in passing that in the Weininger systems (as in the General Electric, Patterson-Moos, and Sprague systems) the active cathodic material is an oxidizing halogen vapor which is substantially free to permeate throughout the cathodic mass to meet and react with silver ions entering through the electrolyte and, furthermore, that the reaction product in each case is the electrolyte itself. These two factors significantly lower cathodic polarization but at a price which involves trading a liquid-containment problem for one involving the long-term containment of a cathodic vapor phase.

As discussed earlier, a high-room-temperature solid-electrolyte resistivity could lose much of its significance through use of extremely thin (and hence low resistance) electrolyte films. This technique, however, cannot be used with impunity with halogen-vapor cathodic materials since such vapors could easily permeate through the thin films to react directly with the anode. The Mallory and Union Carbide "all-solid" systems are

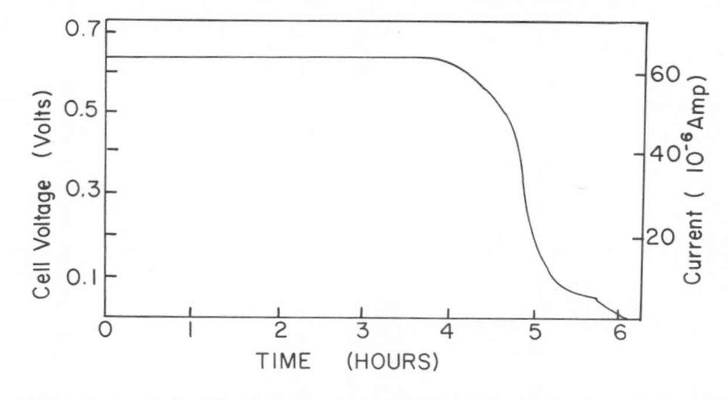

Figure 11.7 Behavior of the Weininger $Ag/AgI/CsI_4$ Tube Cell across 10,000 ohms at 300°C where conductivity of solid AgI and mobility of I_2 vapor are high.

amenable to effective use of thin-film electrolytes since they are not subject to the problem of long-term containment of a halogen vapor cathode.

The "all-solid" system still represents the ideal. Its realization depends on development of better solid ionic electrolytes free of electronic transport and, of equal if not greater importance, of better cathodic materials that can be coupled with such electrolytes. Such improvements will occupy our attention in the following sections.

Recent Progress in Improving the Room-Temperature Conductivity of Solid Electrolytes. In the discussion of the Sprague system it was pointed out that 0.02 percent cadmium was added to the silver of the can to yield a more conductive AgCl through creation of an increased number of Ag^+ ion vacancies in the final AgCl electrolyte. This was an early intimation that it is possible to improve the room-temperature conductivity of solid ionic conductors. In the next section basic concepts of the mechanism of ionic charge transport through solids are developed as background for evaluation of other possible methods to enhance such ionic transport.

IONIC CONDUCTIVITY AND LATTICE DEFECTS. The subject of ionic conduction through solid compounds has received considerable attention during the past several decades. Historically, the voluminous literature can be divided into two time periods: work up to about 1930 and work since then. The early work was summarized in 1932 by Tubandt (61), who cited some 300 references. For our purposes, probably the most important feature of this early work was the experimental development of an empirical relationship between ionic conductivity and temperature. Barrer (62) in 1951 briefly sketched the history of the development of this relationship. He cites (p. 257) Foussereau's (63) work in which an attempt was made to express the temperature dependence of resistivity p in terms of power series of the form $\log p = a - bT + cT^2$. Konigsberger (64) suggested that a better empirical fit between conductivity s and temperature was given by the expression $\log s = c - E/RT$, where c is a constant, E is an energy term, R is the gas constant, and T is the absolute temperature. This expression was confirmed by Phipps, Lansing, and Cook (65) for sodium halides at high temperature.

Although different workers suggested other expressions, we single out the Konigsberger equation because (in a slightly modified form of $s = s_0 \exp(-E/RT)$,[1] where s_0 is a constant, E is an energy term in calories, and R is the gas constant) it can be derived theoretically, has been confirmed by many workers, and is now in general use. Friauf

[1]The term "$\exp(-E/RT)$" should be read as "the base e of natural logarithms raised to the power-E/RT."

(50), citing over 400 references for the period 1930–1960, used the $s = s_0 \exp(-E/RT)$ expression to develop a concise and comprehensive tabulation of modern ionic conductivity data through solids.

Despite the tremendous amount of work done before 1930, in many respects the subject of ionic conductivity through solids was scarcely more than a descriptive study. As Lidiard (66) pointed out, the reason for this was that one could not easily visualize how ions could diffuse from one lattice point of a perfect crystal to another if the effect of increasing temperature involved merely an increase in the amplitude of thermal vibrations of ions about fixed lattice points. Furthermore, the drift of ions through an ionic crystal under the influence of even small applied electric fields was inconceivable if the crystals were as perfect as early x-ray crystallographers depicted and as stoichiometrically perfect as Dalton's Law of Definite Proportions predicted.

This impasse was resolved by Frenkel (67), who developed the idea of intrinsic disorder, or defects, in crystalline materials. Specifically, he proposed that a Boltzmann-type distribution governed the thermal vibrations of ions about the fixed lattice points of stoichiometric crystals and that sometimes some ions statistically received enough thermal energy W to leave their normal lattice points, in effect being pushed into the interstices between lattice points. Such an interstitial ion under further thermal agitation could receive enough energy to surmount a potential energy barrier U to "jump" to an adjacent interstitial position until it eventually found a new normal lattice vacancy previously occupied by a different ion which had left for a different interstitial site. Under the influence of an applied electric field, the potential barrier U would be lowered in the direction of the field and raised by the same amount in the direction against the field. Statistical migration of the interstitials in the direction of the field would therefore be favored and the crystal would exhibit ionic conduction.

Shapiro and Kolthoff (68) have given a simple (but nonrigorous) development of the Frenkel model as follows:

1. The electrical conductivity s of a solid is given by

$$s = neu \tag{2}$$

where n is the number of charge carriers, e is the charge on each carrier, and u is the mobility of the charge carrier.

2. Assuming Frenkel (i.e., interstitial) disorder in cations only, the value of n at an absolute temperature T is given by

$$n = N \exp\left(-\frac{W}{2kT}\right) \tag{3}$$

where N is the total number of lattice positions, W is the energy, in electron volts, required to move a cation from a normal lattice position into an interstitial position, and k is the Boltzmann constant in electron volts.

3. The mobility u is given by

$$u = B \exp\left(-\frac{U}{kT}\right) \tag{4}$$

where B is a constant, U is an energy term in electron volts representing the potential barrier for interstitial "jumps."

4. Letting $s_0 = eBN$, a constant, and $E = \frac{1}{2}W + U$, and combining Eqs. (3) and (4) into (2) yields

$$s = s_0 \exp\left(-\frac{E}{kT}\right) \tag{5}$$

which corresponds to the modified form of the experimental Konigsberger equation discussed previously.

In order to explain simultaneous anion-cation conductivity found in some solids, Schottky (69) in 1935 developed a model based on the assumption that real crystals possessed an electrically equivalent number of anion and cation lattice vacancies but no interstitial ions. The theoretical treatment based on the Schottky model is similar to that used for the Frenkel model and the final Schottky equation for the total specific conductance is of the same exponential form. As one might expect, the Frenkel model can be applied to some materials, the Schottky model to others, and a combination of the two to still others.

Both models account for the straight-line relationship between log s and $1/T°K$ in accord with the experimental equation $s = s_0 \exp(-E/kT)$ at high temperatures. Conductivity of single crystals and of many compressed powders within this high temperature range, the so-called "intrinsic" range, is reproducible from sample to sample since the number of thermally induced defects dominates the scene. As the temperature drops (with a corresponding drop in the number of thermally induced defects), the conductivities of different samples break away from this "intrinsic" linearity, tending to become less dependent on temperature with further temperature drop. The breakpoint and subsequent decreased dependence of conductivity on temperature gives rise to an "intrinsic" or "structure sensitive" family of curves (Fig. 11.8) whose behavior is presumed to be due to defects associated with sample impurity, physical structure, surface distortions, dislocations, previous history, etc.; i.e., to lattice defects which are not primarily associated with thermal agitation. The room-temperature "structure-sensitive" conductivity

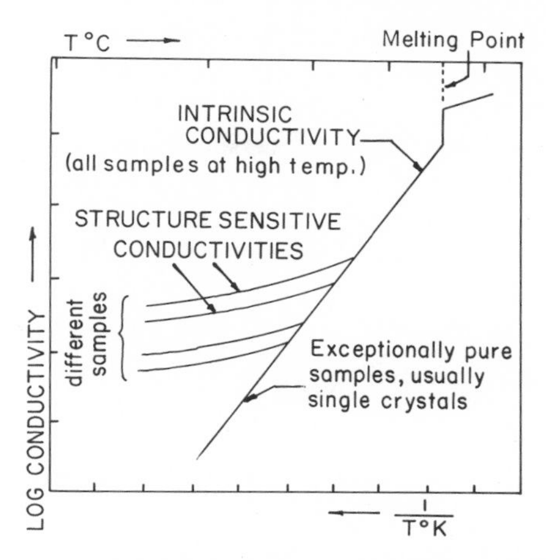

Figure 11.8 Idealized sketch illustrating "intrinsic" and "structure-sensitive" conductivity ranges of most solid ionic conductors. (Structure-sensitive conductivity is always higher than the corresponding intrinsic conductivity and is always less temperature-dependent.)

can sometimes be several orders of magnitude higher than that obtained by extrapolation of the linear "intrinsic" curve to the same room temperature. These "structure sensitive" conductivities are so discordant and intractable to theoretical treatment that, until recently, they received little attention.

Both the Frenkel and Schottky models are concerned with thermally induced, conductivity-imparting lattice defects and are applicable to study of the mechanism of ionic conductivity at the relatively high temperatures (seldom less than 200°C) required for thermal generation of lattice defects in dominant numbers, i.e., enough to overwhelm the effect of nonthermally induced defects. In addition, it is implicit in both models that the sample under study be in the form of a single crystal of known purity.

Of primary interest, however, are ionic materials of high conductivity at room temperature where there are few thermally induced defects. Furthermore, flexibility of battery manufacture and design would be vastly enhanced if multicrystalline solid electrolytes could be used in the form of compressed powders, evaporated films, chemical deposits, painted or sprayed layers, etc. The Frenkel-Schottky discussion, therefore, is not immediately relevant to this interest; it was given here to illustrate a single, crucially important point: an approach to the synthesis of solid electrolyte exhibiting high conductivity at room temperature would be the development of defect structures (i.e., lattice vacancies) by means other than thermal agitation.

Although several alternatives are available (e.g., by addition of appropriate impurities as in the Sprague example, or by heating our material to generate thermal defects followed by quenching to "freeze-in" these defects), the two approaches which seem most promising at present are: (*a*) use of "open" lattices where the crystal structure is such that a cation, for example, can occupy any of several vacant lattice positions, or (*b*) reliance upon the inherently distorted structure of the surface of a solid or the atomic disarrangements known to exist at grain boundaries. These possibilities are discussed in the following two subsections.

DEFECTS ASSOCIATED WITH "OPEN" LATTICE STRUCTURES

Silver iodide. A classic example of an "open" lattice ionic solid can be found in the high-temperature α-phase of AgI, which is stable between 145.8°C and its melting point. In this temperature range the ionic conductivity is very high, rising from 1.3 ohm^{-1} cm^{-1} at 146°C to 2.64 ohm^{-1} cm^{-1} at 550°C (just below the melting point). This conductivity is ascribed solely to movement of Ag^+ ions. As Strock (70) has shown, the crystal structure of the α-phase is body-centered cubic, the distance between I^- ions being 2.18 Å. Each Ag^+ ion, with an ionic radius of 1.26 Å, can be in any one of the 30 largest holes in the rigid I^- ion lattice. The presence of such a large number of available sites, together with the small size of the Ag^+ ions, can be used to explain the Tubandt-Lorenz (71) observation that the conductivity of solid AgI near the melting point (2.64 ohm^{-1} cm^{-1} at 550°C) is actually greater than that of molten AgI (2.36 ohm^{-1} cm^{-1} at 554°C).

A sharp and reversible phase transformation occurs at 145.8°C accompanied by a sharp decrease in conductivity of the AgI, although it still remains much higher than that of most solids. Just below 145.8°C, the AgI forms the so-called β-phase consisting of hexagonal crystals with the wurtzite structure. The iodine atoms in β-AgI now form a close-packed lattice, each atom being equidistant from 12 others. An excellent discussion of the polymorphism of AgI is given by Wells (72). Helmholtz (73) was able to prepare single crystals of the β-phase and reported that the Ag^+ ions in this phase were distributed at random among *four* equivalent lattice sites tetrahedrally surrounding the iodide ions. It seems likely that the reduction from Strock's *30* possible lattice positions for the highly conductive α-phase to Helmholtz's *4* possible positions for the less conductive hexagonal β-phase can be used to account for some of the conductivity drop, but the decrease in Ag^+ ion mobility associated with the drop in temperature cannot be ignored. Donnay, Donnay, Cox, Kennard, and King (74), citing the literature, reported that the hexagonal β-phase was stable in the temperature range 138–146°C but was metastable at lower temperatures. They also reported that the phase stable at room temperatures and up to 138°C was face-centered cubic; this is the so-called γ-phase. Since this γ-phase has not yet been grown in the form of a single crystal large enough for crystallographic study, there is no information concerning the number of vacant sites available to the Ag^+ ions.

Silver mercuric iodide. Another example of high conductivity due to an "open" lattice is the compound Ag_2HgI_4 studied by Ketelaar (75). The structure is similar to that of room-temperature AgI except that the cations (two Ag^+ ions and one Hg^{+2} ion) are distributed at random among *four* available lattice positions. He reported a conductivity of 3×10^{-3} ohm^{-1} cm^{-1} at 50°C. Suchow and Pond (76) observed somewhat lower ionic conductivities than those found by Ketelaar. More recently, Weil and Lawson (77) confirmed the Suchow-Pond lower values and also reported some electronic transport at 80°C. The latter, if not repressed, argues against the use of Ag_2HgI_4 in a solid electrolyte battery. Another complicating factor is that there is apparently some Hg^{+2} ion transport.

Ag_3SI. Reuter and Hardel (78) reported that the high-temperature α-phase of Ag_3SI, stable above 235–245°C, had a face-centered cubic structure with a lattice constant of 4.993 Å in which 3 Ag^+ ions were statistically distributed over 42 equivalent lattice sites. Takahashi and Yamamoto (79) carried out extensive studies of this material as the solid electrolyte in cells of the type $Ag/Ag_3SI/I_2$ operating at room temperature. They prepared their material by heating equimolar amounts of Ag_2S and AgI at 550°C for 17 hr in sulfur vapor at 1 ~ 2 atm. Using a pellet (0.15 cm thick, 1.2 cm diameter) made by compressing powdered Ag_3SI under 1800 kg/cm^2, a silver anode 0.005 cm thick and an iodine-graphite mixture as a cathode, it was found that the open-circuit voltage was about 0.68 V, but that the cell output was limited by anode polarization. The substitution of amalgamated silver electrodes, prepared by rubbing mercury onto polished silver electrodes by Mrgudich (80), reduced this polarization but the open-circuit voltage dropped to 0.60 V, probably due to the lower activity of silver dissolved in mercury. Improvement of the cathodic mix by addition of electronically conductive acetylene black resulted in cells whose flash current was as high as 10 ma/cm^2. These cells could deliver a constant current of 1 ma/cm^2 for 1 hr at 25°C, during which time the closed-circuit cell voltage drop was only 100 mV. This excellent performance, however, may suffer on shelf stand in view of the degradation of the electrolyte according to the reaction

$$I_2(g) + Ag_3SI(s) \rightarrow 3\,AgI(s) + S(s)$$

recently reported by Owens, Argue, Groce, and Hermo (81).

$RbAg_4I_5$. Bradley and Greene (82, 83, 84, 85), in England and Owens and Argue (86) in the United States synthesized a series of solids of the general formula. MAg_4I_5, where M could be Rb, K, or NH_4. These materials have an ionic conductance in the vicinity of 0.1–0.2 ohm^{-1} cm^{-1} at room temperature. The most promising of these is $RbAg_4I_5$. Its crystal structure was studied by Geller (87), who found it to be cubic with a lattice constant of 11.24 Å and a unit cell content of four $RbAg_4I_5$ molecules. His data show three sets of crystallographic nonequivalent sites for the 16 Ag^+ ions; one set is 8-fold and the other two are 24-fold each. Thus there would be 56 available sites at room temperature for the 16 Ag^+ ions. The ionic conductivity due solely to Ag^+ ions is very high at about 0.2 ohm^{-1} cm^{-1}. In a later paper, Bradley and Greene also carried out a structure analysis of $RbAg_4I_5$. They found it to be cubic with 11.14 Å as the lattice constant and 4 molecules per unit cell. They present data indicating that there are 72 sites available to the 16 Ag^+ ions and conclude that "the Ag^+ ions are virtually in a liquid state inside a crystalline lattice formed by I^- and M^+ ions." Bradley and Greene found no evidence of significant electronic transport.

Summing up the essence of the work of both groups, it was found that three compounds, KAg_4I_5, $RbAg_4I_5$, and $NH_4Ag_4I_5$, exhibited room-temperature silver-ion conductivities of about 0.2 ohm^{-1} cm^{-1}, which is some 3 to 4 orders of magnitude greater than the structure-sensitive conductivity of the best fine-particle AgI powders. Further, both groups presented data indicating that the conductivity was due almost entirely to silver-ion transport with only trace electronic flow. The Hittorf transport technique used to demonstrate the virtual absence of electronic conductivity, however, is not so sensitive as to completely resolve the matter of electronic conductivity. The very high level of solid-ionic conductivity, on the other hand, is comparable to the conductivities of *aqueous* electrolytes used in conventional batteries (e.g., 0.2 ohm^{-1} cm^{-1} for the $ZnCl_2$-NH_4Cl electrolyte of Leclanché cells, 0.5 for the KOH electrolyte of Zn/MnO_2, Zn/Ag_2O, Zn/HgO and Cd/NiOOH cells, and 0.7 for the sulfuric acid of Pb/PbO_2 storage batteries).

We shall here devote our attention to $RbAg_4I_5$ since both KAg_4I_5 and $NH_4Ag_4I_5$ are more sensitive to moisture-catalyzed disproportionation at room temperature into AgI and a nonconductive M_2AgI_3 phase. Even $RbAg_4I_5$, according to Topol and Owens (88), is thermodynamically unstable at temperatures below 27°C. The transformation is

$$2\ RbAg_4I_5 = 7\ AgI + (\text{nonconductive})\ Rb_2AgI_3$$

The rate at which the transformation occurs at 27°C and below has yet to be established. It is presumed that this rate is very slow, as long shelf life has been predicted for the $Ag/RbAg_4I_5/RbI_3$ system. At temperatures above 27°C, the reaction is reversed and AgI will react with nonconductive Rb_2AgI_3 to form the conductive $RbAg_4I_5$ compound. (We shall discuss the use of solid $RbAg_4I_5$ as a battery electrolyte in a later section.)

Mellor (89), citing many references, reported that AgI readily forms a very large number of double salts; he lists over 30 such salts. It is quite possible that further work could uncover materials better for our purposes than even $RbAg_4I_5$.

DEFECTS ASSOCIATED WITH SURFACE DISORDER. Returning to a discussion of how to improve room-temperature ionic conductivity, there is ample evidence in the literature that diffusion along crystal surfaces proceeds with greater ease than diffusion through the bulk crystal. Although most of the work has been with metals (and especially along grain boundaries in multicrystalline metals), it is probably valid to assume that ions will diffuse faster along the surface than through the bulk of the crystal. Lidiard (66) considers that enhanced ionic diffusion along surfaces is a factor applicable to a better underatanding of the abnormally high ionic conductivities observed in the "structure-sensitive" region. He concludes, however, that "work on surface conduction is still in a qualitative stage." Welch (90) also showed that surface mobility is considerable at temperatures too low to permit extensive diffusion through the body of the lattice. He further states that factors concerning particle size, clustering, porosity, surface irregularities, previous sample history, and changes in the surface as the diffusing species move along the surface make it impossible to establish a general theoretical treatment applicable to the "structure-sensitive" region.

Because of this complexity and resultant lack of theoretical interest in the structure-sensitive region, there is not much in the literature concerning ionic conductivity along the surfaces of ionic crystallites. The work of Shapiro and Kolthoff (68) seems definitive and can be used to illustrate improvement of low-temperature ionic conductivity. Working under inactive red light, they prepared pure AgBr powders precipitated by reaction of KBr and $AgNO_3$ solutions. The washed and air-dried powders were then compressed between silver electrodes at 3000 atm. Conductivities were then measured at 925 atm pressure over a range of temperatures. These materials were found to obey Ohm's law (i.e., no interfering blocking layers were observed under high pressure). Their results are shown in Fig. 11.9. The observed 3–5 order-of-magnitude improvement over the extrapolated "intrinsic" conductivities could not be attributed to impurities and was interpreted as originating because of increased surface mobility.

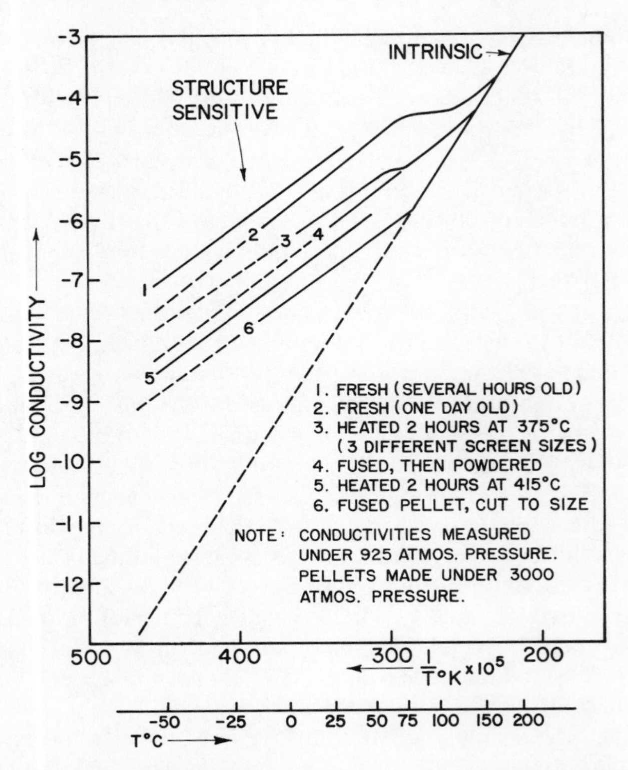

Figure 11.9 Experimental conductivity data for compressed pellets of AgBr. After (68).

An interesting experiment illustrating increased ionic mobility along grain boundaries (a form of internal surface) was carried out by Goddard and Urbach (91). Starting with molten AgBr, they cast rods consisting of individual crystals measuring a few millimeters in size. Next, they machined a thin slab of this rod and prepared an array consisting of

$$\text{Pt/AgNO}_3 \text{ solution/AgBr slab/KBr solution/Pt}$$

Finally, they passed a d-c current through the array, making the Pt in the $AgNO_3$ positive (i.e., Ag^+ ions from the $AgNO_3$ solution moved through the slab into the KBr solution). On reaching the KBr solution, the diffusing Ag^+ ions reacted to form AgBr. It was observed that the AgBr thus formed was concentrated at the grain boundaries, visible as ridges which were often twice as high as the grains were wide. The photograph they show is graphic proof of the much greater mobility of Ag^+ ions along AgBr grain boundaries.

The enhanced conductivity along grain surfaces has also been observed with room-temperature AgI where, as discussed earlier there is reason to believe that high bulk conductivity is due to an "open" lattice. In unpublished work, Paul Bramhall of USAECOM precipitated AgI by simultaneous dropwise addition of normal solutions of $AgNO_3$ and KI from separate burettes into a large volume of boiling water. The precipitated AgI was recovered by centrifuging and washed with distilled water until free of both silver and iodide ions and then air dried. The sample was then pressed under a 3180-kg load into 1-g pellets (1 mm thick, 1.27 cm diameter). The conductivity was measured using amalgamated silver electrodes. The specific conductance was 4.6×10^{-6} ohm^{-1} cm^{-1}. The experiment was repeated using 95 percent ethyl alcohol instead of boiling water. The measured specific conductance of pellets of this ethanolic AgI was 128 times higher at 588×10^{-6} ohm^{-1} cm^{-1}. Electron micrographs of the two materials are shown in Fig. 11.10. It is apparent that the more conductive ethanolic AgI had a much higher specific surface.

A PRECAUTIONARY NOTE: AVOIDANCE OF ELECTRONIC CONDUCTIVITY IN SYNTHESIZED SOLID ELECTROLYTES. Although the two preceding subsections forecast promising approaches to solution of the previous major drawback of solid electrolyte batteries (low ionic conductivity) through synthesis of significantly more conductive materials, it is possible that synthesis of structures heavily laden with internal or surface defects to improve ionic conductivity can easily lead to nonstoichiometry. As Stone (92) pointed out, many chemical compounds show small deviations from stoichiometry while others (in a minority) can withstand large deviations without x-ray evidence of a change in structure. The δ-phase of TiO, for example, has the same rock-salt structure over the entire range of $TiO_{0.66}$ (anion deficiency) to $TiO_{1.33}$ (cation deficiency). Such behavior is undoubtedly connected with the very high degree of Schottky disorder present in stoichiometric $TiO_{1.00}$ where, on the average, every sixth ion is missing in the lattice at room temperature.

Suppose, now, that we synthesize Ag_3SI for use as a solid electrolyte, but that the synthesis resulted in a deficiency of sulfur anions (e.g., the final composition is $Ag_3S_{1-x}I$); the absolute requirement that the molecule be electrically neutral might be met if a number of electrons electrically equivalent to the S^{-2} ion deficiency are present within the atomically nonstoichiometric lattice. The presence of such electrons could impart electronic conductivity to the $Ag_3S_{1-x}I$ and thereby detract from its usefulness as a solid battery electrolyte. A similar argument could be advanced to explain the Evans (38) observation that AgI exhibits some electronic conductivity in the presence of excess iodine vapor. It would

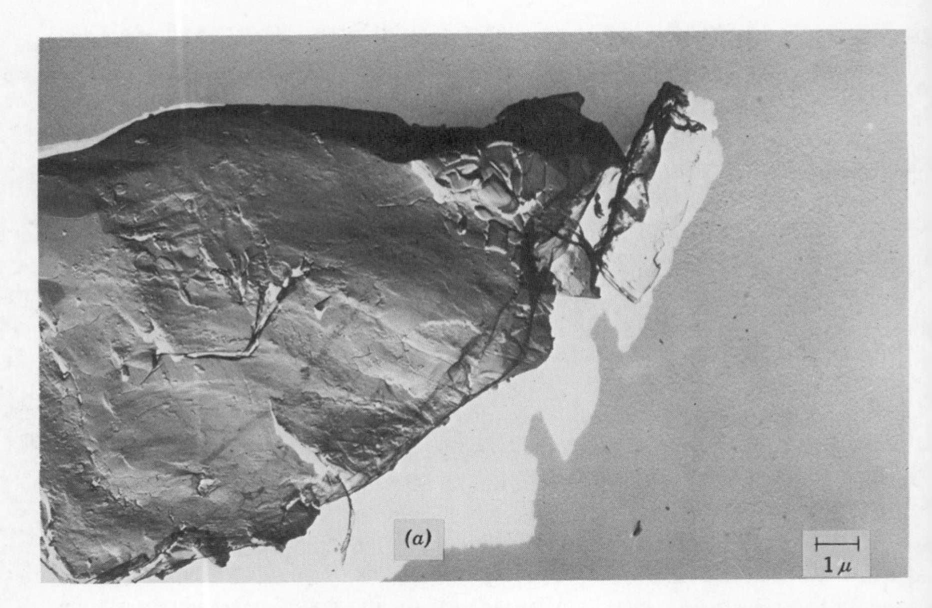

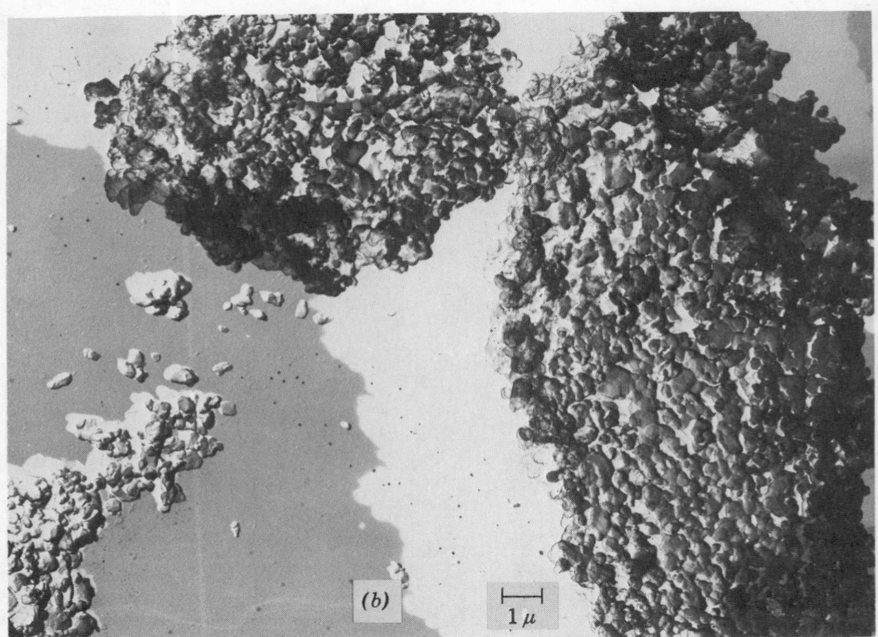

Figure 11.10 (*a*) Electron micrograph of AgI precipitated in hot water. (The conductivity of compressed pellets of this "steaklike" material was 4.6×10^{-6} ohm^{-1} cm^{-1}.) (*b*) Electron micrograph of AgI precipitated in hot ethyl alcohol. (The conductivity of compressed pellets of this "hamburgerlike" material was 583×10^{-6} ohm^{-1} cm^{-1}.)

appear that special precautions should be taken in the synthesis of the more complex halide salts such as the $RbAg_4I_5$ electrolyte where a total of 10 atoms is involved.

A SUGGESTION: IMPROVEMENT OF SOLID CATHODIC MATERIALS THROUGH NONSTOICHIOMETRY. It has been noted that electronic conductivity associated with nonstoichiometry is undesirable in the electrolyte. Such electronic conductivity, however, is desirable in the solid cathodic material when one wants electronic conductivity to bring the electrons entering from the external circuit during cell discharge to the site of reaction (i.e., with the cations that are brought through the electrolyte by virtue of its ionic conductivity). In this connection, Clark (93) has pointed out that the lead peroxide of the positive plate of a storage battery is always deficient in oxygen. The same reference (p. 365) points out that the "oxidation or rusting of metal surfaces proceeds as a result of varying ratio of metal to oxygen and by actual diffusion of cations and electrons from the metallic phase toward the oxygen phase." Further confirmation that solid cathodic materials may benefit by nonstoichiometry is given by the well-known fact that stoichiometric $MnO_{2.00}$ is not good Leclanché cell depolarizer, the preferred composition being somewhere between $MnO_{1.8}$ and $MnO_{1.9}$.

It seems quite likely that battery engineers have not given this matter of control of electronic (and possibly ionic) conductivity of cathodic mixes through deliberate synthesis of cathodic nonstoichiometry the attention it potentially deserves. Such nonstoichiometry can be achieved in several ways: one promising approach could involve doping of the basic cathodic material with multivalent elements such as Sn, Sb, P, Fe, Co, Ni; another might involve heating to drive off part of the anionic constituent followed by quenching to "freeze" nonstoichiometry into the resultant mass. The area is both rich in possibilities and ripe for using techniques well known in the field of electronic semiconductors.

Recent Pellet-Type Solid Electrolyte Cell Systems

THE USAECOM Ag/AgI/Pt SYSTEM. For several years the U. S. Army Electronics Command, through its Institute for Exploratory Research, has sponsored a modest in-house research program in the general area of ionic charge transport through solids. During 1964 in work on the adaptive memory characteristics of all-solid arrays using thin film (about 2000 Å thick) of Ag and Pt vacuum-deposited on opposite faces of a pellet of compressed Mallinckrodt AgI (1.27 cm diameter, 1.0 mm thick, compressed under a 3170 kg load), the batterylike behavior of an array of the type Ag/AgI/Pt was first noticed. The initial open-circuit voltage of such arrays was usually about 0.35–0.40 V. Of particular interest, the flash current was about 100 μa/cm^2 of electrode area. This is about three times higher than the specific flash current of the best of the previous solid electrolyte systems given in Table 11.2. Such encouraging evidence of a sharp decrease in over-all internal resistance gave impetus to further study of the Ag/AgI/Pt system as a possible solid electrolyte battery.

Initial results. Preliminary work [Mrgudich, Schwartz, Bramhall and Schwartz (94)], indicated that the initial discharge curve of a freshly prepared, uncharged Ag/AgI/Pt array showed a closed-circuit voltage which dropped exponentially. Figure 11.11 illustrates a 500,000-ohm discharge curve typical of these early cells. Such cells were also found to be capable of accepting repeated recharging under a 0.40 V charging voltage. In one case the cell was subjected to a total of 42 shallow discharge-charge cycles over a period of some 4 weeks; total input for these 42 cycles was 727 μa-min while total output was some 732 μa-min. The experimental correspondence of input versus output indicates both a high coulombic efficiency and the absence of significant electronic conductivity through the AgI pellet during the 4-week test at room temperature. It was also found that substituting thin-film palladium for platinum was feasible; the Ag/AgI/Pd cell exhibited a somewhat higher flash current at 160 μa/cm^2. Data were presented indicating that the palladium cells were operable, but with sharply diminished output, at temperatures as low as -73°C (-100°F) and with sharply increased output at temperatures as high as 93°C (200°F).

Subsequent progress in improving output capacity. A second paper by Mrgudich, Bramhall, and Finnegan (95) used thinner pellets (0.25 mm thick, 1.27 cm diameter, 0.032 cm^3 volume) and sputtered films of Ag and Pt. Observed flash currents increased to about 600 μa/cm^2, but there was

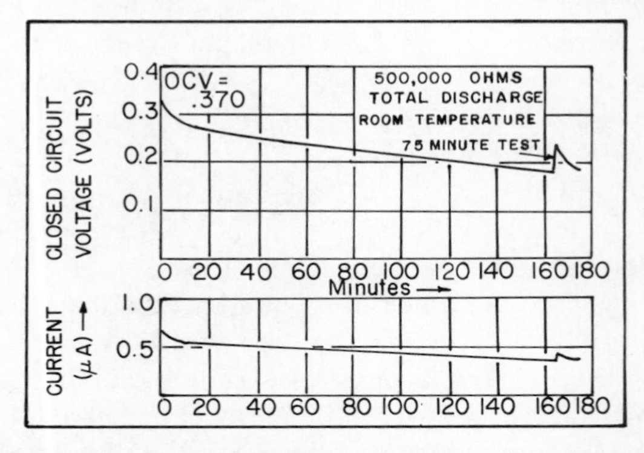

Figure 11.11 Discharge behavior of an early Ag/AgI/Pt cell. [Using a 1-g (1 mm) pellet of compressed AgI powder and vacuum-deposited films of Ag and Pt about 2000 Å thick.]

no significant increase in output capacity per cell. This paper described recharge characteristics under 0.60 V charging voltages. Cell output under a 0.5 megohm discharge showed a slight but steady increase during the first 6 recharge cycles and then became remarkably reproducible for the next 6 cycles (see Fig. 11.12).

A third and more comprehensive paper by Mrgudich, Bramhall, and Schwartz (96) extended and confirmed the recharge capabilities of the cell (Cell # 317). During this testing of recharge capabilities, Cell # 317 was discharged under different loads (see Fig. 11.13). On the heaviest drain (through 0.25 megohm), the cell delivered some 7 min of service to a 0.3 V cutoff (i.e., $\frac{1}{2}$ the open-circuit voltage after charging to 0.6 V). The "average" voltage during this 7-min discharge (calculated by graphical integration of the VT area under the curve and dividing by $T = 7$) was 0.36 V. The corresponding "average" current was 1.44 μ amp. The VIT product yielded an output of 0.061 μw-hr/cell. At a cell volume of 0.032 cm^3, this corresponds to a specific output of 1.9 μw-hr/cm^3. Similarly calculated specific outputs for the 0.5, 1.0, and 2.0 megohm loads to a 0.3 V cutoff yielded values of 3.9, 5.1, and 6.1 μw-hr/cm^3.

Even at 6.1 μw-hr/cm^3, the Ag/AgI/Pt cell output was very low compared to a good liquid electrolyte battery at some 300,000 μw-hr/cm^3 or the best solid electrolyte battery (the Sprague cell) at about 3000 μw-hr/cm^3. It was obvious that efforts should be made to increase output capacity.

One important reason why Cell # 317 with its thin-film electrodes showed such low specific output is that the thin sputtered Pt film cannot

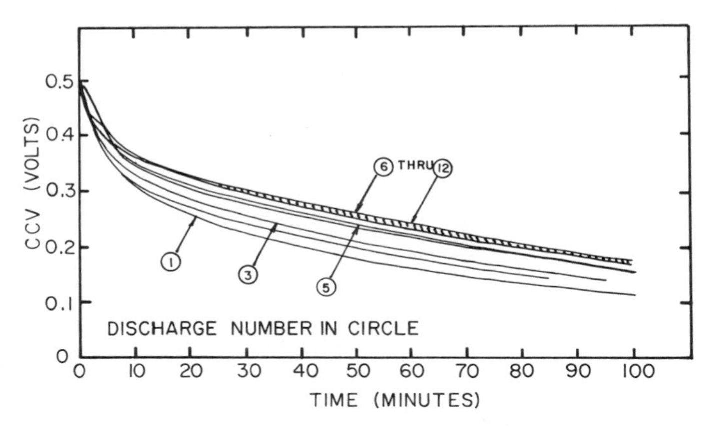

Figure 11.12 Recharge capability of the Ag/AgI/Pt system. (Using a 0.25 mm thick AgI pellet and vacuum deposited electrodes of Ag and Pt. All discharges were through 500,000 ohms after a minimum of 3 days charge under 0.60 V.)

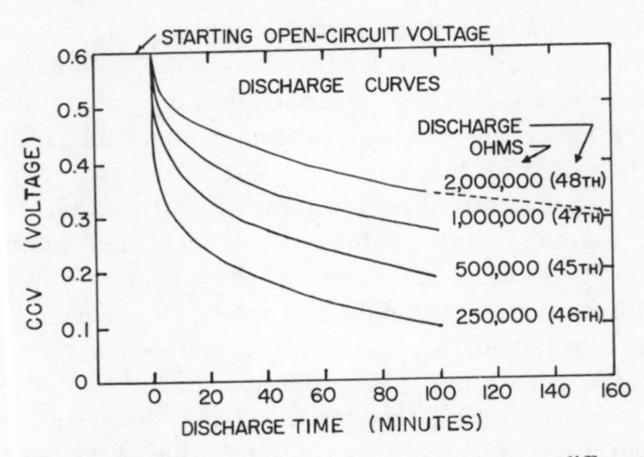

Figure 11.13 Discharge curves of cell of Fig. 11.12 across different loads.

absorb much incoming silver. In principle, all we need do to increase cell output is to increase the amounts of Ag and Pt used. To this end, Bramhall of USAECOM made an Ag/AgI/Pt cell by compressing 0.25 g finely divided Ag powder against one face of a 1.27 cm diameter, 1.0 mm thick pellet of AgI and 0.25 g of platinum black powder against the other face. The over-all thickness was 1.25 mm for a total cell volume of 0.158 cm^3. Its flash current, after charging to 0.58 V, was 450 μ amp. Its discharge curve across a 250,000 ohm load is illustrated in Fig. 11.14. This "more Ag/more Pt" cell gave 39 hr of service to a 0.3 V cutoff. Specific output was about 79 μw-hr/cm^3, or some 40 times better. Use of a thinner pellet of AgI (say, 0.25 mm) should at least triple this specific output, placing the capacity into a range of some 250 μw-hr/cm^3. Use of even larger amounts of electrode materials would probably result in further increases since both the Ag and Pt powders compact rather well. Material costs, however, would become prohibitively large unless palladium is found to be an adequate substitute for platinum.

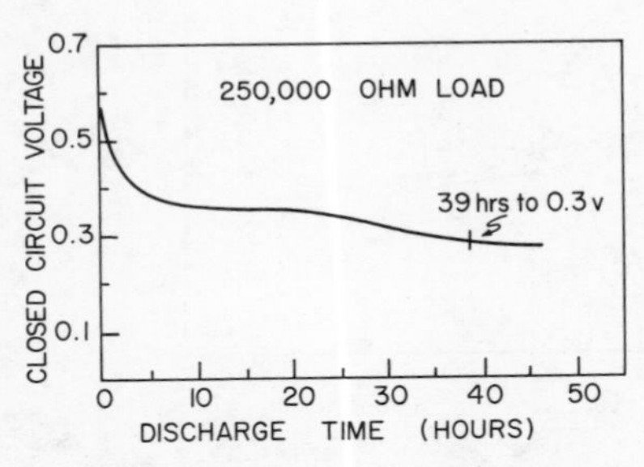

Figure 11.14 Capacity improvement of an Ag/AgI/Pt cell using thicker Ag and Pt electrodes.

Material costs may be lowered appreciably by adopting a somewhat more complicated cell array assembly procedure designed to enhance easier departure of Ag^+ ions from the Ag anode and easier penetration of Ag^+ ions into the Pt cathode. Such (unpublished) improvement is illustrated by the performance of a "laminated" array made in 1967. This was formed by compression in a 1.27 cm diameter pressure cavity of a series of layers composed (in sequence) of 0.1 g of powdered Ag, 0.05 g of a 2 : 1 mixture of Ag and AgI powders, 0.05 g of a 1 : 2 mixture of Ag and AgI powders, 0.50 g of AgI powder, 0.05 g of a 1 : 2 mixture of Pt and AgI powders, 0.05 g of a 2 : 1 mixture of Pt and Ag powders, and, finally, a layer of 0.1 g of Pt powder. Total array weight was 0.90 g, consisting of 0.15 g Ag, 0.60 g AgI, and 0.15 g Pt. Its thickness was about 0.7 mm for a volume of 0.089 cm^3. Charged to 0.6 V, its flash current was about 1800 μa. When discharged across a 50,000 ohm load (see Fig. 11.15), the array yielded 12.8 hr of service to a 0.3 V cutoff for a specific output of about 420 μw-hr/cm^3. Use of thinner AgI (0.25 g) and optimizing the electrolyte : electrode ratios could probably increase this value into the 1000 μw-hr/cm^3 range.

The potential output, taken in conjunction with the established recharge capability and the possibility of a cost-saving Pd-for-Pt substitution, sharply increases the economics of the system. The complicated 6-layer configuration described above, however, requires simplification; we know of no commercially available pelletizing equipment which could fabricate such arrays.

Other characteristics (and problems) of the Ag/AgI/Pt system. The open circuit voltage of charged Ag/AgI/Pt cells stored open to air under moderate clamp pressure reaches a stable value of 0.54 V in 1 month. For example, a series-stack of 3 cells was charged to 0.6 V/cell. In 3 days this dropped to 0.58 V, in 2 weeks to 0.55 V, and in 1 month to 0.54 V, where it remained constant for 2 years. Voltage stability over a period of 2 years is not normally considered entirely satisfactory for the prediction of long-term shelf life. However, remembering that in this system the voltage drops exponentially with discharge, a 2-year stability is strong evidence of the absence of significant internal charge leakage.

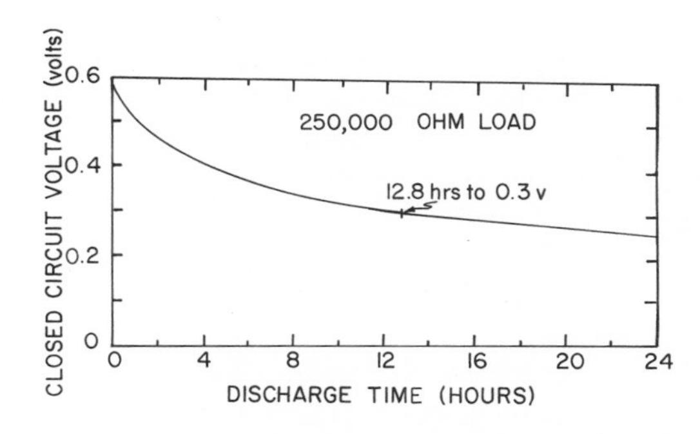

Figure 11.15 Capacity improvement using 'laminated" construction.

Open-circuit voltage stability, although a necessary condition for good shelf life, is not sufficient alone. We must, in addition, insure that secondary effects do not detract from delayed *capacity* output. Time has not been available to carry out these delayed capacity tests, but there is some indirect evidence of potential problem areas. For example, the temperature coefficient of expansion of AgI, although small, is negative, it contracts slightly with increasing temperature [see Mellor (89)] while the coefficients for Ag and Pt are positive. Thus small temperature fluctuations repeated many times over a period of time may disturb the electrode/electrolyte contact and increase over-all cell resistance. This probably accounts for the observed fact that thin 0.25 cm cells with sputtered Ag and Pt films stored uncharged and unclamped for 2 years are definitely inferior to equivalent cells stored under pressure in the test clamps. Incorporation of AgI powder in the electrode powders as described above should alleviate this deficiency. Prompt assembly of fabricated cells into suitably pressure-crimped tubular containers is an obvious manufacturing precaution. Another problem area is that the conductivity of an unclamped pellet (without electrodes) of compressed AgI decreases exponentially with time, leveling off at $\frac{1}{2}$ to $\frac{1}{4}$ of its initial fresh conductivity as shown by Mrgudich (80). Note that Shapiro and Kolthoff (68) (compare Curves 1 and 2 of Fig. 11.9) observed the same effect with AgBr pellets. Still another problem, uncovered by M. N. Hull of USAECOM, is that there is some evidence of iodine evolution at the Pt when Ag/AgI/Pt cells are charged to 0.6 V. It is possible that the drop in open-circuit voltage from 0.60 to 0.54 V in 1 month just described is due to loss of an iodine film generated by the charging process. It seems likely that the maximum *safe* recharging voltage will turn out to be in the range of 0.50–0.52 V.

Summary of the Ag/AgI/Pt(Pd) system. Most of the problem areas described in the preceding subsection, once identified, are thought to be amenable to partial if not complete alleviation. Furthermore, there are several unexplored but promising avenues for further improvement (see, e.g., Fig. 11.10, with its implication that further increase of specific surface area may result in further conductivity improvement). Summing all relevant factors, it appears that the Ag/AgI/Pt(Pd) pellet system is a promising approach to feasible manufacture of reasonably priced, miniature, stable, rugged, rechargeable batteries operable over an extended temperature range for use in applications requiring up to about 10 μa/cm^2 drains per discharge cycle.

THE ATOMICS INTERNATIONAL $Ag/RbAg_4I_5/RbI_3$ SYSTEM. Reference was made earlier to recent work on $RbAg_4I_5$ where an "open" lattice led to an exceptionally high silver ion conductivity. North American Rockwell Corporation, through its Atomics International Division (AI), has conducted an effort to utilize this property of $RbAg_4I_5$ to develop solid electrolyte batteries of exceptionally low internal resistance. This effort is now being continued by Gould Ionics, a new company formed in November, 1968, by North American Rockwell and Gould National Batteries. Other objectives of Gould Ionics involve development of low resistance solid electrolyte electrochemical devices and sensors [see Mrgudich (97)].

In the following discussion, which is based on three publications by Argue, Owens, and Groce [(98), (99), (100)], it must be remembered that the AI system is, in fact, a new system and that many questions, especially those involving reliability, reproducibility and shelf life, cannot be answered at this time.

The basic cell reaction. At first glance one would guess that the function of the cathodic RbI_3 is to supply free iodine by means of the reaction: $RbI_3 = RbI + I_2$; this will react with incoming Ag^+ ions generated by the anode reaction: $Ag = Ag^+ + \text{electrons}$; the discharge products are RbI and AgI. On this basis the over-all cell reaction, similar to that of the Sprague system, would be

$$2Ag + I_2 = 2AgI$$

The AI group, however, using x-ray diffraction methods for product identification, reported that at a discharge temperature of 30°C the discharge products were Rb_2AgI_3 and $RbAg_4I_5$. At a discharge temperature of 0°C they found "no evidence of appreciable $RbAg_4I_5$ whereas AgI and Rb_2AgI_3 were present in the discharged cathode. They related these observations to the Topol-Owens (88) conclusion that $RbAg_4I_5$ is stable above 27°C but slowly transforms into AgI and Rb_2AgI_3 at temperatures below 27°C. Thus above 27°C they propose that the over-all cell reaction is:

$$14\,Ag + 7\,RbI_3 = 3\,RbAg_4I_5 + 2\,Rb_2AgI_3 \qquad (0.66\text{ V})$$

Below 27°C, the proposed cell reaction is

$$4\,Ag + 2\,RbI_3 = 3\,AgI + Rb_2AgI_3$$

Material preparation and battery fabrication. None of the three papers gives complete details for the preparation of $RbAg_4I_5$ or RbI_3. The brief directions given state that the $RbAg_4I_5$ was prepared by melting a 1:4 molecular ratio of RbI and AgI, followed by quenching and grinding. The

RbI_3 was prepared by heating a 1:1 ratio of RbI and I_2 to 125–150°C followed by cooling.

In all cases the $Ag/RbAg_4I_5/RbI_3$ cell consisted of three compressed pellets. The anode pellet was an appropriate mixture of powdered Ag, electronically conductive carbon, and $RbAg_4I_5$. The electrolyte pellet was compressed $RbAg_4I_5$. The cathode pellet was an appropriate mixture of RbI_3, carbon, and $RbAg_4I_5$. The deliberate addition of electronically conductive carbon and ionically conductive $RbAg_4I_5$ to both the silver anode pellet and the RbI_3 cathode pellet served to permit easier flow of electrons and ions throughout the entire anode and cathode pellets with subsequent decrease of polarization effects on discharge.

All pellets were made by compression in a Carver laboratory press. Construction of a typical unit cell is shown in Fig. 11.16. Specifications for several cells and batteries are shown in Table 11.3. Construction of a rugged multicell battery is straight forward since it involves a simple stacking of interlocking unit cells in a plastic-lined stainless steel can, attachment of suitable pickoff leads, and final hot pressing of the entire assembly to seal the plastic rings of the cells to each other and to the can.

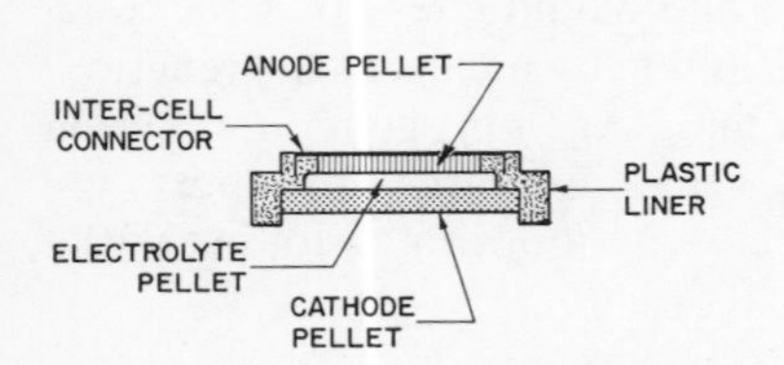

Figure 11.16 Atomics International solid-state cell.

Electrical output. Typical 75°F discharge curves of 2.54 cm diameter cells with a "built-in" cathode capacity of 100 ma-hr are shown in Fig. 11.17. Under a continuous drain across a 130-ohm resistance, the first cell delivered an estimated average current of 5 ma for 1000 min, or

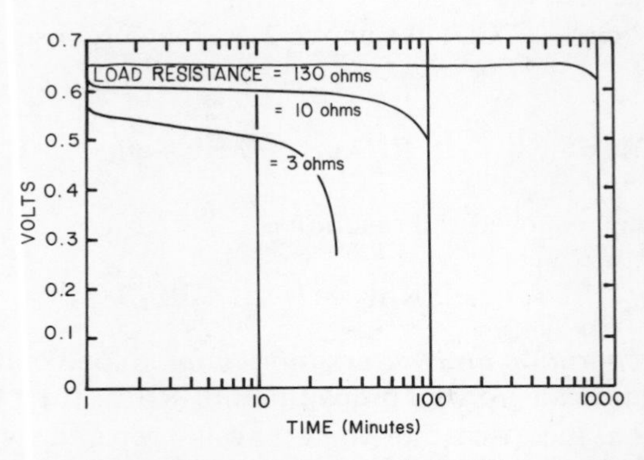

Figure 11.17 Constant-load discharge of AgI solid-state cells at 75°F.

TABLE 11.3. ATOMICS INTERNATIONAL SYSTEM SPECIFICATIONS, CELL AND BATTERY CHARACTERISTICS, PERFORMANCE

Specifications	Low Current Drain[a]	Medium Current Drain[b]	High Current Drain[c]
Cell characteristics			
Cell voltage (V)	0.66	0.66	0.66
Anode wt (g)	0.1	0.07	0.60
Thickness (in.)	0.019	0.007	0.009
Cathode wt (g)	0.2	0.11	0.68
Thickness (in.)	0.040	0.013	0.029
Electrolyte wt (g)	0.3	0.16	1.10
Thickness (in.)	0.020	0.019	0.017
Cell wt (g)	0.7	0.5	3.2
Thickness (in.)	0.085	0.044	0.065
Diameter (in.)	0.485	0.479	0.960
Resistance (ohm)	2 ± 1	0.8 ± 0.2	0.2 ± 0.05
Battery characteristics			
Number of cells	5	10	13
Open-circuit voltage (V)	3.3	6.6	8.58
Battery performance	72 hr at −55°C	After test:	Average of 4 batteries
	144 hr at +25°C	ccv − 4.9 at − 40°C	Temp. (°C) / Calc. bat. resis. (ohms) / Calc. I (amps)
	152 hr at +74°C	ccv − 5.5 at − 18°C	
	After shock:	ccv − 5.8 at 0°C	
	144 hr at +25°C	ccv − 6.0 at +24°C	
	After vibration:		0 / 4.1 / 1.3
	165 hr at +25°C		25 / 2.7 / 1.7

[a] In a battery volume of 0.5×0.5 in., deliver at least 28 μa across 100,000 ohms for 72 hr (−55 − +74°C)

[b] In a battery volume of 0.5×0.5 in., deliver at least 10 ma across 600 ohms for 20 sec (−40− +71°C)

[c] In a battery volume of 1×1 in., deliver at least 800 ma across 2.43 ohms for 8 sec (0−+71°C)

83 ma-hr. Across a 10-ohm resistance, a second cell delivered an estimated average current of 56.8 ma for 100 min, or 95 ma-hr. Across a 3-ohm resistance and to a cutoff of 0.35 V, a third cell delivered an estimated average current of 155 ma for 30 min, or 78 ma-hr. These high coulombic output values, taken in conjunction with the fact that current delivery was at levels some 2–3 orders of magnitude higher than the flash current of the best of previously reported solid electrolyte systems, indicate that "The Common Limitation of Early Systems" can be circumvented.

With respect to energy density in terms of watt-hours per pound of active material, Argue, Owens, and Groce (98) have pointed out that at present the AI system exhibits a value of about 2.2 w-hr/lb compared to a calculated theoretical maximum value of 22 w-hr/lb. Similarly calculated, the theoretical energy density based upon the active anode and cathode alone for the Leclanché system and its alkaline counterpart is 153 w-hr/lb, 131 for the Zn/Ag_2O system, and 116 for Zn/HgO. In terms of watt-hours per cubic inch, the maximum theoretical value for the AI system is 4.2, and 28.6, 28.6, 33.7, and 40.8 for the Leclanché, alkaline Zn/MnO_2, Zn/Ag_2O and Zn/HgO, respectively. Even if the specific outputs of the Atomics International cell were comparable with or even considerably better than those of available aqueous systems, material costs would be so high they would prevent consideration of this system as economically competitive in the broad primary battery market either at present or in the foreseeable future.

Thus it would appear that the most promising application areas for the AI system will revolve about the following inherent advantages of the solid electrolyte approach:

1. Potentially long shelf life without recourse to reserve-battery activation.
2. Freedom from electrolyte leakage.
3. Wider operating temperature range.
4. Rugged construction leading to high resistance to shock and vibration.
5. Improved potential for mechanization of manufacturing process.
6. Improved potential for fabrication of miniature cells.

From the standpoint of continuing interest in the AI system and general impetus to research in solid ionic conductors, it is fortunate that there is a present need for a special family of low voltage, relatively small batteries capable of operation over the range −55 to 71°C under rather severe conditions of shock (13,000 G).

After nearly two decades, solid electrolyte batteries seem at last to be assuming a role—perhaps minor and of limited scope, but nevertheless a role—as new members of the family of modern electrochemical power sources.

Thin-Film Batteries

An intriguing and potentially attractive process for making miniature batteries would be one in which all components of the battery (anode, electrolyte, cathode, intercell connector, strap connectors, external leads, and encapsulating materials) can be brought together through feasible application of a sequence of thin-film vacuum-depositions. Many types of vacuum-deposition equipment are commercially available for special research use or for routine large-scale production. Film thicknesses can be controlled to within 5 percent over a range as low as 50 Å up to 10–20 μ. In addition, many clever masking techniques and film-deposition sequences have been developed so that design configuration can be quite flexible, permitting deposition of thin film batteries in a multiplicity of shapes and onto flat or curved surfaces. Assuming successful thin-film fabrication of batteries, one can visualize the eventual replacement of a battery "compartment" with a battery "surface."

Such long-range, "blue-sky" projections imply that thin-film batteries of the future should be rechargeable since it would be economically unrealistic, except in very special cases, to accept the general thesis that an electronic device can be discarded once its battery has been depleted after but a single discharge. However, one can argue, with some justification, that it hardly makes sense to develop a rechargeable thin-film battery since a larger battery would still be needed for charging purposes and that one might just as well use the larger battery directly. This argument loses much of its validity if we assume, as we did very early in this chapter, that the future will bring with it a need for microwatt power supplies. At these power levels, charging could be from a wound spring or an oscillating weight such as that used to operate certain wrist watches. Another approach to this problem of microwatt charging can be found in recent reports [see Mrgudich (96, 97)] that simple oscillatory bending of a solid Ag/AgI/Ag array the size of a quarter can be used to generate a low impedance (200 ohms) alternating voltage. It is also interesting to note that a very recent announcement by General Electric engineers (101) that electrodes implanted in a rabbit have been used, with no signs of biological damage, to generate 50 μa at 0.8 V for 9 months.

THE SATOR-PERROT $Pb/PbCl_2/AgCl/Ag$ SYSTEM. The earliest attempt to develop all-thin-film solid electrolyte batteries seems to have been made by Sator (29) and Perrot and Sator (102). Several different techniques were used; each culminated as an array of the type $Pb/PbCl_2/AgCl/Ag$ some 10^{-4} to 10^{-3} cm thick. In order to understand the discharge reaction of this "double electrolyte" system it should be pointed out that $PbCl_2$ at room temperature is an ionic conductor for Cl^- ions while AgCl, like AgI, is a conductor for Ag^+ ions [see Friauf (50), pp. 9–70]. On this basis, anodic Pb metal liberates electrons to the external circuit and Pb^{+2} ions to the $PbCl_2$ electrolyte. These Pb^{+2} ions react with mobile Cl^- ions in the $PbCl_2$ to form more $PbCl_2$. The Cl^- ions required for this combination with the Pb^{+2} ions are made available at the $PbCl_2/AgCl$ interface by virtue of the Ag^+ ion conductivity of AgCl which permits Ag^+ ions to move from the interface (leaving it rich in Cl^- ions) toward the Ag cathode where they combine with incoming electrons to form metallic Ag. The over-all discharge reaction, therefore, is probably

$$Pb + 2AgCl = PbCl_2 + 2Ag$$

Sator reported that this system was rechargeable, exhibiting an open-circuit voltage of 0.523 immediately after charging under a 0.5 μa charging current. This open-circuit voltage dropped quite rapidly to a stable value of about 0.44 V. After 7 discharge-charge cycles, initial (charged) open-circuit voltage was 0.44 V, dropping to 0.4 V after 1.5 hr. No discharge data were given in the paper.

THE MELPAR $Pb/PbCl_2/AgCl/Ag$ SYSTEM. A much more comprehensive study of this original Sator system was carried out by Moulton, Hacskaylo, and Feldman (103) using evaporations from heated tungsten boats. The deposition sequence and the final overlapping configuration of their cells is shown in Fig. 11. 18. It will be noted that each final sample consisted of four $Pb/PbCl_2/AgCl/Ag$ cells with a common Ag cathode. The two larger center cells had an effective area of 20 mm^2 while the two smaller outer cells had an effective area of 10 mm^2. Deposition rates for each of the materials were nearly uniform at about 2400–3000 Å (0.24–0.30 μ/min). Each layer thickness was in the range 2.2–2.6 μ for an over-all cell thickness of about 10 μ. The larger central cells therefore had a volume of about 2×10^{-4} cm^3.

Some cells, as formed, had voltages in the range 0.47–0.50 V. The thermodynamically predicted voltage is 0.49 V. The majority of the cells, however, were found to develop little or no voltage. In most of these initially inert cells an electrical activation consisting of passage of a small charging current (in the nanoampere range) for 2 to 7 hr was sufficient to yield cells exhibiting the anticipated voltage.

Cells of the type $Pb/PbBr_2/AgBr/Ag$ were also prepared in the same configuration. Charging under an applied voltage of 0.4 V for about an hour resulted in "bromide" cells whose subsequent open-circuit voltages were in the range 0.33–0.35 V. The theoretical voltage is 0.355 V.

Attempts were made to deposit thin-film $Pb/PbI_2/AgI/Ag$ cells, but cells of near theoretical open-circuit voltage (0.21 V) could not be obtained even after charging activation.

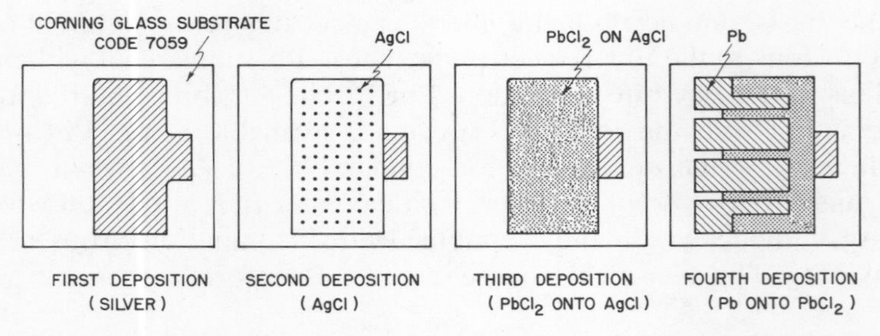

Figure 11.18 Deposition sequence of an all-thin-film $Pb/PbCl_2/AgCl/Ag$ Sator-Melpar cell.

Smyth [see Sprague Electric (56) references] had previously examined the Sator $Pb/PbCl_2/AgCl/Ag$ system, but abandoned this approach when it was observed that some silver deposition took place at the $PbCl_2/AgCl$ interface with subsequent internal shorting. This deposition of Ag metal at the interface may have been due to trace electronic conductivity in AgCl (i.e., some electrons entering the Ag cathode during discharge could move into and through the AgCl to react with Ag^+ ions near the $PbCl_2/AgCl$ interface).

THE HACSKAYLO-FOLEY $Al/CeO_2/Al$ SYSTEM. While investigating the electrical properties of vacuum-deposited dielectric films, Hacskaylo and Foley (104) observed the battery-like behavior of arrays of the type $Al/CeO_2/Al$. They found that such nominally inactive thin-film arrays produced a voltage in the presence of trace moisture. For example, moisture generated by cupping a hand over the cell was sufficient to generate a voltage of 0.1–0.2 V and a short-circuit current of about 30 $\mu a/cm^2$. Research has been described as an environment of fascinating suspense: one can easily imagine the excitement created by this simple experiment.

Further study showed that a drop of water placed on the top Al electrode resulted in generation of an open-circuit voltage of 0.8 V and a flash current of some 350 $\mu a/cm^2$. The moist top Al electrode was always positive with respect to the bottom Al electrode. Substituting a gold film for the top Al electrode resulted in similar moisture-generated voltage and current effects. Substituting a gold film for the bottom Al film, however, resulted in arrays exhibiting no galvanic response to moistening of the top electrode whether it was gold or aluminum.

The authors explained these and other results by the logical postulate that atmospheric oxygen and moisture yielded a cathodic reaction at the top electrode given by

$$2H_2O + O_2 + 4e = 4OH^-$$

Thus the cathodic reaction takes place whether the top metal film is Au or Al. The anodic reaction at the bottom electrode (substantially protected from air by the glass substrate and the covering CeO_2 film) was the oxidation of Al to Al_2O_3 or $Al(OH)_3$. The CeO_2 film acts merely as a moist separator. (Substitution of a $PbCl_2$ film for the CeO_2 yielded an array whose open circuit voltage was 0.6 V when water was placed on the top electrode and delivered a short-circuit current of about 1000 $\mu a/cm^2$ for a few minutes.) This cell is really not a solid electrolyte cell: it could appropriately have been classified as a "thin-film trace liquid electrolyte" system discussed earlier.

It might be mentioned in passing that this system, or one like it, possibly might be developed into a practical humidistat with convenient current or voltage readout.

THE TECHNICAL OPERATIONS Ag/Ag HALIDE/Pt SYSTEMS. Vouros, Clune, and Masters (105) carried out a one-year intensive effort to demonstrate feasibility of vacuum-deposition fabrication of Ag/Ag halide/Pt cells. Vouros and Masters (106) have further reported of these thin-film silver-halide electrolyte cells. It was anticipated, and later confirmed, that the major difficulty involved the deposition of a thin film of solid halide which would be free of pinholes while exhibiting acceptably high Ag^+ ion conductance and substantially infinite electronic resistance.

Deposition techniques and deposition sequence: Early work on Ag/AgI/Pt cells. Platinum films were deposited using a Consolidated Vacuum Corporation AST-100 low energy sputtering unit. Deposition rate was at about 50 Å/min. Final film thicknesses were in the range 0.1–0.2 μ. Electron beam evaporation of platinum would have been much faster but was avoided because of possible harmful effects of the high temperatures involved with such evaporations. Silver films 0.1–0.2 μ thick were deposited by evaporation from heated tantalum filament coil-cups. Silver halide films 5–10 μ thick were deposited by evaporation from heated tungsten boats. Initial work showed that deposition onto glass substrates yielded cells of erratic voltage. To avoid the possibility that this erraticism was due to substrate conductivity, all reported data are for depositions on thin quartz substrates. Initial work also showed that evaporation of AgI films was invariably preceded by evolution of iodine vapor. Since such iodine vapors would react with previously deposited silver films, a deposition sequence of inert platinum-evaporated halide-evaporated silver was adopted for all reported work. Figure 11.19 illustrates the adopted deposition sequence. It also shows the cell configuration used for most of the work.

Initial work attempted to fabricate Ag/AgI/Pt cells. Using auxiliary leads to monitor cell voltage during deposition of the final silver film onto

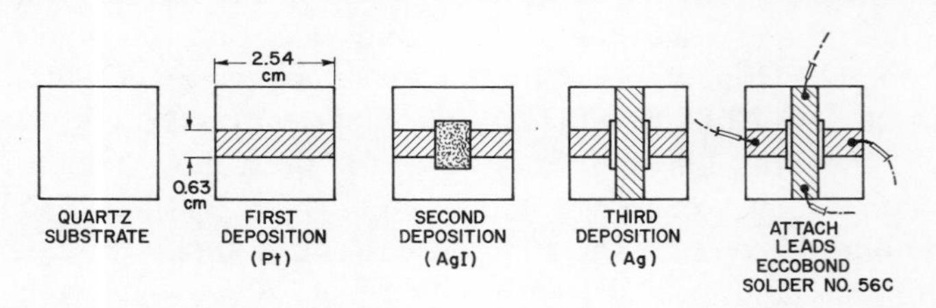

Figure 11.19 Deposition sequence of an all-thin-film Ag/AgI/Pt Technical Operations cell. (Pinholes in evaporated AgI films prevented fabrication of stable cells.)

the deposited AgI film, it was found that only a very small amount of silver could be evaporated before a slowly rising voltage (usually not in excess of 0.3 V) suddenly dropped to zero and stayed there. This was interpreted as indicative of pinholes in the AgI film through which evaporated silver threads could reach the underlying Pt film with resultant internal shorting.

Ag/AgBr/Pt cells. Following other preliminary, but unsuccessful, efforts to avoid pinholes in evaporated AgI films, the work turned to a study of Ag/AgBr/Pt cells, since previous experience had shown that AgBr could be evaporated to yield pinhole-free films as thin as 0.2 μ. Although the Ag^+ ion conductivity of AgBr is some two or three orders of magnitude less than that of AgI and although the literature reports some trace electronic conductivity in AgBr (especially in the presence of free bromine), the advantage of pinhole-free electrolyte films was considered sufficiently important to warrant some exploratory effort using AgBr. After evaporation of a 5–10 μ AgBr cell using the deposition sequence and configuration of Fig. 11.19, the cell (0.4 V initial open-circuit voltage) was charged to a potential of 0.5 V and subjected to a series of discharge-charge cycles. The resultant discharge curves are shown in Fig. 11.20. Using initial voltage drop data (see Eq. (1)), the cell resistance was estimated at about 4×10^4 ohms/cm^2.

Ag/AgI-AgCl/Pt cells. Since it was known that AgCl, like AgBr, could be deposited as pinhole-free thin film, cells were prepared using a deposition sequence of sputtered Pt/0.625 μ AgCl-5 μ AgI/Ag. Such cells exhibited a very high room-temperature resistance estimated at about 4×10^7 ohms/cm^2. Since cell resistance should drop with increasing tem-

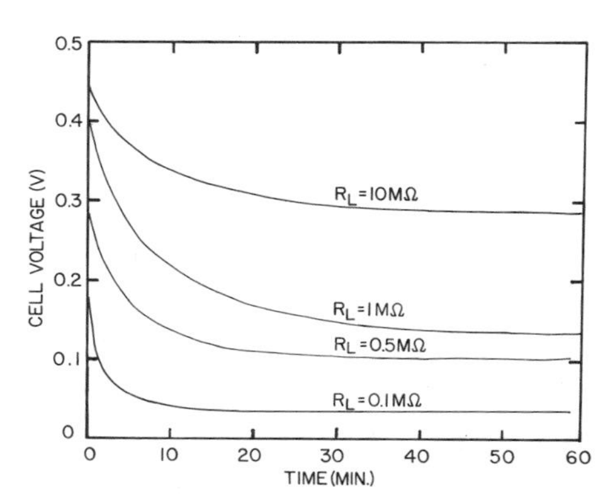

Figure 11.20 Discharge curves of a rechargeable all-thin-film Ag/AgBr/Pt Technical Operations cell.

perature, the AgCl-AgI cell was discharged at higher temperatures across a 1-megohm load. The resulting discharge curves are shown in Fig. 11.21. Each discharge was preceded by a 1-hr charge at 0.5 V.

Ag/evaporated AgI-iodized Ag/Pt cells. The long-term stability of the Union Carbide $Ag/AgI/V_2O_5$ system has already been described. The AgI film used in this process was prepared by iodine vapor conversion of the surface of a silver layer to AgI. Such chemically prepared AgI is apparently quite adherent and, in view of subsequent cell stability during storage, apparently pinhole-free. Technical Operations' approach to a thin-film equivalent of the Union Carbide iodized-Ag film involved deposition of a 1000-Å film of Ag directly onto the sputtered Pt film. This thin film of Ag was then at least partially (if not completely) converted to AgI by a 24-hr storage in an iodine chamber. A 5-μ layer of evaporated AgI was then deposited on the chemically prepared AgI film, followed by a film of Ag evaporated in the configuration of Fig. 11.19. Cell voltage was monitored during deposition of the final Ag film: it never rose above 0.05–0.06 V, presumably due to the presence of some free, unconverted Ag either on or in the Pt film. The cell was then charged at 0.5 V and subjected to a series of discharge-charge cycles. Resultant discharge data are shown in Fig. 11.22. The cell exhibited excellent voltage stability in the charged state. After storage at room temperature for 2 months its open-circuit voltage was 0.47 V, identical with the ocv immediately after charging to a potential of 0.50 V. The cell was then subjected to a resistance measurement involving estimation of the external resistance required to drop cell voltage to one-half its starting ocv (in this case from

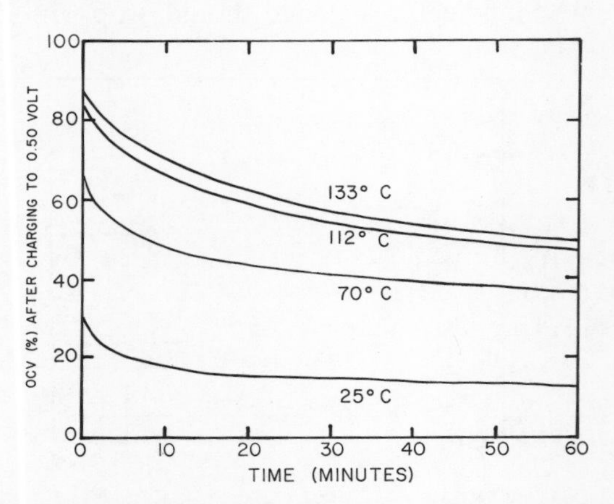

Figure 11.21 Discharge curves of a rechargeable all-thin-film Ag/AgI-AgCl/Pt Technical Operational cell. (Showing high-temperature discharge possibilities.)

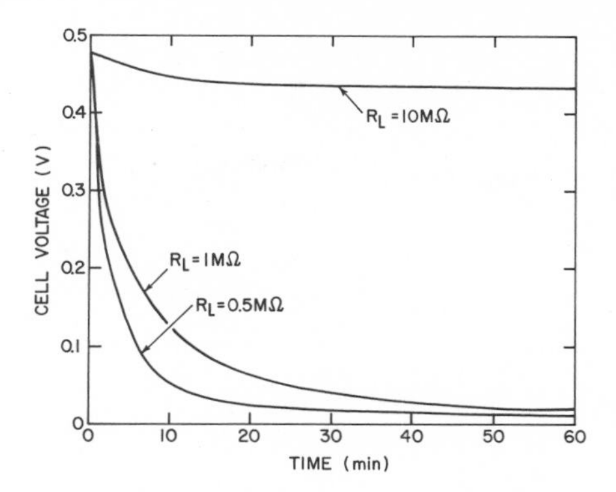

Figure 11.22 Discharge curves of rechargeable and apparently stable Ag/evaporated AgI-iodized AgI/Pt Technical Operations cell.

0.47 to 0.235 V). The resistance was of the order of 10^4–10^5 ohms. Such resistance measurement imposes a momentary but nevertheless heavy drain on the cell and its subsequent ocv dropped to 0.40–0.41 V where it remained for approximately 1.5 months.

Elimination of AgI pinholes by evaporation onto heated substrates. Through electron microscope studies, later work showed that AgI films deposited on substrates heated above 150°C (i.e., above the 145.8°C transition temperature of AgI to the highly conductive α-phase) resulted in pinhole-free films when the film and substrate were subsequently cooled to room temperature. One possible explanation of this is that α-AgI when cooled below its transition temperature converts to the β- and γ-phases with a 5.4 percent volume expansion [see Perrott and Fletcher (107)]. Such expansion on cooling could contribute to a closing of voids (if indeed they exist at higher substrate temperatures). Time was not available to pursue this direct and promising approach to the formidable AgI pinhole problem.

POSSIBLE DEPOSITION SEQUENCES FOR PRODUCTION-SCALE FABRICATION OF SERIES-CONNECTED THIN-FILM BATTERIES. All work to date has been concentrated on fabrication of a satisfactory thin-film *cell.* Assuming ultimate successful development of such cells, the question arises concerning possible ways by which such cells could be deposited to yield series-connected multicell batteries. Figure 11.23 represents an early proposal (Mrgudich, 1965), but this is subject to certain disadvantages, chief of which is that a slight misalignment of any of the several masking steps could ruin the entire assemblage. A simpler and somewhat more foolproof deposition sequence is illustrated in the "ribbon" battery of

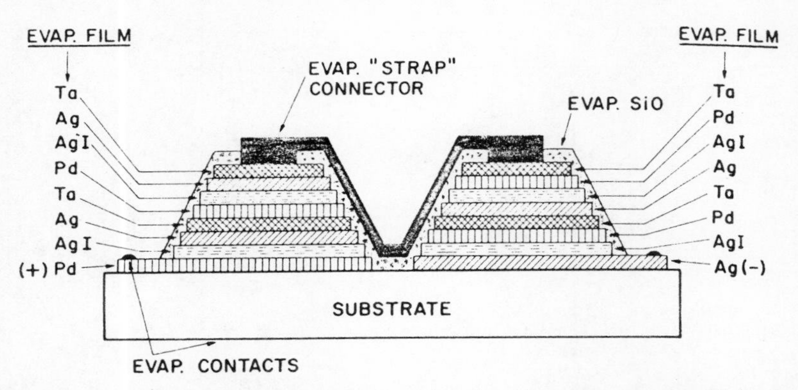

Figure 11.23 A proposed but probably impractical deposition sequence for fabrication of series-connected all-thin-film batteries.

Fig. 11.24 (96). Still another sequence, and possibly the simplest, is illustrated in Fig. 11.25, proposed by Mrgudich and Wilcox (unpublished).

Summary. The substantially immediate promise of the Atomics International $RbAg_4I_5$ battery and the longer-range promise of the Technical Operations work on "printable" batteries are evidence that the foundations for future miniature batteries have been laid. The nature and configuration of these batteries of tomorrow cannot, of course, be detailed now, but it seems valid to assume that the electrochemical imagination and creativity required to meet the needs of an evolving technology can build successfully upon these foundations.

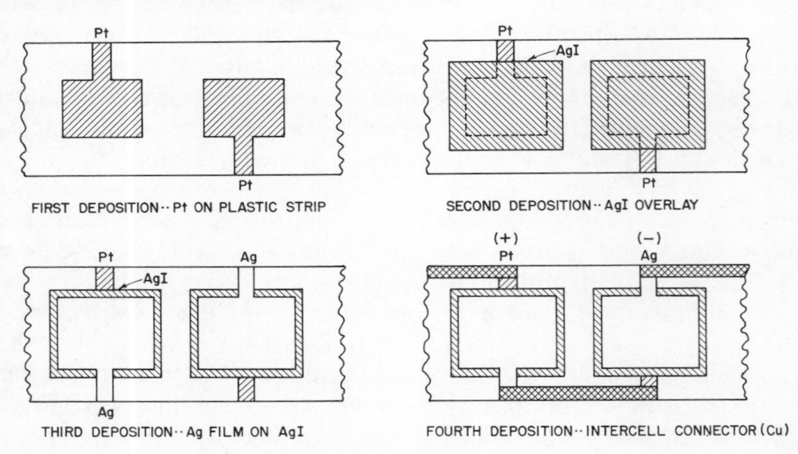

Figure 11.24 A proposed deposition sequence for fabrication of a series-connected all-thin-film "ribbon" battery.

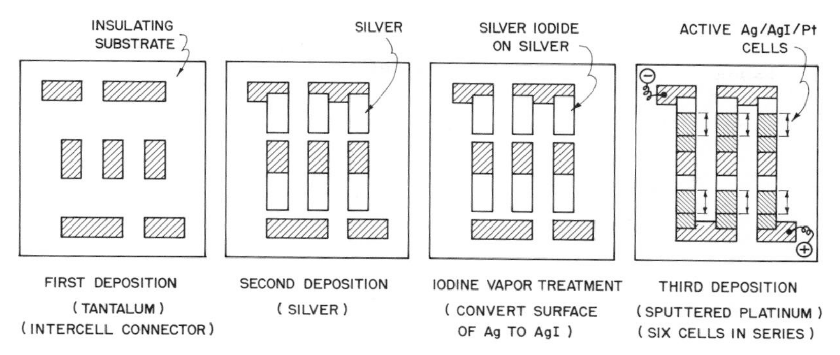

Figure 11.25 A proposed deposition sequence for fabrication of a series-connected all-thin-film Ag/iodized AgI/Pt (Pd) battery on a fixed substrate.

It would be quite appropriate to conclude the discussion of miniature batteries on this optimistic note. To do so, however, would be to ignore recent developments [see Mrgudich (97)] which indicate that electrochemical systems—and especially those employing solid electrolytes—can be used in a variety of applications, none of which involves generation of power as the primary function.

SOLID ELECTROLYTE SENSORS: A NEW HORIZON

Deformation sensing

Consider a simple, dime-sized Ag/AgI/Ag solid electrolyte array consisting of thin powdered silver electrodes compressed onto opposite faces of a pellet of compressed AgI powder. Its voltage is obviously zero. However, when such an array is subjected to even a slight bending force, it generates a small, substantially constant voltage. The compressed Ag electrode is negative with respect to the Ag electrode in tension. It is as if compression tends to "squeeze" positive silver ions into the AgI, thus leaving the electrode negative. Voltage output is proportional to the bending force at about 1 μV/μin. strain up to 1 mV (pellet breaking points). The output increases with array thickness and can be further increased by series connection of simultaneously responding arrays. It is substantially independent of temperature providing that both electrodes are at the same temperature. Array impedance is quite low at about 100 ohms so that the output is amenable to low-noise amplification and subsequent display on a recorder or oscilloscope. Arrays will respond to frequencies at least as high as 5000 cps. In short, the array behaves like a strain gage with a unique advantage—it generates its own voltage. Because of this, power requirements are lowered so that the device can permit simplification of unattended, long-term monitoring of remote terrestrial, or even lunar deformable elements such as diaphragms, bellows, bourdon tubes, accelerometers, lightweight seismographs, microphone pickups, and covert alarm systems.

Temperature sensing

It has long been known that if one electrode of an Ag/AgI/Ag array is warmed, the array generates a voltage whose magnitude is related to the temperature difference between the two electrodes. The warmer electrode is negative and the temperature coefficient of voltage is about 0.8 mV/°C. This is some 10–20 times higher than that of the best conventional thermocouples at 0.04–0.08 mV/°C. The authors know of no reported study to exploit this high coefficient. For such exploitation, consider an Ag/AgI/Ag array in which one Ag electrode is extremely thin while the AgI and other Ag electrode are very much thicker. Mount the array in a suitable insulating cylinder so that only the thin electrode surface is exposed. The heat capacity of this thin electrode is quite low so that even a small heat input will reflect as a relatively high temperature change and hence as a relatively high voltage change. It has been found that array response is both sensitive and remarkably fast. A mere whiff of warm breath directed toward the thin exposed electrode resulted in a 1-msec response of 1000–2000 μV. "Talking" and indeed inaudible "whispering," to the electrode generated a rapidly fluctuating voltage as the electrode responded to the varying gusts of warm breath. The array can easily detect a difference between fingertip temperatures or a change in fingertip blood circulation associated with holding the arm upward or downward. Cooling a fingertip in ice water followed by estimation of the rate of temperature recovery may be a measure of circulatory defficiencies. Other skin surface temperature measurements can be easily made through momentary physical contact. The area of the exposed sensing electrode can be made quite small for precision plotting of surface isotherms. (Note that a cancerous growth is some 1–2°C warmer than normal surrounding tissue.) In principle, an ionic temperature sensor could be mounted on the tip of a catheter for measurement of internal temperatures. The diagnostic value of such temperature sensing, where voltage outputs can be easily read into a computer for storage or analysis, is but one of many unexplored application areas.

An ionic balancing device

Passage of direct current through an Ag/AgI/Ag array will transport silver from one electrode to the other with resultant change in the over-all center of gravity. Sensitivity can be very high; 1 μa-sec of easily controlled external current will transport 1.12×10^{-9} g silver. Exploitation of this simple concept of easily controlled balancing awaits the efforts of some enterprising group of development engineers.

An ionic adaptive memory element

Consider a three-terminal array of the type $Ag/AgI/Pt_1/AgI/Pt_2$. The voltage between the Ag and Pt_1 can be adjusted by suitable charging (see previous discussion of the USAECOM pellet battery). The voltage between Pt_1 and Pt_2 can be made equal to zero by preliminary shorting. Now momentarily discharge across Ag and Pt_1. The discharge delivers silver from Ag to Pt_1 which, being now richer in silver than Pt_2, becomes negative with respect to Pt_2. The device can be viewed as an adaptive memory element in which "read-in" is current passing through $Ag\text{-}Pt_1$ and "read-out" is a voltage change between Pt_1 and Pt_2. Erasure can be accomplished by connecting Pt_1 and Pt_2 in parallel and charging against the Ag electrode. Use of a four-terminal array of the type $Ag_1/AgI/Pt_1/AgI/Pt_2/AgI/Ag_1$ and alternate discharge between Ag_1 and Pt_1 and between Ag_2 and Pt_2 can yield a positive or negative voltage readout between Pt_1 and Pt_2.

Summary

The preceding brief and far from complete recitation of but a few examples of solid electrolyte ionic sensors and responders is given primarily to tease the reader's imagination and ingenuity. We will conclude by pointing out that all of the many electrical effects known to occur in living systems originate because of the interplay of *ionic* charge transport between biological fluids separated by membranes of variable ionic permeability. The God of living things—of birds and bees and cabbages and kings—uses ionic, not electronic, charge transport in His systems.

REFERENCES

1. Radio Corporation of America, *Scientific American*, **218**, No. 3, 17 (March 1968).
2. deLuc, J. A., *Nicholson's J.*, **27**, 81, 161, 241 (1810); **28**, 5 (1811): through *New Edinburgh Ency.*, Vol. 8, p. 293, Amer. Ed., David Brewster, Ed., Edward Parker, Philadelphia (1816).
3. Mottelay, P. F., *Bibliographical History of Electricity and Magnetism*, Griffin and Co., Ltd., London, pp. 404–5 (1922).
4. Zamboni, G., *Della Pila Elettrica a Secco*, Verona (1812); *Ann. chim. phys.*, **11**, 190 (1812); cf. Mottelay (3), pp. 420 ff.
5. Elliott, A., *Electronic Eng.*, **20**, 317 (1948).
6. Howard, P. L., *J. Electrochem. Soc.*, **99**, 333 (1952).
7. Victoreen, J. A., U.S. Pat. 2,666,801 (1954).
8. Reiner, I., U.S. Pat. 2,701,272 (1955).

9. Bjorksten, J., U.S. Pat. 2,688,649 (1954).
10. Wood, R. E., *J. Electrochem. Soc.*, **103**, 417 (1956); U.S. Pat. 2,762,858 (1956); Proc. Twelfth Annual Battery R & D Conf., Fort Monmouth, N.J., p. 62 (1958).
11. Kirkwood, J. G., and West, F. W., U.S. Pat. 2,747,009 (1956).
12. Corren, S. A., U.S. Pat. 2,853,537 (1958).
13. Zablocki, H. S., U.S. Pat. 2,918,518 (1958).
14. Louis, A. S., U.S. Pat. 2,843,649 (1958).
15. Thompson, H. S., *J. Royal Agr. Soc. Engl.*, **11**, 68 (1850).
16. Nachod, F. C., and Shubert, J., *Ion Exchange Technology*, Academic Press, New York (1956).
17. Kitchener, J. A., *Ion Exchange Resins*, Methuen, London (1957).
18. Kunin, R., *Ion Exchange Resins*, 2nd ed., John Wiley and Sons, New York (1958).
19. Salman, J. E., and Hale, D. K., *Ion Exchange*, Academic Press, New York (1959).
20. Kunin, R., *Elements of Ion Exchange*, Reinhold, New York (1960).
21. Meyer, K. H., *Helv. Chim. Acta*, **23**, 795 (1940).
22. Juda, W., and McRae, W. A., U.S. Pat. 2,636,851 (1953).
23. Robinson, P., U.S. Pat. 2,786,088 (1957).
24. Pauli, W. J., U.S. Pat. 2,851,510 (1958).
25. Schwartz, M., and Franklin, P. J., U.S. Pat. 2,844,642 (1958); U.S. Pat. 2,895,000 (1959).
26. Harding, M. S., U.S. Pat. 2,831,045 (1958).
27. Grubb, W. T., Proc. Eleventh Annual Battery R & D Conf., Fort Monmouth, N.J., p. 5 (1957); *J. Electrochem. Soc.*, **106**, 275 (1959); U.S. Pat. 2,861,116 (1958); U.S. Pat. 2,933,547 (1960).
28. Ruben S., U.S. Pat. 2,001,978 (1935).
29. Sator, A., *Compt. rend.*, **234**, 2283 (1952).
30. Lehovec, K., and Broder, J., *J. Electrochem. Soc.*, **101**, 208 (1954).
31. van der Grinten, W. J., *J. Electrochem. Soc.*, **103**, 201C (1956).
32. Lehovec, K., U.S. Pat. 2,689,876 (1954); U.S. Pat. 2,696,513 (1954).
33. Broder, J. D., U.S. Pat. 2,690,465 (1954).
34. Bradshaw, B. C., and Shuttleworth, R. A., U.S. Pat. 2,718,539 (1955).
35. Shorr, W., U.S. Pat. 2,778,754 (1957).
36. Hack, A. J., Shapiro, S. J., Shorr, W., U.S. Pat. Reissue 24,408 (1957).
37. Louzos, D. V., U.S. Pat. 2,894,053 (1959).
38. Evans, G. E., U.S. Pat. 2,894,052 (1959).
39. Shapiro, S., Proc. Eleventh Annual Battery R & D Conf., Fort Monmouth, N.J., p. 3 (1957).
40. van der Grinten, W. J., U.S. Pat. 2,793,244 (1957).
41. van der Grinten, W. J., and Mohler, D., U.S. Pat. 2,928,890 (1960).
42. Wagner, B. F., Proc. 1957 Electronic Components Symposium, IRE, Inc., p. 133; *Electronic Design*, p. 44 (Oct. 1, 1957).
43. Luckey, G., and West, E., *J. Chem. Phys.*, **24**, 879 (1956).
44. Weininger, J. L., *J. Electrochem. Soc.*, **105**, 578 (1958).
45. Ruben, S., U.S. Pat. 2,707,199 (1955); U.S. Pat. 2,816,151 (1957); U.S. Pat. 2,852,591 (1958).
46. Buchinski, J. J., et al., U.S. Pat. 2,932,569 (1960).
47. Richter, E. W., et al., U.S. Pat. 3,004,093 (1961).
48. Louzos, D. V., Battery Division Extended Abstracts, Boston Meeting, The Electrochem. Soc., p. 88 (1962).
49. Lieb, H. C., U.S. Pat. 2,930,830 (1960).

50. Friauf, R. J., *Am. Inst. of Physics Handbook*, 2nd ed., D. E. Gray, Ed., Section 9f-1 to 9f-7, McGraw-Hill Book Co., New York (1963).
51. Lehovec, K., Proc. Tenth Annual Battery R & D Conf., Fort Monmouth, N.J., p. 50 (1956).
52. Smyth, D. M., Proc. Twelfth Annual Battery R & D Conf., Fort Monmouth, N.J., p. 64 (1958).
53. Smyth, D. M., and Cutler, M. E., *J.A.C.S.*, **80**, 4462 (1958).
54. Smyth, D. M., *J. Electrochem. Soc.*, **106**, 635 (1959).
55. Smyth, D. M., and Shirn, G. A., U.S. Pat. 2,905,740 (1959).
56. Sprague Electric Company, Final Report, Solid Electrolyte Battery Systems, Contract DA-36-039-SC-63151 (30 Nov. 1956); Contract DA-36-039-SC-72349 (31 Aug. 1957).
57. Koch, E., and Wagner, C., *Z. Phys. Chem. Abt.* **B38**, 295 (1937).
58. Ebert, I., and Teltow, J., *Ann. Physik*, **15**, 268 (1954).
59. Weininger, J. L., *J. Electrochem. Soc.*, **106**, 475 (June, 1959); U.S. Pat. 3,003,017 (1959).
60. *Id.*, *J. Electrochem. Soc.*, **105**, 439 (August, 1958).
61. Tubandt, C., *Handbuch der Experimentalphysik*, Vol. 12, Part 1, Akad. Verlag, Leipzig, pp. 383–470 (1932).
62. Barrer, R. M., *Diffusion in and through Solids*, Cambridge Univ. Press, London, Chap. VI (1951).
63. Foussereau, *Ann. Chim. Phys.*, **5**, 241, 371 (1885); cited by Barrer (62), p. 257.
64. Konigsberger, J., *Physik. Z.*, **8**, 833 (1907).
65. Phipps, T. E., Lansing, W. D., and Cook, T. G., *J.A.C.S.*, **48**, 112 (1926).
66. Lidiard, A. B., *Encyclopedia of Physics*, Vol. *22*, Springer, Berlin, p. 246–349 (1957).
67. Frenkel, J., *Z. Physik*, **35**, 652 (1926).
68. Shapiro, I., and Kolthoff, I. M., *J. Chem. Phys.*, **15**, 41 (1947).
69. Schottky, W., *Z. phys. Chem.*, Abt **B29**, 335 (1935).
70. Strock, L. W., *Z. phys. Chem.*, **B25**, 441 (1934).
71. Tubandt, C., and Lorenz, F., *Z. phys. Chem.*, **87**, 513 (1914).
72. Wells, A. F., *Structural Inorganic Chemistry*, Oxford, 134 and 271 (1950).
73. Helmholtz, L., *J. Chem. Phys.*, **3**, 740 (1935) .
74. Donnay, J. D. H., Donnay, G., Cox, E. G., Kennard, O., and King, M. V., *Crystal Data*, 2nd ed., Amer. Crystal. Assn., Williams and Heintz, Washington, D.C., pp. 751, 867, 907 (1963).
75. Ketelaar, J. A. A., *Trans. Faraday Soc.*, **34**, 874 (1938).
76. Suchow, L., and Pond, G. R., *J.A.C.S.*, **75**, 5242 (1953).
77. Weil, R., and Lawson, A. W., *J. Chem. Phys.*, **41**, 832 (1964).
78. Reuter, B., and Hardel, K., *Naturwiss.*, **48**, 161 (1961).
79. Takahashi, T., and Yamamoto, O., *Denki Kagaku* (Japan), **31**, 42 (1963); **32**, 610,664 (1964); **33**, 346, 518, 733 (1965). See *Chem. Abstracts*, **61**, 15458; **62**, 14173; **63**, 7697; **64**, 1622, 10465. (An article in English can be found in *Electrochemica Acta*, **11**, 911 (1966)—*Chem. Abstracts*, **65**, 5016 (1966). The authors are indebted to Dr. Gary Argue for an English translation of the **32**, 610 (1964) article. Also Dr. Takahashi kindly supplied us with English translations of several of his articles.)
80. Mrgudich, J. N., *J. Electrochem. Soc.*, **107**, 475 (1960).
81. Owens, B., Argue, G. R., Groce, I. J., and Hermo, L. D., *J. Electrochem. Soc.*, **116**, 312 (1969).
82. Bradley, J. N., and Greene, P. D., *Trans. Faraday Soc.*, **62**, 2069 (1966).
83. *Ibid.*, **63**, 424 (1967).

84. *Ibid.*, **63**, 1023 (1967).
85. *Ibid.*, **63**, 2516 (1967).
86. Owens, B. B., and Argue, G. R., *Science*, **157**, 308 (1967).
87. Geller, S., *Science*, **157**, 310 (1967).
88. Topol, L. E., and Owens, B. B., *J. Phys. Chem.*, **72**, 2106 (1968).
89. Mellor, J. W., *Comp. Survey of Inorg. and Phys. Chem.*, Vol. 3, Longmans-Green pp. 426–436 (1922).
90. Welch, A. J. E., *Chemistry of the Solid State*, W. E. Garner, Ed., Academic Press, New York, Chap. 12, pp. 302–3 (1955).
91. Goddard, P. E., and Urbach, F., *J. Chem. Phys.*, **20**, 1975 (1952).
92. Stone, F. S., In (90), Chap. 2, pp. 44–49.
93. Clark, G. L., *Applied X Rays*, 3rd ed., McGraw-Hill Book Co., New York, p. 363 (1940).
94. Mrgudich, J. N., Schwartz, A., Bramhall, P. J., and Schwartz, C. M., Proc. Nineteenth Annual Battery R & D Conf., Fort Monmouth, N. J., p. 86 (1965).
95. Mrgudich, J. N., Bramhall, P. J., and Finnegan, J. J., IEEE Trans. on Aerospace and Electronic Systems, AES-1, p. 290 (1965).
96. Mrgudich, J. N., Bramhall, P. J., and Schwartz, C. M., 1966 Army Science Conference Proceedings, Vol. 2, p. 127 (available through Clearinghouse of Federal Scientific and Technical Information, Dept. of Commerce, Springfield, Va., 22151).
97. Mrgudich, J. N., "Batteries as Temperature and Pressure Sensing Devices," Proc. Twenty-first Annual Power Sources Conference, Fort Monmouth, N.J. (May 1967); also "Ionic Sensors," 1968 Army Science Conference Proceedings (to be available through Clearinghouse of Federal Scientific and Technical Information, Dept. of Commerce, Springfield, Va., 22151); also "Electronically Active Ionic Devices," *Army R & D Newsmagazine*, **9**, 26 (November 1968) available through Supt. of Documents, U.S. Govt. Printing Office, Washington, D.C. 20402.
98. Argue, G. R., Owens, B. B., and Groce, I. J., Proc. Twenty-Second Annual Battery R & D Conf., Fort Monmouth, N.J. (May 1968).
99. *Id.*, Proc. Sixth International Power Source Symposium, Brighton, Sussex, England (Sept. 1968).
100. *Id.*, Preprint 22D, 61st Annual Mtg. Am. Inst. Chem. Eng. (Dec. 1968).
101. (General Electric), "Artificial Hearts May Use Body Power," *Electronics*, **41**, 33–4 (Oct. 28, 1968).
102. Perrot, M., and Sator, A., *Compt. rend.*, **234**, 1883 (1952).
103. Moulton, C. W., Hacskaylo, M., and Feldman, C., "The Electrochemistry of Thin Films," presented at the Toronto Meeting, Electrochemical Society (May, 1964). Note: A copy of this presentation was made available to us by Mr. Hacskaylo.
104. Hacskaylo, M., and Foley, R. T., *J. Electrochem. Soc.*, **113**, 1231 (1966).
105. Vouros, P., Clune, J., and Masters, J. I., Technical Operations, Inc., Burlington, Mass., Final Report, "Fabrication of Ultrathin Solid-Electrolyte Batteries," Contract DAAB-07-67-C-0339 (Aug. 1968).
106. Vouros, P., and Masters, J. I., *J. Electrochem. Soc.*, **116**, 880 (1969).
107. Perrott, C. M., and Fletcher, N. H., *J. Chem. Phys.*, **48**, 2143 (1968).

12

Standard Cells

Walter J. Hamer

As physical representations of the unit of electromotive force (emf), standard cells fulfill a unique role and serve a special purpose. They serve in the maintenance of that unit, and they are used as the standards with which emf and *i-r* drop of other cell systems can be compared. Quality has been all-important, cost a secondary consideration. Open-circuit voltage, as measured to five decimal places, must be reasonably reproducible and stable over long periods of time. The cells are not intended to supply current and therefore differ from conventional primary batteries in function as well as in design. Investigators of academic and industrial background have so thoroughly studied the subject that the modern standard cell probably represents an application of the most highly developed technology in the primary battery field.

Standard cells serve in the maintenance of the unit of emf in various national laboratories, such as the National Physical Laboratory (Teddington, England), Laboratoire Central des Industries Électriques (Fontenay-aux-Roses, France), Physikalisch-Technische Bundesanstalt (Braunschweig, Germany), the Electrotechnical Laboratory (Tokyo, Japan), the Institute of Metrology of U.S.S.R. (Leningrad, U.S.S.R.), and the National Bureau of Standards (Washington, D.C., U.S.A.). The units of emf of the various nations involved are compared every third year at the Bureau International des Poids et Mesures, Sèvres, France. These comparisons are effected by means of standard cells maintained by the participating countries and by the International Bureau.

Owing to their special use, standard cells are required to meet certain performance criteria and, for precise measurements, to have certain inherent characteristics. They must be reasonably reproducible, exhibit long lifetimes, possess low emf-temperature coefficients, be relatively insensitive to minute current drains, have an emf of convenient magnitude, and have a low or moderately low internal resistance. Very few primary cells, as will be seen later, meet these severe requirements.

Historically, primary cells have played a significant role in the evolution of the electrical units. They were used alone or in series as early *arbitrary* standards of potential difference; results were expressed in terms of the effects the cells produced, rather than in units, and could be stated with sufficient accuracy merely by stating the number of cells in the circuit. Faraday's law of electrolysis was widely used to report results; thus, in a Faraday experiment two or three Daniell cells in series were stated to give twice or three times the effect, through the same resistor, of a single cell, and so on. Together with standards of resistance, standard cells were used in establishing magnitudes of electric current.

The word "standard" in the expression "standard cell" has its roots in this history. Since a standard is a physical representation of a unit, it is obvious why permanency and high reproducibility are of prime importance in a standard cell. The precision with which standard cells are measured, accordingly, exceeds that normally required for other types of primary cells.

ELECTRICAL UNITS

The practical unit of emf, the volt, is not an arbitrary one but like the other electrical units is derived from the basic mechanical units of length, mass, and time using the principles of electromagnetism with the value of the permeability of free space taken as $4\pi/10^7$ on the Giorgi rationalized mksa (meter-kilogram-second-ampere) system of units. Other systems for the basic units of length, mass, and time were used previously but the mksa system, which is a part of the Système Internationale d'Unités, was recommended for universal adoption by a resolution of the General Conference on Weights and Measures, Paris, 1960. With the permeability of free space taken as $4\pi/10^7$, the mksa system is the same as the practical volt-ohm-ampere system.

It has been customary, following the first use of the term by Gauss,[1] to refer to electrical units based on the basic units of length, mass, and time as absolute electrical units.[1] The transition from arbitrary to absolute units began with the work of Gauss (1) in 1833 and of Weber (2) in 1851; Weber showed that it was possible to measure electrical quantities in terms of mechanical units. However, the electrical units determined in Weber's mms (millimeter-milligram-second) or in the cgs (centimeter-gram-second) electromagnetic (e.m.) systems are of inconvenient size

[1]It is unfortunate that the name "absolute" has persisted. It is sometimes wrongly interpreted to imply that there are no errors involved in the measurements or that perfection has been attained.

for practical use. For example, Thomson (Lord Kelvin) (3) in 1851 showed that the emf of a Daniell cell was about 1×10^8 cgs e.m. units. Even so, a Committee of the British Association for the Advancement of Science in 1873 recommended the cgs e.m. system for use in both science and in practical engineering. The electricians and telegraphists, however, although they agreed with the desirability of relating electrical units to mechanical ones, objected to the use of the cgs e.m. system in practice because of the magnitudes involved and since they had been using such terms as ohms and volts for their physical standards. The practitioners' view steadily gained favor and in 1881 the International Congress of Electricians meeting in Paris adopted the cgs e.m. system of units as the fundamental system and the volt-ohm-ampere system for practical use, with the practical units being made larger or smaller than the corresponding cgs e.m. units by an appropriate power of 10. As time went on the practical volt-ohm-ampere units gradually became the real entities, and their forebears, the cgs e.m. units, became their shadows.

The following factors were chosen for emf, resistance, and current:

1 volt (practical unit) = 10^8 cgs electromagnetic units of emf
1 ohm (practical unit) = 10^9 cgs electromagnetic units of resistance
1 ampere (practical unit) = 0.1 cgs electromagnetic unit of current

The factor 10^8 was chosen for the emf since the emf of a Daniell cell, widely used at that time as a rough standard of emf, became approximately 1 volt. The factor 10^9 was chosen for the resistance for in this way the value of the Siemens mercury column, already used as a resistance standard, especially on the European continent, became approximately 1 ohm (actually about 0.94 ohm). The factor for the ampere was then fixed as 0.1 by the requirement of Ohm's law. These factors are now academic since, as has been stated, the use of the mksa system of units makes the theoretical (or absolute) and the practical systems identical.

To date, no direct absolute measurement of the unit of emf in the e.m. system has been found feasible. Instead its value is established experimentally through Ohm's law by the measurement of the fall of potential produced in a resistance by a current, with resistance and current determined in absolute measure[2] as follows:

[2]The unit of emf may also be determined in cgs electrostatic (e.s.) units using absolute electrometers. The accuracy of the methods, however, is much lower than the e.m. approach, the order of magnitude being 1 part in 10,000. The unit of emf in e.s. units is therefore arrived at indirectly by converting e.m. to e.s. values using the experimental factor $(2.997925 \pm 0.000003) \times 10^{10}$ cm sec^{-1}, the speed of propagation of an electrical disturbance (within experimental error, the speed of light in free space). As an example, Thomson (Lord Kelvin) (4) found the emf of a Daniell cell to be 0.00374 cgs e.s. unit from the attraction between two parallel discs connected to the opposite poles of a Daniell cell. This value of Thomson is equivalent to 1.121×10^8 cgs e.m. units or 1.121 V.

The *ohm in absolute measure* may be obtained in terms of length and time by means of inductance and frequency (5), in which a mutual inductor of known dimensions, and thus of calculable inductance, is placed in a suitable circuit containing the resistor whose value is to be determined, a battery, a galvanometer, and two rotary reversing switches. The primary of the mutual inductor is placed in series with the battery and resistor; the secondary is connected to the potential terminals of the resistor through the galvanometer. The rotary switches serve to reverse the connections to the primary and to the secondary. When balance is attained, as indicated by the galvanometer, the induced emf ($4nIM$) in the secondary equals the IR drop in the resistor, the current I having the same magnitude in both instances. Since M, the mutual inductance, may be calculated from dimensions referred to length standards and from the permeability of the medium; and n, the frequency of commutation is measured indirectly in terms of the unit of time, the resistance R is given in terms of the basic units of length and time and of the permeability of the space surrounding the windings of the inductor. The over-all accuracy of the method is about 5 ppm (parts per million). The ohm may also be obtained with a similar accuracy using a self-inductor (6) or with an accuracy of about 2 ppm using a computable capacitor (7).

The *ampere in absolute measure* is obtained in terms of length, mass, and time by means of a current balance (8), in which the electrodynamic force of attraction or repulsion, in the x (vertical) direction, between two coils (one movable and one stationary) through which electricity is flowing, is balanced against the gravitational attraction of a known mass m having a weight mg where g is the acceleration due to gravity. From this force and the measured dimensions of the coils (from which the rate of increase, dM/dx, of the mutual inductance between the stationary and movable coils may be calculated), together with the permeability of space in which the coils act, the current is expressed in terms of units of length, mass, and time, where time enters into the calculations through the acceleration of gravity. The over-all accuracy of the method is about 5 ppm. The ampere may also be obtained with similar accuracy by an electrodynamometer in which a torque is measured (9).

The *volt in absolute measure* is obtained at the same time a measurement of the absolute ampere is made. The usual procedure is to connect in series with the coils of the current balance a resistor having a value, R, previously obtained in absolute measure. The voltage drop, IR, in this resistor is opposed to the emf, E, of a standard cell. When the current, I, is held at such a value that IR balances E and at the same time the mass, m, of the weight has been adjusted to balance the electrodynamic force, the relation

$$E = R\left[\frac{mg}{(dM/dx)}\right]^{1/2} = RI \tag{1}$$

follows and gives emf of the standard cell in absolute measure. Considering the uncertainties in the absolute determinations of R and I, the value of the volt in absolute measure should not be in error by more than 7 parts per million (rms).

The emf of any primary cell could be determined in a similar manner. Obviously, such a procedure would be cumbersome and if required would be very undesirable, since the absolute measurements involved require painstaking and time-consuming work and therefore are unsuitable for frequent or routine measurements of emf. The circumvent the necessity for frequent absolute measurements, standard cells are used and their emfs are determined in relation to current and resistance in absolute measure. These cells are then used to maintain the volt in the interval of five or ten years, between absolute determinations conducted at national laboratories. In other words, the results of absolute measurements are preserved in a physical object, the standard cell.[3] The validity and realization of this approach depends on the possibility of constructing standard cells, the emfs of which are independent (or nearly so) of time; otherwise, the unit of emf would be lost or drift in value in the interim and a repetition of absolute measurements would be imperative at frequent intervals.

It is obvious that this matter was and is of critical importance. Over the years extensive work was conducted to find that electrochemical system which would exhibit constant emfs for long periods of time, say decades, or in battery vernacular have the perfect or nearly perfect shelf life. Fortunately, and anticipating what is to follow, such an electrochemical system has been found.

It is essential that the dissemination of the unit of emf, i.e., the transfer

[3]Standard resistors, usually hermetically sealed coils of annealed manganin wire, are used similarly to maintain the unit of resistance. Owing to the transitory nature of an electric current, no physical standard, to date, has been found feasible. The ampere in absolute measure, however, is embodied in coulometers wherein the current of electricity is found to be that steady current which in a specified time will produce a specified chemical change; specifically that steady current which will cause the dissolution of 1.117972 mg of silver per second (10). The current producing this amount of dissolution is known in terms of the ratio of an absolute emf and an absolute resistance or is obtained directly by an absolute method such as a current balance.

of the unit from a standardizing laboratory, be carried out with high precision involving a minimum of uncertainty. In this case, comparisons are made between reference cells—standard cells whose emf have been determined directly in absolute units—and unknown cells—other standard cells whose emf have not been so determined. These comparisons are nearly one-to-one ratio measurements and therefore may be made with high accuracy with potentiometers by the *difference method.* In this method the two cells (reference and unknown) are connected in series with their emfs in opposition and the difference in emf between the two is measured with a potentiometer using a galvanometer of high sensitivity. The difference in emf between the two cells is small, and only a moderate accuracy in its determination is required to give the emf of the unknown cell in terms of the known reference cell. Thus, if the difference between the two cells were 100 μV the accuracy of the measurements need be only 1 percent to give the difference to 1 μV or only 0.1 percent to give an accuracy of 0.1 μV in the unknown cell.

The difference method has the disadvantage that the operator must determine which of the two cells has the higher emf, i.e., the method is not direct reading. In addition to this disadvantage the accuracy of the method may be limited by the presence of parasitic thermal emfs in the galvanometer circuit. At the National Bureau of Standards a special comparator, designed by Brooks (11), is used; it is read directly and compensates for parasitic thermal emfs. Other types of potentiometers that may be used are cited by Vincent (12) and Miller (13). Other details associated with the dissemination and the maintenance of the unit of emf are given in reference (14).

DEVELOPMENT OF STANDARD CELLS

Several primary cells have been proposed as standards of emf even though many did not meet the criteria of a standard cell; these criteria, however, became evident *a posteriori.* Although Faraday and others of his time used various cells such as Grove and Bunsen cells in their investigations, the Daniell cell (15) of 1836

$$(-)\mathrm{Zn(s)}|\mathrm{ZnSO_4,H_2SO_4(aq)}\|\mathrm{CuSO_4(aq)}|\mathrm{Cu(s)}(+)$$[4]

was the first one used seriously as a standard of emf, as it exhibited less gassing than its predecessors. However, being of the two-fluid type, it

[4]In this chapter a single vertical line is used to indicate the interface of two distinct phases, a double vertical line to indicate a liquid junction, and s = solid, l = liquid, aq = aqueous solution, and for cells described later c = crystals and sat aq = saturated aqueous solution.

did not exhibit a constant emf with time. Somewhat better results were obtained when saturated solutions (no acid) and amalgamated zinc were used, but even these cells did not show the permanency required in a standard. Nevertheless, for over 35 years, Daniell cells were used as rough standards of emf, thus accounting for a large part of the delay in placing electrical units on a sound basis. Absolute electrical measurements gave 1.07–1.14 V for the emf of new cells, the actual value depending on the concentration and acidity of the solution used. Absolute measurements were not needed to show the lack of constancy in the emf of the cells; this behavior was evident from comparing new cells with old ones.

Clark Cell

In 1872 Latimer Clark (16) proposed a cell which had a profound effect on electrical measurements and on work pertaining to the electrical units. His cell,

$$(-)\mathrm{Zn(s)|ZnSO_4{\cdot}7H_2O(c)|ZnSO_4(sat\ aq)|ZnSO_4{\cdot}7H_2O(c)|Hg_2SO_4(s)|Hg(l)}(+)$$

was a one-fluid type without a liquid junction. The over-all cell reaction is

$$\mathrm{Zn(s)+Hg_2SO_4(s)}+\frac{7}{m-7}(\mathrm{ZnSO_4}{\cdot}m\mathrm{H_2O})(\mathrm{sat\ aq}) = \frac{m}{m-7}(\mathrm{ZnSO_4{\cdot}7H_2O})(\mathrm{c})+2\mathrm{Hg(l)} \qquad (2)$$

where m denotes the number of moles of water associated with 1 mole $ZnSO_4$ in the saturated solution. Clark used a very simple construction (Fig. 12.1*a*). Various modifications of this general construction were used in commercial cells; one such modification is the Board of Trade (British) cell illustrated by Fig. 12.1*b*. In all of these the zinc anode was in contact with a paste consisting of mercurous sulfate, a saturated solution of zinc sulfate, and crystals of $ZnSO_4{\cdot}7H_2O$.

Clark determined the emf of his cells in absolute units with (*a*) a sine galvanometer and (*b*) an electrodynamometer with a B.A. resistor.[5] With these he obtained 1.45735 V and 1.45761 V, respectively, at 15.5°C, the mean of which he rounded to 1.457 V as the absolute emf of the Clark cell. Clark's cell, although much better as a standard than its predecessors,

[5]British Association for Advancement of Science coil of silver-platinum alloy (66.6 percent Ag, 33.4 percent Pt). The B.A. ohm = 0.9866 international ohm = 0.9871 absolute ohm.

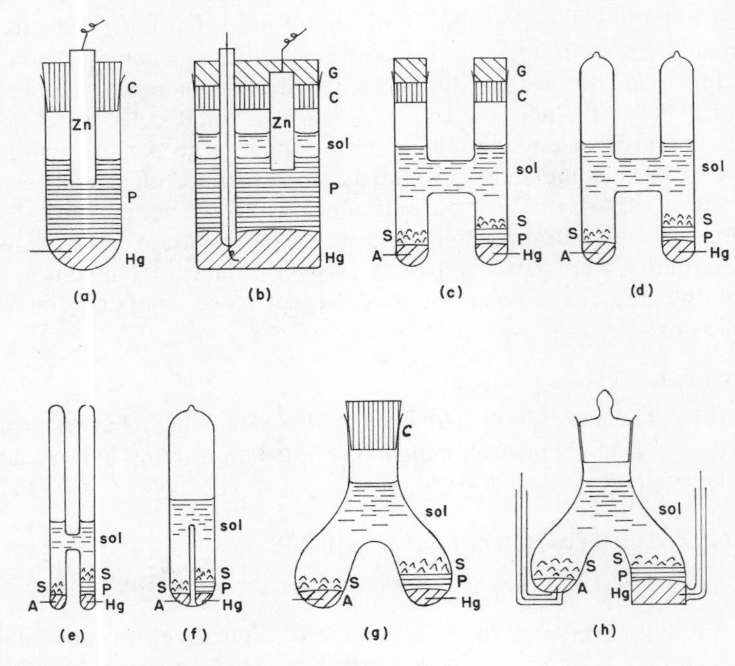

Figure 12.1 Forms of Clark or Weston standard cells.
(a) Original design of Clark
(b) Board of Trade (British) commercial Clarke cell
(c) Rayleigh-Sidgwick unsealed H-shaped cell
(d) Callendar-Barnes sealed H-shaped cell
(e) Hulett short crossarea H-shaped cell
(f) Single tube cell with partition
(g) Kahle's open-stoppered Wright-Thompson inverted γ-form cell
(h) Cooper modific of Kahle's cell.

did not show the permanency in emf hoped for. The cell tended to gas at the zinc anode and the emf showed large variations owing mainly to concentration gradients that developed within the compact pasto during slight changes in ambient temperature.

In 1884 Rayleigh and Mrs. Sidgwick (17) introduced two modifications in the Clark cell that resulted in substantial improvements. They substituted zinc amalgam, (percentage composition not stated)[6] for zinc and

[6]Rayleigh and Mrs. Sidgwick presented their paper in June, 1884; in the paper published later (Dec., 1884) they added the note "Some H-cells have been set up . . . with amalgams of known composition, varying from $\frac{1}{32}$ zinc to $\frac{1}{5}$ zinc by weight. The duration of the test has as yet been scarcely adequate, but it appears that the smaller quantity of zinc is sufficient."

used an H-shaped container in which the zinc anode could be kept entirely under solution and out of contact with the mercurous sulfate paste (Fig. 12.1*c*.) The zinc amalgam reduced the rate of gassing at the anode and by being under solution and out of contact with the mercurous sulfate paste the zinc anode exhibited more nearly constant potentials. They obtained 1.453 V for the absolute emf of their cells at 15°C using a current balance (now known as the Rayleigh balance) and a B.A. resistor. These cells tended to leak in time and Callendar and Barnes (18) were the first to recommend the hermatic sealing of H-shaped cells (Fig. 12.1*d*).

Other types of cell containers, also illustrated in Fig. 12.1, have been proposed. Hulett (19) suggested a shorter cross-arm (Fig. 12.1*e*) and today some cells are made with no cross-arm but with a partition at the base of a single tube (Fig. 12.1*f*). These forms when used in the construction of unsaturated cells give cells of somewhat shorter life because of the shortness of the diffusion path between the two electrodes. Wright and Thompson (20) proposed the inverted-Y form, hermetically sealed, and Kahle (21) proposed the same form but with a cork or ground glass stopper (Fig. 12.1*g*). These forms are more difficult to fill. Cooper (22) proposed a modified Kahle type (Fig. 12.1*h*) which required no support in a thermostatically controlled bath; his form could rest on a flat surface without support. The Cooper cell could also be used partially submerged in water as well as in oil because the cell terminals were not exposed but protruded above the bath fluid. Today, the H-shaped container, hermetically sealed, is used most widely, although some single-tube types are made either like (Fig. 12.1*f*) or in a modified form of the original Clark design (Fig. 12.1*a*).

With increases in temperature the emf of the Clark cell decreases about 0.01 percent/°C. For the range 0–28°C Callendar and Barnes (18) gave

$$E_{t^\circ} = E_{15} - 0.0012(t - 15^\circ) - 0.0000062(t - 15^\circ)^2 \tag{3}$$

for the relation between emf and temperature. At 15 and 25°C, this relation gives -0.00120 and -0.001076 V/°C, respectively, for dE/dt. Kahle (23) and Jaeger and Kahle (24) also studied the temperature dependence of the emf of Clark cells, obtaining results in substantial agreement with Eq. (3). On the other hand, Giauque, Barieau, and Kunzler (25), by plotting $[(E_t - E_{15^\circ}) + 0.00127t]$ against t, concluded that the data of Callendar and Barnes give -0.001340 V/°C for dE/dt at 25°C.

The Clark cell played an important role in the formulation of international and national standards of emf. In 1893 the International Electrical Congress meeting in Chicago chose the Clark cell as the basic standard of emf to which they assigned a value of 1.434 international[7] volts at 15°C. The Clark cell and this value for its emf were legalized as standards in the United States by an act of Congress, July 12, 1894 (14). The value 1.434 international volts at 15°C followed from the determinations of Rayleigh and Sidgwick mentioned earlier (17) and the determinations of Carhart (26), Kahle (23), and Glazebrook and Skinner (27), who used current balances or silver coulometers (the electrochemical equivalent of Ag having been determined by absolute methods) to determine current and B.A. or British legal ohms (known in cgs e.m. units) as standards of resistance. Clark cells served as standards of emf until 1908 (*vide infra*).

WESTON CELLS

Following the success obtained with the Clark cell many other one-fluid cells were proposed as standards including the Weston *unsaturated* form (28) of Clark cell, which was found to have an emf-temperature coefficient of about one-half that of the saturated type of cell (unsaturated and saturated refer to the electrolyte). For a cell made with a solution having a concentration equivalent to one saturated at 0°C, Carhart (29) gave

$$E_t = E_{15°}[1 - 0.000387(t - 15°) + 0.0000005(t - 15°)^2] \quad (4)$$

Carhart also found that the emf-temperature coefficient was practically independent of the density of the solution providing no crystals of $ZnSO_4 \cdot 7H_2O$ were present. An unsaturated cell (electrolyte concentration equivalent to one that is saturated at 0°C) has an emf which is about 6.5 mV higher than the saturated type of cell at normal temperatures.

Other one-fluid cells proposed as standards were those of de la Rue (30), Helmholtz (31), Gouy (32), and Weston (33). The de la Rue cell, (−) Zn(s)|$ZnCl_2$(aq)|AgCl(s)|Ag(s) (+), has an emf of 1.03 V at 15°C and a dE/dt of about −0.001 V/°C. The Helmholtz cell, (−) Zn(s)|$ZnCl_2$(aq)|HgCl(s)|Hg(l) (+), has an emf of 1.00 V at 15°C and a dE/dt of 0.0000857 V/°C and thus, unlike other standard cells, has a positive dE/dt. The Gouy cell, (−) Zn(s)|$ZnSO_4$(10% aq)|HgO(s)Hg(l), has an emf of 1.39 V at 12°C and a dE/dt of about −0.0014 V/°C. The Weston cell, (−) Cd(s)|$CdSO_4$(aq)|Hg_2SO_4(s)|Hg(l) (+), is similar to the Clark unsaturated cell

[7]The values were referred to as "international" because international agreement had been obtained. The international ohm was defined in terms of the resistance offered to an unvarying electric current by a column of mercury at the temperature of melting ice, of constant cross-sectional area, 106.3 cm in length, and weighing 14.4521 g. The international ampere was defined as that steady current which deposits from a solution of silver nitrate 1.118 mg silver/sec. The international volt was defined in terms of the international ohm and ampere or as 1000 ÷ 1434 of the electromotive force of the Clark cell.

except that Cd replaces Zn. It was stated by Weston to have an emf of 1.019 V over a temperature range of 200°F, i.e., to have a negligible dE/dt (Weston was referring to an unsaturated cell).

For all of these cells, amalgamated anodes were recommended and H-shaped containers were generally used. Of these four cells, the first three are inferior to the Clark cell as standards while the last one is superior. The first three exhibit larger drifts in emf with time and have higher internal resistance than the Clark cell, whereas the Weston cell has a much smaller dE/dt (about 1/30 that of the Clark cell), tends to gas at a slower rate, and has an emf closer to unity than the Clark cell. The Weston cell is also made with saturated solutions of cadmium sulfate; crystals of $CdSO_4 \cdot \frac{8}{3}H_2O$ are added to assure saturation. A small amount of sulfuric acid is also added to some Weston cells (*vide infra*).

The saturated Weston cell made with amalgamated anodes may be represented by

$$(-)\ \mathrm{Cd,Hg(2p)}|\mathrm{CdSO_4 \cdot \tfrac{8}{3}H_2O(c)}|\mathrm{CdSO_4(sat\ aq)}|\mathrm{CdSO_4 \cdot \tfrac{8}{3}H_2O(c)}|\mathrm{Hg_2SO_4(s)}|\mathrm{Hg(l)}\ (+)$$

where 2p = two phases (liquid and solid) and the other symbols have the meaning given in discussing the Daniell cell. The cell reaction is

$$x\mathrm{Cd},(y\mathrm{Hg})(2\mathrm{p}) + \mathrm{Hg_2SO_4(s)} + \frac{8/3}{m-8/3}(\mathrm{CdSO_4} \cdot m\mathrm{H_2O})(\mathrm{sat\ aq}) = \frac{m}{m-8/3}(\mathrm{CdSO_4 \cdot \tfrac{8}{3}H_2O})(\mathrm{c}) + 2\mathrm{Hg(l)} + (x-1)\mathrm{Cd}, y\mathrm{Hg}(2\mathrm{p\ or\ 1}) \qquad (5)$$

where x moles of Cd are associated with y moles of Hg in the amalgam and m is the number of moles of water associated with 1 mole of $CdSO_4$ in the saturated solution. At 25°C, $m = 15.089$. At the end of the reaction the amalgam may be two phases (solid and liquid) or a liquid phase only, depending on the extent of the reaction or the relative amounts of amalgam and mercurous sulfate used in preparing the cell. When not discharged, as is normal when the cell is used as an emf standard, the amalgam remains in two phases. This cell made with a 10 percent amalgam has a nominal emf of 1.018636 V at 20°C or 1.018410 V at 25°C. The saturated cell is also referred to as the Weston Normal Cell (or Element).

Two equations have been proposed relating the emf of saturated Weston cells with temperature. Wolff (34) working with $12\frac{1}{2}$ percent (2-phase) amalgams, obtained

$$E_t = E_{20°} - 0.00004060(t-20°) - 0.000000950(t-20°)^2 + 0.000000010(t-20°)^3 \qquad (6)$$

in international volts where E_t is the emf at temperature t and $E_{20°}$ is the emf at 20°C. It is valid from 12 to 40°C but may be used to 0°C as long as the liquid and solid phases of the amalgam are both present. Vigoureux and Watts (35), using 6 and 10 percent amalgams to extend the temperature range, obtained

$$E_t = E_{20°} - 0.00003939(t-20°) - 0.000000903(t-20°)^2 + 0.00000000660(t-20°)^3 - 0.000000000150(t-20°)^4 \qquad (7)$$

in international volts. These two equations are also applicable in absolute volts. Equation (7) is valid from −20 to 40°C; Vigoureux and Watts used the 6 percent amalgam for the lower temperatures but the 10 percent amalgam gives the same results as long as the amalgam remains in two phases; thereafter the emf decreases below that given by Eq. (7). Both equations show that the emf exhibits a maximum value at 3°C, the temperature of the minimum solubility of $CdSO_4 \cdot \frac{8}{3}H_2O$.

The unsaturated Weston cell differs from the saturated type only in that an unsaturated solution of cadmium sulfate and no crystals of $CdSO_4 \cdot \frac{8}{3}H_2O$ are used. It is customary to use a solution having a concentration equivalent to one that is saturated at 3 or 4°C; the solution is then unsaturated at higher temperatures since the solubility of $CdSO_4 \cdot \frac{8}{3}H_2O$ increases with temperature (36). The emf of the unsaturated cell is about 0.05 percent higher than that of the saturated cell. The cell reaction is simply

$$x\text{Cd},(y\text{Hg})(2\text{p}) + \text{Hg}_2\text{SO}_4(\text{s}) = \text{CdSO}_4(\text{in sol}) + 2\text{Hg}(\text{l}) + (x-1)\text{Cd},(y\text{Hg})(2\text{p or }1) \qquad (8)$$

unless the reaction is continued until crystals of $CdSO_4 \cdot \frac{8}{3}H_2O$ are formed; then the reaction is the same as the saturated cell. Again, when the cell is not discharged, as is normal when the cell is used as an emf standard, the solution remains unsaturated and the amalgam in two phases. The effect of temperature on the emf of unsaturated cells will be discussed later.

Because of its advantage over the Clark cell, the Weston Normal Cell was given increased attention, and in 1908 the International Conference on Electrical Units and Standards (37) meeting in London officially accepted it as the standard of emf. This Conference adopted 1.0184 international V as the emf of the Weston Normal Cell at 20°C but recommended additional studies. As a result, scientists from France, Germany, and Great Britian met with American scientists at the National Bureau of Standards in Washington, D.C., in 1910, and as a result of joint measurements with mercury resistance columns and silver coulometers found the emf of the Weston Normal Cell at 20°C to be 1.0183 int. V (38). Values derived from this were later assumed to be significant to the fifth, sixth, or seventh decimal as a basis of measurement (14). This value of 1.0183 int. V served as the fundamental basis of all voltage measurements throughout the world from 1911 to 1948. By 1948, improved techniques led to a final realization of the electrical quantities in cgs. e.m. units, and on January 1, 1948, changes from international to absolute units officially went into effect.[8] The changes (39) for the United States were:

1 international volt = 1.00033 absolute volts
1 international ohm = 1.000495 absolute ohms
1 international ampere = 0.999835 absolute ampere

Other electrical quantities, such as the coulomb and farad, had corresponding changes (39). Additional absolute electrical measurements from 1948 to 1969 showed that better factors for the volt and ampere were, respectively, 1.0003384 and 0.9998435, and that no change in the factor for the ohm was necessary. These new factors were adopted officially in the United States on January 1, 1969. The conversion factors in other countries were nearly the same as those given above and those adopted in the United States on January 1, 1969. The magnitude of the changes from international to absolute units is insignificant for many purposes but is of importance in scientific work of the highest precision and accuracy and is of importance when data of one era are compared with those of another. All emfs in this chapter are based on the 1969 factor.

Miscellaneous types

Although the Weston Normal Cell proved to be a very satisfactory standard of emf, other cells have, from time to time, been proposed as standards. As early as 1911, Brönsted (40) proposed the cell

$$(-)\ \mathrm{Pb, Hg(1)|K_2SO_4 \cdot PbSO_4(c)|K_2SO_4(sat\ aq)|Hg_2SO_4(s)|Hg(l)}\ (+)$$

as a standard, stating that it had an emf of 1.0481 V at 22°C and an emf-temperature coefficient of −0.0001 V/°C. Henderson and Stegeman (41), however, stated that the Brönsted cell did not exhibit steady emfs with

[8]Although "international" and "absolute" are frequently used in referring to the units, it is now best to consider these terms of historical interest only. There is only one kind of volt in the e.m. system of units, and the original terms had significance only during the time when efforts were being made to achieve the theoretical unit. No adjective is now necessary for the volt; the unit is now, within experimental limitations, the theoretical one.

time and that better results were obtained when Na_2SO_4 was used as the electrolyte. However, their cell at 25°C had a lower emf of 0.96463 V and a higher emf-temperature coefficient of +0.000174 V/°C. These cells have not been used as standards probably because of their high emf-temperature coefficients.

In 1937, Vosburgh, Guagenty, and Clayton (42) proposed the cell

$$(-)\ \underset{(3p)}{\text{Cd,Bi,Hg}} \left| \begin{matrix} CdSO_4 \cdot \tfrac{8}{3}H_2O(c), \\ \\ CdSO_4 \cdot Na_2SO_4 \cdot 2H_2O(c) \end{matrix} \right| \begin{matrix} \text{sat aq sol of the two} \\ \text{salts in 0.1 } M \text{ acetic} \\ \text{acid or in 0.01 } M\ H_2SO_4 \end{matrix} \left| \ Hg_2SO_4(s) \right| Hg(l)\ (+)$$

which is a modified Weston cell in which a 3-phase (3p) amalgam (8.9 per^ent Cd, 11.1 percent Bi, 80 percent Hg) and a 0.1 *M* acetic acid or 0.01 *M* H_2SO_4 solution saturated with the two salts, as indicated, are used. At 25°C, Vosburgh et al. reported an emf of 1.0184 V and a dE/dt of +0.000013 V/°C for their cell. This dE/dt is about 2 to 10 times that of the unsaturated Weston cell but less than $\frac{1}{3}$ that of the saturated Weston cell at 20–30°C but of opposite sign. Data available on this cell indicate that it does not have the long-term stability exhibited by saturated Weston cells but compares favorably with the unsaturated type (43).

Dry primary cells having flat discharge curves, such as the mercury dry cell (see Chap. 5), have been proposed as secondary standards of emf. They must be considered as secondary standards since they do not exhibit the steadiness in emf with time shown by Weston or Clark cells.

CRITERIA FOR STANDARD CELLS

Gibbs phase rule applied to primary cells determines why some primary cells are suitable as standards of emf while others are not. The phase rule expresses the relation between the number of independent variables V which it is necessary to specify in order to fix the state of a system and the number of components C and phases P constituting the system. For primary cells the number of surfaces S at which there is an electrical potential must also be considered. The relation between those is given by $V = C - P + 2 + S$ (where the number 2 is valid only if there are two variables, commonly temperature and pressure, in addition to concentration).

For the *saturated* type of Weston cell there are 4 components: Cd, Hg, SO_3, and H_2O ($CdSO_4 \cdot \frac{8}{3}H_2O$ comes from Cd, SO_3, and H_2O; Hg_2SO_4

from Hg, SO_3, and H_2O); 6 phases; solid Cd amalgam, liquid Cd amalgam, $CdSO_4 \cdot \frac{8}{3}H_2O$ crystals, saturated solution of cadmium sulfate, Hg_2SO_4, and Hg; and 2 surfaces, the 2 electrodes, at which an electrical potential exists. Therefore, $V = 4 - 6 + 2 + 2 = 2$. Hence if the temperature and pressure of the system is fixed, the system is invariant and the emf remains unaltered.

For the *unsaturated* Weston cell there is one less phase since the cell contains no $CdSO_4 \cdot \frac{8}{3}H_2O$ crystals and $V = 3$. Therefore, at constant temperature and pressure, $V = 1$ and the emf of the unsaturated cell is a function of the concentration of the solution in the cell. If the concentration is fixed (or remains constant), the unsaturated cell is invariant like the saturated one.

If the components of the electrode diffuse to the other electrode and interact, or if the components of the individual electrodes interact, the phase rule no longer applies since the system is not at equilibrium. More important, the cell under such conditions would not be a standard cell unless the interactions did not disturb the number of components and phases constituting the electrodes or alter the composition of a phase. In fact, Van Ginneken and Kruyt (44) defined a standard cell as a cell composed of two parts, both invariant systems of the *same* components, when the temperature and pressure are constant. The emf of such a cell would not be expected to be altered by a flow of electricity through the cell in either direction, by diffusion, or by internal action.

Of the preceding cells, only the Clark and Weston cells of the saturated type meet the definition of Van Ginneken and Kruyt. In each, the two electrodes consist of the same components. Also each electrode in each cell consists of 4 components and 4 phases; thus an open circuit $V = 2$ and each electrode is an invariant system when the temperature and pressure are held constant. In the saturated Weston cell, Hg_2SO_4, which is sparingly soluble, diffuses slowly from the cathode to the anode where it is reduced by Cd with a resultant decrease in the amount of Cd in the amalgam and the formation of crystals of $CdSO_4 \cdot \frac{8}{3}H_2O$. However, since the solution is already saturated with cadmium sulfate, the formation of more cadmium sulfate is without effect on the emf of the cell. Also, as will be shown later (see Fig. 12.2), the emf of the Weston cell is very insensitive to changes in the composition of the Cd amalgam, and therefore the slight reduction in the amount of Cd in the amalgam caused by the interaction of Cd and Hg_2SO_4 is without significant effect on the emf of the cell. In the unsaturated cell, however, the production of additional cadmium sulfate at the anode by interaction of Cd and Hg_2SO_4 increases the concentration of cadmium sulfate with a resultant decrease in the emf of the cell; this decrease continues until the cell becomes saturated with cadmium sulfate (discussed later). In some cells, a small amount of sulfuric acid is added to prevent the hydrolysis of Hg_2SO_4. This does not increase the number of components since sulfuric acid is produced in the hydrolysis of cadmium sulfate and therefore is not an independent component; or it comes from SO_3 and H_2O, already counted as components. However, if sulfuric acid is added in excessive amount, above 0.1 N, it will have an adverse effect on the cell by chemically reacting with the cadmium in the amalgam, causing an appreciable decrease in the cadmium concentration of the amalgam, the production of hydrogen gas, and a decrease in the emf of the cell. This reaction represents an example of interactions of the components of any one electrode leading to variation in the emf of a cell.

A less restrictive definition of a standard cell than the one given by Van Ginneken and Kruyt, although requiring that both electrodes be invariant when the temperature and pressure are constant, would not require that both electrodes be of the same components. However, the components of any one electrode, as a result of diffusion or migration under a potential gradient, cannot interact with the components of the other electrode. The cell of Vosburgh et al. would meet this definition. Both electrodes are invariant under constant temperature and pressure but are not made with the same components; Bi is part of the anode but not of the cathode. However, Bi does not take part in the cell reaction, acting merely as part of the medium in which Cd is produced or consumed, and is so inert that it apparently does not reach the cathode by diffusion or migration. If Bi were to reach the cathode compartment it would have to do so as a separate phase; otherwise the cathode would not remain an invariant system under constant temperature and pressure.

THERMODYNAMICS OF STANDARD CELLS

The changes in Gibbs energy (free energy), enthalpy, entropy, and heat capacity for the cell reaction in standard cells are given, respectively, by: $\Delta G = -nFE$, $\Delta H = -nFE + nFT(dE/dT)$, $\Delta S = nF(dE/dT)$, and $\Delta C_p = d(\Delta H)/dT = nFT(d^2E/dT^2)$, where F is the faraday and n is the number of equivalents involved in the cell reaction; in the present case, for Weston cells, $n = 2$. The value of F is 96,487 coul/gram equivalent on the ^{12}C scale of atomic weights (10, 45). Thus, if E is expressed in volts, ΔG is given in volt-coulombs (or joules) per gram equivalent (46); these may be converted to the thermochemical calorie (defined) using the relation

$$1 \text{ thermochemical calorie (defined)} = 4.1840 \text{ joules}$$

and are so reported for a saturated Weston cell made with a 10 percent amalgam, for a series of temperatures in Table 12.1.

TABLE 12.1. THERMODYNAMIC DATA FOR THE REACTION IN WESTON NEUTRAL, SATURATED STANDARD CELLS MADE WITH 10 PERCENT CADMIUM AMALGAM, AS OBTAINED FROM ELECTROCHEMICAL MEASUREMENTS

Temperature (°C)	Emf, E (V)	Gibbs Energy Change, ΔG (cal/mole)	Enthalpy Change, ΔH (cal/mole)	Entropy Change, ΔS (cal/mole/°C)	Heat Capacity Change, ΔC_p (cal/mole/°C)
0	1.0189899	−46,997.9	−46,878.8	+0.436	−41.81
3	1.0190084	−46,998.5	−46,998.8	−0.001	−38.21
5	1.0190026	−46,998.2	−47,072.7	−0.268	−36.00
10	1.0189402	−46,995.4	−47,240.3	−0.865	−31.12
15	1.0188182	−46,989.7	−47,385.6	−1.374	−27.24
20	1.0186446	−46,981.7	−47,514.4	−1.817	−24.43
25	1.0184258	−46,971.6	−47,631.7	−2.214	−22.74
30	1.0181654	−46,959.6	−47,743.8	−2.587	−22.24
35	1.0178649	−46,945.9	−47,856.4	−2.955	−22.99
40	1.0175240	−46,930.0	−47,975.9	−3.340	−25.05

The data of Table 12.1 at 25°C may be compared with thermal data as follows. The cell reaction Eq. (5) may be considered as the sum of five reactions:

$$\mathrm{Cd(s) + Hg_2SO_4(s) = CdSO_4(c) + 2Hg(l)} \tag{a}$$
$$\mathrm{CdSO_4(c) + H_2O(l) = CdSO_4 \cdot H_2O(c)} \tag{b}$$
$$\mathrm{CdSO_4 \cdot H_2O(c) + \tfrac{5}{3}H_2O(l) = CdSO_4 \cdot \tfrac{8}{3}H_2O(c)} \tag{c}$$
$$\frac{8/3}{m-8/3}\mathrm{CdSO_4} \cdot m\mathrm{H_2O\ (sat\ aq)} = \tfrac{8}{3}\mathrm{H_2O(l)} + \frac{8/3}{m-8/3}\mathrm{CdSO_4 \cdot \tfrac{8}{3}H_2O(c)} \tag{d}$$
$$\mathrm{Cd(10\ percent\ amalgam) = Cd(s)} \tag{e}$$

and for each, reaction e excepted, we have thermal data. These are given in Table 12.2 and compared with the electrochemical values derived from the emf and the dE/dT (Eq. (7)). The agreement between the thermal and electrochemical data is remarkably good, even for ΔC_p, considering that the electrochemical value depends on a second derivative. The good agreement between the thermal and electrochemical data for the Weston saturated cell shows that the cell must be highly reproducible and that the electrodes must be of the reversible type (47). The thermal data in each case, in order, were obtained as outlined in the following paragraphs.

For reaction a, Cohen, Helderman, and Moesveld (48) found the difference between the heats of formation of $CdSO_4$ and Hg_2SO_4 to be −45,346 cal/mole at 18°C. This value becomes −45,386 cal/mole at 25°C in terms of the defined thermochemical calorie used here. Accepted values for the entropies of Cd, Hg_2SO_4, $CdSO_4$, and Hg at 25°C are 12.37 (49), 47.98 (50), 29.41 (51), and 18.19 (49) cal/mole/°C, respectively; hence

TABLE 12.2. COMPARISON OF THERMAL AND ELECTROCHEMICAL DATA ON THE WESTON NEUTRAL SATURATED STANDARD CELL MADE WITH 10 PERCENT CADMIUM AMALGAM, AT 25°C

Reaction	ΔG	ΔH	ΔS	ΔC_p
(a)	(−47,008)[a]	−45,386	+5.44	−0.63
(b)	−2,160	(−4,933)	−9.30	−9.69
(c)	−308	(−3,224)	−9.78	−11.26
(d)	184	224	(0.13)	−1.00
(e)	2,329	5,680	11.24	0.00
Sum	−46,963	−47,634	−2.27	−22.58
Emf	−46,971.7	−47,631.9	−2.214	−22.74

[a]Values in parenthesis calculated from the relation $\Delta G = \Delta H - T\Delta S$.

ΔS_a for reaction a is 5.44 cal/mole/°C and ΔG_a is therefore −47,008 cal/mole. The heat capacities at 25°C of Cd, Hg_2SO_4, $CdSO_4$, and Hg are, respectively, 6.19 (50), 31.55 (50), 23.81 (51), and 6.65 (50) cal/mole/°C and accordingly $\Delta C_{p(a)}$ is −0.63 cal/mole/°C.

The Gibbs energies for reactions b and c are given, respectively, by

$$\Delta G_b = RT \ln \frac{p^*}{p^\circ_{H_2O}} = RT \ln \frac{0.62}{23.756} = -2160 \text{ cal/mole}$$

and

$$\Delta G_c = \tfrac{5}{3} RT \ln \frac{p^{**}}{p^\circ_{H_2O}} = \tfrac{5}{3} RT \ln \frac{17.4}{23.756} = -308 \text{ cal}$$

where $p^\circ_{H_2O}$ is the vapor pressure of water at 25°C (52), $p^*_{H_2O}$ the vapor pressure of the system $CdSO_4 \cdot H_2O$—H_2O, and $\rho^{**}_{H_2O}$ the vapor pressure of the system $CdSO_4 \cdot \frac{8}{3}H_2O$—$CdSO_4 \cdot H_2O$. Ishikawa and Murooka (53) obtained 0.62 mm for $\rho^*_{H_2O}$ and the combined results of Ishikawa and Murooka (53) and Carpenter and Jette (54) give 17.4 mm for $p^{**}_{H_2O}$. The entropies of $CdSO_4 \cdot H_2O$, H_2O(l), and $CdSO_4 \cdot \frac{8}{3}H_2O$ are, respectively, 36.82, 16.71, and 54.89 cal/mole/°C (51), and ΔH_b and ΔH_c are, accordingly, −4933 and −3223 cal/mole, respectively. The heat capacities of H_2O(l), $CdSO_4 \cdot H_2O$, and $CdSO_4 \cdot \frac{8}{3}H_2O$ are, respectively, 18.04 (55), 32.16 (51), and 50.97 (51) cal/mole/°C; therefore, $\Delta C_{p(b)}$ and $\Delta C_{p(c)}$ are −9.69 and −11.26 cal/mole/°C, respectively.

The Gibbs energy for reaction d is given by

$$\Delta G_d = \frac{8}{3} RT \ln \frac{p^\circ_{H_2O}}{p_{H_2O\,(ss)}} = \frac{8}{3} RT \ln \frac{23.756}{21.17} = 183.8 \text{ cal}$$

where $p_{H_2P(ss)}$ is the vapor pressure of a saturated solution of $CdSO_4 \cdot \frac{8}{3} H_2O$ and $p^{\circ}_{H_2O}$ is again the vapor pressure of water. Ishikawa and Murooka (53) obtained 21.17 mm for the vapor pressure of the saturated solution at 25°C. Holsboer (56) found that 1044 cal of heat were evolved when 1 mole of $CdSO_4 \cdot \frac{8}{3} H_2O$ was dissolved in enough water to form the saturated solution at 25°C; for $(8/3)/(m-8/3)$ mole of salt the heat evolved would then be 224.1 cal. Combining these values of ΔG_d and ΔH_d gives 0.13 cal/°C for ΔS_d. Cohen and Holsboer (57) obtained 0.5823 cal/g for the specific heat of a saturated solution of $CdSO_4 \cdot \frac{8}{3} H_2O$ at 25°C; accordingly, $(8/3)/(m-8/3)$ mole of a saturated solution has a heat capacity of 60.05 cal/°C and with the heat capacities of $H_2O(1)$ and $CdSO_4 \cdot \frac{8}{3} H_2O$, given above, gives -1.00 cal/°C for ΔC_p for reaction d. Hulett (58) and Getman (59) measured the emf of the cell

Cd (metal) | cadmium sulfate solution | Cd amalgam (8–10%)

at a series of temperatures and obtained

$$E = 0.0504871 - 0.0002436(t - 25°\text{C}) \qquad (9)$$

for the emf as a function of temperature. Their results, expressed in international volts, are converted to absolute volts here. Their emf results give 2329 cal/mole, 5680 cal/mole, 11.24 cal/mole/°C, and 0 cal/mole/°C for ΔG_e, ΔH_e, ΔS_e, and $\Delta C_{p(e)}$, respectively, for reaction e. For some cells, Goetman obtained lower emfs than those given above, in agreement with earlier observations of Cohen (60). They attributed these lower emfs to different allotropic forms of Cd. However, these allotropic forms, if they exist, give emfs that are not consistent with the emfs observed for saturated Weston cells.

Giauque, Barieau, and Kunzler (25) have shown by similar calculations that the thermal and electrochemical data on the Clark cell are consistent.

PRACTICAL CONSIDERATIONS

The Weston (or Cadmium Sulfate) Cell

The cadmium sulfate cell is, for all intents and purposes, the only electrochemical system used today as a standard cell. The cell, as noted above, is made in two general types; saturated and unsaturated. It is the saturated type that has the nearly perfect shelf life and is the one used in the international comparisons of the unit of emf. Both are made commercially in the United States by the Eppley Laboratory, Inc., Newport,

R. I., and Weston Instruments, Inc., Newark, N. J. Both companies issue cells with their calibration reports of the emf. Such calibration reports may also be obtained from the National Bureau of Standards when a direct comparison with the national standard is desired. Most saturated cells made in the United States must be hand-carried, whereas the unsaturated type may be shipped by common carrier. The methods used to render unsaturated cells transportable are discussed later. Standard cells made abroad are usually of the saturated type, and some may be shipped by common carrier. Some miniature types recently made in the United States also can be shipped by common carrier.

Although standard cells in use today are based on the general proposal of Weston, research has shown that for a reliable standard cell attention must be given to (*a*) the acidity of the solution, (*b*) the composition of the amalgam, (*c*) the concentration of the solution, and (*d*) the nature of the solid electrolyte phase. Weston was not specific on these factors nor did he state that crystals of cadmium sulfate were added to assure saturation at all temperatures; in fact, his emf corresponded to that for an unsaturated cell. A discussion of these factors follows.

Acidity of the Solution. Although reactions 5 and 8 are the main ones in the cadmium sulfate cell, a side reaction tends to occur at the cathode. Mercurous sulfate hydrolyzes

$$2\,Hg_2SO_4(s) + H_2O(l) = Hg_2O \cdot Hg_2SO_4(s) + H_2SO_4(aq) \qquad (10)$$

to form the basic sulfate and an equilibrium concentration of sulfuric acid. Gouy (61) and Hager and Hulett (62) found the equilibrium concentration of H_2SO_4 to be 0.001 *M* and Craig, Vinal, and Vinal (63) showed that it did not change appreciably with temperature being 0.00099 *M* at 0°C and 0.00108 *M* at 28°C. Cadmium sulfate also hydrolyzes to form H_2SO_4 and $Cd(OH)_2$ or CdO. The molarity of H_2SO_4 in a saturated solution of cadmium sulfate is only 0.00046 (14), however, and insufficient acid is produced in the hydrolysis of cadmium sulfate to prevent the formation of the basic mercurous sulfate. Accordingly, sulfuric acid is added in small amount to the cells when they are constructed. Although, theoretically, a concentration slightly exceeding the equilibrium concentration could be used, in practice a higher concentration, 0.03–0.04 *M*, is used. Hulett (64) and Sir Frank Smith (65) found that mercurous sulfate has a minimum solubility at approximately 0.02 *M* H_2SO_4 while Craig, Vinal, and Vinal (63), in a more detailed study, found the minimum solubility to be at 0.03 *M* for 28°C and at 0.04 *M* for 0°C. It is for this reason that 0.03–0.04 *M* acid is now generally used in cadmium sulfate cells since it is advantageous to keep the amount of soluble mercurous sulfate at a minimum (see section on Criteria for Standard Cells). Higher concentrations of acid are to be avoided since excess acid promotes corrosion of the cadmium amalgam anode. These cells are customarily called "acid cells" to distinguish them from the so-called neutral cells to which no acid has been deliberately added. Acid cells have been found to exhibit greater permanency in emf than *strictly* neutral ones.

The addition of acid decreases the emf of the cell slightly. For the saturated cell several equations have been proposed relating the change in emf to acid concentration. For acid concentrations up to 4 normal ($4\,N$) Sir Frank Smith (66) gave

$$\Delta E(\mathrm{V}) = -(0.00060x + 0.00005x^2) \tag{11}$$

where ΔE is the difference in emf of acid cells from the neutral cell and x is the normality of the sulfuric acid solution before it is saturated with $CdSO_4 \cdot \frac{8}{3} H_2O$. For acid concentrations up to $0.4\,N$ only, the National Physical Laboratory (67) gave the linear relation

$$\Delta E(\mu\mathrm{V}) = -615x \tag{12}$$

Obata (68) and Ishibashi and Ishizaki (69) also gave linear relations with the coefficient being, respectively, -855 and -833, and Vosburgh (70) found Obata's equation to be valid to $1.49\,N$. In these last two equations, x refers to the normality of the acid in a saturated solution of $CdSO_4 \cdot \frac{8}{3} H_2O$. All of these formulas agree closely if applied properly. For low acidities, the acidity of a solution of sulfuric acid after saturated with $CdSO_4 \cdot \frac{8}{3} H_2O$ is 0.767 of that before saturation.

Sir Frank Smith (66) investigated the effect on the emf of cadmium cells if the acid were confined to one or the other of the electrode compartments. For these effects, valid to $4.0\,N$, he gave

$$\Delta E_{\mathrm{negative}}(\mathrm{V}) = 0.01090x - 0.00125x^2 \tag{13}$$

$$\Delta E_{\mathrm{positive}}(\mathrm{V}) = -0.01150x + 0.00120x^2 \tag{14}$$

the summation of which gives his equation (Eq. (11) above) for the effect of acid on the emf of the cell as a whole. These relations show that more acid at the negative electrode increases the emf while more acid at the positive electrode decreases the emf. For high reproducibility and stability in emf the acid concentration must be the same and remain the same at both electrodes.

Composition of the Amalgam. A 10 or $12\frac{1}{2}$ percent amalgam is generally used in the construction of cadmium sulfate standard cells. The choice of these percentages rests mainly on the comprehensive work of Sir Frank Smith (71), who studied the relation between the emf of saturated cells at various temperatures and the cadmium content of the amalgam. His results converted to absolute volts (given in Fig. 12.2) show that the useful range of amalgam composition is from about 8 to 14 percent cadmium for normal temperatures. The useful range of amalgam composition is limited to the horizontal part of the curves. In this region the emf is not critically dependent on the exact percentage of cadmium in the amalgam; the amalgam consists of two phases, solid and liquid, with the solid phase being a solid solution of mercury and cadmium. For low and high percentages of cadmium the emf is very sensitive to the exact amount of cadmium in the amalgam; this sensitivity is more marked at lower temperatures. The range of use for a $12\frac{1}{2}$ percent amalgam is 12.1–62°C, while for a 10 percent amalgam the range is about −8–51°C. For cells for use below −8°C a 6 percent amalgam is preferred; these cells may be used from −24 to 28°C.

Concentration of the Solution. The emf of cadmium sulfate cells varies inversely with the electrolyte concentration. Over a limited range of concentration near saturation the emf increases about 0.0017 V for a 1-percent decrease in the cadmium sulfate content. For more dilute solutions the change is somewhat greater. In Table 12.3 the emfs at 25°C corresponding to various concentrations of cadmium sulfate, near saturation, are given, as are the differences in emf for several temperature

TABLE 12.3. EFFECT OF CONCENTRATION ON ELECTROMOTIVE FORCE AND TEMPERATURE COEFFICIENTS OF CADMIUM SULFATE STANDARD CELLS (NEAR SATURATION)

$CdSO_4$ (Percent)	Emf, at 25°C (V)	15–25°C	25–35°C (μV/°C)	35–40°C	40–45°C
41.84	1.02132	−13.0	−13.0	−10.0	−8.0
42.39	1.020033	−5.5	−5.5	−3.2	−0.6
42.63	1.019688	−3.5	−3.6	−0.9	+0.9
42.77	1.019328	−1.3	−1.4	+0.7	+3.5
42.90	1.019133	−0.1	−0.3	+2.1	+4.5
42.94	1.019056	+0.1	0.0	+2.5	+5.4
43.06	1.018893	+1.0	+1.1	+3.4	+5.9
43.12	1.018719	+2.1	+2.0	+4.7	+7.1
43.22	1.018615	+2.8	+2.7	+5.0	+8.1
43.35[a]	1.018399	−40.0	56.0	−32.6	−35.1

[a]Saturated solution at all temperatures.

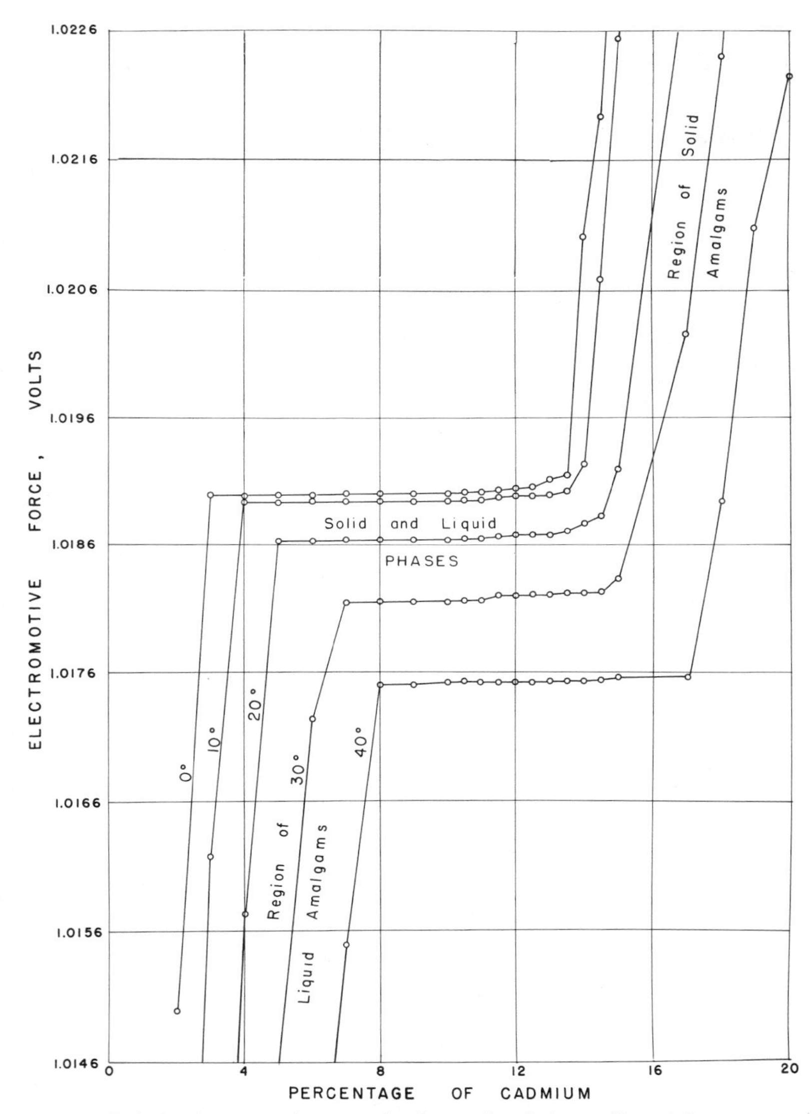

Figure 12.2 Relation between electromotive force of cadmium cells and the percentage of cadmium in the amalgam at various temperatures.

intervals. These data are based on the results of Vosburgh and Eppley (72) for cells with $12\frac{1}{2}$ percent amalgam and 0.023 N H_2SO_4. The data reported in international volts have been converted to absolute volts here. Interpolation of the data of Table 12.3 shows that a cell made with a

42.93 percent solution will have a zero or negligible temperature coefficient. Vosburgh and Eppley showed that this percentage corresponded closely to a solution saturated at 4°C. Mrs. Brickwedde (36) found the percentage of a solution saturated at 4°C to be 43.13, which is slightly higher than that found by Vosburgh and Eppley, who used acidified solutions. A lower value for the solubility of cadmium sulfate would be expected for acidified solutions due to the common-ion effect. Since solutions saturated at 4°C lead to cells with a negligible emf-temperature coefficient, they are frequently used in the construction of unsaturated cells. At 25°C such cells have an emf of 1.01907 V, which incidentally agrees with 1.019 V given by Weston in his patents. Unsaturated cells are also made with somewhat more dilute solutions so that the initial emf at 25°C is about 1.01940 V; these cells, when new, have a higher emf-temperature coefficient (about $-5\,\mu V/°C$) at 25°C but a longer life (*vide infra*).

Solid Electrolyte Phase. In saturated standard cells crystals of $CdSO_4 \cdot \frac{8}{3} H_2O$ must be used in excess to assure saturation of the solution over a wide range of operating temperatures. The upper limit of temperature is set by the transition point, 43.6°C (36), of cadmium sulfate from $CdSO_4 \cdot \frac{8}{3} H_2O$ to $CdSO_4 \cdot H_2O$ and the lower limit at -17°C where freezing of the electrolyte begins (cell becomes completely frozen at -24°C). The hydrates, $CdSO_4 \cdot \frac{8}{3} H_2O$ and $CdSO_4 \cdot H_2O$, may exist in a metastable state above and below 43.6°C, respectively, and may be maintained in this state for considerable time if only one crystal form is present since the transition in phase is a slow process. The solubility of $CdSO_4 \cdot \frac{8}{3} H_2O$ increases with temperature while that of $CdSO_4 \cdot H_2O$ decreases; the solubilities of the two forms are equal at the transition point (see Table 12.4). When a cell made with a particular hydrate is carried over into the supersaturated solutions may form.

The emfs of standard cells made with the two hydrates differ and are equal only at the transition point. In each case, cells made with the more stable hydrate have the higher emf at a particular temperature, (see Table 12.4). When a cell made with a particular hydrate is carried over into the metastable range, its emf will at first correspond to that for the supersaturated solution of the metastable hydrate and then slowly rise in value as the hydrate is converted to the more stable form. Accordingly, cells with one or the other hydrate should be used only in the temperature range in which the hydrate is stable.

It has been reported (74) that $CdSO_4 \cdot \frac{8}{3}H_2O$ exhibits a transition to $CdSO_4 \cdot 7H_2O$ at 4°C; Mylius and Funk (75) believed the transition point was at or below 0°C. However, the emf of cells near 4°C does not exhibit

TABLE 12.4. ELECTROMOTIVE FORCES OF CELLS[a] MADE WITH $CdSO_4 \cdot \frac{8}{3}H_2O$ AND $CdSO_4 \cdot H_2O$ (AND SOLUBILITIES OF THE TWO HYDRATES) AT VARIOUS TEMPERATURES

	Solubility		Electromotive Force	
Temperature (°C)	$CdSO_4 \cdot \frac{8}{3}H_2O$	$CdSO_4 \cdot H_2O$	$CdSO_4 \cdot \frac{8}{3}H_2O$	$CdSO_4 \cdot H_2O$
	(mole $CdSO_4$ per mole H_2O)		(V)	
0	0.06546	–	–	–
5	0.06555	–	–	–
10	0.06566	–	–	–
15	0.06580	–	–	–
20	0.06600	–	1.01867	1.01334
25	0.06627	–	1.01845	1.01418
30	0.06663	–	1.01818	1.01502
35	0.06710	–	1.01786	1.01588
40	0.06773	–	1.01753	1.01673
43.6	0.06833	0.06833	1.01730	1.01730
45	0.06857	0.06797	1.01717	1.01758
50	0.06966	0.06659	1.01684	1.01843
60	–	0.06385		

[a]Cells contained 10 percent amalgams and 0.032 N H_2SO_4 (73).

an abrupt change or inflection corresponding to a phase change but exhibits a maximum at 3°C.

Since unsaturated standard cells do not contain crystals of cadmium sulfate their behavior is not dependent on state of hydration of the solid electrolyte phase. Unsaturated standard cells with $12\frac{1}{2}$ percent amalgams may be used as high as 62°C, but will decrease in emf at this temperature at a rate of about 0.03 percent/year and should be frequently recalibrated. By increasing the cadmium content of the amalgam unsaturated standard cells may be used to even higher temperatures, but these will also exhibit rapid decreases in emf with time. The lower temperature limit for unsaturated cells is the same as for saturated cells, i.e., they become completely frozen at −24°C.

STRUCTURAL FEATURES

Today the majority of saturated standard cells are made in soft-glass, H-shaped containers with platinum leads sealed in the bottom of each limb. Pyrex containers with tungsten leads and silica containers with molybdenum leads in Houskeeper seals (76, 77) may also be used. Although it has generally been thought that interactions between the container and the electrolyte cause "aging" or drifts in the emf of standard cells, available data have not shown significant differences between cells made in soft glass, Pyrex, or silica on a long-term basis. Plastic containers (78) have also been proposed. These must prevent vapor transport or a drying out of the cells may occur.

Saturated standard cells as made at the National Bureau of Standards are illustrated in Fig. 12.3. The container is of the H-form and of soda-lime glass. The height of the cell is 9.2 cm, the diameter of the vertical limbs 1.6 cm, the diameter of the cross-arm 1.1 cm, and the distance between limbs 2.2 cm. A constriction is made in both limbs of the container, as shown, to lock-in part of the crystals. The constriction may be a complete circumferential indentation or may consist of several knobs directed inward. The platinum leads, which have a tendency to break at the seal if not securely held in place, are secured by cotton thread to the side of each limb at the constriction.

Mercury is first placed in the bottom of one limb and the amalgam in the bottom of the other limb, each to a depth of about 0.6 cm. A 10 percent amalgam is added while warm and in a single phase; on cooling it becomes two phases. Mercurous sulfate, moistened with a saturated solution of $CdSO_4 \cdot \frac{8}{3}H_2O$, is then placed over the mercury to a depth of 1.3 cm and finely-divided crystals of $CdSO_4 \cdot \frac{8}{3}H_2O$ are added in both limbs to a depth of 1 cm at the anode and 0.8 cm at the cathode. The crystals used are of a size that will pass through a tube of 4-mm bore. Finally, a saturated solution of $CdSO_4 \cdot \frac{8}{3}H_2O$ is added to a level slightly above the cross-arm,

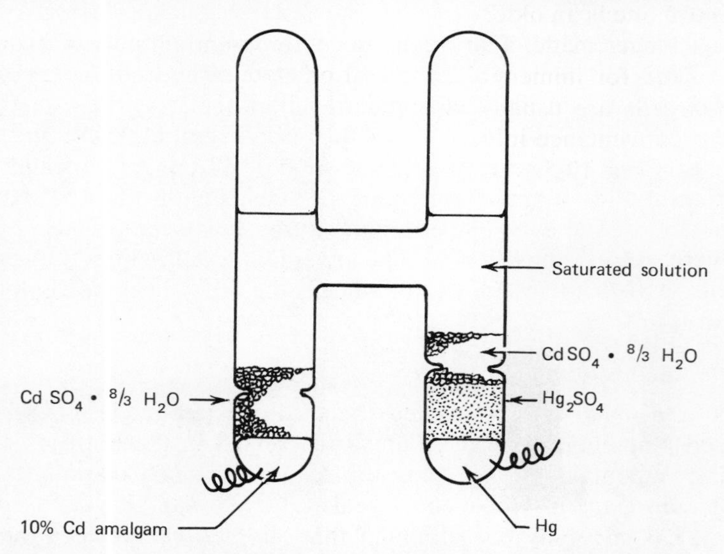

Figure 12.3 Cross section of a National Bureau of Standards saturated cadmium sulfate cell.

and the cell is hermetically sealed. The mercurous sulfate should be mixed with a small amount of mercury and washed with a saturated solution of $CdSO_4 \cdot \frac{8}{3}H_2O$ before introduction to the cell. Finely-divided crystals of $CdSO_4 \cdot \frac{8}{3}H_2O$ also are mixed in small amount with the mercurous sulfate paste. In some cells, especially of larger size, larger crystals (about 1–1.5 cm in diameter) of $CdSO_4 \cdot \frac{8}{3}H_2O$ are used. Larger crystals have an advantage over smaller crystals in that any gas formed at the electrodes will not be entrapped by the crystals whereby an open circuit might result. However, cells with large crystals tend to come to equilibrium, after a temperature change, more slowly than those made with small crystals.

Unsaturated cells are made similarly except that no crystals of $CdSO_4 \cdot \frac{8}{3}H_2O$ are used and they are made shippable by inserting cork or plastic rings, covered with linen, over the elements of the two electrodes. In some cells ceramic disks, either locked in place or supported by a ceramic rod which protrudes through stoppers in each limb, are used (79, 80). The unsaturated cell is the commercial type used widely in the United States for work requiring no greater accuracy than 0.01 percent. It is used unmounted for pyrometer work, in *p*H meters, recording instruments, etc., but is usually mounted in nontransparent copper-shielded cases for laboratory work. A copper-shielded case is employed to assure temperature equilibrium at both limbs of the cell (*vide infra*). Various types of unsaturated cells made by manufacturers are shown in Fig. 12.4. The central one is an older type, similar to the original design of Weston, and is no longer made. The saturated cell is not mounted in cases since it is intended for immersion in temperature-controlled oil or air baths. Saturated cells are usually mounted in groups of 3, 4, or 6 on special racks for convenience in use. Two types of commercial racks with cells are shown in Fig. 12.5.

Figure 12.4 Several types of unsaturated standard cells.

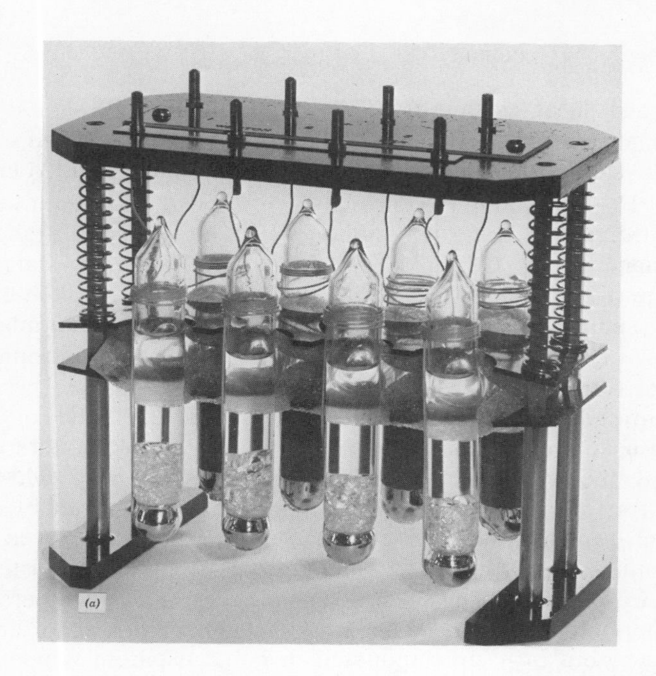

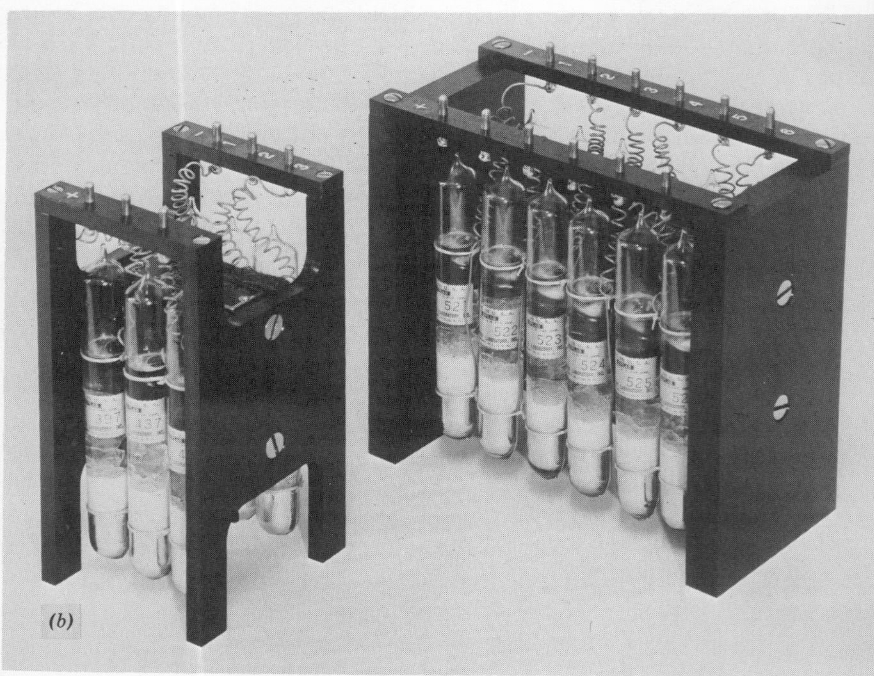

Figure 12.5 Typical racks of saturated standard cells.

For the preparation of precision cells, only materials of the highest purity should be used. The four principal materials needed (mercury, cadmium, sulfuric acid, water) may all be purified by distillation. The following procedures are used at the National Bureau of Standards. Mercury, after washing in dilute nitric acid then in distilled water, is dried and distilled in a Hulett (81) still in a stream of dry air. Cadmium of electrolytic grade is sublimed under reduced pressure. Sulfuric acid of reagent grade is twice distilled in Pyrex. Conductivity water having a conductivity of 10^{-6} ohm^{-1} cm^{-1} or less, prepared in a Barnstead still, is used. The mercurous sulfate is prepared electrolytically in a darkened room (see effect of light, discussed later) by the Wolff and Waters (82) or Hulett (83) method using mercury anodes, platinum cathodes, and sulfuric acid diluted with conductivity water. Cadmium sulfate is made from recrystallized cadmium nitrate (prepared from the metal and redistilled nitric acid) and redistilled sulfuric acid. Cadmium sulfate of high purity may also be obtained by several recrystallizations of reagent grade cadmium sulfate. The recrystallization must be conducted below 43.6°C. All materials should be free of manganese, which has a deleterious effect on standard cells (84). Moreover, all metals that reduce hydrogen overvoltage on cadmium must be absent or present only in minute amounts.

CHARACTERISTICS OF STANDARD CELLS

Effect of temperature

Unsaturated standard cells have a very low emf-temperature coefficient; the value depends on the concentration of cadmium sulfate in the cell (see Table 12.3). As unsaturated cells age, their emf-temperature coefficients increase. *Saturated standard cells* exhibit a much larger emf-temperature coefficient than the unsaturated type because of the changes in solubility of $CdSO_4 \cdot \frac{8}{3}H_2O$ with temperature. The relation between emf and temperature for saturated cells has already been considered (see Eqs. (6) and (7) except for the separate limbs of the cell. The temperature coefficient of each electrode (or limb) differs greatly. For the anode and cathode Sir Frank Smith (71) obtained -0.00035 V/°C and $+0.00031$ V/°C for dE/dt at 20°C, respectively, the summation of which gives the emf-temperature coefficient of the cell as a whole. Since these temperature coefficients are large and of opposite sign, it is obvious that all parts of the cell must be kept at the same temperature, otherwise large errors will ensue. It is for this reason that copper shields are used on unsaturated shippable cells (*vide supra*).

In practice, saturated standard cells are maintained at a constant temperature in thermostatically controlled oil baths or in portable thermostatically controlled air boxes. The latter are generally made on the design first proposed by Mueller and Stimson (85). The cells are housed in a thin-walled aluminum box which rests in a larger, thick-walled aluminum box, whose temperature is controlled by a mercury-in-glass thermoregulator. The aluminum boxes are thermally insulated and are enclosed in a wooden box which also contains an a-c relay, a transformer, and a pilot light. The box is operated on 100 V-60 cycle a-c line. The leads from the individual cells are brought to binding posts on the outside of the box. These boxes are designed to operate at temperatures above room temperature; the choice of temperature depends on the location where the boxes will be used or on the size of the box. A higher temperature is generally chosen for smaller boxes. As a rule these boxes operate at some temperature between 28 and 37°C. New types of boxes that operate at about 3°C are now available. These utilize thermoelectric cooling. A temperature of 3°C is chosen since this is the temperature at which cadmium sulfate standard cells have a maximum emf. The emf of standard cells for use above 43.6°C, i.e., of cells containing crystals of $CdSO_4 \cdot H_2O$, increase linearly with temperature at the rate of 0.000173 V/°C. These cells are not made commercially but the usual saturated type may be converted to the monohydrate type by heating and keeping them above 43.6°C. The long-range stability of such cells is not known, however.

Hysteresis

In general, if saturated standard cells are subjected to slowly changing temperatures, their emfs will follow closely the emf-temperature relations previously given. If, however, the cells are subjected to abrupt temperature changes, deviations from the expected emf will occur. These deviations are generally referred to as hysteresis. On cooling, the cells show first too high an emf and then a slow decrease in emf to equilibrium values. On heating, the cells show first too low an emf and then a slow rise in emf to equilibrium values. The magnitude of the hysteresis is given by the percentage deviation from the equilibrium value and is usually greater when the temperature is decreased than when it is increased. Unsaturated cells also show hysteresis. The magnitude of the hysteresis in modern unsaturated cells is usually less than 0.01 percent for an abrupt temperature change of 5°C.

The magnitude and duration of hysteresis depend on the type of construction, age, and composition of the cell and also on the rate, frequency, and extent of the temperature change. Accordingly, it is not possible to give quantitative data that may be applied as corrections to the emf of cells under diversified conditions involving abrupt temperature changes. Furthermore, the magnitude of hysteresis for unsaturated cells increases with age; a cell 10 years old exhibits about 10 times the hysteresis of a new cell. In general, for a 5°C abrupt change in temperature the hysteresis of new unsaturated cells ranges from 0.01 to 0.02 percent (0.0001 to 0.0002 V) on cooling and from 0.005 to 0.01 percent (0.00005 to 0.0001 V) on heating. The cells usually recover their original emf within 1 or 2 days after cooling and within 10 or 12 hours after heating. For older cells, hysteresis may persist for days or even months. Proportionate hysteresis is to be expected for larger or smaller temperature intervals than 5°C.

Many explanations have been given for hysteresis, but no single factor is responsible. Vosburgh and Eppley (72, 86) found that the size of the mercurous sulfate grains and the acidity of the electrolyte affected hysteresis, small grains and acidity being favorable to small hysteresis. As mentioned earlier, cells with small grains of cadmium sulfate also show less hysteresis than cells made with large grains. Wold (87) attributed hysteresis to the lag in the solution or precipitation of cadmium sulfate. Vinal and Brickwedde (88) by comparing cells with made with $CdSO_4 \cdot \frac{8}{3}H_2O$ crystals and metastable ones made with $CdSO_4 \cdot H_2O$ crystals, likewise showed that hysteresis is associated with the precipitation of the crystalline salt. Niederhauser and Hulett (89) and Vosburgh and Elmore (90) gave evidence that hysteresis was the result of the formation of supersaturated solutions of mercurous sulfate when the temperature is reduced. Vosburgh and Eppley (91) and Eppley and Vincent (92) showed that extracts from the cork septa in unsaturated cells increased hysteresis but that an increase in the acidity of the electrolyte reduces the hysteresis. Modern cells made with plastic septa obviously are free of this cause of hysteresis. In any case, standard cells should be maintained at a constant temperature if at all possible. Temperature fluctuations may be kept at a minimum for portable (or shippable) unsaturated cells by placing them in temperature-lag boxes or in Dewar flasks.

Temperature range

The temperature range over which saturated and unsaturated cells may be used has been given earlier in connection with discussions of the cadmium amalgam and the solid phases of cadmium sulfate.

Effect of pressure

The effect of pressure on the emf of an open galvanic cell at constant temperature is given by

$$\left(\frac{dE}{dP}\right)_T = -\frac{(\Delta V)k}{nF} \qquad \text{(V/atm)} \tag{15}$$

where ΔV is the volume change in cubic centimeters at atmospheric pressure per faraday, n and F have the significance given previously, and k is a conversion factor, 0.101325, which converts cubic centimeter-atmosphere to joules, *i.e.*, volt-coulombs. For Weston saturated cells the volume change may be calculated from the atomic or molecular weights and the densities of the reactants and products of the cell reaction. At 20°C the accepted densities (93) of Cd, Hg_2SO_4, $CdSO_4 \cdot \frac{8}{3}H_2O$, Hg, and saturated solution of $CdSO_4 \cdot \frac{8}{3}H_2O$ are, respectively, 8.648, 7.56, and 3.090 g/cm³ and 13.5463 and 1.6119 g/ml. Using these values, the accepted atomic and molecular weights, and the value of F given above (10, 45), $(dE/dP)_T = 6.4\ \mu V/atm$, which agrees excellently with the experimental value, 6.1 μV/atm, of Ramsey (94). For sealed cells at pressures below that required to crack the cell container, the emf does not change with pressure (14).

Internal Resistance

The internal resistance ranges from 100 to 500 ohms for unsaturated cells and from 500 to 750 ohms for saturated cells made in the United States and is, therefore, much higher than that of most primary cells. The internal resistance increases at a rate of about 2 percent for a decrease in temperature of 1°C, and increases slightly as the cells age. The resistance of the positive and negative limbs of the symmetrical H-shaped cells are nearly equal.

The internal resistance of a standard cell may be estimated by placing a 10 megohm resistor across the cell terminals and reading the emf, E_R. The internal resistance R is then given by

$$R = 10^7 \frac{(E_0 - E_R)}{E_0} \tag{16}$$

where E_0 is the open-circuit value. The change in internal resistance during the passage of electricity through the cell is the same whether the passage is in the forward or reverse direction. The cell recovers its initial emf within a few minutes after the 10-megohm resistor is removed. If the cell exhibits a high internal resistance or insensitivity it may contain a gas bubble at the anode, which may be removed by tapping the cell when inclined 45°. If this treatment is ineffective the cell should be discarded.

Effect of current

Normally, standard cells should not be subjected to more current than is required with potentiometers. Even so, the question frequently arises about the effect of electric current on standard cells, especially the unsaturated type. When standard cells are charged or discharged the rise or drop in emf is initially dictated by IR and subsequently by a voltage change, V_c, associated with the chemical changes in the cell, and by electrode polarization and changes in internal resistance. For the unsaturated cell, V_c is made up of two parts: V_a, the change in emf associated with the change in composition of the amalgam during charge or discharge and V_e, the change in emf associated with the change in electrolyte content of the solution during charge or discharge. When unsaturated cells have been discharged until their solutions are saturated this no longer applies. For saturated cells, V_c consists of V_a only.

The magnitudes of V_a and V_e on discharge at 25°C are given, respectively, by (14)

$$V_a = -\frac{0.001067\, It}{FW_a} = -1.106 \times 10^{-8} \frac{It}{W_a} \quad \text{(V)} \tag{17}$$

and

$$V_e = -\frac{0.161\, It}{FW_e} = -1.67 \times 10^{-6} \frac{It}{W_e} \quad \text{(V)} \tag{18}$$

where I is the current in amperes, t the time in seconds, F the faraday, W_a the weight of the amalgam in grams, and W_e the weight of the solution in grams. For commercial cells having internal resistances of 100 and 500 ohms, W_a is, respectively, 67 and 20 g and W_e is 100 and 200 g.

The magnitude of electrode polarization, including any change in internal resistance that may occur during discharge, is given by

$$\log \Delta E = 1.6048 + 0.786 \log \left(\frac{C}{A}\right) \quad (\mu\text{V}) \tag{19}$$

where ΔE is the change in emf arising from electrode polarization and internal resistance, C is the quantity of electricity in coulombs, and A is apparent (geometric) electrode area in square centimeters (14). For unsaturated cells having internal resistances of 100 and 500 ohms, A is about 5.5 and 1.4 cm^2, respectively. The length of time standard cells will sustain a discharge depends on the amount of material in the cells; most unsaturated cells made in the United States contain sufficient material to yield from 700 to 5000 coul if the current is kept below about 2×10^{-5} amp/cm^2 of electrode area. In Table 12.5, the calculated changes in emf caused by various currents of low magnitude are given for cells having cross-sectional areas of 5.5 and 1.4 cm^2 and internal resistances of 100 and 500 ohms (14).

When an external load is removed, the cells recover their initial emf,

TABLE 12.5. EFFECT OF CURRENT ON THE EMF OF TWO SIZES OF UNSATURATED STANDARD CELLS AT 25°C

	Internal Resistance (initial change)	
Current (amp)	100 ohms (μV)	500 ohms (μV)
10^{-10}	0.01	0.05
10^{-9}	0.1	0.5
10^{-8}	1.0	5.0
10^{-7}	10.0	50.0
10^{-6}	100.0	500.0
10^{-5}	1000.0	5000.0

CHANGES IN EMF OWING TO CHANGES IN AMALGAM (V_a) AND ELECTROLYTE (V_e) COMPOSITION (AFTER ONE YEAR)

	Internal Resistance 100 ohms		Internal Resistance 500 ohms	
Current (amp)	V_a (μV)	V_e (μV)	V_a (μV)	V_e (μV)
10^{-10}	—[a]	—	—	—
10^{-9}	—	—	—	—
10^{-8}	—	—	—	0.03
10^{-7}	—	0.05	—	0.26
10^{-6}	—	0.5	0.02	2.6
10^{-5}	0.05	5.3	0.17	26.3

CHANGES IN EMF ARISING FROM ELECTRODE POLARIZATION (AFTER ONE YEAR)

Current (amp)	Internal Resistance 100 ohms (μV)	Internal Resistance 500 ohms (μV)
10^{-10}	0.1	0.3
10^{-8}	0.7	2.0
10^{-8}	4.3	12.5
10^{-7}	26.	76.
10^{-6}	159.	466.
10^{-5}	970.	2844.

[a] Dash means that the change is less than 0.01 μV.

providing the discharge has not been a prolonged one. The time required for recovery depends on the severity of the discharge. For example, if an unsaturated cell (internal resistance = 100 ohms) is discharged for 5 min at 6×10^{-6} amp/cm^2 it will recover its original emf within 5 μV in 30 min and will have completely recovered in 6 hr. At a higher current density of 6×10^{-4} amp/cm^2 the emf will be about 180 μV below its original value after 30 min, 5 μV after 6 hr, and several days will be required for complete recovery. If the cell were discharged at a current density of 6×10^{-4} amp/cm^2 to a very low cutoff voltage of 0.001 V, recovery will be exceedingly slow, requiring several months, and full recovery will not be attained after this prolonged period because the normal emf will have declined (*vide infra*).

Short Circuit. Standard cells may be short-circuited momentarily without permanent damage to the cells. The cells will recover their original emf within a few minutes after taken off short circuit. If kept on short circuit they will be completely discharged within ½–2 days depending on the size and internal resistance of the cell and will *not* recover their initial emf. The short-circuit current is given by the ratio of the open-circuit emf and the internal resistance of the cell. For cells having internal resistances of 100 and 500 ohms, the short-circuit (flash) current will therefore be 1×10^{-2} and 2×10^{-3} amp, respectively.

EFFECT OF LIGHT AND RADIATION

Mercurous sulfate is sensitive to light and changes in color at a slow rate through tan, gray-brown, dark brown, to black. Although standard cells having discolored mercurous sulfate may have normal emfs (95), they exhibit slower approach to equilibrium values after temperature or other changes than cells not exposed. Standard cells therefore should be mounted in nontransparent cases or kept in the dark and used only for short periods at a time under diffuse light.

Windus (96) has reported that high-energy radiation (neutron flux of 5×10^{11} n/cm^2 and gamma flux of 3×10^7 rad) as produced in a nuclear reactor also causes the mercurous sulfate in a standard cell to turn black. He attributed this to the presence of mercury sulfides (detected by chemical analysis) as well as small globules of mercury. He also showed that unsaturated cells act as a heat source when so irradiated and, unless means are provided for rapid heat conduction, their temperature will rise about ½°C/cm^2 of surface area within 2 hr, with attendant changes in emf. After corrections were made for the temperature rise Windus found that the emf was about 0.2 percent below its original value and that three weeks were required for total recovery.

EFFECT OF SHOCK

Mechanical shocks insufficient to fracture or break or scramble the components of unsaturated cells have no lasting effects on the cells. When subjected to shocks of 10–40 g for durations of 6–18 msec, unsaturated cells exhibit large transient changes in emf ranging from 4200 to 31,000 μV (97). After the shock the cells immediately recover their initial emf within 2 μV.

The saturated cells usually found in the United States should not be subjected to sudden shock, should not be shipped by common carrier, should be transferred by messenger, and should not be tilted more than 45°. Some new saturated standard cells of novel design are stated to be shippable but extensive studies will be required to determine their long-range stability.

EFFECT OF VIBRATION

Vibrations at frequencies from 10 to 1000 Hz (c/s)[9] with accelerations of 1–10 g have no lasting effects on the emf of unsaturated standard cells (97). During the vibration, however, rather large a-c voltages of the same frequency are generated. For an unsaturated cell having an internal resistance of 500 ohms this a-c voltage ranges from about 25 μV at 1 g and 1000 Hz to 9900 μV at 10 g and 50 Hz. Furthermore, there is a decrease in the d-c emf ranging from 3 μV at 1 g and 1000 Hz to about 200 μV at 10 g and 100 Hz. Generally, at frequencies above 100 Hz the waveform of the a-c voltage is sinusoidal, whereas below 100 Hz it is nonsinusoidal owing to the resonance of the various components of the cell. In most cases the a-c and d-c effects of the vibration appear and disappear instantaneously when vibration is started or stopped. In some instances the d-c change may be rapid in the initial moments of vibration and then build up slightly in an exponential manner for 2 or 3 min. In these instances when the vibration is stopped the d-c emf decays in the same fashion as it was built up.

LIFE AND DRIFT IN EMF OF STANDARD CELLS

Saturated standard cells have an exceedingly long life. Some cells at the National Bureau of Standards have been in use for nearly 65 years, and they have retained their emfs within a few microvolts; at least absolute electrical measurements have not shown drifts in emf exceeding the uncertainties in absolute measurements (14). On the other hand, unsaturated cells at room temperature decrease in emf at a rate of about 20–40 μV/

[9]Hz means hertz, which is now used to designate cycles per second, defined in the parenthesis (c/s).

year (12). Since the emfs of new unsaturated standard cells generally range from 1.01900 to 1.01940 V, depending on the concentration and acidity of the electrolyte, these cells on the average reach an emf of 1.01830 V within 23 to 37 years, providing they are maintained at around 25°C and are not subjected to abuse, such as discharging or charging current. A practical lifetime for these cells is about 12 to 18 years. When unsaturated standard cells reach an emf of 1.01830 V or lower the cells generally behave erratically (largely because the electrolyte may become supersaturated on cooling), have large emf-temperature coefficients, and show excessive emf-temperature hysteresis. The life of the cell is considerably reduced if the cell is stored at higher temperatures. The rate of decrease in emf is approximately doubled for every 12°C increase in temperature; thus at 37°C the life of an unsaturated cell would be half that given above for 25°C.

Earlier unsaturated standard cells decreased in emf at room temperature at a rate of 70–85 μV/year (98, 99, 100) and therefore had an average theoretical life of 10–15 years or a practical life of 8–12 years. The improvement in the life noted in modern cells has resulted mainly, if not entirely, from the use of improved septa in the cells.

THE HULETT STANDARD BATTERY

Although standard cells are normally not intended to supply current, Hulett (101) designed a large-sized Weston cell, of simple construction, as a power supply for potentiometer circuits. Since he designed the cell as a replacement for storage batteries in potentiometer circuits, he called it a "standard battery" although it consisted of one cell only. The cell is suitable for low currents only, but it gives an exceedingly constant current at steady emfs and is little effected by variations in ambient temperature.

A cross section of the Hulett battery, in simple form, is shown in Fig. 12.6. A 500 cc bottle stoppered with a paraffined cork serves as the container. Mercury, 50 cm² in area, is placed on the bottom to a depth of $\frac{1}{2}$ cm, and mercurous sulfate is formed on its surface to complete the cathode *C*, by passing 1 amp for 20 min through the cell after assembly. The anode *A* is a $12\frac{1}{2}$ percent cadmium amalgam, 20 cm² in area, housed in a flat-bottom dish which is suspended in the solution 2 cm above the cathode by a glass tube sealed to the center of the bottom of the dish. A platinum wire is included in the seal so that the tube serves also to make electrical connections to the anode. The bottle is half filled with an acidified solution of cadmium sulfate (200 g $CdSO_4 \cdot \frac{8}{3}H_2O$ in 200 cc of 0.28 *M* H_2SO_4).

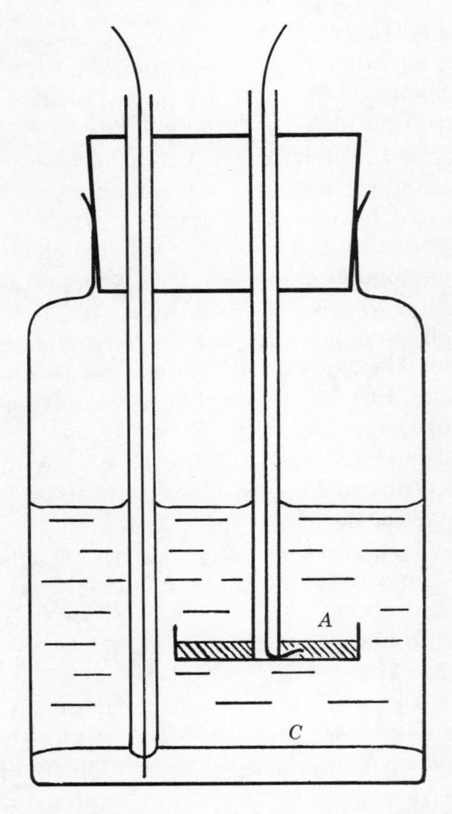

Figure 12.6 Cross section of a Hulett "standard battery."

This cell gives an emf of 1.02218 int. V(1.02253 ab. V) at normal temperatures, has an internal resistance of 6.2 ohms, exhibits very little polarization, and recovers in emf after short-term discharges at low current and even after short-circuiting if not prolonged. In going from high to low currents, or the reverse, the cells exhibit very little hysteresis in emf. Hulett stated that the cell gave satisfactory service with potentiometers for about 2 years and required very little attention. He also designed smaller cells with cathode areas of 20 cm^2 and anode areas of 8 cm^2; Spence (102) made larger cells with cathode and anode areas, respectively of 340 and 78 cm^2. After use these cells may be recharged or replenished by current reversal; the cell is rotated during the charging.

Elliot and Hulett (103) later altered the design. In the new design they used anodes of 10 percent cadmium amalgam, a solution of cadmium sulfate saturated at 4°C and 0.08 *M* in H_2SO_4, and reagent or electrolytic mercurous sulfate as "depolarizer" (it was not prepared *in situ* as in the earlier design). A glass crystallizing dish, 30 cm in diameter and 15 cm in depth, is covered on the bottom first with 5 kg of mercury and then with a layer of 300 g of mercurous sulfate. The anode is housed in 4 Petri dishes, 19 cm in diameter and 1.5 cm in height, stacked on a tripod, and separated by glass triangles. The tripod supports the Petri dishes 2 to 3 cm above the cathode. The amalgam in all four Petri dishes is connected to one binding post and a glass-covered wire which extends through the solution to a terminal in a greased and tight-fitting glass cover. Another glass-covered wire is also used as a terminal connector to the cathode. The total area of the anode is 1300 cm^2 and that of the cathode 666 cm^2. For this design Elliot and Hulett did not recommend recharging but suggested that the cell could be restored to its original condition by additions of water and mercurous sulfate. The amounts of water and mercurous sulfate required may be determined from records kept on the extent of the discharge.

In Table 12.6 the emf and average working voltage after 20 or 30 min of operation on various loads are given for the Elliot-Hulett cell of dimensions listed above (the emfs have been converted to absolute volts). In the last column the average variations in the emf during the discharge are listed. The cell maintains a constant emf to within 10 ppm during $\frac{1}{2}$ hr of discharge. This "battery" is suitable only for stationary service but offers an answer to those end uses in which steady currents and constant emfs are required. When so used it should not be depended upon as an ultimate standard of emf.

TABLE 12.6. VOLTAGE OF ELLIOTT–HULETT "STANDARD BATTERY"

Initial Open-Circuit (emf) (V)	Resistance (ohms)	Current (ma)	Closed-Circuit Voltage (for one-half hour) (V)	Variation in Closed-Circuit Voltage During One-Half Hour (μV)
1.018032	5000	0.2	1.017835	± 3.8
1.018023	3000	0.33	1.017716	± 3.1
1.018026	1000	1.0	1.017151	± 7.0
1.018033	500	2.0	1.016345	± 5.0
1.018012	250	4.0	1.014716	± 3.2
1.018056	150	6.67	1.012647	± 1.5
1.018043	100	10.0	1.010130	± 3.0
1.018061	50	20.2	1.002797	± 10.0

ZENER DIODES

In recent years solid-state devices known as zener diodes have appeared on the market as d-c reference standards of emf. A zener diode is a variant of the silicon-junction diode which exhibits a constant voltage when electricity of proper magnitude is caused to flow in the reverse direction across its junction. Therefore they differ fundamentally from standard cells (or primary cells) for their operation. Unlike standard cells, which have emfs of the order of 1 V, zener diodes currently being considered as standards have operating voltages ranging from 5 to 12 V. For operation these require a d-c current ranging from 5 to 15 ma. Recent studies (46, 104) have shown that the better temperature-compensated[10] diodes, after stabilization (about 5 days required), are stable in emf to 3 ppm over a 3-year period.

Zener diodes show a much wider spread in voltage than do standard cells, i.e., their construction has not yet been standardized. Despite this factor and their higher voltage, they may, with proper circuitry (resistors, etc.), be used to provide d-c reference voltages corresponding to the emf of standard cells. When so used they have made considerable impact on the use of *unsaturated* standard cells in electrical equipment. Although zener diodes require a current source for operation, they have the advantage over unsaturated standard cells of being rugged and compact and are therefore suitable for use under many conditions where unsaturated cells would be unsuitable. However, in maintaining the unit of emf, zener diodes do not yet compete with *saturated* standard cells, which may be made in a highly reproducible form and which show remarkable emf stability with time (14). Zener diodes therefore now must be considered as secondary standards of emf.

SUMMARY

Table 12.7 summarizes the various primary cells that have been proposed as standard cells. Of these, only the Weston or cadmium sulfate cell is used today as a standard of electromotive force since it meets the criteria for a standard cell better than all of the others. It is made in two types: unsaturated and saturated, with the latter having the higher stability and the longer life but the higher emf-temperature coefficient. In the future either the saturated cadmium sulfate cell will be modified

[10]A temperature-compensated zener diode is one consisting of a series package of a zener diode and one or more diodes that operate in the normal or forward direction. The negative emf-temperature coefficients of the added diodes are balanced against the positive coefficient of the zener diode.

TABLE 12.7. A SUMMARY OF STANDARD CELLS

Inventor	Date	Electrochemical system	Nominal Voltage	Temperature Range (°C)
Daniell	1836	(−) Zn(s) | $ZnSO_4$(s) | $ZnSO_4 \cdot H_2SO_4$(aq) || $CuSO_4$(aq) | Cu(s)(+)	1.07–1.14	18–30
Clark	1872	(−) Zn(s) | $ZnSO_4 \cdot 7\,H_2O$(c) | $ZnSO_4$(sat aq) | $ZnSO_4 \cdot 7\,H_2O$(c) | Hg_2SO_4(s) | Hg(1)(+)	1.457	15.5
de la Rue	1878	(−) Zn(s) | $ZnCl_2$(aq) | AgCl(s) | Ag(s)(+)	1.03	18–30
Helmholtz	1882	(−) Zn(s) | $ZnCl_2$(aq) | HgCl(s) | Hg(l)(+)	1.00	15
Weston	1884	unsaturated form of Clark	1.46	18–30
Raleigh-Sidgwick	1884	Clark system with improvement in design	1.453	15
Gouy	1887	(−) Zn(s) | $ZnSO_4$ (10% aq) | HgO(s) | Hg(l)(+)	1.39	12
Weston	1892	(−) Cd(s) | $CdSO_4$(aq) | Hg_2SO_4(s) | Hg(l)(+) saturated	1.018636	18–30
Weston	1892	same system as above unsaturated	1.019	20
Hulett	1908	a larger version of the Weston cell	1.02252	18–30
Brönsted	1911	(−) Pb, Hg(l) | $K_2SO_4 \cdot PbSO_4$(s) | K_2SO_4(sat aq) | Hg_2SO_4(s) | Hg(l)(+)	1.0481	22
Henderson-Stegeman	1918	(−) Pb, Hg(1) | Na_2SO_4(c) | Na_2SO_4(sat aq) | Hg_2SO_4(s) | Hg(1)(+)	0.96463	25
Vosburgh et al.	1937	(−) Cd, Bi, Hg | $CdSO_4 \cdot \frac{8}{3}\,H_2O$(c) or $CdSO_4 \cdot Na_2SO_4 2\,H_2O$(c) | Sat aq sol of 2 salts in 0.1 M acetic acid or in 0.1 M H_2SO_4 | Hg_2SO_4(s) | Hg(1)(+)	1.0184	25

to reduce its temperature coefficient or other electrochemical systems will be developed with lower emf-temperature coefficients. This line of research is necessary if the emf of standard cells is to be measured with higher precision; the present emf-temperature coefficient limits their measurement to 0.1 μV. It is also to be expected that standard cells of lower emf of the order of 0.1 V will ultimately be produced. These would be of value as standards in electrical measurements where voltages of the order of 0.1 V rather than 1 V are involved, e.g., electrical measurements of temperature and temperature differences. Finally, a need exists for standard cells for low- and high-temperature use. Efforts to develop such cells will no doubt be made in the future.

REFERENCES

1. Gauss, C. F., *Pogg. Ann. Physik.*, **28**, 241 (1833), German translation of a paper in Latin read before the Royal Scientific Society of Göttingen (1833).
2. Weber, W. E., *Pogg. Ann. Physik.*, **82**, 337 (1851).
3. Thomson, Sir William (Lord Kelvin), *Phil. Mag.*, **2**, (4), 429 (1851); Ayrton, W. E., and Mather, T., *Practical Electricity*, Cassell and Co., Ltd., London, p. 205 (1911); *Brit. Assoc. Reports on Electrical Standards*, Cambridge Univ. Press, p. 137 (1911).
4. *Id., Papers on Electrostatics and Magnetism*, Macmillan and Co., London, p. 246 (1884).
5. Wenner, F., *Science*, **29**, 475 (1909), by title only; details of the method are given by Thomas, J. L., Peterson, C., Cooter, I. L., and Kotter, F. R., *J. Research Natl. Bur. Standards*, **43**, 291 (1949).
6. Curtis, H. L., Moon, C., and Sparks, C. M., *J. Research Natl. Bur. Standards*, **21**, 375 (1938).
7. Cutkosky, R. D., *J. Research Natl. Bur. Standards*, **65A** (Phys. and Chem.), 147 (1961).
8. Driscoll, R. L., and Cutkosky, R. D., *J. Research Natl. Bur. Standards*, **60**, 297 (1958); Curtis, R. W., Driscoll, R. L., and Critchfield, C. L., *J. Research Natl. Bur. Standards*, **22**, 485 (1939).
9. Driscoll, R. L., *J. Research Natl. Bur. Standards*, **60**, 287 (1958).
10. Craig, D. N., Hoffman, J. I., Law, C. A., and Hamer, W. J., *J. Research Natl. Bur. Standards*, **64A** (Phys. and Chem.), 381 (1960).
11. Brooks, H. B., *J. Research Natl. Bur. Standards*, **11**, 211 (1933).
12. Vincent, G. D., *IRE Trans. Instrumentation*, **I-7**, 221 (1958).
13. Miller, J. H., AIEE Communications and Electronics, No. 14, 413 (Sept. 1954).
14. Hamer, W. J., *Standard Cells, Their Construction, Maintenance, and Characteristics*, Natl. Bur. Standards Monograph 84 (1965).
15. Daniell, J. F., *Phil. Mag. III*, **8**, 421 (1836).
16. Clark, Latimer, *Proc. Roy. Soc. (London)*, **20**, 444 (1872); *Phil. Trans.*, **164**, 1 (1874).
17. Lord Rayleigh and Mrs. Henry Sidgwick, *Phil. Trans., II*, **175**, 411 (1884).
18. Callendar, H. L., and Barnes, H. T., *Proc. Roy. Soc. (London)*, **62**, 177 (1897).
19. Hulett, G. A., *Phys. Review*, **32**, 257 (1911); *The Electrician*, **67**, 129 (1911).
20. Wright, C. R. A., and Thompson, C., *Phil. Mag. V*, **16**, 28 (1883).
21. Kahle, K., *The Electrician*, **31**, 265 (1893).

22. Cooper, W. R., *Primary Batteries, Their Theory, Construction, and Use,* 2nd ed., D. Van Nostrand, New York, p. 336 (1916).
23. Kahle, K., *Wied. Ann.,* **51**, 197 (1894).
24. Jaeger, W., *Die Normalelemente und ihre Anwendung,* Wilhelm Knapp, Halle a.s., 1902.
25. Giauque, W. F., Barieau, R. E., and Kunzler, J. E., *J. Am. Chem. Soc.,* **72**, 5685 (1950).
26. Carhart, H. S., *Am. J. Sci.,* **28**, 374 (1884).
27. Glazebrook, R. T., and Skinner, S., *Proc. Roy. Soc. (London),* **51**, 60 (1892).
28. Weston, Edward, U.S. Pat. 310,004 (1884).
29. Carhart, H. S., *Phil. Mag.,* **28**, 420 (1889); *Am. J. Sci.,* **38**, 402 (1889); *Proc. Am. Ind. Elec. Eng.,* **9**, 615 (1892).
30. de la Rue, W., and Müller, H. W., *J. Chem. Soc.,* **21**, 488 (1868); *Phil. Trans. Roy. Soc., I,* **169**, 55 (1878).
31. Helmholtz, H., *Monatsbericht der Berlin Akad,* 0. 945 (Nov. 1881).
32. Gouy, M., *Compt. rendu,* **104**, 781 (1887).
33. Weston, Edward, German Pat. 75,194 (1892); Brit. Pat. 22,482 (1892); U.S. Pat. 494,827 (1893).
34. Wolff, F. A., *Natl. Bur. Standards Bull.,* **5**, 309 (1908).
35. Vigoureux, P., and Watts, S., *Proc. Phys. Soc. (London),* **45**, 172 (1933).
36. Brickwedde, L. H., *J. Research Natl. Bur. Standards,* **36**, 377 (1946).
37. Report of International Conference on Electrical Units and Standards, 1908, given as Appendix 2 in Natl. Bur. Standard Circular 475, Establishment and Maintenance of the electrical units, F. B. Silsbee (1949).
38. Report to the International Committee on Electrical Units and Standards of a Special Technical Committee Appointed to Investigate and Report on the Concrete Standards of the International Electrical Units and to Recommend a Value for the Weston Normal Cell, U.S. Government Printing Office, Washington, D.C. (Jan. 1, 1912).
39. Announcement of Changes in Electrical and Photometric Units, Natl. Bur. Standards Circ. C459 (1947).
40. Brönsted, J. N., *Z. physik. Chom.,* **77**, 315 (1911).
41. Henderson, W. E., and Stegeman, G., *J. Am. Chem. Soc.,* **40**, 84 (1918).
42. Vosburgh, W. C., Guagenty, M., and Clayton, W. J., *J. Am. Chem. Soc.,* **59**, 1256 (1937).
43. Vosburgh, W. C., and Bates, R. G., *J. Electrochem. Soc.,* **111**, 997 (1964).
44. Van Ginneken, P. J. H., and Kruyt, H. R., *Z. physik. Chem.,* **77**, 758 (1911).
45. Hamer, W. J., and Craig, D. N., *J. Electrochem. Soc.,* **111**, 1434 (1964).
46. Hamer, W. J., *J. Wash. Acad. Sciences,* **54**, 297 (1964).
47. *Id., Encyclopedia of Chemical Technology,* Vol. 3, The Interscience Encyclopedia, Inc., New York, p. 293 (1949).
48. Cohen, E., Helderman, W. D., and Moesveld, A. L. Th., *Z. physik. Chem.,* **96**, 259 (1920).
49. Stull, D. R., and Sinke, G. C., *Thermodynamic Properties of the Elements,* Am. Chem. Soc., Washington, D.C. (1956).
50. Rossini, F. D., Wagman, D. D., Evans, W. H., Levine, S., and Jaffe, I., *Selected Values of Chemical Thermodynamic Properties*, Natl. Bur. Standards Circ. 500 (1952).
51. Papadopoulos, M. N., and Giauque, W. F., *J. Am. Chem. Soc.*, **77**, 2740 (1955).
52. International Critical Tables, **3**, 210 (1928).
53. Ishikawa, F., Murooka, T., *Bull. Inst. Phys. Chem. Res* (*Tokyo*), **9**, 781 (1933).
54. Carpenter, C. D., and Jette, E. R., *J. Am. Chem. Soc.*, **45**, 578 (1923).

55. Kelley, K. K., *Contributions to the Data on Theoretical Metallurgy*, Vol. XIII., *High-Temperature Heat-Content, Heat-Capacity, and Entropy Data for the Elements and Inorganic Compounds*, Bull. 584, Bur. Mines (1960).
56. Holsboer, H. B., *Z. physik. Chem.*, **39**, 691 (1902).
57. Cohen, E., and Holsboer, H. B., *Z. physik. Chem.*, **95**, 305 (1920).
58. Hulett, G. A., *Trans. Am. Electrochem. Soc.*, **50**, 435 (1909).
59. Getman, F. H., *J. Am. Chem. Soc.*, **39**, 1806 (1917).
60. Cohen, E., *Z. physik. Chem.*, **85**, 419 (1913); **87**, 409, 419, 426, 431 (1914); *Trans. Faraday Soc.*, **10**, 216 (1915).
61. Gouy, M., *Compt. rendu*, **130**, 1399 (1900).
62. Hager, O. B., and Hulett, G. A., *J. Phys. Chem.*, **36**, 2095 (1932).
63. Craig, D. N., Vinal, G. W., and Vinal, F. E., *J. Research Natl. Bur. Standards*, **17**, 709 (1936).
64. Hulett, G. A., *Trans. Am. Electrochem. Soc.*, **6**, 109 (1904); *Trans. Inst. Elec. Cong., St. Louis*, **2**, 109 (1904).
65. Smith, Sir Frank, *A Dictionary of Applied Physics*, Glazebrook, Sir Richard, Ed., Macmillan and Co., Ltd., London, p. 267 (1922).
66. Ref. (65), p. 268.
67. Report of National Physical Laboratory, *Electrician*, **75**, 463 (1915).
68. Obata, J., *Proc. Math. Phys. Soc.* (*Japan*), **2**, (3), 232 (1920).
69. Ishibashi, V., and Ishizaki, T., *Research Electrotechnical Lab.*, No. 318 (1931).
70. Vosburgh, W. C., *J. Am. Chem. Soc.*, **47**, 1265 (1925).
71. Smith, Sir Frank, *Proc. Phys. Soc.* (*London*), **22**, 11 (1910); *Phil. Mag.*, **19**, 250 (1910); *National Physical Laboratory Collected Researches*, **6**, 137 (1910).
72. Vosburgh, W. C., and Eppley, M., *J. Am. Chem. Soc.*, **45**, 2271 (1923).
73. Vinal, G. W., and Brickwedde, L. H., *J. Research Natl. Bur. Standards*, **26**, 455 (1941).
74. International Critical Tables, **1**, 120 (1926).
75. Mylius, F., and Funk, F., *Ber. deut. chem. Ges.*, **30**, 824 (1897).
76. Vinal, G. W., Brickwedde, L. H., and Hamer, W. J., *Compt. rend. de la Quinzieme Conference*, Int. Union Pure and App. Chem., Amsterdam, p. 92 (1949).
77. Hamer, W. J., Brickwedde, L. H., and Robb, P. R., *Electrochemical Constants*, Natl. Bur. Standards Circ. 524, Chap. 12 (1953).
78. Dyer, C. A., U.S. Pat. 2,647,155 (1953).
79. Heinrich, R. O., U. S. Pat. 631,044 (1899).
80. Eicke, H., and Kessner, A., *Electrochem. Z.*, (A), **74**, 623 (1953).
81. Hulett, G. A., *Z. physik. Chem.*, **33**, 611 (1900).
82. Wolff, F. A., and Waters, C. E., *Natl. Bur. Standards Bull.*, **4**, 15 (1907).
83. Hulett, G. A., *Phys. Rev.*, **32**, 257 (1911).
84. Eppley, M., *J. Franklin Inst.*, **201**, 24 (1926).
85. Mueller, E. F., and Stimson, H. F., *J. Research Natl. Bur. Standards*, **13**, 699 (1934).
86. Vosburgh, W. C., and Eppley, M., *J. Am. Chem. Soc.*, **46**, 109 (1924).
87. Wold, P. I., *Phys. Rev.*, **27**, 329 (1908).
88. Vinal, G. W., and Brickwedde, L. H., *J. Research Natl. Bur. Standards*, **36**, 377 (1946).
89. Niederhauser, W. S., and Hulett, G. A., *J. Am. Chem. Soc.*, **51**, 2345 (1929).
90. Vosburgh, W. C., and Elmore, K. L., *J. Am. Chem. Soc.*, **53**, 2819 (1931).
91. Vosburgh, W. C., and Eppley, M., *J. Optical Soc. Am. and Rev. Sci. Instr.*, **9**, 67 (1924).
92. Eppley, M., and Vincent, G. D., *Electrochemical Constants*, Natl. Bur. Standards Circ. 524, Chap. 10 (1953).
93. *International Critical Tables*, **1**, 120, 121 (1926); **2**, 456, 458 (1927); **3**, 66 (1928).
94. Ramsey, R. R., *Phys. Rev.*, **XVI**, No. 2, 105 (1902).

95. Hulett, G. A., *Phys. Rev.*, **22**, 321 (1906).
96. Windus, W. B., M.S. dissertation, Washington State University (1962).
97. Brodd, R. J., and Eicke, W. G., Jr., *J. Research Natl. Bur. Standards*, **66C** (Eng. and Instr.), 85 (1962).
98. Vinal, G. W., Craig, D. N., and Brickwedde, L. H., *Trans. Electrochem. Soc.*, **68**, 139 (1935).
99. Vinal, G. W., *Primary Batteries*, John Wiley and Sons, New York, p. 211 (1950).
100. Lamb, F. X., *Electrochemical Constants*, Natl. Bur. Standards Circ. 524, Chap. 11 (1953).
101. Hulett, G. A., *Phys. Rev.*, **27**, 33 (1908).
102. Spence, B. J., cited in (101), p. 37.
103. Elliott, R. B., and Hulett, G. A., *J. Phys. Chem.*, **37**, 489 (1933).
104. Eicke, W. G., Jr., *ISA Trans.*, **3**, 93 (1964); *Natl. Bur. Standards Tech. News Bull.*, **48**, 1 (1964).

Author Index

Subject Index